Engineering Materials

This series provides topical information on innovative, structural and functional materials and composites with applications in optical, electrical, mechanical, civil, aeronautical, medical, bio- and nano-engineering. The individual volumes are complete, comprehensive monographs covering the structure, properties, manufacturing process and applications of these materials. This multidisciplinary series is devoted to professionals, students and all those interested in the latest developments in the Materials Science field, that look for a carefully selected collection of high quality review articles on their respective field of expertise.

Indexed at Compendex (2021) and Scopus (2022)

Vijay Kumar · Pardeep Singh ·
Devendra Kumar Singh
Editors

Green Carbon Quantum Dots

Environmental Applications

Springer

Editors
Vijay Kumar
Ramanujan College
University of Delhi
New Delhi, Delhi, India

Devendra Kumar Singh
Department of Chemistry
Government P.G. College
Amarpatan, Madhya Pradesh, India

Pardeep Singh
Department of Environmental Studies
PGDAV College
University of Delhi
New Delhi, Delhi, India

ISSN 1612-1317 ISSN 1868-1212 (electronic)
Engineering Materials
ISBN 978-981-97-6205-7 ISBN 978-981-97-6203-3 (eBook)
https://doi.org/10.1007/978-981-97-6203-3

This Springer imprint is published by the registered company Springer Nature Singapore Pte Ltd.
The registered company address is: 152 Beach Road, #21-01/04 Gateway East, Singapore 189721, Singapore

Contents

An Introduction to Carbon Quantum Dots

Neeru Rani, Permender Singh, Sandeep Kumar, Vinita Bhankar, Dinesh Kumar, and Krishan Kumar

Abstract Carbon quantum dots (CQDs), a novel family of carbon dots (CDs) are one of the emerging materials among several carbon-based nano-ranged materials. They have attained much interest and become a profound competitor to traditional semiconductor quantum dots for several applications. Since their discovery, CQDs are widely used for several applications like chemical sensing, bioimaging, biosensing, drug delivery, photocatalysis because of several characteristics associated with them such as tiny sizes (<10 nm), low toxicity, bio-compatibility, optical properties, water solubility, photocatalytic properties, ease of surface modifications, simple and eco-friendly synthesis routes by making usage of waste biomass that ultimately made the process cost-effective. Methods widely used for synthesis purpose are hydrothermal, microwave, ultrasonication, laser ablation, and pyrolysis among which hydrothermal is more common. Green CQDs are basically CDs synthesised by making usage of biomass. Availability and abundance of biomass makes the process eco-friendly and cost effective by providing a variety of precursors. Along with, these materials are becoming promising candidates for environmental applications such as sensing and absorption of toxic metal ions, water remediation through degradation of harsh textile effluents. This chapter provides the basic details regarding the material including synthesis, properties and basic applications.

N. Rani · P. Singh · D. Kumar · K. Kumar (✉)
Department of Chemistry, Deenbandhu Chhotu Ram University of Science & Technology, Murthal, Sonepat, Haryana 131039, India
e-mail: krishankumar.chem@dcrust.org

S. Kumar (✉)
J. C. Bose University of Science & Technology, YMCA, Faridabad, Haryana 121006, India
e-mail: sandeepkumar@jcboseust.ac.in

V. Bhankar
Department of Biochemistry, Kurukshetra University, Kurukshetra, Haryana 136119, India

V. Kumar et al. (eds.), *Green Carbon Quantum Dots*, Engineering Materials,
https://doi.org/10.1007/978-981-97-6203-3_1

1 Introduction

CQDs also termed as CDs are novel member of zero-dimensional carbon based nano-ranged materials with size generally below 10 nm [1]. These are accidentally discovered in the course of the refinement of single walled CNTs in 2004 [2]. CQDs are becoming an interesting material for industrial and scientific community because of several unique properties associated with them including high chemical stability, excellent solubility in aqueous media, low toxicity, good optical characteristics, biocompatibility and unique physiochemical properties [3]. Existence of abundant functional moieties such as hydroxyl, carboxylic, phenolic groups on the surface are responsible for these properties and provides ease of surface modifications. Because of these associated properties CQDs are used for several applications such as photocatalysis [4], detection of heavy metal ions [5], electrocatalysis [6], bioimaging [7], drug delivery [8]. One of the benefits associated with these materials is that they can be synthesised from natural renewable, easy available precursors that makes the process cost efficient and eco-friendly at the same time, termed as green synthesis [9]. CQDs fabricated from green synthesis are generally termed as green quantum dots. These green CQDs are efficient over chemical based CQDs in terms of biocompatibility, non-toxicity which make them as efficient nanomaterials for application in health sector such as bioimaging, biomedicine, bioimaging [10]. Variety of available precursors, especially the utilization of waste biomass, e.g., plant [11] fruit and vegetable peels [12], flower [13], leaves [14], plastic [15] makes the synthesis efficient and also provides an alternative to get rid of domestic waste efficiently. Along with superiorities, these nanostructures are associated with some limitations too such as low yield, purification issues, size and morphology control, less knowledge of mechanisms [7, 10]. We have summarized all the basic details of CQDs regarding their synthesis, characterization techniques, properties, applications and related challenges associated in this book chapter with the aim of providing basic background to readers.

2 Synthesis Methods of Green CQDs

Fabrication strategies of CQDs can be generally classified into two categories: top –down and bottom-up. Among these the common one is bottom-up strategy as mentioned in Fig. 1. Top-down strategy comprises the conversion of large bulk structures into nano-size counterparts by utilizing some energy source [16, 17]. The fabrication methods under this approach includes arc discharge [18], laser ablation [19], electrochemical oxidation [20] which basically involves the fabrication of CQDs from large size carbon related materials like graphene membranes, activated carbon, carbon nanotubes and graphite. Requirement of expensive equipment, specific steps and harsh experimental conditions limits their practical and wide scale applicability

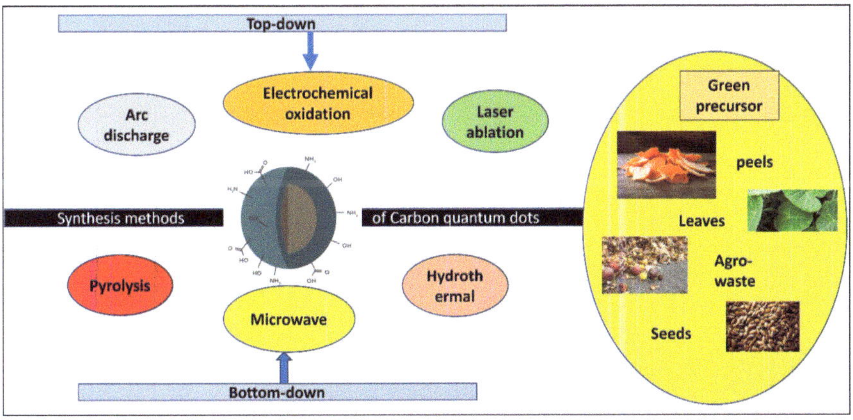

Fig. 1 Several synthesis methods of CQDs and related green precursor

[21]. Bottom-up approach is widely utilized to fabricate CQDs because of its simple application and ease to tune the properties [7].

This strategy basically involves the formation of CQDs through self-assembly of small size components (generally organic molecules) by aid of physical and chemical procedures such as hydrothermal [22], pyrolysis [23], solvothermal [24], microwave assisted [25] and combustion [26] etc. Among this hydrothermal method is quite famous because of its various advantages such as eco-friendly, non-expensive and non-toxic processes. Hydrothermal method generally utilize aqueous medium wherein precursors are tend to carbonized at high pressure and temperature conditions for fabrication of water soluble CQDs [7].

Nowadays, green precursors are widely used for these methods for their several characteristics such as non-toxic nature, eco-friendliness, renewability and ease of providing cost effectiveness to the reaction procedure. Green precursors such as orange peels [27], rice husk [28], sugarcane bagasse [29], wood, plant seeds [30], leaves [14], etc., are widely employed for synthesis purpose. Xia et al., fabricated fluorescent CQDs through hydrothermal treatment of flowers of winter sweet, and later on utilize this sensor for sensing of Cr^{6+} and Fe^{3+} with huge sensitivity and selectivity by exhibiting LOD of 0.07 and 0.15 μM, sequentially as shown in Fig. 2 (Vi, 2022). Solvothermal method requires the utilization of solvents such N, N-diethylformamide (DMF) and ethanol [31]. Using biomass based green, eco-friendly precursors such as honey, bamboo leaves and corn bract. Using electromagnetic radiation (about 2.45 GHz) CQDs can also be synthesised through microwave method using green precursors such as lotus roots [32], sucrose [33] and potatoes [34]. A brief comparison of all the methods is summarized in Table 1.

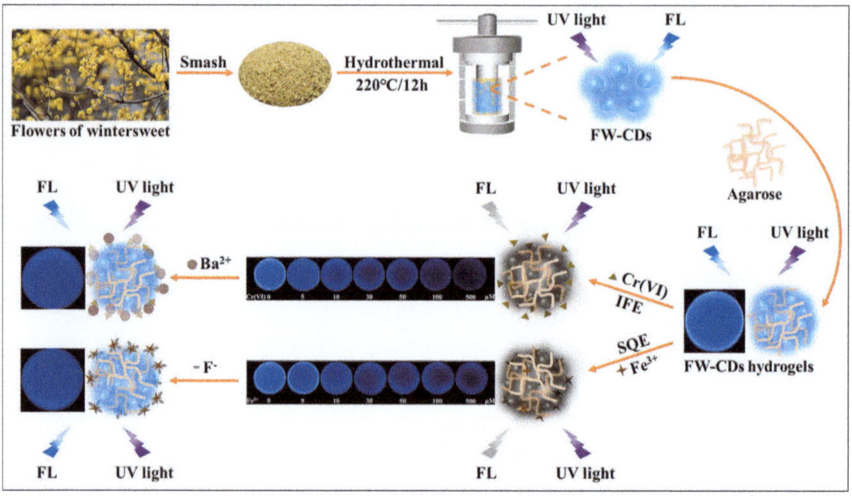

Fig. 2 Fluorescent CQDs fabricated from winter sweet flowers utilized as sensors [35]

3 Characterizations of CQDs

After fabrication, characterizations of CQDs become crucial for applicability in several fields. "X-ray powder diffraction (XRD), X-ray photoelectron spectroscopy (XPS), Fourier transform infrared spectroscopy (FTIR), transmission electron microscopy (TEM), photoluminescence spectrophotometry (PL) and UV–vis spectroscopy are some of the common characterization techniques used for analysing CQDs as shown in Fig. 3".

3.1 XRD

Crystalline and amorphous nature of CQDs is determined through XRD techniques. Figure 4. Represent XRD pattern of highly fluorescent N-doped CQDs by utilizing grass as precursor and hydrothermal as synthesis method with a broad diffraction peak around 22° owing to the clear amorphous nature of CQDs [49]. Magnetic carbon dots synthesised by Singh et al., using *Allium Cepa* exhibited highly crystalline nature and degraded hazardous RhB and MB dye up to 83.05% and 92.76%, respectively [16]. CQDs fabricated using pyrolysis of peanut shell, demonstrated a wide peak around 23.5° ascribed to graphene structure of resultant CQDs [50]. Similarly, there are several reports available in literature demonstrating crystalline and amorphous nature if CQDs using green precursors.

Table 1 Entries of several methods of synthesis and their associated advantages and disadvantages

Method	Advantage	Disadvantage	Refs
Laser ablation (Top-down)	Reproducibility possible, purity achieved, ease of control over size and morphology	Large scale production not feasible, low yield, complicated operations, expensive	[36, 37]
Arc discharge (Top-down)	Graphene CQDs with doping can be fabricated	Sometimes unwanted materials generated, purification required, high energy required	[38–40]
Electrochemical oxidation (Top-down)	Good reproducibility and purity can be achieved, normal temperature and pressure required	Purification process is complex,	[39, 41]
Hydrothermal (Bottom-up)	Non-toxic, ease over control in size, purity of sample achieved	Requirement of high-pressure conditions	[42, 43]
Solvothermal (Bottom-up)	Slow process, solvent might have influence on properties of CQDs	Problem of low yield and purity, uncontrollable morphology and size	[42]
Microwave (Bottom-up)	Fast process, direct solid-state synthesis, direct heating of the sample possible	Large scale production not feasible due to usage of small reactors	[44, 45]
Pyrolysis (Bottom-up)	Simple, cost-effective process, reliable for mass production	Requirement of high temperature, difficult to separate CQDs with raw materials	[46]
Thermal decomposition (Bottom-up)	Easy operations, low cost, less time required	Non-fluorescent intermediates can be obtained	[47, 48]

Table 2 Green synthesized CQDs and their composite in degradation of several dyes

Precursor	Photocatalyst	Dye degraded	Reaction condition	Light source	Shape and size	Efficiency	Ref
Pear juice	CQDs	MB	Hydrothermal 180 ◦C, 36 h	Visible	Spherical 3–6 nm	99.5% in 130 min	[67]
Orange peel	CQDs/ZnO	NBB	Hydrothermal 180 °C, 12 h	UV	Sphere 2–7 nm	100% in 45 min	[27]
Avocado seeds	CQDs	MB	Pyrolysis 400 °C	UV-A	Spherical 4.2 ± 1.5 nm	45% in 2 h	[30]
Sugarcane bagasse	WO_3/NCQDs	MB	Hydrothermal 190 °C, 24 h			84.6%	[68]

Table 3 Several biomass derived green CQDs in metal sensing

Precursor	Synthesis method	Metal sensing and LOD	Ref
Lemon peel	Hydrothermal	Cr^{6+} (73 nM)	[72]
Broccoli	Hydrothermal	Ag^{+} (0.5 μM)	[73]
Table sugar	Microwave assisted	Pb^{2+}(14 ppb)	[33]
Human urine	Pyrolysis	Hg^{2+}(2.7 nM)	[74]
Chocolate	Hydrothermal	Pb^{2+}(12.7 nM)	[75]

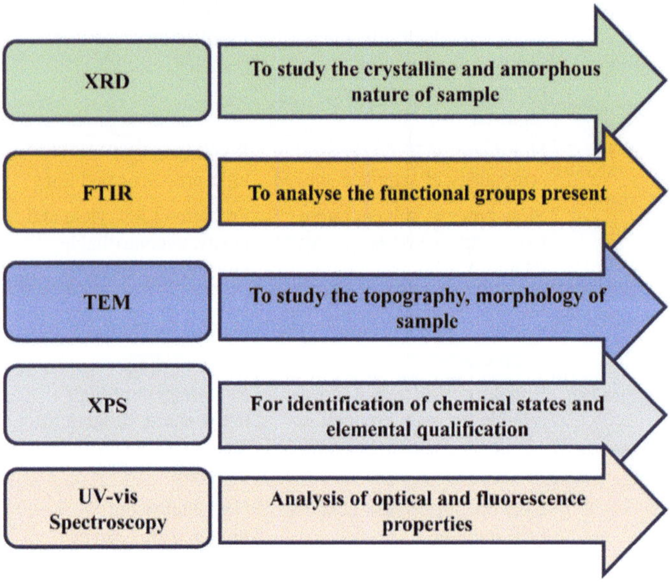

Fig. 3 Different characterization techniques of CQDs

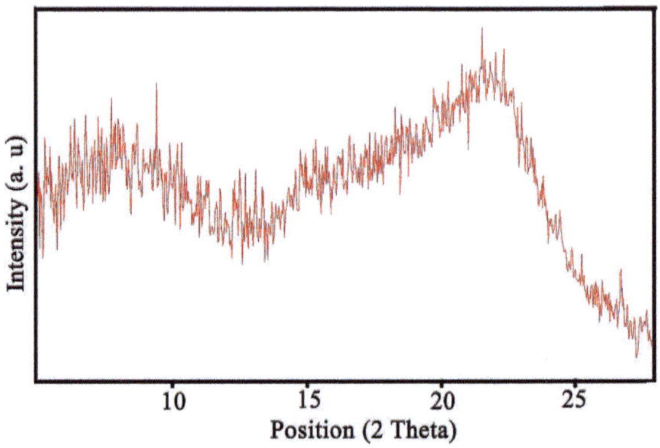

Fig. 4 XRD pattern of N-CQDs synthesised using grass. Reprinted with permission copyright © 2019 Elsevier Ltd, (License number—5652540827237]. All rights reserved [49]

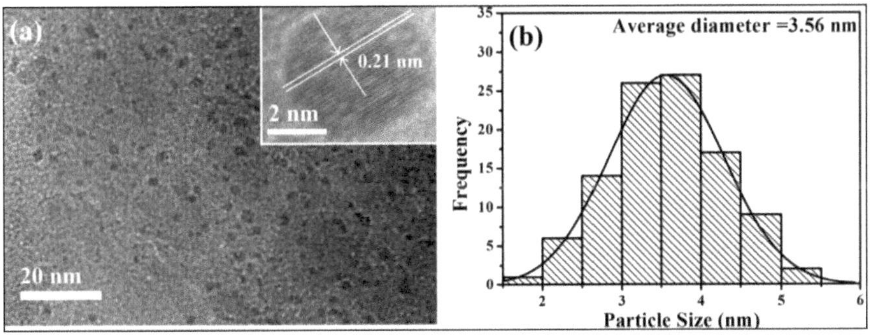

Fig. 5 TEM image of particles of CQDs with circular shape and average diameter 3.56 nm. Reprinted with permission copyright © 2020 Elsevier Ltd, (License number—5652580070538]. All rights reserved [51]

3.2 TEM

Texture and morphological analysis of the CQD samples are done through TEM. Figure 5a, b represents shape and mean diameter of the particles. as shown in Fig. 5a, the CQDs are observed to be circular with fringe spacing of 0.21 nm and average diameter of 3.56 nm when calculated for 100 particles [51]. In different study by Xue et al., CQDs with mean size of 1.62 nm were synthesised using peanut shells, whereas nanospheres of size range 10–40 nm were obtained in another study using peanut skin as precursor [50, 52].

3.3 FTIR

Highly fluorescent Barley derived green CQDs by Xie et al., was used for the detection for Hg^{2+}[53]. FTIR spectra was used for the determining the abundant functional moieties in the sample as displayed in Fig. 6.

Absorption band close to 3380 cm^{-1} corresponds to N–H and O–H bonds. Peak at 2922 cm^{-1} attributed to C-H bonding. Absorption bands around 1440 and 1640 cm $^{-1}$ was ascribed for COO^{-}vibrations. The distinctive absorption band 1574 cm^{-1} was observed for C-N. Furthermore, a band around 1340 cm^{-1} and 1032 cm^{-1} ascribed for C-H and C-O mode vibrations. FTIR data clearly confirms the availability of several hydrophilic functional moieties in the CQDs which ultimately leads to increment in the aqueous solubility and sensing application in aqueous media. Allium Cepa derived CDs exhibited excellent decontamination efficiency for MB and RhB dye with a decomposition efficiency of 92.76% and 83.05%. FTIR spectra of these CDs exhibited several peaks corresponds to different functional groups (3118–3420 cm^{-1} ascribed to stretching vibration of -OH and N–H, 2888-2930 cm^{-1} for C-H group,

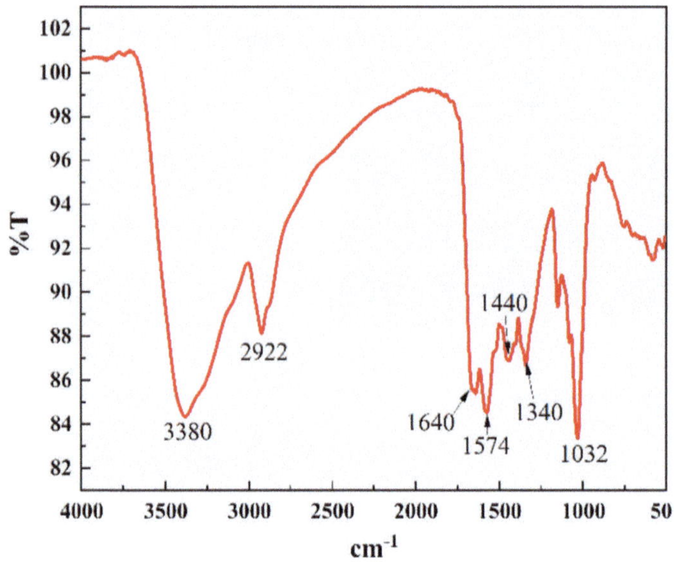

Fig. 6 FTIR image of CQDs for analysing functional groups [53]

2165–2171 corresponds to C-H overtone, peaks around 1520–1567 cm^{-1} for NH bending/C = C [16].

3.4 UV–vis Spectroscopy

Fluorescent nature of CQDs played a crucial role for applicability in several sectors. Optical properties are found to be very vital for determining the fluorescent based properties. Figure 7. Represents the UV–vis absorption and fluorescence emission spectra of green CQDs synthesised from waste orange peels [27]. A weak absorption band around 268 nm contributes to $\pi - \pi^*$ electronic transition. Along with, CQDs displayed a fluorescence emission band around 431 nm for excitation wavelength of 340 nm as illustrate in Fig. 7b. Similarly, in another study by Permender et al., UV–vis absorption spectra showed peaks among range of 268–273 nm ascribed to $\pi - \pi^*$ transition, along with peaks around 323–327 nm corresponds to n- π^* transitions for C-O, C = O. These are believed to be the common bands demonstrated by CQDs and later used for calculation of optical band gap [16].

Such kind of fluorescent behaviour was also observed for several other biomass based green CQDs. It was believed that these unique emission characteristics arises because of size, accessibility of sp^2 sites, conjugating aromatic structure and defects of the material.

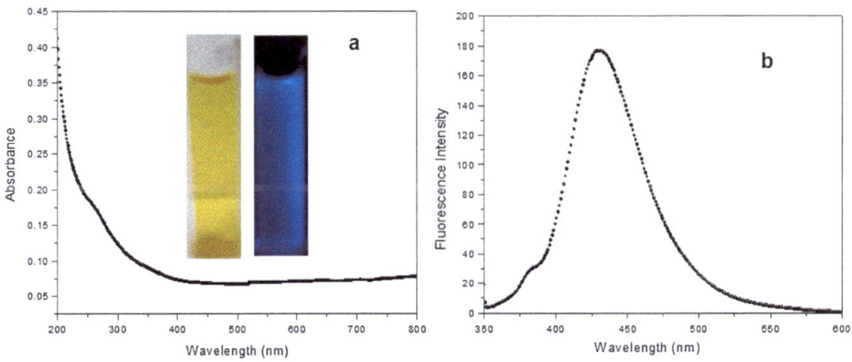

Fig. 7 A UV – visible absorption spectrum of green CQDs synthesised from orange peels, inset, fluorescence behaviour under daylight (left) and UV radiations (right). **b** Photoluminescence behaviour of CQDs. Reprinted with permission from reference [27]. Copyright {2013} American Chemical Society

3.5 XPS

Chemical and elemental analysis of the green CQDs is determined through XPS. Analysis of XPS survey spectrum of green CQDs synthesised using milk as precursor exhibit three intense peaks located at 286.0, 400.0 and 533.0 eV ascribed to elements C1s, N1s and O1s sequentially as displayed in Fig. 8a [54]. High resolution analysis of these peaks was further shown in Fig. 8b, c, and d. In specific Fig. 8b represents high resolution spectra of C1s, reveals four distinct peaks at values of 285.7, 286.8, 287.5 and 288.9 eV corresponding to groups C-H, C-O, C-N, and C = O, correspondingly. Figure 8c corresponds to XPS of O1 s, exhibiting two peaks at 532.8 and 535.6 eV corresponding to C = O/C–O–C, and N–O groups sequentially. Similarly, XPS N1s spectrum of is displayed in Fig. 8d, here, two distinct peaks around 400.7 and 402 eV corresponds to C-N, N–O correspondingly [54]. Similarly, using Bran as precursor, showed three intense peaks at 284.5, 399.7 and 531.9 eV associated with C1s, N1s and O1sequentially, which additionally confirms the results [55]. Similarly, *Allium Cepa* derived magnetic CDs having average size of 13.37 nm exhibited peaks of elements N, O, C, S in XPS spectrum which further demonstrated the role of green precursors in synthesis of CQDs [16].

4 Characteristics of CQDs

Because of the excellent characteristics exhibited by CQDs, CQDs are utilized in various fields such as biosensing, biomedical applications, bioimaging, drug delivery and wastewater remediation. Several properties of CQDs are discussed below which make them ideal probes for application in several sectors.

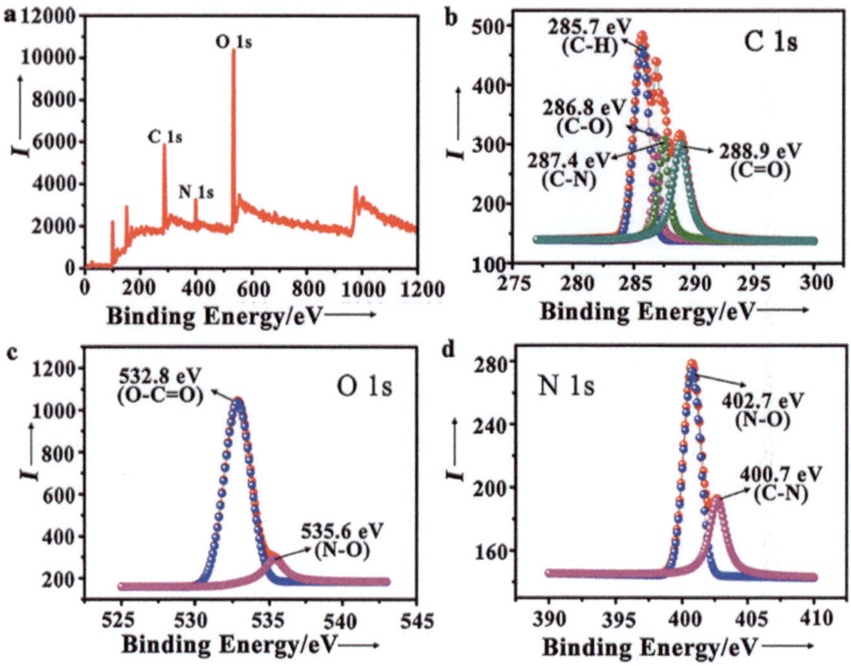

Fig. 8 XPS spectra of CQDs synthesised from milk. **a** XPS, **b** C 1 s, **c** O 1 s, **d** N 1 s. Reprinted with permission copyright © 2014 WILEY–VCH Verlag GmbH & Co. KGaA, Weinheim Ltd, (License number—5653451071398]. All rights reserved [54]

4.1 Optical Characteristics

Because of the remarkable optical features of CDs (fluorescent nature) they have been broadly employed for numerous applications including biosensing, bioimaging that have a huge importance in health sector. Prior to applicability of these materials in these sectors, it is important to have some basic information about optical properties of CQDs.

4.1.1 Absorbance

$\Pi - \pi^*$ transitions arise in the CQDs due to available sp^2 $c = c$ framework present in the structure of CQDs. Due to this sp^2 nature of carbon framework, absorbance is generally observed in short wavelength region ranging from 260–320 nm. This absorbance range may vary depending upon type of modification introduced (heteroatom doping, introduction of surface functional groups). As a result of doping, n-π^*also arises which further alter the absorption wavelength [7, 10].

4.1.2 Fluorescence Characteristics

(a) Up-conversion fluorescence

A fluorescence behaviour is said to be up-conversion when excitation wavelength is greater than wavelength of emission. CQDs fabricated using ultrasonic approach tends to exhibit such type of fluorescence. It is observed that larger excitation wavelength tends to reduced background auto fluorescence, which plays a significant role in bioimaging applications [7, 37].

(b) Down-conversion fluorescence

Fluorescent mechanism of CQDs are still under investigation, but currently it is believed that factors basically responsible for fluorescent nature may be due to vacant zig-zag sites, multi-emissive centres, conjugated structure, quantum confinement effects, trapped excitons and special edge deformities. As CQDs are zero dimensional nanomaterials therefore fluorescence can be aroused because of availability of e^-/h^+pair in these nanostructures [56]. Energy of band gap decreases with the increment in the size of CQDs. Therefore, fluorescent behaviour of these nanomaterials can be tuned by altering in their quantum confinement effects. Surface oxidation (due to varieties of functional moieties on surface having oxygen) is also one of the possible mechanism of fluorescent behaviour, arises due to creation of surface deformities by oxygen enriched functional moieties on edge of CQDs [7].

(c) Emission characteristics

Various fluorescent emissions can be achieved by having proper control over excitation wavelength which may be accompanied through regulating synthesis parameters such as pH, concentration, temperature [7]. "Protonation and deprotonation of the surface functional groups" results in pH dependent emission, surface state emissions results in concentration dependent fluorescence, whereas temperature dependent emissions are a result of non-radiative decay occurs on surface [57].

(d) Chemical stability

Fluorescence biosensing and bioimaging requires stable fluorescence signals and long emission lifetimes. This can be possible by the utilization of CQDs, because of their ability to produce stable signals [58]. Furthermore, they have the ability to emit fluorescence for long durations (up to a year). Also CDs have stability over wide range of pH values (generally from 3–12), that leads to result in remarkable impedance for photobleaching [57].

4.1.3 Phosphorescence

CQDs demonstrate phosphorescence characteristics. "Based on the CQDs, it has been possible to fabricate an organic room temperature phosphorescent material, with enhancement in phosphorescent lifetime of about 380 MS". When CDs are dispersed in polyvinyl alcohol matrix, it is possible to observe a clear phosphorescence under UV light at room temperature. Initial investigations revealed that triplet excited states of surface carbonyl groups were mainly responsible for phosphoresce. Basically, hydrogen bonding among the matrixes preserve triplet excited energy states from vibrational and rotational loss [7, 59, 60].

4.1.4 Chemiluminescence

Chemiluminescence (CL) generally refers to a process in which light emitted due to occurrence of any chemical reaction. Under suitable environment in redox reactions, CQDs exhibited CL in aqueous medium, with the generation of some unstable products from intermediate radicals. CL generation through CDs result of either excitation after oxidation or enhancement/inhibition of luminescence [58]. CL is initially arising due to mixing of CQDs with several oxidants like Cerium (IV) and $KMnO_4$. These oxidants generate holes into CDs reveals by EPR spectrum. The increase in holes leads to speed up e^-/h^+annihilation process causing CL emission. Also, CL intensity and CQDs concentration is interrelated [7, 59].

4.1.5 Photoluminescence

Photoluminescence (PL) is one of the efficient properties of CQDs which played a major role in photocatalysis. This is very exciting property of CQDs, attaining a continuous interest from the past years. "Emission pattern of PL resembles with stokes type emission patterns where emission wavelength of PL is longer than excitation wavelength of laser. Several reports analysis the PL behaviour of CQDs from diverse sources [59]. Detailed analysis of several spectrum reveals that PL emissions fall into one of two groups. One is associated with band gap transitions associated with conjugated p-domains, other one is due to defects". Both are inter-related as manipulation of defects creates p-domains [59].

4.2 Electrical Properties

CQDs are currently being utilized in the field of electrocatalysis and electrochemistry by virtue of several factors:

(a) Comparison to several different carbon based nanostructures, CQDs are associated with several characteristics such as good charge transfer ability, large surface area, enhanced conductivity, less toxicity along with synthesis procedure is comparatively cost-effective [4, 7, 59].

(b) Numerous surface functional moieties provides variety of sites for surface modification along with increase electrocatalytic activity through enhancing the intermolecular electroconductivity [7, 25, 31, 61].
(c) Doping of CQDs with heteroatoms including B, S, N etc., enhanced the electronic characteristics because of intramolecular charge transferability [58, 61, 62].
(d) CQDs are effective at improving the electrocatalysis technique for electrochemical processes, including the hydrogen evolution reaction, oxygen evolution reaction, alcohol oxidation reaction and oxygen reduction reaction [10, 59, 61].

These aforementioned characteristics of CQDs make them probable candidates for serving in electrochemical applications as electrocatalytic agents.

4.2.1 Electrical Conductivity

CQDs as electrocatalysts displayed good electrical conductivity by removing the Schottky barrier at electrolyte-catalyst junction, which further enhanced the effective energy transformations [58, 61]. Also due to exhibition of good electrical conductivity, they are used for rapid electron transfer during electrochemical reactions [62].

4.2.2 Electronic Structural Arrangement Cause by Doping of Heteroatoms

Doping of heteroatoms in CDs results in alteration of chemical framework, electric charge can be remarkably moved from several adjacent carbon atoms [63]. These modified nanomaterials exhibit good electrochemical performance because of enhanced intrinsic performance of functional sites, distortion of electronic configuration and acceleration of adsorption–desorption process [58, 61, 62].

4.2.3 Stability Enhancement

Abundance of various types of surface functional moieties on CQDs, also their good long term stability (chemical) in variety of solvents, made them potential candidates for refining the chemical stability associated with hybrid catalysts [61]. Along with, the production of CQDs nanocomposites with metals, metal oxides result in enhancement in electrocatalytic activity by preventing the agglomeration. Also, stability of CQDs in aqueous media reinforce the steadiness of these catalysts. Electrostatic interactions of CQDs with catalysts enhance the stability of hybrid catalysts [10].

4.2.4 Active Centre and Defect Sites

CQDs can proficiently behaves as active centres due to their several characteristics such as several defect moieties, active edges, good electroconductivity and high surface area to volume ratio. Henceforth, fusion of CQDs with some conductive materials leads to facilitate their electrochemical performances [61, 64].

5 Applications of Green CQDs

5.1 CQDs as Photocatalyst: Dye Degradation

Photocatalysis is termed as "Change in the rate of a chemical reaction or its initiation under the action of ultraviolet, visible or infrared radiation in the presence of a substance (the photocatalyst) that absorbs light and is involved in the chemical transformation of the reaction partners"[65], 66, 16. CQDs can yield light through absorption. "The absorbed photons excite electrons from their HOMO to their LUMO by leaving a hole in HOMO. The photo-induced electrons and holes cause chemical transformation of reaction partners rendering CQDs as photocatalysts. CQDs have a strong capacity to absorb light in the UV–vis range due to π–π* transition of sp^2 C = C bonds present in the inner core whereas n-–π* transition arises because of the numerous functional groups (-COOH, -OH, NH_2) on the outer surface".

This property of CQDs serves in water remediation through degradation of harmful dyes in wastewater from textile industries. There are various reports on role of green CQDs in decontamination of dyes such as MB, RhB, MG, CV etc. Ease of surface modification in CQDs also enhanced the results through doping and composites formation. For instance, degradation efficiency of composite of CQDs for naphthol blue black (NBB) dye, derived utilizing orange peels as precursor was found to be very high. The resultant composite CQD/ZnO exhibited 100% efficiency in 45 min which is 84.3% comparative to ZnO alone as illustrated in Fig. 9 [27].

5.2 Metal Sensing

CQDs are extensively employed for sensing of dangerous heavy metal ions specially ions like Fe^{2+},Zn^{2+}, $Hg^{+2}$and radioactive ions which cause negative impacts on environment and human body even in trace amounts [17] 16. These metals can be quantitatively sensed by synthesised CQDs.

Presence of these metal ions in aquatic system possess a risk to aquatic animals and human beings by generating the risk of several lethal diseases such as kidney failure, risk of cancer and damage of central nervous system [69]. Green synthesised CQDs are intensively interesting materials for sensing of metal ions in terms of their

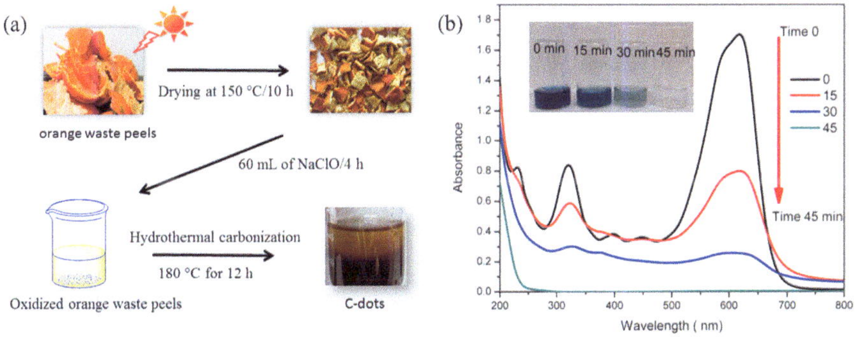

Fig. 9 **a** Fabrication of CQDs from orange waste peels making use of hydrothermal method. **b** Complete degradation of NBB dye by composite of CQDs with ZnO in 45 min. Reprinted with permission from reference [27]. Copyright {2013} American Chemical Society

ease of synthesis, inexpensive and eco-friendly precursors, ease of surface modifications. "The procedure of sensing these heavy metal ions is initiated by the selective quenching of metal ions, which is further directly related to the concentration of that metal ion which is to be sensed". There are several reports available in literature where green synthesised CQDs are widely employed for sensing several metal ions successfully. Using *Borassus flabellifer* (ice apple) as precursor, Nagaraj et al., synthesised green CQDs via hydrothermal approach that were further employed for sensing Fe^{+3} with a detection limit of 2.01 μM. This sensor was very efficient that can be used for tap and potable water [70]. Using bamboo leaves as starting material and solvothermal method, green CDs were synthesised which were further analysed for sensing of two metal ions $Hg^{2+}$and Pb^{2+} with LOD of 0.22 and 0.14 nM, sequentially as demonstrated in Fig. 10a, b, c, d [71]. These results clearly demonstrated that green CQDs can be widely employed for act as efficient, eco-friendly and cost-effective sensors for heavy metal ions.

5.3 *Bioimaging*

CQDs are frequently employed for bioimaging because of their comparatively low cytotoxicity and its resistive nature to photobleaching [7]. CQDs have been employed for bioimaging for "in vivo and in vitro imaging". CQDs are considered as ideal probes for bioimaging due to their several properties like nanosized, high aqueous stability and strong fluorescent nature. Using watermelon peel as precursor, Zhou et al., fabricated high quality fluorescent CQDs having size 2 nm emitting intense blue fluorescence and good stability over a widespread range of pH (2–11) as shown in Fig. 11. [76].

The as synthesised CQDs were later on analysed for imaging of Hela cells successfully as illustrate in Fig. 12 [76]. These results revealed the role of CQDs

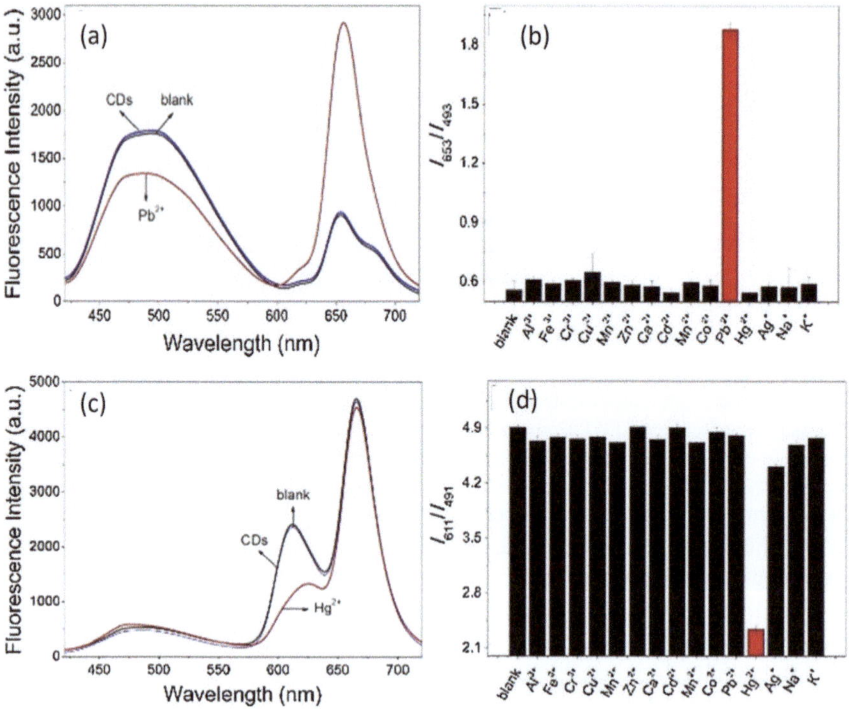

Fig. 10 Sensing of two metal ions (**a** and **b**) Hg^{+2} and (**c** and **d**) Pb^{+2}. Reprinted with permission [71]

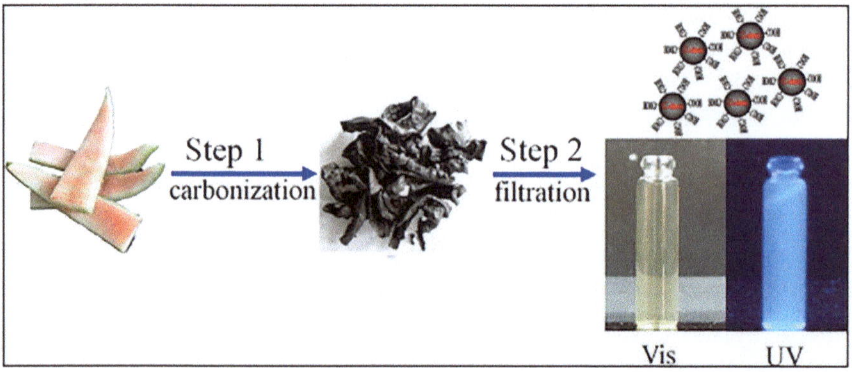

Fig. 11 Schematic illustration of fluorescent CQDs from watermelon peels. Reprinted with permission [76]

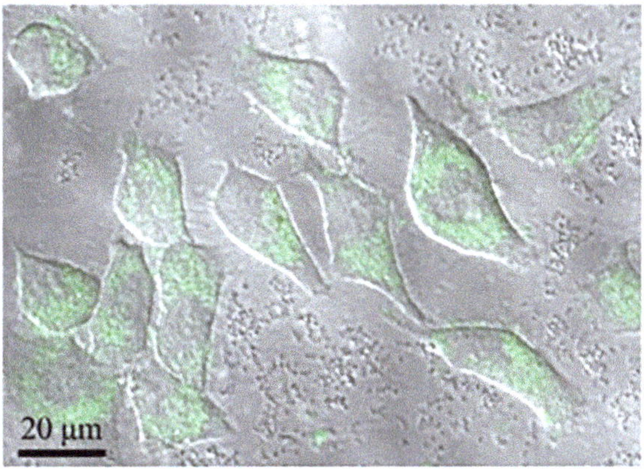

Fig. 12 Confocal microscopy picture of Hela cell incubated with CQDs. Reprinted with permission copyright © 2011 Elsevier Ltd, (License number—5653070746557). All rights reserved [76]

as bioimaging agents and their biocompatible nature and could be utilized as efficient optical imaging probes [76]. Excitation dependent fluorescence behaviour of some CQDs allow their use as multicolour fluorescence imaging agents. For instance, highly stable and biocompatible fluorescent CQDs exhibiting quantum yield of 9.91% derived from pyrolysis of peanut shell [50].

These CQDs were later analysed for imaging of HepG2 cells resulting in blue, green and red emissions corresponding to excitation wavelengths of 405, 488 and 514 nm sequentially. Similarly, *Allium fistulosum* derived CQDs have been utilized for multicolour imaging of MCF-7 and K562 cells by utilizing excitation wavelengths of 405, 488 and 561 nm associated to blue, green, and red emissions respectively [77]. Surface passivation and functionalization of CQDs leads to have an impact on their related cytotoxicity and potential imaging in several cells. Dehvari et al., using shells of crab, fabricated CQDs and used folic acid for their functionalization and analysed for bioimaging of HeLa cells that exhibited significantly extra folate receptors compared to healthy cells [78]. Along with, these CQDS demonstrated enhanced uptake for cancerous cells [78]. Green CQDs can also be used for in vivo imaging. In an investigation by Wei et al., embryos of Zebra fish exhibited survival rate of 93% after 94 h of exposure to greater than 200 $\mu g\ mL^{-1}$ of CQDs in comparison to 97% in control [77]. Results revealed that *gynostemma* derived CQDs can safely utilized for bioimaging applications.

5.4 *Biomedicine*

CQDs plays a predominant role in biomedical applications because of the abundance of various surface functional moieties which results in active targeting [7]. Wang et al., synthesised hollow CQDs using bovine serum albumin having size range of 6.8 nm and further loaded it with doxorubicin [79]. This complex upon 90 min of incubation with A549 cells results in red emission from drug in the nucleus of

A549 cells. In terms of mechanism, author proposed that this complex (doxorubicin-CQDs) through endocytosis enters the cell and in the environment of low pH lysosome complex releases the drug which further arrives the nucleus. Similarly, Shao et al., synthesised CQDs utilizing mulberry leaves and further loaded with lycorine (anti-cancer drug) [80]. Compared to lycorine alone, this lycorine-CQDs complex demonstrated enhanced cell death in cancerous HepG2 cell which reveals the role of CQDs in biomedicine. CQDs are also been known for their biomedical applications. There are several studies available in the literature which reveals the antibacterial nature of CQDs. For instance, CQDs synthesised using henna leaves were found to be very efficient against "Gram-negative (*Escherichia coli)* and Gram-positive bacteria (*Staphylococcus aureus)*" as compared to bulk henna alone [81].

5.5 Ink

One of the emerging applications of CQDs is their use as fluorescent ink. Cost-effective fluorescent ink for anti-counterfeiting applications such as invisible inks for security purposes have been developed by green CQDs. For instance, utilizing oriental plane leaves as precursor for fabrication of CQDs, and then utilized these for developing fluorescent ink that is invisible in the day light and become visible in UV-light [82]. Later on milk was also utilized as precursor for developing fluorescent ink in another study [54]. Researchers used this ink for printing over commercial paper. Printed patterns demonstrated green and red fluorescence in response to excitation wavelengths of 455 and 523 nm sequentially as shown in Fig. 13.

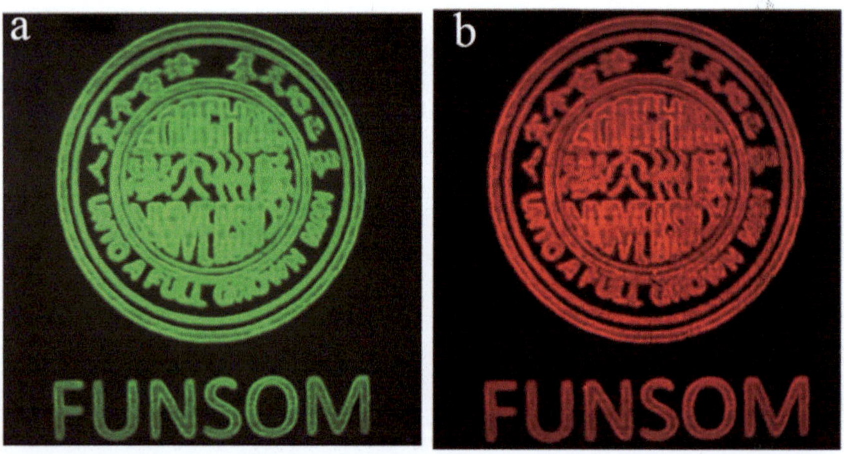

Fig. 13 Green CQDs synthesised from milk exhibited green and red fluorescence in response to excitation wavelengths of 455, 523 nm. Reprinted with permission copyright © 2014 WILEY–VCH Verlag GmbH & Co. KGaA, Weinheim Ltd, (License number—5653451071398]. All rights reserved [54]

Qu et al., developed CQDs with fluorescent efficiencies by utilizing citric acid and urea [83]. They also demonstrated that these CQDs are non-toxic to mice and bean sprouts. After testing its biocompatibility, they had successfully applied it to human skin for obtaining fingerprints. These CQDs exhibited extreme stability in printed patterns and showed same fluorescent still 3 months [84]. However, for high security purposes the stability should be for some years, therefore more investigations are required to analyse the long-term stability of green CQDs.

6 Associated Problems and Future Outlooks

In this chapter, we have introduced CQDs, a new emerging nanomaterial having size of generally 1–10 nm, associated with several characteristics such as low toxicity, biocompatibility, easy and eco- friendly synthesis, small size, high surface to volume ratio. Synthesis methods, related properties, characterization techniques and applications are being summarized in order to provide a basic background about the nanomaterial to the readers. Applications in various sectors (sensing, bioimaging, biomedicine, ink, photocatalysis), probable method of synthesis especially green synthesis using hydrothermal, microwave, ultrasonic and pyrolysis are mentioned. Despite having several advantages, they are associated with several challenges too. Fluorescent quantum yield of CQDs is generally low in comparison to semiconductor CQDs. This point out that further research should be required in this area to deeply explore the fluorescent mechanism of these nanostructures. Also, there are several challenges still reside in synthesis procedures of CQDs. Reproducibility and size uniformity is still a challenge to be addressed for applicability in several sectors. Since fluorescence properties and toxicity are dependent on distribution of particle size. Therefore, inappropriate distribution of size may be one of the reasons for their limitations in some bio-based applications specifically in vivo. Similarly, inadequate reproducibility of these materials in terms of quantum yield, size, fluorescence-based intensity limits their largescale construction. Therefore, efficient and cost-effective strategies should be required for bulk production. Green synthesis of CQDs can be an alternative for eco-friendly, cost-effective CDs but low yield is still an issue to be addressed. CQDs are diverse class of nanomaterials with several production routes. Narrowing the gap between green and chemically synthesised CQDs requires a deeper and in depth understanding of the interconnections among precursors, synthesis approach, purification techniques, resultant physical and chemical characteristics and the effect of these factors on the role of CQDs in several applications. Proper investigations on the aforementioned challenges will be required to advance the field. However, it is speculated that upcoming generations of CQDs would influentially address these barriers for becoming a prominent candidate for several applications.

First Author: Neeru Rani (neerusihan0867@gmail.com).

Author's Contributions: - The first draft writing, formal analysis, data collection and investigations were performed by [Neeru Rani]. Review and editing of this chapter have been done by [Permender Singh], and [Vinita Bhankar]. Editing by [Dinesh Kumar] Conceptualization, investigation, and resources by [Krishan Kumar] and [Sandeep Kumar].

References

1. Atchudan, R., et al. (2020). Sustainable synthesis of carbon quantum dots from banana peel waste using hydrothermal process for in vivo bioimaging. *Physica E: Low-Dimensional Systems and Nanostructures, 126*, 114417. https://doi.org/10.1016/j.physe.2020.114417
2. Chandra, S., et al. (2012). Tuning of photoluminescence on different surface functionalized carbon quantum dots. *RSC Advances, 2*(9), 3602–3606. https://doi.org/10.1039/c2ra00030j
3. Zhao, D. L., & Chung, T. S. (2018). Applications of carbon quantum dots (CQDs) in membrane technologies: A review. *Water Research, 147*, 43–49. https://doi.org/10.1016/j.watres.2018.09.040
4. Han, M., et al. (2018). Recent progress on the photocatalysis of carbon dots: Classification, mechanism and applications. *Nano Today, 19*, 201–218. https://doi.org/10.1016/j.nantod.2018.02.008
5. Sun, X., & Lei, Y. (2017). Fluorescent carbon dots and their sensing applications. *TrAC - Trends in Analytical Chemistry, 89*, 163–180. https://doi.org/10.1016/j.trac.2017.02.001
6. Feng, T., et al. (2020). Recent advances in energy conversion applications of carbon dots: From optoelectronic devices to electrocatalysis. *Small (Weinheim an der Bergstrasse, Germany), 16*(31), 1–30. https://doi.org/10.1002/smll.202001295
7. Chahal, S., et al. (2021). Green synthesis of carbon dots and their applications. *RSC Advances, 11*(41), 25354–25363. https://doi.org/10.1039/d1ra04718c
8. Feng, T., et al. (2016). Charge-convertible carbon dots for imaging-guided drug delivery with enhanced in vivo cancer therapeutic efficiency. *ACS Nano, 10*(4), 4410–4420. https://doi.org/10.1021/acsnano.6b00043
9. Zheng, J., et al. (2020). An efficient synthesis and photoelectric properties of green carbon quantum dots with high fluorescent quantum yield. *Nanomaterials, 10*(1), 1–15. https://doi.org/10.3390/nano10010082
10. Mansuriya, B. D. (2021). Carbon dots: Classification, properties, synthesis, characterization, and applications in health care—An Updated Review (2018—2021)'. *Nanomaterials, 11*(10), 2525. https://doi.org/10.3390/nano11102525.
11. Rani, N., Singh, P., Kumar, S., Kumar, P., Bhankar, V., & Kumar, K. (2023). Plant-mediated synthesis of nanoparticles and their applications: A review. *Materials Research Bulletin, 163*(January), 112233. https://doi.org/10.1016/j.materresbull.2023.112233
12. Himaja, A. L., et al. (2014). Synthesis of carbon dots from kitchen waste: Conversion of waste to value added product. *Journal of Fluorescence, 24*(6), 1767–1773. https://doi.org/10.1007/s10895-014-1465-1
13. Feng, Y., et al. (2015). Carbon dots derived from rose flowers for tetracycline sensing. *Talanta, 140*, 128–133. https://doi.org/10.1016/j.talanta.2015.03.038
14. Singh, J., et al. (2020). Highly fluorescent carbon dots derived from Mangifera indica leaves for selective detection of metal ions. *Science of the Total Environment, 720*, 137604. https://doi.org/10.1016/j.scitotenv.2020.137604
15. Kuang, T., et al. (2023). Functional carbon dots derived from biomass and plastic wastes. *Green Chemistry, 25*(17), 6581–6602. https://doi.org/10.1039/d3gc01763j
16. P, Singh, K. S., Kumar, K., & Kumar, P. B. V. (2023). Photocatalytic technologies in the removal of pharmaceuticals from aquatic systems. *Pharmaceuticals in Aquatic Environments*, pp. 173–201. https://doi.org/10.1201/9781003436607-9.

17. Rani, N., Singh, P., Kumar, S., Kumar, P., Bhankar, V., Kamra, N., et al. (2023). Recent advancement in nanomaterials for the detection and removal of uranium : A review. *Environmental Research, 234*(July), 116536. https://doi.org/10.1016/j.envres.2023.116536
18. Dey, S., et al. (2014). Luminescence properties of boron and nitrogen doped graphene quantum dots prepared from arc-discharge-generated doped graphene samples. *Chemical Physics Letters, 595–596*, 203–208. https://doi.org/10.1016/j.cplett.2014.02.012
19. Yatom, S., et al. (2017). Detection of nanoparticles in carbon arc discharge with laser-induced incandescence. *Carbon, 117*, 154–162. https://doi.org/10.1016/j.carbon.2017.02.055
20. Bao, L., et al. (2011). Electrochemical tuning of luminescent carbon nanodots: From preparation to luminescence mechanism. *Advanced Materials, 23*(48), 5801–5806. https://doi.org/10.1002/adma.201102866
21. Kang, C., et al. (2020). A review of carbon dots produced from biomass wastes. *Nanomaterials, 10*(11), 1–24. https://doi.org/10.3390/nano10112316
22. Yuan, M., et al. (2015). One-step, green, and economic synthesis of water-soluble photoluminescent carbon dots by hydrothermal treatment of wheat straw, and their bio-applications in labeling, imaging, and sensing. *Applied Surface Science, 355*, 1136–1144. https://doi.org/10.1016/j.apsusc.2015.07.095
23. Tan, X. W., et al. (2014). Carbon dots production via pyrolysis of sago waste as potential probe for metal ions sensing. *Journal of Analytical and Applied Pyrolysis, 105*, 157–165. https://doi.org/10.1016/j.jaap.2013.11.001
24. Zhang, W., et al. (2021). Insights into fluorophores of dual-emissive carbon dots derived by naphthalenediol solvothermal synthesis. *Journal of Physical Chemistry C, 125*(9), 5207–5216. https://doi.org/10.1021/acs.jpcc.0c11409
25. Deng, J., et al. (2021). Eco friendly synthesis of fluorescent carbon dots for the sensitive detection of ferric ions and cell imaging. *Arabian Journal of Chemistry, 14*(7), 103195. https://doi.org/10.1016/j.arabjc.2021.103195
26. Sarswat, P. K., & Free, M. L. (2015). Light emitting diodes based on carbon dots derived from food, beverage, and combustion wastes. *Physical Chemistry Chemical Physics, 17*(41), 27642–27652. https://doi.org/10.1039/c5cp04782j
27. Prasannan, A., & Imae, T. (2013). One-pot synthesis of fluorescent carbon dots from orange waste peels. *Industrial and Engineering Chemistry Research, 52*(44), 15673–15678. https://doi.org/10.1021/ie402421s
28. Thongsai, N., et al. (2019). Real-time detection of alcohol vapors and volatile organic compounds via optical electronic nose using carbon dots prepared from rice husk and density functional theory calculation. *Colloids and Surfaces A: Physicochemical and Engineering Aspects, 560*, 278–287. https://doi.org/10.1016/j.colsurfa.2018.09.077
29. Eslami, A., et al. (2018). Preparation of activated carbon dots from sugarcane bagasse for naphthalene removal from aqueous solutions. *Separation Science and Technology (Philadelphia), 53*(16), 2536–2549. https://doi.org/10.1080/01496395.2018.1462832
30. Monje, D. S., et al. (2021). Carbon dots from agroindustrial residues: A critical comparison of the effect of physicochemical properties on their performance as photocatalyst and emulsion stabilizer. *Materials Today Chemistry, 20*,. https://doi.org/10.1016/j.mtchem.2021.100445
31. Choi, Y., et al. (2018). Carbon dots: Bottom-up syntheses, properties, and light-harvesting applications. *Chemistry - An Asian Journal, 13*(6), 586–598. https://doi.org/10.1002/asia.201701736
32. Gu, D. et al. (2016). Green synthesis of nitrogen-doped carbon dots from lotus root for Hg(II) ions detection and cell imaging. *Applied Surface Science, 390*(Ii), pp. 38–42. https://doi.org/10.1016/j.apsusc.2016.08.012.
33. Ansi, V. A., & Renuka, N. K. (2018). Table sugar derived Carbon dot—a naked eye sensor for toxic Pb^{2+} ions. *Sensors and Actuators, B: Chemical, 264*, 67–75. https://doi.org/10.1016/j.snb.2018.02.167
34. Gupta, A., et al. (2016). Paper strip based and live cell ultrasensitive lead sensor using carbon dots synthesized from biological media. *Sensors and Actuators, B: Chemical, 232*, 107–114. https://doi.org/10.1016/j.snb.2016.03.110

35. Xia, L., Li, X., Zhang, Y., Zhou, K., Yuan, L., Shi, R., Zhang, K., Fu, Q. (2022). Sustainable and green synthesis of waste-biomass-derived carbon dots for parallel and semi-quantitative visual detection of Cr(VI) and Fe^{3+} Molecules, *27*, 1258. https://doi.org/10.3390/molecules27041258.
36. Atchudan, R., et al. (2017). Facile green synthesis of nitrogen-doped carbon dots using Chionanthus retusus fruit extract and investigation of their suitability for metal ion sensing and biological applications. *Sensors and Actuators, B: Chemical, 246*, 497–509. https://doi.org/10.1016/j.snb.2017.02.119
37. Li, Z., et al. (2019). Frontiers in carbon dots: Design, properties and applications. *Materials Chemistry Frontiers, 3*(12), 2571–2601. https://doi.org/10.1039/c9qm00415g
38. Mishra, V., et al. (2018). Carbon dots: Emerging theranostic nanoarchitectures. *Drug Discovery Today, 23*(6), 1219–1232. https://doi.org/10.1016/j.drudis.2018.01.006
39. Pan, M. et al. (2020). Fluorescent carbon quantum dots-synthesis, functionalization and sensing application in food analysis. *Nanomaterials, 10*(5). https://doi.org/10.3390/nano10050930.
40. Wang, J., Wang, C. F., & Chen, S. (2012). Amphiphilic egg-derived carbon dots: Rapid plasma fabrication, pyrolysis process, and multicolor printing patterns. *Angewandte Chemie - International Edition, 51*(37), 9297–9301. https://doi.org/10.1002/anie.201204381
41. Pang, Y. L. et al. (2020). Biomass-based photocatalysts for environmental applications. In: Inamuddin, Asiri, A., Lichtfouse, E. (eds) Nanophotocatalysis and Environmental Applications. Environmental Chemistry for a Sustainable World, vol. 30. Springer, Cham. https://doi.org/10.1007/978-3-030-12619-3_3.
42. Lou, Y., et al. (2021). Recent advances of biomass carbon dots on syntheses, characterization, luminescence mechanism, and sensing applications. *Nano Select, 2*(6), 1117–1145. https://doi.org/10.1002/nano.202000232
43. Zhang, J., & Yu, S. H. (2016). Carbon dots: Large-scale synthesis, sensing and bioimaging. *Materials Today, 19*(7), 382–393. https://doi.org/10.1016/j.mattod.2015.11.008
44. Cao, M., et al. (2019). A novel method for the preparation of solvent-free, microwave-assisted and nitrogen-doped carbon dots as fluorescent probes for chromium(VI) detection and bioimaging. *RSC Advances, 9*(15), 8230–8238. https://doi.org/10.1039/c9ra00290a
45. De Medeiros, T. V., et al. (2019). Microwave-assisted synthesis of carbon dots and their applications. *Journal of Materials Chemistry C, 7*(24), 7175–7195. https://doi.org/10.1039/c9tc01640f
46. Lai, C. W., et al. (2012). Facile synthesis of highly emissive carbon dots from pyrolysis of glycerol; Gram scale production of carbon dots/mSiO 2 for cell imaging and drug release. *Journal of Materials Chemistry, 22*(29), 14403–14409. https://doi.org/10.1039/c2jm32206d
47. Ludmerczki, R., et al. (2019). Carbon dots from citric acid and its intermediates formed by thermal decomposition. *Chemistry - A European Journal, 25*(51), 11963–11974. https://doi.org/10.1002/chem.201902497
48. Sharma, A., & Das, J. (2019). Small molecules derived carbon dots: Synthesis and applications in sensing, catalysis, imaging, and biomedicine. *Journal of Nanobiotechnology, 17*(1), 1–24. https://doi.org/10.1186/s12951-019-0525-8
49. Sabet, M., & Mahdavi, K. (2019). Green synthesis of high photoluminescence nitrogen-doped carbon quantum dots from grass via a simple hydrothermal method for removing organic and inorganic water pollutions. *Applied Surface Science, 463*, 283–291. https://doi.org/10.1016/j.apsusc.2018.08.223
50. Xue, M., et al. (2016). Green synthesis of stable and biocompatible fluorescent carbon dots from peanut shells for multicolor living cell imaging. *New Journal of Chemistry, 40*(2), 1698–1703. https://doi.org/10.1039/c5nj02181b
51. Zhao, S., et al. (2020). Green production of fluorescent carbon quantum dots based on pine wood and its application in the detection of Fe^{3+}. *Journal of Cleaner Production, 263*, 121561. https://doi.org/10.1016/j.jclepro.2020.121561
52. Saxena, M., & Sarkar, S. (2012). Synthesis of carbogenic nanosphere from peanut skin. *Diamond and Related Materials, 24*, 11–14. https://doi.org/10.1016/j.diamond.2012.01.035

53. Xie, Y. et al. (2019). Green hydrothermal synthesis of N-doped carbon dots from biomass highland barley for the detection of Hg2+. *Sensors (Switzerland), 19*(14). https://doi.org/10.3390/s19143169.
54. Wang, J., et al. (2015). Large-scale green synthesis of fluorescent carbon nanodots and their use in optics applications. *Advanced Optical Materials, 3*(1), 103–111. https://doi.org/10.1002/adom.201400307
55. Xu, J., et al. (2020). Synthesis of green-emitting carbon quantum dots with double carbon sources and their application as a fluorescent probe for selective detection of Cu^{2+} ions. *RSC Advances, 10*(5), 2536–2544. https://doi.org/10.1039/c9ra08654d
56. Ding, H. et al. (2020). Surface states of carbon dots and their influences on luminescence. *Journal of Applied Physics, 127*(23). https://doi.org/10.1063/1.5143819.
57. El-Shafey, A. M. (2021). Carbon dots: Discovery, structure, fluorescent properties, and applications. *Green Processing and Synthesis, 10*(1), 134–156. https://doi.org/10.1515/gps-2021-0006
58. Jorns, M., & Pappas, D. (2021) A review of fluorescent carbon dots, their synthesis, physical and chemical characteristics, and applications. *Nanomaterials, 11*(6). https://doi.org/10.3390/nano11061448.
59. Jung, H., et al. (2022). Recent progress on carbon quantum dots based photocatalysis. *Frontiers in Chemistry, 10*(April), 1–28. https://doi.org/10.3389/fchem.2022.881495
60. Liu, Y., et al. (2017). The synthesis of water-dispersible zinc doped $AgInS_2$ quantum dots and their application in Cu2+ detection. *Journal of Luminescence, 192*(March), 547–554. https://doi.org/10.1016/j.jlumin.2017.07.018
61. Tian, L., et al. (2021). Carbon quantum dots for advanced electrocatalysis. *Journal of Energy Chemistry, 55*, 279–294. https://doi.org/10.1016/j.jechem.2020.06.057
62. Miao, S., et al. (2020). Hetero-atom-doped carbon dots: Doping strategies, properties and applications. *Nano Today, 33*, 100879. https://doi.org/10.1016/j.nantod.2020.100879
63. Singh, P. A., et al. (2023). Assessment of biomass-derived carbon dots as highly sensitive and selective template for sensing of hazardous ions. *Nanoscale* [Preprint]. https://doi.org/10.1039/d3nr01966g.
64. Gayen, B., Palchoudhury, S., & Chowdhury, J. (2019). Carbon dots: A mystic star in the world of nanoscience. *Journal of Nanomaterials, 2019*,. https://doi.org/10.1155/2019/3451307
65. Madaan, V. et al. (2022). Metal-decorated CeO_2 nanomaterials for photocatalytic degradation of organic pollutants. *Inorganic Chemistry Communications, 146*(October). https://doi.org/10.1016/j.inoche.2022.110099.
66. Singh, P. et al. (2022). Nanomaterials photocatalytic activities for waste water treatment : a review, Environmental Science and Pollution Research. Springer Berlin Heidelberg. https://doi.org/10.1007/s11356-022-22550-7.
67. Das, G. S., et al. (2019). Biomass-derived carbon quantum dots for visible-light-induced photocatalysis and label-free detection of Fe(III) and ascorbic acid. *Scientific Reports, 9*(1), 1–9. https://doi.org/10.1038/s41598-019-49266-y
68. Wahyu, M. et al. (2021). Chemosphere synthesis of tungsten oxide / amino-functionalized sugarcane bagasse derived-carbon quantum dots (WO_3/N-CQDs) composites for methylene blue removal, *277*. https://doi.org/10.1016/j.chemosphere.2021.130300.
69. Yoo, D. et al. (2019). Carbon dots as an effective fluorescent sensing platform for metal ion detection. *Nanoscale Research Letters, 14*(1). https://doi.org/10.1186/s11671-019-3088-6.
70. Nagaraj, M. et al. (2022). Detection of Fe^{3+} ions in aqueous environment using fluorescent carbon quantum dots synthesized from endosperm of Borassus flabellifer. *Environmental Research*, 212, p. Nagaraj, M. et al. (2022) 'Detection of Fe^{3+} ions. https://doi.org/10.1016/j.envres.2022.113273.
71. Liu, Z., et al. (2019). Ratiometric fluorescent sensing of Pb^{2+} and Hg^{2+} with two types of carbon dot nanohybrids synthesized from the same biomass. *Sensors and Actuators, B: Chemical, 296*(February), 126698. https://doi.org/10.1016/j.snb.2019.126698
72. Tyagi, A., et al. (2016). Green synthesis of carbon quantum dots from lemon peel waste: Applications in sensing and photocatalysis. *RSC Advances, 6*(76), 72423–72432. https://doi.org/10.1039/c6ra10488f

73. Arumugam, N., & Kim, J. (2018). Synthesis of carbon quantum dots from Broccoli and their ability to detect silver ions. *Materials Letters, 219*(February), 37–40. https://doi.org/10.1016/j.matlet.2018.02.043
74. Essner, J. B., et al. (2016). Pee-dots: Biocompatible fluorescent carbon dots derived from the upcycling of urine. *Green Chemistry, 18*(1), 243–250. https://doi.org/10.1039/c5gc02032h
75. Liu, Y., et al. (2016). Selective and sensitive chemosensor for lead ions using fluorescent carbon dots prepared from chocolate by one-step hydrothermal method. *Sensors and Actuators, B: Chemical, 237*, 597–604. https://doi.org/10.1016/j.snb.2016.06.092
76. Zhou, J., et al. (2012). Facile synthesis of fluorescent carbon dots using watermelon peel as a carbon source. *Materials Letters, 66*(1), 222–224. https://doi.org/10.1016/j.matlet.2011.08.081
77. Wei, X., et al. (2019). Green synthesis of fluorescent carbon dots from gynostemma for bioimaging and antioxidant in zebrafish. *ACS Applied Materials and Interfaces, 11*(10), 9832–9840. https://doi.org/10.1021/acsami.9b00074
78. Dehvari, K., et al. (2019). Sonochemical-assisted green synthesis of nitrogen-doped carbon dots from crab shell as targeted nanoprobes for cell imaging. *Journal of the Taiwan Institute of Chemical Engineers, 95*, 495–503. https://doi.org/10.1016/j.jtice.2018.08.037
79. Wang, Q., et al. (2013). Hollow luminescent carbon dots for drug delivery. *Carbon, 59*, 192–199. https://doi.org/10.1016/j.carbon.2013.03.009
80. Shao, Y. et al. (2020). Green synthesis of multifunctional fluorescent carbon dots from mulberry leaves (Morus alba L.) residues for simultaneous intracellular imaging and drug delivery. *Journal of Nanoparticle Research, 22*(8). https://doi.org/10.1007/s11051-020-04917-4.
81. Shahshahanipour, M., et al. (2018). An ancient plant for the synthesis of a novel carbon dot and its applications as an antibacterial agent and probe for sensing of an anti-cancer drug. *Materials Science and Engineering C, 98*, 826–833. https://doi.org/10.1016/j.msec.2019.01.041
82. Zhu, L., et al. (2013). Plant leaf-derived fluorescent carbon dots for sensing, patterning and coding. *Journal of Materials Chemistry C, 1*(32), 4925–4932. https://doi.org/10.1039/c3tc30701h
83. Qu, S., et al. (2012). A biocompatible fluorescent ink based on water-soluble luminescent carbon nanodots. *Angewandte Chemie - International Edition, 51*(49), 12215–12218. https://doi.org/10.1002/anie.201206791
84. Bhatt, S. et al. (2018). Green route for synthesis of multifunctional fluorescent carbon dots from Tulsi leaves and its application as Cr(VI) sensors, bio-imaging and patterning agents. *Colloids and Surfaces B: Biointerfaces, 167*(Vi), 126–133. https://doi.org/10.1016/j.colsurfb.2018.04.008.
85. Singh, P., & Kumar, S. K. K. (2023). Biogenic synthesis of Allium cepa derived magnetic carbon dots for enhanced photocatalytic degradation of methylene blue and rhodamine B dyes. *Biomass Conversion and Biorefinery*]. https://doi.org/10.1007/s13399-023-05047-2.
86. Singh, P., Rani, N., et al. (2023). Assessing the biomass-based carbon dots and their composites for photocatalytic treatment of wastewater. *Journal of Cleaner Production, 413*(January), 137474. https://doi.org/10.1016/j.jclepro.2023.137474
87. Wei, Z., et al. (2019). Green synthesis of nitrogen and sulfur co-doped carbon dots from Allium fistulosum for cell imaging. *New Journal of Chemistry, 43*(2), 718–723. https://doi.org/10.1039/c8nj05783d

Green Synthesis of Carbon Quantum Dots Through Various Strategies

Sarita Shaktawat, Surendra K. Yadav, Diksha Singh, and Jay Singh

Abstract Carbon quantum dots (CQDs) are nano-sized particles made of carbon with distinct optical, electronic, and chemical features. These particles can be produced using a range of environmentally friendly techniques, including the hydrothermal process, microwave-assistance, electrochemical methods, laser ablation, and the use of green precursors, solar energy, plant-based extracts, biomass, eco-friendly ligands, and solvents. These sustainable approaches minimize environmental harm while often resulting in CQDs with superior qualities for a wide range of uses. Their unique properties make them highly valuable in diverse sectors, including biomedical imaging, drug delivery systems, sensing technologies, photocatalytic processes, light-emitting devices, energy storage solutions, solar energy conversion, catalysis, optical labelling, and environmental surveillance. Ongoing research efforts are dedicated to finding more eco-conscious ways to manufacture CQDs.

Keywords Carbon quantum dots · Green synthesis · GQDs

1 Introduction

Green CQDs (G-CQD) represent an advanced synthesis of QDs, a field recognized by the Nobel Prize in Chemistry 2023, awarded to Ekimov, Brus, and Bawendi for their pioneering work in developing and discovering QDs. These entities are a prime demonstration of quantum mechanics in action, encapsulating matter in its most fundamental form within a quantum box [1]. QDs are semiconductor particles, mere nanometers in size, known for their unique optoelectronic properties. This new wave of inorganic, nanoscale materials possess a variable number of electrons in quantum states, marking a significant stride in semiconductor technology [2]. Employed across a spectrum of fields including medical, technological, and therapeutic applications,

S. Shaktawat · S. K. Yadav · D. Singh · J. Singh (✉)
Department of Chemistry, Institute of Sciences, Banaras Hindu University, Varanasi, Uttar Pradesh 221005, India
e-mail: jaysingh.chem@bhu.ac.in

V. Kumar et al. (eds.), *Green Carbon Quantum Dots*, Engineering Materials,
https://doi.org/10.1007/978-981-97-6203-3_2

as well as in imaging, these artificial semiconductors have broadened the horizons of research [3].

CQDs stand out as exemplary zero-dimensional semiconductor QDs, noted for their solubility, low toxicity, eco-friendly nature, affordability, and straightforward synthesis [4]. The scientific community is increasingly leaning towards renewable biomass to curtail the use of poisonous substances, safeguarding both animals and the environment. In this light, G-CQDs have been developed from renewable sources like agricultural waste, plants, and organic biomaterials [5], embodying an eco-conscious synthesis approach. G-CQDs excel in the nanomaterial realm due to their optoelectronic properties, biocompatibility, non-toxicity, consistent particle size, high photostability, cost-effectiveness, and adjustable photoluminescence. These attributes have propelled G-CQDs into the forefront of applications in bio-imaging, medical diagnostics, sensing, photocatalysis, metal ion detection, and drug delivery [6]. The production of G-CQDs utilizes numerous methods such as laser ablation, hydrothermal processes, electrochemical oxidation, reflux, microwave irradiation, and ultrasonication, which are classified into bottom-up or top-down approaches. The creation of CQDs necessitates several chemical reactions and functionalization with carbonic and hydroxyl groups, among others [7] etc. However, synthesizing G-CQDs poses a unique challenge: achieving controlled growth at the quantum mechanical level using renewable biomass, a task that continues to challenge scientists.

Recent studies have highlighted the potential of renewable biomass for G-CQD synthesis. For instance, Hou et al. (2020) reported the creation of fluorescent G-CQDs from grapefruit peel using a one-pot, eco-friendly hydrothermal method [8]. Similarly, [9] produced G-CQDs from lemon peel waste, demonstrating their application in sensitive fluorescent detection of Cr^{6+} ions in water and the photocatalytic degradation of dye [9]. Furthermore Gonzalez et al. [10] utilized agave bagasse, a by-product of tequila production, to synthesize G-CQDs through a simple drying and grinding process, showcasing their optical properties and photoluminescence [11]. This discussion aims to concisely review the diverse methods of G-CQD synthesis, exploring the strategies, challenges, and solutions within this burgeoning field.

1.1 CQDs and G-CQDs

QDs are minuscule semiconductor particles, just a few nanometers across, whose optical and electronic characteristics diverge from larger particles due to the principles of quantum mechanics. They stand at the forefront of nanotechnology research. The highly customizable nature of quantum dots makes them extremely appealing for a broad spectrum of applications.

QDs can be used as qubits, the basic units of information in quantum computing, offering a pathway toward powerful quantum computers. Their adjustable optical properties make quantum dots excellent for use in biological imaging, where they can be used to tag or highlight specific parts of a biological system in vivid colours. QDs can be employed in solar cells to improve efficiency. They can be engineered to absorb a wider range of wavelengths from the solar spectrum compared to traditional solar cell materials. Quantum dot Light Emitting Diodes (QLEDs) offer brighter, more saturated colours than traditional LEDs, making them attractive for displays in televisions, smartphones, and other devices. Beyond LEDs, quantum dots are used in a variety of optoelectronic devices due to their unique electronic properties, including lasers and photodetectors. The unique light-emitting properties of quantum dots make them suitable for use in security inks and as tracers for tracking the distribution of products or substances.

The size of a quantum dot gives it a specific quantum confinement effect, where the energy levels that the electrons can occupy are quantized. By changing the size of the quantum dot, you can control the wavelengths of light they absorb and emit. This tunability is what makes them so useful in a wide range of applications, as you can "dial in" the desired optical properties by simply changing the size of the dots during synthesis.

CQDs were unexpectedly discovered in 2004 by researchers who were purifying single-walled carbon nanotubes through the arc discharge method, according to Xu and her team [12]. These nanoparticles are carbon-based, spherical in shape, and larger than 10 nm, exhibiting distinctive fluorescent properties akin to semiconductors. They are functionalized with groups such as hydroxyl, amino, carboxyl, and aromatic, making them highly appealing for various applications. CQDs stand out due to their high quantum yield, tunable emission wavelengths, excellent photostability, minimal cytotoxicity, outstanding biocompatibility, simple surface modification, and exceptional chemical stability, as noted [13]. CQDs can be manufactured through numerous methods, including the environmentally friendly approach of using renewable biomass to produce what are known as G-CQDs. Although utilizing renewable biomass and waste at the quantum level presents significant challenges for researchers, G-CQDs offer a sustainable alternative. These CQDs are eco-friendly, biodegradable, compatible with biological systems, non-toxic, and safe, finding applications across diverse scientific, medical, and technological domains. Figure 2.1 describe the quantum confinement effect that have been visualized, as the levels get wider as size below 10 nm.

1.2 Importance of Green Synthesis

Synthesis process were minimize hazardous chemical, negligible environment impact and fabricate sustainable product is briefly called green synthesis [14]. Today, numerous studies focus on creating nanometer-sized particles using chemical, physical, and eco-friendly synthesis techniques [15]. However, green synthesis methods are becoming more favoured over physical and chemical methods due to their lower energy consumption, reduced toxicity, and minimal use of hazardous chemicals [16]. Physical method such as aerosols, ultraviolet thermal decomposition etc. require high temperature (aerosols method, 2400 K) and pressure [17–19]. Currently green synthesis highlight due to environmental impact reduction, resource conservation,

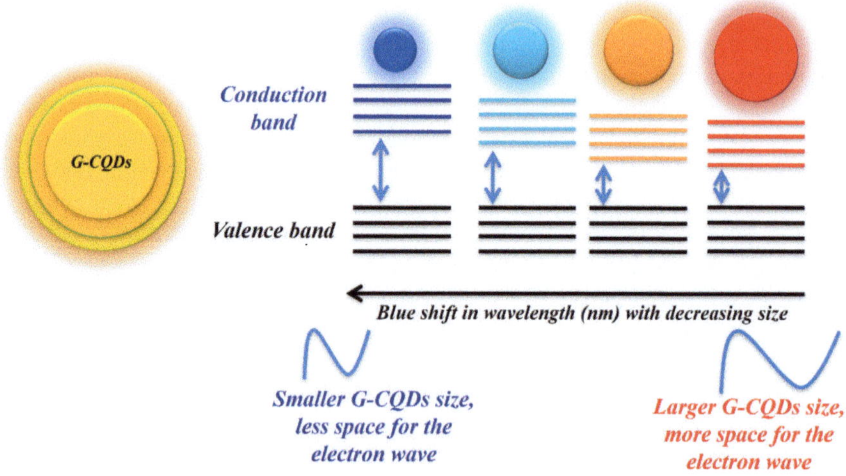

Fig. 2.1 The quantum confinement effect has been visualized, as the levels get wider as size ceases below 10 nm

nontoxic, cost-effective, regulatory compliance, sustainable, biodiversity preservation, technological advancements, market demand, biodegradable, biocompatible etc. properties has been shown [14]. Green synthesis process has been utilized mainly microorganism (fungi, bacteria and algae), plant extract (leaf, flower, roots, peeling, fruit, seeds of various plant), industrial waste (hazardous chemicals etc.) [20].

1.3 Various Method for Synthesis of G-CQDs

The discovery of CQDs has presented new challenges to the quantum mechanics of matter, specifically regarding the control of growth at the nanometer scale for spherical particles. The synthesis of Graphene -CQDs is principally categorized into two approaches: "top-down" and "bottom-up," [4]. Each approach encompasses a variety of methods for synthesizing CQDs, employing diverse chemical and physical parameters. In contrast to the time-intensive, hazardous, and complex nature of chemical and physical synthesis methods, green synthesis offers a more appealing alternative. Green synthesis methods are distinguished by their sustainability, non-toxicity, biodegradability, biocompatibility, and chemical reduction capabilities, making them a preferred choice for the synthesis of CQDs. Figure 2.2 shows the schematic representation of various methods for preparation of GQDs.

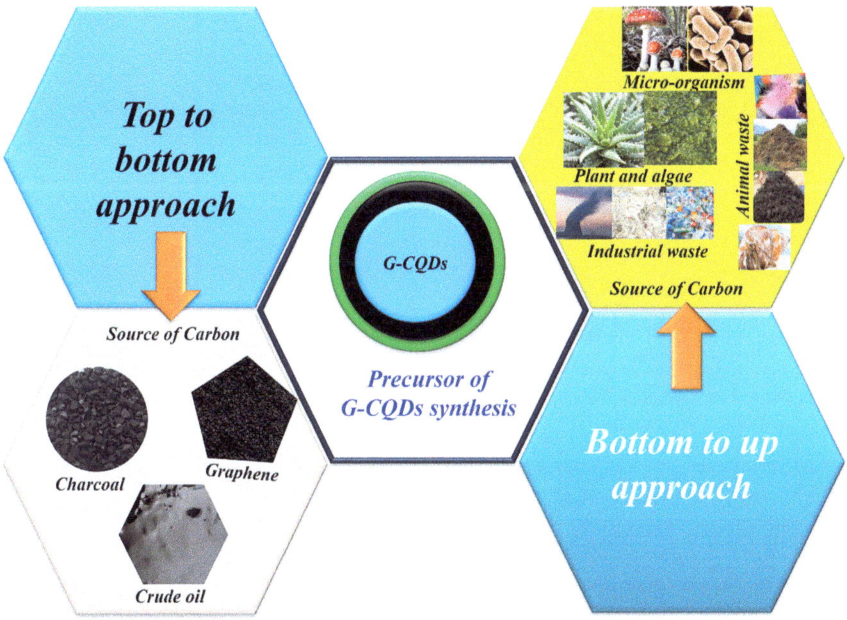

Fig. 2.2 The synthesis methods for desired C-QDs

1.3.1 Top-Down Approach

Top-down approaches involve physical and mechanical techniques for creating nano-sized particles through processes such as cutting, etching, grinding, ball milling, and lithography, starting from bulk raw materials. This conventional approach has been employed to create CQDs from larger carbon sources, including graphite, graphene, and polysaccharides, among others [21]. Despite its utility, this approach faces several challenges, including spectral inefficiency, low yields, poor size control, and the necessity for high temperatures during experimentation. These challenges have led to a shift towards developing new strategies that are more straightforward, environmentally friendly, and sustainable, aiming to reduce global warming while being renewable and cost-effective in the synthesis of CQDs [22]. This discussion primarily centers on the synthesis of G-CQDs employing various top-down techniques.

Arc Discharge

The Arc discharge method operates at low temperatures within an inert gas chamber, equipped with a graphite cathode and an anode made from powdered carbon precursors. Initially, the precursor is evaporated and consumed, leading to the sublimation of the resultant carbon precursor. This sublimated carbon then travels towards the cathode, where it is deposited on the graphite surface as nano-sized CQDs [23].

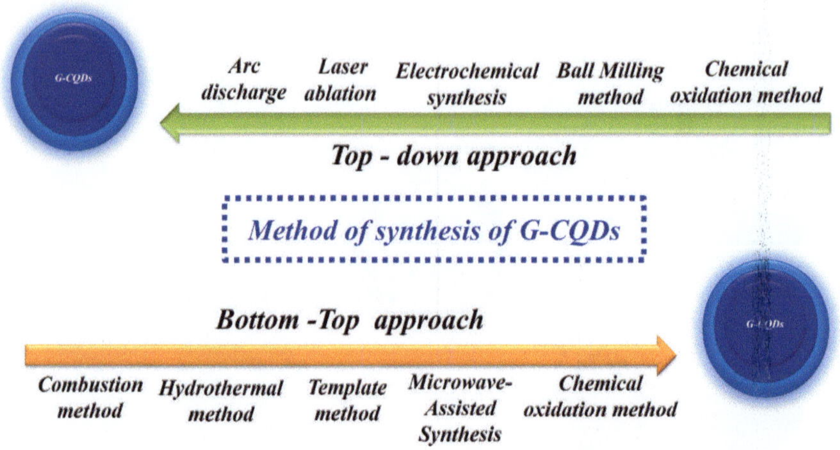

Fig. 2.3 Various physical Methods had been mentioned that ultimately used for QDs synthesis

Chao-Mujica and their team have devised a technique for creating fluorescent CQDs through submerged arc discharge in water (SADW). They emphasize the method's straightforwardness, inherent phase separation, and its scalability potential [24].

Laser Ablation

Laser ablation represents a distinctive and promising method that, despite being time-intensive, operates at high temperatures and pressures to cut materials into CQDs, albeit with concerns regarding residue impurity and contamination [25] 26. Cui et al. exploited a process for producing homogeneous CQDs using a cost-effective carbon cloth and an ultrafast dual beam pulsed laser ablation technique. The CQDs synthesized via this method demonstrated a notable 35.4% quantum yield, enhancing their application in photoluminescence (PL) emission mechanisms and cellular bio-imaging [27]. Similarly, [28] adeptly blended a composite of gold nanoparticles and CQDs through the laser ablation method, aiming to amplify the fluorescence properties. Initially, CQDs were produced hydrothermally using biochar from empty fruit bunches. Subsequently, these CQDs were employed to create a Gd/CQDs nanocomposite through laser ablation in solution, which was then used for detecting pyrene through photoluminescence spectroscopy. The laser ablation method is highlighted for its simplicity, speed, and green approach in producing highly pure CQDs [28]. Figure. 2.3 represents the various physical methods used for QDs synthesis.

Electrochemical Synthesis

Electrochemical synthesis is a method used to create chemical compounds through the application of electrical potentials or currents. This setup involves the use of electrochemical cells, such as potentiostats or galvanostats, which contain electrodes, solvents, supporting electrolytes, and precursor chemicals [29]. The process, which involves multi-electron transfers from metal and non-metal ionic melts, occurs at very high temperatures [30]. [10] reviewed the electrochemical making of CQDs and discussed how varying electrochemical parameters can adjust the properties of CQDs. They highlighted that the electrochemical method is an environment friendly, practical, and efficient way to produce high-quality CQDs [10]. Similarly, Ding et al. emphasized the simplicity, mild reaction conditions, low cost, and scalability of the electrochemical synthesis method for CQDs [31]. The electrochemical synthesis of Nitrogen-doped CQDs (N-CQDs) with Copper(I) Oxide (Cu_2O) and their application in the sensitive and selective detection of non-steroidal anti-inflammatory drugs in berries. Initially, N-CQDs were synthesized from citric acid and ethylenediamine in deionized water with microwave treatment. Subsequently, N-CQDs were in-situ deposited on Cu microcubes using an electrochemical synthesis method, cycling between -0.5 V and +0.3 V for 40 cycles on a glassy carbon electrode (GCE). The resulting N-CQDs/Cu_2O/GCE platform demonstrated excellent sensitivity, with a linear detection range of 1 to 907 μM, a limit of detection (LOD) of 0.002 μM, and a sensitivity of 21.87 $\mu A/\mu M.cm^2$ [32], 33 created Iron-Nitrogen doped CQDs (Fe-N-CQDs) using an electrochemical synthesis technique, establishing a fluorescent probe that is both sensitive and selective for detecting Cu^{2+} ions. The production of Fe-N-CQDs involved electrochemical oxidation of a Fe-N-C layer on a carbon cloth electrode. This method demonstrated exceptional sensitivity, exhibiting a linear response to Cu^{2+} concentrations between 100 nM and 1000 nM ($R^2 = 0.997$), with a detection limit of 59 nM. [33].

Ball Milling Method

The grinding process is recognized as one of the modest and most proficient mechanical methods among top-down approaches for material synthesis. This method enables the diminution of nanoparticles by conveying kinetic energy from the grinding medium to the material, under precise temperature and pressure conditions. The powder obtained from the ball milling process exhibits significant effects on crystal size and surface properties [34].

An article on the synthesis of carbon nanoparticles via a green mechanochemical approach. This method, followed by careful thermal annealing for structural refinement, successfully produced nitrogen-doped carbon nanoparticles. These nanoparticles were used as an electrocatalyst for the oxygen reduction reaction, demonstrating a kinetic restricting current density of 5.2 $mA.cm^{-2}$, which surpasses the performance of conventional Pt/C catalysts. The findings suggest that the mixture of ball milling and annealing represents a scalable production technique for developing highly efficient and practical electrolysis systems [35]. Furthermore, through the mechanical ball milling method, synthesized magnesium-doped carbon quantum dots using a mixture of cellulose and magnesium powder. This approach led to the successful creation of a fluorescent sensor for Fe^{3+} detection, highlighting the ball milling method's suitability for the leading industrial production of CQDs [36].

Chemical Oxidation Method

Feng et al. developed G-CQDs from coke, a byproduct of coal, through a chemical oxidation process. These eco-friendly G-CQDs exhibit blue fluorescence, with a fluorescence quantum yield of 9.2% and a spectral configuration that includes 48% blue-green-red spectra. Such characteristics make these CQDs particularly suitable for use in the production of cool white light zone light-emitting diodes (LEDs), highlighting their significant potential for applications in lighting technology [37].

Other

In a novel strategy, S. et al. derived intensely fluorescent CQDs from sugarcane bagasse pulp through a top-down technique that includes chemical oxidation followed by the subsequent exfoliation of sugarcane carbon. This process, which leverages a sustainable and renewable basis, was validated through techniques such as X-ray diffraction (XRD), Fourier-transform infrared spectroscopy (FTIR), and high-resolution transmission electron microscopy (HR-TEM) (S. and D., 2016). The resulting CQDs have found applications in fields like biosensors, bioimaging, and drug delivery, showcasing the versatility and environmental friendliness of this synthesis approach [37]. Navarro-Badilla et al. introduced a green synthesis technique for CQDs using extracts from Opuntia ficus-indica and Agave Maximiliano. This method employed mesoporous zeolite 4A as a catalyst for the thermal treatment of the extracts, leading to the production of CQDs with narrow size distributions of 2 nm and 3 nm, and an absorption band at 220 nm. These CQDs were also evaluated as a surface-enhanced Raman scattering (SERS) substrate for the detection of methylene blue (MB), demonstrating the potential of such materials in environmental and analytical applications [38].

1.3.2 Bottom-Up Approach

The bottom-up approach encompasses both physical and chemical strategies that leverage the self-assembly of atoms, molecules, and clusters through a nucleation process, which can be finely tuned at the nanoscale. Physical methods involve the condensation of vapour-phase species through techniques such as evaporation, sputtering, plasma arcing, and laser ablation. On the chemical side, condensation is achieved through processes like combustion, hydrothermal synthesis, templating, microwave-assisted synthesis, and electrochemical methods [39] 40 4 41 42. This approach is favoured for its precise control over nanoparticle size, higher yields, reduced toxicity, and its environmentally friendly nature. Modern adaptations of the bottom-up approach emphasize reducing toxicity and the use of dangerous substances by incorporating biomass and waste materials. This focus on sustainable sources has led to the development of CQDs that are biodegradable, biocompatible, non-toxic, sustainable, and eco-friendly, marking a significant advancement in materials science for applications requiring environmentally conscious materials.

Chemical Oxidation Method

The said method involves processes where electron transfer occurs from an oxidizing agent to the chemical species being oxidized. Ding et al. showed the production of G-CQDs from agricultural biomass using this method. The G-CQDs produced through this approach exhibited strong fluorescence and a consistent particle size when dissolved in common solvents. The synthesized G-CQDs displayed an excitation wavelength range from 350 to 390 nm and were employed as fluorescent probes for the precise and careful detection of Fe^{3+} ions within a concentration range of 0 to 500 μM. This study effectively showcased the transformation of agricultural waste into high-value products [43]. Similarly, Baweja et al. undertook a sustainable and cost-effective approach to synthesize graphene and CQDs by chemically oxidizing and then exfoliating sugarcane bagasse waste. This process yielded graphene and CQDs characterized by their bright blue luminescent properties, highlighting an innovative use of agricultural waste for producing valuable nanomaterials [44].

Combustion Method

The combustion technique, often referred to as self-propagating high-temperature synthesis (SHS), utilizes a series of exothermic reactions between a fuel and an oxidizer to maintain a chemical reaction [45]. This process is typically carried out in two phases: the initial self-diffusion high-temperature synthesis followed by a thermal explosion mode. Using the combustion method, Liu et al. were the first to synthesize CQDs with aqueous solubility and tunable fluorescence properties [46]. In a separate study, Raikwar utilized the carbonization technique, which involves breaking down complex carbonaceous materials like wood or agricultural residues into elemental carbon and chemical compounds through heating. This method was employed using Aloe Vera gel extract to produce CQD@$LaPO4:Eu^{3+}$ nanocomposites. The inclusion of CQDs in the nanocomposite was shown to significantly enhance its luminescence intensity, a finding supported by photoluminescence excitation and emission spectra measurements at 385 nm and 485 nm, respectively [47].

Hydrothermal Method

The said method is recognized globally as a sustainable and environmentally friendly approach for nanotechnology manufacturing. It involves processing at high temperatures (200°C to 400 °C) and pressures, with specific flow rates, composition, and mixing turbulence, akin to natural geological processes that occur over millions of years. This method allows for controlled nucleation, resulting in nanoparticles of uniform size, and employs surfactants to ensure their dispersion in solvents [48]. It is particularly effective for producing CQDs through the carbonization of biomass and waste materials [49], representing a green approach to utilizing renewable carbon sources. Innovative strategies have been developed to convert biomass and waste into valuable nanomaterials. For instance, Sabet et al. reported the synthesis of green nitrogen doped CQDs via hydrothermally treated grass, which were effective in eliminating both pollutants from water. These N-CQDs demonstrated high photoluminescence and could decompose various dyes while adsorbing heavy metals like Cd^{2+} and Pb^{2+} from water [50]. Tyagi et al. utilized lemon peel waste in a hydrothermal process to create CQDs for photoluminescence sensing of Cr^{6+} and the photocatalytic degradation of dye [9]. Photoluminescent G-CQDs, characterized by their non-toxicity, stability, water solubility, biodegradability, small size, and biocompatibility, have been synthesized from natural products using hydrothermal methods. Mindivan et al. produced RC-CQDs from rosehip fruit and CA-CQDs from citric acid, incorporating them into a polycaprolactone (PCL) matrix through ultrasonic mixing to create biodegradable polymer composites. These composites were tested for biodegradability, with RC-CQDs showing distinct properties such as an amorphous structure and high hydrophobicity compared to the more hydrophilic CA-CQDs films [51]. Researchers have successfully created G-CQDs and activated carbon from banana peels through a single-step hydrothermal process. This development highlights the promise of G-CQDs for use in catalysis, biosensing, and electronic applications. Meanwhile, the activated carbon has proven to be an effective supercapacitor material, attributed to its extensive surface area (294.6 m^2/g) and porous architecture. [52]. Furthermore, Mohammadi et al. embraced a green chemical approach that avoids organic solvents by synthesizing multifunctional amphiphilic CQDs using a hydrothermal process. These carbon quantum dots (CQDs), when paired with an anion-exchange method, acted as an innovative catalytic system for the targeted oxidation of alcohols into aldehydes and ketones in water. This approach demonstrated outstanding selectivity, high turnover numbers, and impressive yields [53]. This breadth of research underscores the hydrothermal method's versatility and potential for green synthesis in nanotechnology.

Template Method

The template method utilized as an established framework for controlled production of nanoparticles and materials with nanoscale precision. This approach is divided into categories such as colloidal, soft, and hard templates, each leading to nanomaterials with distinct structures and morphologies as noted by Liu and colleagues [54]. An example of innovative application is the work by Shi and associates, who crafted a three-dimensional porous carbon structure from economical CQDs using an ice-template method. By transforming 0-D G-CQDs obtained from coal tar pitch into 3-dimensional porous carbon frameworks, the researchers created materials that are well-suited for use in high-performance lithium-ion batteries (LIBs). These LIBs, when incorporated with the porous carbon framework as the anode, demonstrated superior long-term stability, showcasing a reversible capacity of up to 273 $mA.g^{-1}$ after 1000 charging discharging cycles at a current of 2000 $mA.g^{-1}$, as reported in 2023 [55].

Further, Kwon and team's research into the precise size control during the soft template synthesis of carbon nanodots (CNDs) revealed a blue-shift in photoluminescence (PL), attributing to the unique properties of oleylamine-capped CNDs. These CNDs were synthesized utilizing an emulsion strategy, acting as a self-assembled flexible template, achieving a remarkable 60% quantum yield [56]. Wang and colleagues pursued a template-free approach, creating oxygen-rich ultrathin porous CQDs on $g\text{-}C_3N_4$ substrates. Their synthesis method was highlighted for its photocatalytic efficiency in degrading pharmaceutical and personal care products (PPCPs), marking a significant stride in environmental remediation technology [57].

Microwave-Assisted Synthesis

Zhou and colleagues achieved the successful synthesis of environment friendly and efficient CQDs, focusing on their luminescent characteristics. They employed citric acid as a starting material and glutathione as a dopant providing nitrogen and sulfur, to create N/S-doped CQDs within an hour using a microwave-assisted oil bath technique [58]. In a similar vein, Padron and his team developed fluorescent CQDs from lignocellulosic waste using a catalyzed microwave-assisted method. The resultant CQDs were thoroughly characterized through numerous analytical methods, including HR-TEM revealing an average diameter of 17.5 nm, UV-visible spectroscopy showing a peak at 325 nm, and fluorescence spectroscopy with excitation and emission peaks at 360 nm and 435 nm, respectively. These findings underscore the microwave-assisted approach as a viable and effective strategy for generating green CQDs [59]. Furthermore, Archita and her team innovated in the synthesis of biomass-derived CQDs using leaves from Plectranthusamboinicus, commonly known as Mexican Mint, employing a microwave-assisted reflux method. These CQDs exhibited fluorescent properties and were applied in the detection of various metal ions via a quenching mechanism. The sensor developed from these CQDs demonstrated high selectivity and a strong linear response, with an R^2 value of 0.9111, in the concentration range of 0 to 15 μM for Fe^{3+} ions. The report highlights the potential of plant derived CQDs in environmental monitoring and metal ion detection applications [60].

1.4 Various Strategies for Synthesis of G-CQDs

The inquiry focuses on "How matter is manipulated at the quantum level." Quantum science employs a variety of strategies, techniques, and parameters—including choice of precursor, solvent, temperature, pressure, and ambient conditions—to fabricate CQDs. G-CQDs are highlighted for their biodegradability, biocompatibility, sustainability, eco-friendliness, cost-effectiveness, efficiency, and their capability to minimize the carbon source to the quantum scale. This discussion covers several reported methodologies for generating G-CQDs. Notably, G-CQDs have been produced from a range of traces, including biomass waste, industrial by-products, and microorganisms like bacteria and fungi. Examples include grass [50], banana peel waste [61], lemon juice [62], corn stalk shell [63], Curcuma zedoaria [64], red cabbage [65], Calotropis gigantea [66] among others. Figure 2.4 represents the variety of renewable, locally available materials can serve as precursors for the eco-friendly production of carbon dots (CDs).

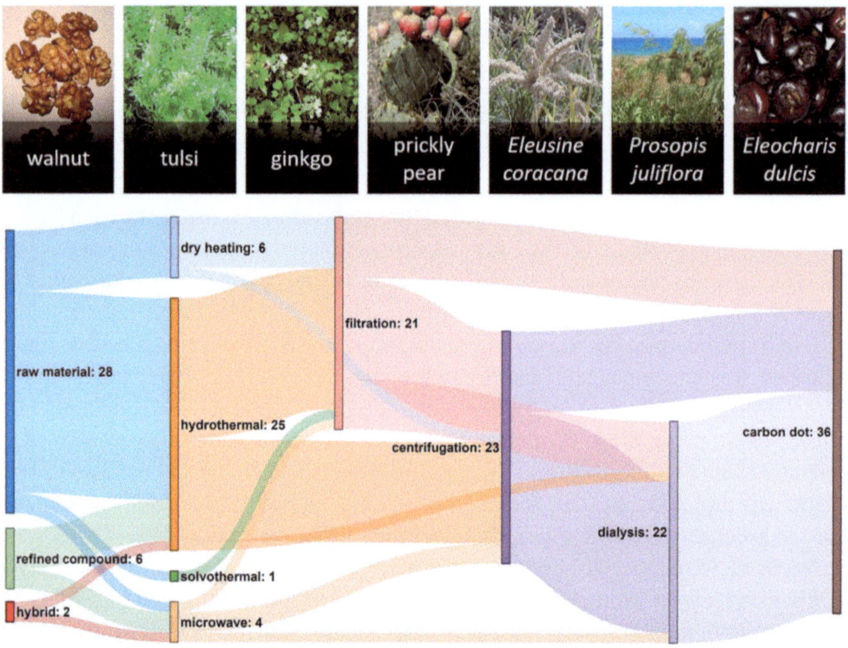

Fig. 2.4 A variety of renewable, locally available materials can serve as precursors for the eco-friendly production of carbon dots (CDs). A Sankey diagram illustrates the quantity of publications focused on the green synthesis of CDs, depicting the tr transition from carbon sources to the final CDs. For the purposes of this diagram, it is hypothesized that the purification process adheres to a sequence of (1) filtration, (2) centrifugation, and (3) dialysis. However, it's important to note that this sequence was not universally applied across all studies and is proposed here merely for the sake of visualization [67] (Copyright with permission under **CC BY-NC 3.0** 2021, RSC)

1.4.1 Green Precursor

Green precursors to produce CQDs are materials that are environment friendly and sustainable, used in the preparation of these nanoscale particles. The use of green precursors associates with the realm of green chemistry, aiming to minimize the environmental impact and ensure safety in material production. The use of green precursors in the synthesis of CQDs is a promising approach that aligns with the principles of sustainable and environmentally friendly chemistry, offering a wide range of applications for these versatile nanomaterials. Here are some commonly used green precursors for CQD synthesis, mentioned below.

Plant Extract

The exploitation of plant extracts for the production of CQDs represents an intriguing and eco-friendly method. The presence of natural products, such as carbohydrates and polyphenols within these extracts, is crucial for the formation of CQDs, serving primarily as a rich carbon source [68]. These natural compounds play a prominent role in the precise fabrication of CQDs through several mechanisms:

(1) Providing a carbon source, for instance, glucose, citric acid, or other organic compounds.
(2) Facilitating surface passivation through functional groups like carboxyl, amino, and hydroxyl groups.
(3) Introducing tamper with groups such as sulfur (S), nitrogen (N), and phosphorus (P).
(4) Acting as reducers and stabilizers for metal precursors during CQD synthesis.
(5) Allowing for the tuning of CQD properties as different plant extracts contain unique compositions of carbon, heteroatoms, and functional groups. This variability can influence the size, surface morphology, and optical characteristics of the produced CQDs [6].
(a) Fruit

Ramezani et al. successfully synthesized green CQDs employing quince fruit (*Cydonia oblonga*) powder as a precursor in one pot microwave radiation. G-CQDs are highly stable, luminescent, high Photoluminescent intense, multicolour 4.85 ± 0.07 nm size and maximum emission intensity at 450 nm. These G-CQDs are exploited in cell tomography and As^{3+} determination [69]. Saini et al. fabricated unique green, fluorescent nanoaminocatalyst NAT-CQDs (Natural amine-targeted CQDs) for production of therapeutically significant heteroaromatic methane via Knovenagel/(aza) Michael addition reaction of 1,3 – Di carbonyls and pungent aldehydes with nitrogen and carbon-based nucleophiles in water. Herein synthesis NAT-CQDs by *Trichosanthes diocia (T. dioica;* Parwal) via Carbonization method was utilized and minimized waste for formulation of medical drug warfain and chemotherapy short of any metal infection. These reactions are characterized by high atom economy, minimal E-factor, reduced process mass intensity (PMI), enhanced reaction mass efficiency (RME), superior carbon efficiency, and excellent catalyst reusability, collectively contributing to significantly high sustainability values [70] 69, adeptly produced eco-friendly CQDs from quince fruit (Cydonia oblonga) powder via a straightforward one-pot microwave irradiation method. These green CQDs are characterized by their high stability, bright luminescence, intense photoluminescence with multicolour emission, an average size of 4.85±0.07 nm, and a peak emission strength at 450 nm. Such attributes make these G-CQDs suitable for applications in cell imaging and the detection of arsenic (As^{3+}), as documented in their study. In their innovative work, Ramezani et al. adeptly synthesized eco-friendly CQDs using quince fruit (Cydonia oblonga) powder through a one-step microwave irradiation process. These green CQDs exhibited remarkable stability and luminosity, with an intense photoluminescence feature, showcasing a multicolour emission, a precise average size of 4.85±0.07 nm, and a peak emission at 450 nm. The unique properties of these G-CQDs have been effectively applied in the grounds of cellular tomography and arsenic (As^{3+}) detection, as highlighted in their 2018 study. Following this, Saini et al. introduced an innovative green, fluorescent nano amino catalyst known as NAT-CQDs (Natural amine-targeted CQDs) aimed at synthesizing medically significant heteroaromatic methane. The creation of NAT-CQDs was achieved using Trichosanthes dioica (T. dioica; Parwal) through a carbonization method, significantly reducing waste in the production of clinical drugs like warfarin and chemotherapy agents, devoid of metal contaminants. Such advancements underscore the potential of using green precursors and sustainable methodologies in the synthesis of CQDs, paving the way for their application in crucial areas like drug synthesis, bioimaging, and environmental monitoring.

(b) Leaf

Phoenix dactylifera (date palm) leaves were employed as a carbon source in the hydrothermal synthesis of green carbon quantum dots (G-CQDs) ranging in size from 5 to 15 nm. Athinarayanan and colleagues synthesized these fluorescent CQDs, demonstrating their application in bio-imaging for both normal and cancerous cells [71]. Ren and his team transformed biomass waste from peach trees into luminous CQDs through an eco-friendly, one-step hydrothermal process. These CQDs exhibited photoluminescent features, including a quantum yield of 7.71%, and fluorescence lifetime of 5.96 ns rendering them apt for fluorescence tomography applications in both in vivo zebrafish and in vitro cell cultures studies. Remarkably, they demonstrated extremely low toxicity and proved to be efficient in biological imaging.

[72]. Hemmati et al. innovatively synthesized sulfur and nitrogen-doped CQDs using feijoa leaves through a microwave-assisted pyrolysis method. The resulting fluorescent S, N-CQDs were integrated into a smartphone fluorimeter device, enabling the visual determination of levodopa (L-DOPA), showcasing a novel application of these materials [73].

(c) Flower

Rheumatoid arthritis (RA), a chronic condition characterized by joint inflammation and pain, has been the focus of innovative therapeutic strategies. Qiang et al.

developed CQDs from the herbal materials of safflower (Carthamus tinctorius L.; flower) and Angelica sinensis (root), creating nano agents aimed at RA treatment. These CQDs exhibited exceptional lubrication and anti-inflammatory properties, offering a novel approach to managing the disease [74]. Beyond their traditional medicinal uses, flowers are increasingly recognized in the field of medical chemistry and technology for their rich phytochemical content. Building on this concept, Padmapriya and colleagues utilized the biomass extract from the bract of the banana flower (Musa acuminata) to create nitrogen and phosphorus-doped carbon quantum dots (N, P-CQDs) using a hydrothermal process. These N, P-CQDs exhibited the ability for selective and sensitive dopamine detection, achieving a detection limit (LOD) of about 500 pM across a linear range from 6.0 μM to 0.1 mM for electrochemical detection, highlighting their promising applications in the field of biomedicine. [75].

(d) Seed

Xu and his team crafted a fluorescent probe dedicated to the selective sensing of Cu^{2+} ions, employing a one-pot solvothermal technique that leverages bran extract for the synthesis of CQDs. These CQDs, notable for their green fluorescence, have an average diameter of 4.8 nm and exhibit emission at 539 nm upon excitation at 450 nm. Designed specifically for Cu^{2+} ions, the probe demonstrated a linear detection spectrum from 0 to 0.5 mM, with a detection threshold (LOD) of 0.0507 mM [76].

Industrial Waste

Enhancing the sustainability of our environment involves controlling pollution and recycling industrial waste through viable methods. Significant contributors to industrial pollution include waste from oil, paper, dye, and plastic industries, with their degradation and recycling presenting substantial challenges. As an innovative solution, alternative materials like plastic, cat feedstock, and petroleum coke are being explored as carbon supplies for the production of CQDs. This task is increasingly feasible thanks to advances in quantum nano chemistry. For instance, Wei et al. explained the possibility of repurposing wastepaper into CQDs using a straightforward one-step hydrothermal synthesis process. The resulting water-soluble fluorescent CQDs are useful in biomedical applications [77]. In a similar vein, Wang et al. employed a hydrothermal method to convert wastepaper into environment friendly N-CQDs. These single crystalline N-CQDs exhibit intense blue-green luminescence, surpassing that of standard CQDs in terms of brightness [78]. Additionally, Srinivasan et al. ventured into the synthesis of fluorescent CQDs using kerosene fuel soot, employing a simple one-pot oxidative acid treatment. This approach has established that the CQDs' applicability across a diverse range of fields [79], highlighting the potential of repurposing waste materials into valuable nanomaterials for various applications.

Animal Waste and By-Product

Animal waste, including human hair, animal manure, crustacean shells, and dairy products, are viable carbon sources for the synthesis of CQDs. Techniques such as carbonization, chemical oxidation, and hydrothermal treatment are employed to transform these wastes into CQDs for various scientific applications. For instance, sustainable, highly fluorescent CQDs were produced from human hair through thermal treatment at 200 °C for 24 hours, exhibiting strong blue emission with a 10.75% quantum yield and showing effective quenching in the presence of Hg^{2+} ions [80]. Similarly, Gedda et al. utilized prawn shells through a combination of chemical oxidation and hydrothermal treatment to create a selective and responsive fluorescence probe for detecting Cu^{2+} ions [81]. The use of animal waste for CQD synthesis has expanded, showcasing a variety of methods and sources. Examples include CQDs synthesized from crab shells via hydrothermal or microwave-assisted methods [82], sonochemical [83], Cow manure hydrothermal/microwave assisted/chemical oxidation [15, 84, pigen feathers, egg shells via carbonization [85], expired milk through heating [86], and whey thermal treatment [87]. These studies underscore the potential of repurposing animal waste into valuable nanomaterials for wide-ranging applications, contributing to waste reduction and sustainable material science.

Micro-Organism

Microorganisms present a promising bio-precursor for the creation of highly photoluminescent CQDs, thanks to their rich content of carbohydrates, peptidoglycan, proteins, and peptides, which are abundant in carbon and heteroatoms such as N, S, O, and P. Utilizing microbes for CQD synthesis represents a novel, sustainable, and eco-friendly approach, employing various methodologies to produce green CQDs (G-CQDs) [88]. Ghorbani and colleagues have innovatively produced CQDs from baker's yeast through a one-pot microbial method. They have shown that these carbon quantum dots (CQDs) are safe and non-toxic at concentrations up to 3.5 mg/ml in human colon cancer cell lines (HCT-116) and TM4 mouse cells. Furthermore, they have effectively employed these CQDs in creating an antimicrobial bacterial nanocellulose membrane, highlighting the broad utility of these nanomaterials in various applications. [88]. Here tabulated in Table 1 various G-CQDs synthesis by Micro-organism:

1.4.2 Effects of Various Parameter on Size of G-CQDs

The different parameters such as temperature, pressure, solvent, green precursor like biomass, waste raw materials, washing step play a critical role in precise indistinguishable G-CQDs distribution for employing in diverse applications. Temperature and pressure is major parameter has been controlled growth of G-CQDs. Here briefly

Table 1 Green synthesis of G-CQDS using various microorganisms

S.No	G-CQD synthesis microorganism	method	temperature	Size and of G-CQDs	Application	Reference
1	*Forsythia* (anti-wood rot fungus) by *Forsythia@CQDs*	Microwave synthesis	90 °C/5 min/500 W	0.38 nm crystalline size	Antifungal test of *G. trabeum* and *C. versicolor*	[57]
2	CQDs synthesis by three probiotic bacterial strains (*Bifidobacterium breve*, *Lactobacillus fermentum*, *Lactobacillus acidophilus*) and three human pathogens (*Escherichia coli*, *Klebsiella pneumoniae*, *Pseudomonas aeruginosa*)	Pyrolytic carbonization method	160 °C to 220 °C	1.9 nm and 2.5 nm	Temperature effect on quantum yield	[100]
3	CQDs synthesis by Mushroom (fungus)	Hydrothermal method	200 °C/ 6 h	2 to 4 nm	Hyaluronic acid and hyaluronidase sensing	[88]

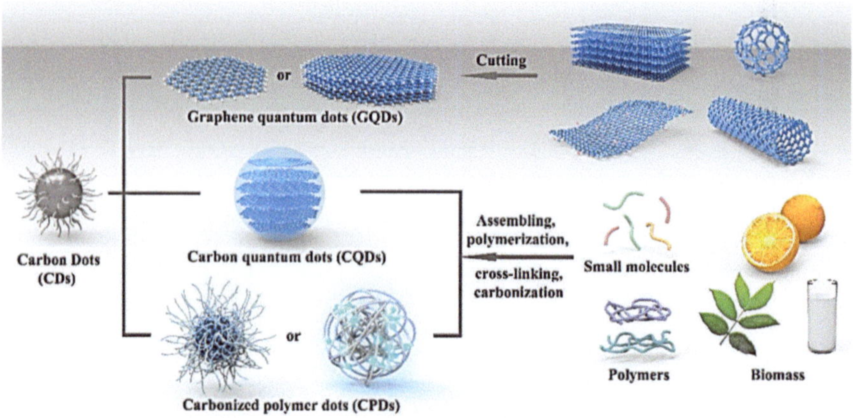

Fig. 2.5 Including the synthesis and main preparation techniques for graphene QDs, CQDs, and carbonized polymer dots (CPDs) [91], (Copyright with permission under CC-BY-NC-ND, 2020, ACS)

discus effect of temperature and pressure on biomass and waste reduced into quantum level. Figure 2.5 shows the synthesis and main preparation techniques for graphene QDs, CQDs, and carbonized polymer dots (CPDs).

Temperature

Temperature is one of the leading parameter for controlled and indistinguishable growth of G-CQDs. Mainly biomass and waste material contain carbon has been reduced at 150 ˚C to 300 ˚C temperature into G-CQDs. Here tabulated in Table 2 various G-CQDs synthesis method depend on temperature.

Pressure

Various techniques are utilized to achieve controlled nucleation of carbon quantum dots (CQDs) through pressure manipulation. The arc discharge method operates as a top-down approach within a low-pressure inert atmosphere, whereas the hydrothermal method represents a bottom-up strategy, facilitating controlled growth of CQDs under high pressure [5]. Baragau and colleagues developed a method to produce blue luminescent nitrogen-doped CQDs (N-CQDs) hydrothermally. This method involves merging a serve of supercritical water (374 °C, 22.1 MPa) with a precursor directly in the response zone for a specified duration. This approach is highlighted for its simplicity, eco-friendliness, cost-effectiveness, and potential for large-scale production, offering a new pathway for synthesizing uniformly distributed N-CQDs with biomass-derived precursors like citric acid [95].

Table 2 Temperature depended synthesis of G-CQDs

S.No	G-CQD synthesis extract	Method	Temperature	Size and of G-CQDs	Application	References
1.	G-CQD by quince fruit (Cydonia oblonga) powder	Hydrothermal method	200 °C/30 min	4.85 ± 0.07 nm	Cell imaging and As^{3+} detection	[68]
2.	G-CQD by Echinops persicus (leaf, stem, root, and flower) powder	Hydrothermal method	200 °C/10 h	3 to 5 nm	Synthesis of G-CQDs via variable extract in phitochemicals	[90]
3.	Narrow bandwidth emission triangular carbon quantum dots (NBE-T-CQDs)	Solvothermal method	200 °C	2 to 4 nm	Multicolour LEDs	[91]
4.	CQDs/Ag by Orange juice	Solvothermal method	120°C/2.5 h	10 nm	Anticancer activity and imaging on colorectal carcinoma	[92]
5.	Natural amine-targeted CQDs (NAT-CQDs) by *Trichosanthes dioica (T. dioica;* Parwal)	Carbonization method	300°C/3 h	2 to 9 nm	Used as nanoaminocatalyticc platform for medicinally important heteroaromatic methanes	[69]

1.5 Future of Synthesis Direction of G-CQDs

The G-CQDs represent a sustainable, economical, and sustainable methodology for transforming biomass and industrial waste into valuable materials, thereby enriching various advanced application domains. Their development is a testament to the ongoing quest for a balance among economic growth, environmental preservation, and societal progress [15]. The burgeoning interest in G-CQDs is steering future innovations and enhancements in their synthesis techniques, which in turn, broadens their applicability across diverse scientific areas due to:

I. The drive to reduce costs and enhance the quantum yield through refined synthesis methods of G-CQDs [96].
II. Tailoring the size, surface functionality, and electronic characteristics of G-CQDs to boost their performance in sensing, bioimaging, and energy conversion applications [23].
III. Leveraging the photoluminescence efficiency of G-CQDs to pioneer new advancements in optoelectronics and bioimaging [97].
IV. Enhancing drug delivery and bioimaging techniques by improving the interaction between G-CQDs' biocompatible surfaces and biological systems [98].
V. Progress in developing hybrid materials through the integration of G-CQDs with various precursors and polymers, paving the way for innovative composite materials [15].
VI. Advancing sensitive and selective environmental monitoring methods for detecting pollutants, heavy metals, and other harmful substances [99].
VII. Pioneering energy conversion and storage solutions, including batteries, supercapacitors, and photovoltaic devices, utilizing the unique properties of G-CQDs [100].
VIII. The multidisciplinary potential of G-CQDs and their widespread scientific interest underscore the importance of achieving uniform, precisely controlled nucleation as a strategic priority for future research and applications.

1.6 Challenges in Victorious Process of G-CQDs

This section delves into the diverse methodologies and tactics employed in the synthesis of G-CQDs, focusing on achieving precise control over their growth. However, this endeavour faces several challenges, including:

I. The complexity of achieving uniform synthesis of G-CQDs, given the variety of precursor materials, reaction environments, and resultant properties, poses a significant challenge [101].
II. Enhancing the photostability and quantum yield of G-CQDs is critical, especially for their relevance in bioimaging and optoelectronics, yet remains a demanding task.

III. Achieving mono dispersity and exact sizing of G-CQDs is challenging but essential for advancing their use across scientific and technological domains [68].
IV. Scaling up G-CQD production to meet industrial demands presents considerable difficulties [22].
V. Navigating the ethical landscape associated with the use of nanomaterials like G-CQDs, ensuring their safe and responsible application in various fields, is imperative [22].
VI. There is a need for rigorous and standardized optoelectronic characterization methods to precisely determine the structural and morphological features of G-CQDs [89].
VII. Addressing the demand for advanced metrics to evaluate the potential environmental and biological impacts of G-CQDs remains a challenge [97].

In reply to such encounters, green chemistry principles are increasingly being applied to the sustainable, eco-friendly, biodegradable, and biocompatible development, design, and production of G-CQDs, aiming to overcome these hurdles while fostering environmental stewardship.

1.7 Environmental and Economic Implications

The global challenges of environmental degradation and economic sustainability have prompted the United Nations to establish a mission for worldwide development, aiming for accord and fortune for both people and the planet, now and into the future [102]. CQDs have been identified as having significant potential across various applications, including biosensing, drug delivery, and bioimaging, contributing to the achievement of the third Sustainable Development Goal [103, 104]. The environmental and economic impacts of G-CQDs largely depend on the methods of synthesis, their applications, and the lifecycle considerations. G-CQDs, synthesized through eco-friendly and sustainable approaches, offer lower environmental impacts compared to traditional CQDs. This advantage arises from their reliance on biomass and industrial waste, which helps reduce the use of hazardous chemicals. Despite their biocompatibility, there's a need to address the toxicity concerns associated with G-CQDs, especially in biomedical applications, to ensure their safe usage.

The recyclability and degradability of G-CQDs are influenced by their specific composition and surface functionalization. Economically, G-CQDs stand out for their cost-effectiveness, utilizing advanced biomass and industrial waste in their production processes [22]. Their versatility allows G-CQDs to find applications in diverse fields, potentially increasing market demand and competitiveness against other materials. Recent uses of fluorescent CQDs include food contamination detection [105], energy and bioremediation [100], dye sensitized solar energy conversion [106], multicolour imaging [107], catalytic reduction of dye and microbial activity [108] etc. Despite these promising applications, challenges remain in scaling up production, controlling nucleation processes, and achieving uniform growth of G-CQD particles [108]. Addressing these challenges is essential for fully realizing the ability of G-CQDs in green development and environmental conservation.

1.8 Conclusion

The search of Sustainable Development Goals underscores the importance of transforming biomass and industrial waste into G-CQDs, driving forward eco-friendly and sustainable innovations for the future. The creation of G-CQDs, similar to other nanoparticles, encounters specific challenges, including inconsistent particle sizes and uncontrolled nucleation. In response to these challenges, several strategies have been devised, broadly categorized into top-down and bottom-up approaches. Methods like hydrothermal synthesis, arc discharge, microwave assistance, laser ablation, electrochemical synthesis, ball milling, chemical vapour deposition, combustion, and templating have been utilized to generate G-CQDs. Despite the success of these methods in generating G-CQDs, achieving a uniform particle size and controlled nucleation growth still poses a considerable challenge. This discussion aims to explore the diverse strategies and parameters that may facilitate the controlled growth of G-CQDs from biomass and industrial waste. It also examines the implementation and future potential of G-CQDs in various applications, acknowledging the complications that synthesis challenges such as uncontrolled nucleation and polydispersity bring to the process. From an environmental and economic perspective, this new wave of CQDs offers a promising and viable option for broad application. The chapter seeks to bridge the gap between the synthesis of G-CQDs and their possible tenders, emphasizing the necessity of overcoming the hurdles in synthesis to unlock their full potential.

Acknowledgements Author S.S; S.K.Y and D.S acknowledge the Banaras Hindu University for financial support. Author J. S. acknowledges BHU for providing a seed grant and BRIDGE grant under MoE Govt. India, Institute of Eminence (IoE), under Dev. Scheme No. 6031 & 6031A respectively.

References

1. Manna, L. (2023). The bright and enlightening science of quantum dots. *Nano Letters, 23*, 9673–9676.
2. Yang, W., Li, X., Fei, L., Liu, W., Liu, X., Xu, H., & Liu, Y. (2022). A review on sustainable synthetic approaches toward photoluminescent quantum dots. *Green Chemistry, 24*, 675–700.
3. Reimann, S. M., & Manninen, M. (2002). Electronic structure of quantum dots. *Reviews of Modern Physics, 74*, 1283–1342.
4. Wang, Y., & Hu, A. (2014). Carbon quantum dots: Synthesis, properties and applications. *J. Mater. Chem. C, 2*, 6921.

5. Jing, H., Bardakci, F., Akgöl, S., Kusat, K., Adnan, M., Alam, M., Gupta, R., Sahreen, S., Chen, Y., Gopinath, S., & Sasidharan, S. (2023). Green carbon dots: synthesis, characterization, properties and biomedical applications. *J. Funct. Biomater., 14*, 27.
6. Manikandan, V., & Lee, N. Y. (2022). Green synthesis of carbon quantum dots and their environmental applications. *Environmental Research, 212*, 113283.
7. Banerjee, A., Pons, T., Lequeux, N., & Dubertret, B. (2016). Quantum dots–DNA bioconjugates: Synthesis to applications. *Interface Focus, 6*, 20160064.
8. Huo, X., He, Y., Ma, S., Jia, Y., Yu, J., Li, Y., & Cheng, Q. (2020). Green synthesis of carbon dots from grapefruit and its fluorescence enhancement. *Journal of Nanomaterials, 2020*, 1–7.
9. Tyagi, A., Tripathi, K. M., Singh, N., Choudhary, S., & Gupta, R. K. (2016). Green synthesis of carbon quantum dots from lemon peel waste: Applications in sensing and photocatalysis. *RSC Advances, 6*, 72423–72432.
10. Rocco, D., Moldoveanu, V. G., Feroci, M., Bortolami, M., & Vetica, F. (2023). Electrochemical synthesis of carbon quantum dots. Chem Electro Chem, p. 10.
11. Guerrero-Gonzalez, R., Vázquez-Dávila, F., Saucedo-Flores, E., Ruelas, R., Ceballos-Sánchez, O., & Pelayo, J. E. (2023). Green approach synthesis of carbon quantum dots from agave bagasse and their use to boost seed germination and plant growth. *SN Appl. Sci., 5*, 204.
12. Xu, X., Ray, R., Gu, Y., Ploehn, H. J., Gearheart, L., Raker, K., & Scrivens, W. A. (2004). Electrophoretic analysis and purification of fluorescent single-walled carbon nanotube fragments. *Journal of the American Chemical Society, 126*, 12736–12737.
13. Li, H., Kang, Z., Liu, Y., & Lee, S.-T. (2012). Carbon nanodots: Synthesis, properties and applications. *Journal of Materials Chemistry, 22*, 24230.
14. Ying, S., Guan, Z., Ofoegbu, P. C., Clubb, P., Rico, C., He, F., & Hong, J. (2022). Green synthesis of nanoparticles: Current developments and limitations. *Environmental Technology and Innovation, 26*, 102336.
15. Mensah, J. (2019). Sustainable development: Meaning, history, principles, pillars, and implications for human action: Literature review. *Cogent Soc. Sci. 5.*
16. Alsammarraie, F. K., Wang, W., Zhou, P., Mustapha, A., & Lin, M. (2018). Green synthesis of silver nanoparticles using turmeric extracts and investigation of their antibacterial activities. *Colloids Surfaces B Biointerfaces, 171*, 398–405.
17. Lassenberger, A., Grünewald, T. A., van Oostrum, P. D. J., Rennhofer, H., Amenitsch, H., Zirbs, R., Lichtenegger, H. C., & Reimhult, E. (2017). Monodisperse iron oxide nanoparticles by thermal decomposition: elucidating particle formation by second-resolved in situ small-angle x-ray scattering. *Chemistry of Materials, 29*, 4511–4522.
18. Smirniotis, P. G., Boningari, T., & Inturi, S. N. R. (2018). Single-step synthesis of N-doped TiO 2 by flame aerosol method and the effect of synthesis parameters. *Aerosol Science and Technology, 52*, 913–922.
19. Wojnarowicz, J., Chudoba, T., Gierlotka, S., & Lojkowski, W. (2018). Effect of microwave radiation power on the size of aggregates of ZnO NPs prepared using microwave solvothermal synthesis. *Nanomaterials, 8*, 343.
20. Aswathy Aromal, S., & Philip, D. (2012). Green synthesis of gold nanoparticles using Trigonella foenum-graecum and its size-dependent catalytic activity. *Spectrochimica Acta Part A: Molecular and Biomolecular Spectroscopy, 97*, 1–5.
21. Zhang, X., Yin, J., & Yoon, J. (2014). Recent advances in development of chiral fluorescent and colorimetric sensors. *Chemical Reviews, 114*, 4918–4959.
22. Desmond, L. J., Phan, A. N., & Gentile, P. (2021). Critical overview on the green synthesis of carbon quantum dots and their application for cancer therapy. *Environmental Science. Nano, 8*, 848–862.
23. Rasal, A. S., Yadav, S., Yadav, A., Kashale, A. A., Manjunatha, S. T., Altaee, A., & Chang, J.-Y. (2021). Carbon quantum dots for energy applications: a review. *ACS Appl. Nano Mater., 4*, 6515–6541.
24. Chao-Mujica, F. J., Garcia-Hernández, L., Camacho-López, S., Camacho-López, M., Camacho-López, M. A., Reyes Contreras, D., Pérez-Rodríguez, A., Peña-Caravaca, J. P., Páez-Rodríguez, A., Darias-Gonzalez, J. G., Hernandez-Tabares, L., Arias de Fuentes,

O., Prokhorov, E., Torres-Figueredo, N., Reguera, E., & Desdin-García, L. F. (2021). Carbon quantum dots by submerged arc discharge in water: Synthesis, characterization, and mechanism of formation. J. Appl. Phys. *129*.
25. De Bonis, A., Lovaglio, T., Galasso, A., Santagata, A., & Teghil, R. (2015). Iron and iron oxide nanoparticles obtained by ultra-short laser ablation in liquid. *Applied Surface Science, 353*, 433–438.
26. Yang, G. W. (2007). Laser ablation in liquids: Applications in the synthesis of nanocrystals. *Progress in Materials Science, 52*, 648–698.
27. Cui, L., Ren, X., Wang, J., & Sun, M. (2020). Synthesis of homogeneous carbon quantum dots by ultrafast dual-beam pulsed laser ablation for bioimaging. *Mater. Today Nano, 12*, 100091.
28. Sadrolhosseini, A. R., Krishnan, G., Safie, S., Beygisangchin, M., Rashid, S. A., & Harun, S. W. (2020). Enhancement of the fluorescence property of carbon quantum dots based on laser ablated gold nanoparticles to evaluate pyrene. *Optical Materials Express, 10*, 2227.
29. Sharma, M. K. (2021). Synthesis of advanced materials by electrochemical methods, pp. 435–466.
30. Koelmel, J., Prasad, M. N. V., Velvizhi, G., Butti, S. K., & Mohan, S. V. (2016). Metalliferous waste in India and knowledge explosion in metal recovery techniques and processes for the prevention of pollution. In: Environmental Materials and Waste. Elsevier, pp. 339–390.
31. Ding, X., Niu, Y., Zhang, G., Xu, Y., & Li, J. (2020). Electrochemistry in Carbon-based Quantum Dots. *Chem. – An Asian J. 15*, 1214–1224.
32. Muthusankar, G., Sasikumar, R., Chen, S.-M., Gopu, G., Sengottuvelan, N., & Rwei, S.-P. (2018). Electrochemical synthesis of nitrogen-doped carbon quantum dots decorated copper oxide for the sensitive and selective detection of non-steroidal anti-inflammatory drug in berries. *Journal of Colloid and Interface Science, 523*, 191–200.
33. Sun, S., Bao, W., Yang, F., Yan, X., Sun, Y., Zhang, G., Yang, W., & Li, Y. (2023). Electrochemical synthesis of FeNx doped carbon quantum dots for sensitive detection of Cu2+ ion. *Green Energy Environ., 8*, 141–150.
34. Abid, N., Khan, A. M., Shujait, S., Chaudhary, K., Ikram, M., Imran, M., Haider, J., Khan, M., Khan, Q., & Maqbool, M. (2022). Synthesis of nanomaterials using various top-down and bottom-up approaches, influencing factors, advantages, and disadvantages: A review. *Advances in Colloid and Interface Science, 300*, 102597.
35. Xing, T., Sunarso, J., Yang, W., Yin, Y., Glushenkov, A. M., Li, L. H., Howlett, P. C., & Chen, Y. (2013). Ball milling: A green mechanochemical approach for synthesis of nitrogen doped carbon nanoparticles. *Nanoscale, 5*, 7970.
36. Han, Y., Chen, Y., Wang, N., & He, Z. (2019). Magnesium doped carbon quantum dots synthesized by mechanical ball milling and displayed Fe 3+ sensing. *Materials Technology, 34*, 336–342.
37. Feng, X., & Zhang, Y. (2019). A simple and green synthesis of carbon quantum dots from coke for white light-emitting devices. *RSC Advances, 9*, 33789–33793.
38. Navarro-Badilla, A., Calderon-Ayala, G., Delgado-Beleño, Y., Heras-Sánchez, M. C., Hurtado, R. B., Leal-Pérez, J. E., Hurtado-Macias, A., & Cortez-Valadez, M. (2023). Green synthesis for carbon quantum dots via opuntia ficus-indica and agave Maximiliana : surface-enhanced Raman scattering sensing applications. *ACS Omega, 8*, 33342–33348.
39. Qiu, Y., Zhou, B., Yang, X., Long, D., Hao, Y., & Yang, P. (2017). Novel single-cell analysis platform based on a solid-state zinc-coadsorbed carbon quantum dots electrochemiluminescence probe for the evaluation of CD44 expression on breast cancer cells. *ACS Applied Materials & Interfaces, 9*, 16848–16856.
40. Shen, J., Zhu, Y., Yang, X., & Li, C. (2012). Graphene quantum dots: Emergent nanolights for bioimaging, sensors, catalysis and photovoltaic devices. *Chemical Communications, 48*, 3686.
41. Wang, X., Qu, K., Xu, B., Ren, J., & Qu, X. (2011). Microwave assisted one-step green synthesis of cell-permeable multicolor photoluminescent carbon dots without surface passivation reagents. *Journal of Materials Chemistry, 21*, 2445.

42. Ming, H., Ma, Z., Liu, Y., Pan, K., Yu, H., Wang, F., & Kang, Z. (2012). Large scale electrochemical synthesis of high quality carbon nanodots and their photocatalytic property. *Dalt. Trans., 41*, 9526.
43. Ding, S., Gao, Y., Ni, B., & Yang, X. (2021). Green synthesis of biomass-derived carbon quantum dots as fluorescent probe for Fe3+ detection. *Inorganic Chemistry Communications, 130*, 108636.
44. Baweja, H., & Jeet, K. (2019). Economical and green synthesis of graphene and carbon quantum dots from agricultural waste. *Mater. Res. Express, 6,* 0850g8.
45. Stojanovic, B. D., Dzunuzovic, A. S., & Ilic, N. I. (2018). Review of methods for the preparation of magnetic metal oxides. In: Magnetic, Ferroelectric, and Multiferroic Metal Oxides. Elsevier, pp. 333–359.
46. Liu, H., Ye, T., & Mao, C. (2007). Fluorescent carbon nanoparticles derived from candle soot. *Angew. Chemie, 119*, 6593–6595.
47. Raikwar, V. R. (2022). Synthesis and study of carbon quantum dots (CQDs) for enhancement of luminescence intensity of CQD@LaPO4:Eu3+ nanocomposite. *Materials Chemistry and Physics, 275*, 125277.
48. Darr, J. A., Zhang, J., Makwana, N. M., & Weng, X. (2017). Continuous hydrothermal synthesis of inorganic nanoparticles: applications and future directions. *Chemical Reviews, 117*, 11125–11238.
49. Das, R., Bandyopadhyay, R., & Pramanik, P. (2018). Carbon quantum dots from natural resource: A review. *Mater. Today Chem., 8*, 96–109.
50. Sabet, M., & Mahdavi, K. (2019). Green synthesis of high photoluminescence nitrogen-doped carbon quantum dots from grass via a simple hydrothermal method for removing organic and inorganic water pollutions. *Applied Surface Science, 463*, 283–291.
51. Mindivan, F., & Göktaş, M. (2023). The green synthesis of carbon quantum dots (CQDs) and characterization of polycaprolactone (PCL/CQDs) films. *Colloids Surfaces A Physicochem. Eng. Asp., 677*, 132446.
52. Nguyen, T. N., Le, P. A., & Phung, V. B. T. (2022). Facile green synthesis of carbon quantum dots and biomass-derived activated carbon from banana peels: Synthesis and investigation. *Biomass Convers. Biorefinery, 12*, 2407–2416.
53. Mohammadi, M., Rezaei, A., Khazaei, A., Xuwei, S., & Huajun, Z. (2019). Targeted development of sustainable green catalysts for oxidation of alcohols via tungstate-decorated multifunctional amphiphilic carbon quantum dots. *ACS Applied Materials & Interfaces, 11*, 33194–33206.
54. Liu, Y., Goebl, J., & Yin, Y. (2013). Templated synthesis of nanostructured materials. *Chemical Society Reviews, 42*, 2610–2653.
55. Shi, F., Xing, B., Zeng, H., Guo, H., Qu, X., Huang, G., Cao, Y., Li, P., & Zhang, C. (2023). Ice template induced assembly strategy for preparation of 3D porous carbon frameworks from low-cost carbon quantum dots for high-performance lithium-ion batteries. *J. Energy Storage, 70*, 107982.
56. Kwon, W., Lee, G., Do, S., Joo, T., & Rhee, S. (2014). Size-controlled soft-template synthesis of carbon nanodots toward versatile photoactive materials. *Small (Weinheim an der Bergstrasse, Germany), 10*, 506–513.
57. Wang, R., Ji, Y., Wu, X., Liu, R., Chen, L., & Ge, G. (2016). Experimental determination and analysis of gold nanorod settlement by differential centrifugal sedimentation. *RSC Advances, 6*, 43496–43500.
58. Zhao, X., Wang, L., Ren, S., Hu, Z., & Wang, Y. (2021). One-pot synthesis of Forsythia@carbon quantum dots with natural anti-wood rot fungus activity. *Materials and Design, 206*, 109800.
59. Rodríguez-Padrón, D., Algarra, M., Tarelho, L. A. C., Frade, J., Franco, A., de Miguel, G., Jiménez, J., Rodríguez-Castellón, E., & Luque, R. (2018). Catalyzed microwave-assisted preparation of carbon quantum dots from lignocellulosic residues. *ACS Sustain. Chem. Eng., 6*, 7200–7205.

60. Architha, N., Ragupathi, M., Shobana, C., Selvankumar, T., Kumar, P., Lee, Y. S., & Kalai Selvan, R. (2021). Microwave-assisted green synthesis of fluorescent carbon quantum dots from Mexican Mint extract for Fe3+ detection and bio-imaging applications. *Environmental Research, 199*, 111263.
61. Atchudan, R., Jebakumar Immanuel Edison, T. N., Shanmugam, M., Perumal, S., Somanathan, T., & Lee, Y. R. (2021). Sustainable synthesis of carbon quantum dots from banana peel waste using hydrothermal process for in vivo bioimaging. *Phys. E Low-dimensional Syst. Nanostructures, 126,* 114417.
62. Hoan, B. T., Tam, P. D., & Pham, V.-H. (2019). Green synthesis of highly luminescent carbon quantum dots from lemon juice. *J. Nanotechnol., 2019*, 1–9.
63. Li, Z., Wang, Q., Zhou, Z., Zhao, S., Zhong, S., Xu, L., Gao, Y., & Cui, X. (2021). Green synthesis of carbon quantum dots from corn stalk shell by hydrothermal approach in near-critical water and applications in detecting and bioimaging. *Microchemical Journal, 166*, 106250.
64. Zhang, Y., Li, P., Yan, H., Guo, Q., Xu, Q., & Su, W. (2023). Green synthesis and multi-functional applications of nitrogen-doped carbon quantum dots via one-step hydrothermal carbonization of Curcuma zedoaria. *Analytical and Bioanalytical Chemistry, 415*, 1917–1931.
65. Sharma, N., Das, G. S., & Yun, K. (2020). Green synthesis of multipurpose carbon quantum dots from red cabbage and estimation of their antioxidant potential and bio-labeling activity. *Applied Microbiology and Biotechnology, 104*, 7187–7200.
66. Sharma, N., Sharma, I., & Bera, M. K. (2022). Microwave-assisted green synthesis of carbon quantum dots derived from calotropis gigantea as a fluorescent probe for bioimaging. *Journal of Fluorescence, 32*, 1039–1049.
67. Chahal, S., Macairan, J.-R., Yousefi, N., Tufenkji, N., & Naccache, R. (2021). Green synthesis of carbon dots and their applications. *RSC Advances, 11*, 25354–25363.
68. de Oliveira, B. P., & da Silva Abreu, F. O. M. (2021). Carbon quantum dots synthesis from waste and by-products: Perspectives and challenges. *Materials Letters, 282*, 128764.
69. Ramezani, Z., Qorbanpour, M., & Rahbar, N. (2018). Green synthesis of carbon quantum dots using quince fruit (Cydonia oblonga) powder as carbon precursor: Application in cell imaging and As3+ determination. *Colloids Surfaces A Physicochem. Eng. Asp., 549*, 58–66.
70. Saini, S., Saini, P., Kumar, K., Sethi, M., Meena, P., Gurjar, A., Dandia, A., Dhuria, T., & Parewa, V. (2023). Unlocking the molecular behavior of natural amine-targeted carbon quantum dots for the synthesis of diverse pharmacophore scaffolds via an unusual nanoaminocatalytic route. *ACS Applied Materials & Interfaces, 15*, 49083–49094.
71. Athinarayanan, J., Periasamy, V. S., Al-Harbi, L. N., & Alshatwi, A. A. (2023). Phoenix dactylifera leaf-derived biocompatible carbon quantum dots: Application in cell imaging. *Biomass Convers. Biorefinery, 13*, 12989–12998.
72. Ren, H., Yuan, Y., Labidi, A., Dong, Q., Zhang, K., Lichtfouse, E., Allam, A. A., Ajarem, J. S., & Wang, C. (2023). Green process of biomass waste derived fluorescent carbon quantum dots for biological imaging in vitro and in vivo. *Chinese Chem. Lett., 34*, 107998.
73. Hemmati, A., Emadi, H., & Nabavi, S. R. (2023). Green synthesis of sulfur- and nitrogen-doped carbon quantum dots for determination of L-DOPA using fluorescence spectroscopy and a smartphone-based fluorimeter. *ACS Omega, 8*, 20987–20999.
74. Qiang, R., Huang, H., Chen, J., Shi, X., Fan, Z., Xu, G., & Qiu, H. (2023). Carbon quantum dots derived from herbal medicine as therapeutic nanoagents for rheumatoid arthritis with ultrahigh lubrication and anti-inflammation. *ACS Applied Materials & Interfaces, 15*, 38653–38664.
75. Padmapriya, A., Thiyagarajan, P., Devendiran, M., Kalaivani, R. A., & Shanmugharaj, A. M. (2023). Electrochemical sensor based on N, P–doped carbon quantum dots derived from the banana flower bract (Musa acuminata) biomass extract for selective and picomolar detection of dopamine. *Journal of Electroanalytical Chemistry, 943*, 117609.
76. Xu, J., Wang, C., Li, H., & Zhao, W. (2020). Synthesis of green-emitting carbon quantum dots with double carbon sources and their application as a fluorescent probe for selective detection of Cu 2+ ions. *RSC Advances, 10*, 2536–2544.

77. Wei, J., Zhang, X., Sheng, Y., Shen, J., Huang, P., Guo, S., Pan, J., Liu, B., & Feng, B. (2014). Simple one-step synthesis of water-soluble fluorescent carbon dots from waste paper. *New Journal of Chemistry, 38*, 906.
78. Wang, R.-C., Lu, J.-T., & Lin, Y.-C. (2020). High-performance nitrogen doped carbon quantum dots: Facile green synthesis from waste paper and broadband photodetection by coupling with ZnO nanorods. *Journal of Alloys and Compounds, 813*, 152201.
79. Venkatesan, S., Mariadoss, A. J., Arunkumar, K., & Muthupandian, A. (2019). Fuel waste to fluorescent carbon dots and its multifarious applications. *Sensors Actuators B Chem., 282*, 972–983.
80. Guo, Y., Zhang, L., Cao, F., & Leng, Y. (2016). Thermal treatment of hair for the synthesis of sustainable carbon quantum dots and the applications for sensing Hg2+. *Science and Reports, 6*, 35795.
81. Gedda, G., Lee, C.-Y., Lin, Y.-C., & Wu, H. (2016). Green synthesis of carbon dots from prawn shells for highly selective and sensitive detection of copper ions. *Sensors Actuators B Chem., 224*, 396–403.
82. Yao, Y.-Y., Gedda, G., Girma, W. M., Yen, C.-L., Ling, Y.-C., & Chang, J.-Y. (2017). Magnetofluorescent carbon dots derived from crab shell for targeted dual-modality bioimaging and drug delivery. *ACS Applied Materials & Interfaces, 9*, 13887–13899.
83. Dehvari, K., Liu, K. Y., Tseng, P.-J., Gedda, G., Girma, W. M., & Chang, J.-Y. (2019). Sonochemical-assisted green synthesis of nitrogen-doped carbon dots from crab shell as targeted nanoprobes for cell imaging. *Journal of the Taiwan Institute of Chemical Engineers, 95*, 495–503.
84. Haryadi, H., Purnama, M. R. W., & Wibowo, A. (2018). C dots derived from waste of biomass and their photocatalytic activities. *Indones. J. Chem., 18*, 594.
85. Ye, Q., Yan, F., Luo, Y., Wang, Y., Zhou, X., & Chen, L. (2017). Formation of N, S-codoped fluorescent carbon dots from biomass and their application for the selective detection of mercury and iron ion. Spectrochim. *Acta Part A Mol. Biomol. Spectrosc., 173*, 854–862.
86. Su, R., Wang, D., Liu, M., Yan, J., Wang, J.-X., Zhan, Q., Pu, Y., Foster, N. R., & Chen, J.-F. (2018). Subgram-scale synthesis of biomass waste-derived fluorescent carbon dots in subcritical water for bioimaging, sensing, and solid-state patterning. *ACS Omega, 3*, 13211–13218.
87. Devi, P., Kaur, G., Thakur, A., Kaur, N., Grewal, A., & Kumar, P. (2017). Waste derivitized blue luminescent carbon quantum dots for selenite sensing in water. *Talanta, 170*, 49–55.
88. Ghorbani, M., Tajik, H., Moradi, M., Molaei, R., & Alizadeh, A. (2022). One-pot microbial approach to synthesize carbon dots from baker's yeast-derived compounds for the preparation of antimicrobial membrane. *Journal of Environmental Chemical Engineering, 10*, 107525.
89. Vyas, Y., Chundawat, P., Dharmendra, D., Punjabi, P. B., & Ameta, C. (2021). Review on hydrogen production photocatalytically using carbon quantum dots: Future fuel. *International Journal of Hydrogen Energy, 46*, 37208–37241.
90. Yang, K., Liu, M., Wang, Y., Wang, S., Miao, H., Yang, L., & Yang, X. (2017). Carbon dots derived from fungus for sensing hyaluronic acid and hyaluronidase. *Sensors Actuators B Chem., 251*, 503–508.
91. Liu, J., Li, R., & Yang, B. (2020). Carbon dots: a new type of carbon-based nanomaterial with wide applications. *ACS Central Science, 6*, 2179–2195.
92. Nasseri, M. A., Keshtkar, H., Kazemnejadi, M., & Allahresani, A. (2020). Phytochemical properties and antioxidant activity of Echinops persicus plant extract: Green synthesis of carbon quantum dots from the plant extract. *SN Appl. Sci., 2*, 670.
93. Yuan, F., Yuan, T., Sui, L., Wang, Z., Xi, Z., Li, Y., Li, X., Fan, L., Tan, Z., Chen, A., Jin, M., & Yang, S. (2018). Engineering triangular carbon quantum dots with unprecedented narrow bandwidth emission for multicolored LEDs. *Nature Communications, 9*, 2249.
94. Mishra, S., Das, K., Chatterjee, S., Sahoo, P., Kundu, S., Pal, M., Bhaumik, A., & Ghosh, C. K. (2023). Facile and green synthesis of novel fluorescent carbon quantum dots and their silver heterostructure: An in vitro anticancer activity and imaging on colorectal carcinoma. *ACS Omega, 8*, 4566–4577.

95. Baragau, I.-A., Power, N. P., Morgan, D. J., Heil, T., Lobo, R. A., Roberts, C. S., Titirici, M.-M., Dunn, S., & Kellici, S. (2020). Continuous hydrothermal flow synthesis of blue-luminescent, excitation-independent nitrogen-doped carbon quantum dots as nanosensors. *J. Mater. Chem. A, 8*, 3270–3279.
96. Mishra, D. M., & Chahal, D. P. (2023). Green, hybrid synthesis and characterization of improved CQD with antioxidant properties for biomedical applications. *Migr. Lett., 20*, 169–178.
97. Lee, H., Song, H.-J., Shim, M., & Lee, C. (2020). Towards the commercialization of colloidal quantum dot solar cells: Perspectives on device structures and manufacturing. *Energy & Environmental Science, 13*, 404–431.
98. Han, J., Hong, J., Lee, H., Choi, S., Shin, K., Gu, M., & Kim, S.-H. (2023). Advances in polyphenol-based carbon dots for biomedical engineering applications. *European Polymer Journal, 197*, 112354.
99. Barton, J., García, M. B. G., Santos, D. H., Fanjul-Bolado, P., Ribotti, A., McCaul, M., Diamond, D., & Magni, P. (2016). Screen-printed electrodes for environmental monitoring of heavy metal ions: A review. *Microchimica Acta, 183*, 503–517.
100. Ahuja, V., Bhatt, A. K., Varjani, S., Choi, K.-Y., Kim, S.-H., Yang, Y.-H., & Bhatia, S. K. (2022). Quantum dot synthesis from waste biomass and its applications in energy and bioremediation. *Chemosphere, 293*, 133564.
101. Ghosh, D., Sarkar, K., Devi, P., Kim, K.-H., & Kumar, P. (2021). Current and future perspectives of carbon and graphene quantum dots: From synthesis to strategy for building optoelectronic and energy devices. *Renewable and Sustainable Energy Reviews, 135*, 110391.
102. Sharma, S. K. (2023). *Green chemistry, its role in achieving sustainable development goals, Volume1*. CRC Press.
103. Ruggerio, C. A. (2021). Sustainability and sustainable development: A review of principles and definitions. *Science of the Total Environment, 786*, 147481.
104. Sugunakara Chary, K. J., Sharma, A., & Singh, A. (2023). Carbon quantum dots in healthcare: A promising solution for sustainable healthcare and biomedical practices. E3S Web Conf. 453, 01017.
105. Kumar, J. V., & Rhim, J. -W. (2024). Fluorescent carbon quantum dots for food contaminants detection applications. J. Environ. Chem. Eng. 111999.
106. Vivekanandhan, S. (2023). Recent developments and emerging opportunities for biomass derived carbon materials in dye sensitized solar energy conversion. *ChemBioEng Rev., 10*, 993–1005.
107. Moniruzzaman, M., Anantha Lakshmi, B., Kim, S., & Kim, J. (2020). Preparation of shape-specific (trilateral and quadrilateral) carbon quantum dots towards multiple color emission. *Nanoscale, 12*, 11947–11959.
108. Gugulothu, Y., Anjaiah, P., Prashanthi, M., & Utkoor, U. K. (2022). Ultra speed synthesis of carbon quantum dots (GCQDs) and Gold (GCQDs-Au) Nano composites, for the Catalytic reduction of MG Dye, Microbial activity and stability studies. *Applied Nanoscience, 12*, 3963–3981.
109. D'Angelis do E. S., Barbosa, C., Corrêa, J. R., Medeiros, G. A., Barreto, G., Magalhães, K. G., de Oliveira, A. L., Spencer, J., Rodrigues, M. O., & Neto, B. A. D. (2015). Carbon dots (C-dots) from cow manure with impressive subcellular selectivity tuned by simple chemical modification. *Chem. – A Eur. J., 21*, 5055–5060.
110. S. T. D. R. S. (2016). Green synthesis of highly fluorescent carbon quantum dots from sugarcane bagasse pulp. *Appl. Surf. Sci., 390*, 435–443.
111. Wang, R., Lu, K.-Q., Tang, Z.-R., & Xu, Y.-J. (2017). Recent progress in carbon quantum dots: Synthesis, properties and applications in photocatalysis. *J. Mater. Chem. A, 5*, 3717–3734.
112. Wang, X., Wang, A., Ma, J., & Fu, M. (2017). Facile green synthesis of functional nanoscale zero-valent iron and studies of its activity toward ultrasound-enhanced decolorization of cationic dyes. *Chemosphere, 166*, 80–88.
113. Wu, H., Xu, H., Shi, Y., Yuan, T., Meng, T., Zhang, Y., Xie, W., Li, X., Li, Y., & Fan, L. (2021). Recent advance in carbon dots: From properties to applications. *Chinese J. Chem., 39*, 1364–1388.

114. Wu, Y., Wei, H., van der Mei, H. C., de Vries, J., Busscher, H. J., & Ren, Y. (2021). Inheritance of physico-chemical properties and ROS generation by carbon quantum dots derived from pyrolytically carbonized bacterial sources. *Mater. Today Bio, 12*, 100151.

Various Properties of Green Synthesized Carbon Quantum Dots

Diksha Singh, Sarita Shaktawat, Ranjana Verma, and Jay Singh

Abstract Due to their unique features and environmentally benign manufacturing methods, carbon quantum dots (CQDs) produced in this way have focused a lot of interest. The many characteristics of CQDs made using green synthesis methods, such as using natural carbon sources, biocompatible precursors, and sustainable processes, are outlined in this chapter. Green-synthesized carbon quantum dots (G-CQDs) optical characteristics display exceptional photoluminescence with controllable emission wavelength. Due to these characteristics, they are appropriate for assistance in bioimaging as well as in light-emitting diodes, together with other optoelectronic applications. G-CQDs created utilizing environmentally friendly methods frequently exhibit great biocompatibility, making them appropriate for biomedical applications such as drug delivery, imaging, as well as therapy. Green synthesis techniques enable precise control of CQD size and morphology, affecting their characteristics and functionality in a variation of function. The remarkable electrical conductivity of G-CQDs makes it possible to use them as electrocatalysts, supercapacitors, and sensors. The higher photostability of CQDs produced in an environmentally responsible way is well recognized and increases their potential for long-term applications. In keeping with the objectives of sustainable development, the green synthesis of CQDs minimizes the environmental impact by using less energy and hazardous waste. For their continuous advancement and incorporation into environmentally friendly technology, green-synthesized CQDs (G-CQDs) must be fully understood and utilized. This opens a promising path for sustainable nanomaterials with a wide range of uses.

D. Singh · S. Shaktawat · J. Singh (✉)
Department of Chemistry, Institute of Sciences, Banaras Hindu University, Uttar Pradesh, Varanasi 221005, India
e-mail: jaysingh.chem@bhu.ac.in

R. Verma
Department of Physics, Institute of Sciences, Banaras Hindu University, Uttar Pradesh, Varanasi 221005, India

V. Kumar et al. (eds.), *Green Carbon Quantum Dots*, Engineering Materials,
https://doi.org/10.1007/978-981-97-6203-3_3

1 Introduction

The element carbon is extremely common on Earth and found in various forms. Thus, the numerous findings of remarkable carbon allotropes have attracted and piqued the interest of expert across research fields. The luminescence, biocompatibility, and solubility of CQDs derived from natural products are higher than those synthesized from other sources [1]. Carbon quantum dots (CQDs), a unique fluorescent carbon nanomaterial, have piqued the interest of scientists throughout the world due to their exceptional water solubility, chemical stability, biocompatibility, hydrophilicity, high conductivity, along with excellent photoluminescence features. Because of its affordability, sustainability, environmental friendliness, along with commercialization, the transformation of biomass leftover into value-added CQDs has been studied as a green synthesis pathway for CQD production [2]. CQDs have gotten an appreciable attention because of their exceptional physicochemical features such as water solubility, biocompatibility, nontoxicity, ease of functionalization, including remarkable chemical stability together with electrical conductivity make them excellent for biosensors, drug administration, gene therapy, bioimaging, besides the development of nanocomposite materials. CQDs have initiate implementation in a various sector, inclusive of tissue engineering, and regenerative medicine, owing to their remarkable chemical stability, along with optical properties [3]. The exceptional impacts of quantum confinement, surface defect, including molecule state, in particular, equip CQDs with a distinctive fluorescence feature [4, 5]. With these benefits, CQDs hold considerable promise in a variety of applications such as photocatalysis, bioimaging, sensing, drug delivery, energy storage, and so on [6, 7]. Biomass, as a benign as well as abundant feedstock, has an excessive carbon and oxygen content [8]. It has 40–50% primary components cellulose, 20–30% hemicellulose, 20–25% lignin, and 1–10% inorganic compounds with remarkable water solubility, making them appealing CQD sources [9]. So far, rice husk [10], wheat straw [11], sugar cane bagasse [12], orange juice [13], watermelon peel [14], soybean [4], onion [5], peanut shell [15], potato [8], durian [6, 16] mango peel [17], lychee [18], bamboo leaves [19], lemon grass herb [20], coffee grounds [21], milk, wool [21], prawn shell, and other agricultural, forestry, and food wastes have been used to make photoluminescent CQDs. Lemon peels are abundant in proteins as well as fibres, which function as a possible carbon resource for the synthesis of CQD [22]. Lu et al. explore chlorogenic acid (ChA), a bioactive component in everyday coffee, for the green synthesis of CQDs, Fig. 4.1 correspond to advantages of the G-CQDs. This method is gaining interest due to its potential for physiologically active CQDs with minimal toxicity [23]. The transformation of biomass into value-added goods achieve a twin purpose of efficiently disposing of solid waste while also relieving the deteriorating environmental, resource, and energy crisis [24]. CQDs have advantages in terms of aqueous solubility, physicochemical, and photochemical stability, good optical performance, and excellent biocompatibility. Due to all these properties, CQD is used in various fields which is shown in Fig. 4.2.

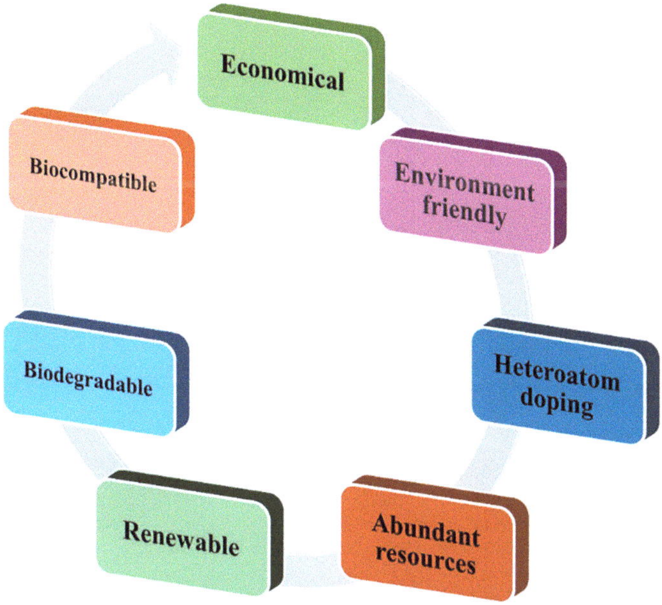

Fig. 4.1 Illustrates the advantages of the Green synthesized Carbon Quantum Dots (G-CQDs)

Fig. 4.2 Illustrating utilization of the G-CQDs in various applications

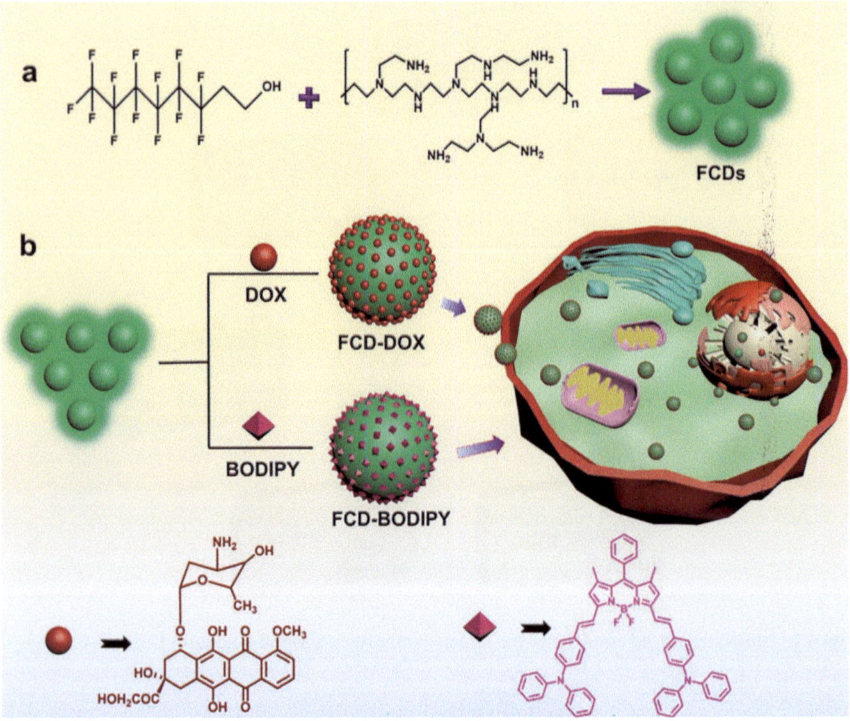

Fig. 4.3 (**a**) Diagram representing fabrication of carbon dots doped with fluorine (FCDs) (**b**) Creating FCDs containing doxorubicin (DOX) and boron dipyrromethene (BODIPY), as well as monitoring cellular uptake of FCD-DOX. Reprint with permission [98]

2 Various Properties of Green Synthesized Carbon Quantum Dots (G-CQDS)-

2.1 Chemical Properties-

(a) ***Surface Functional Groups-*** C-dots from chia seeds had an average dimension of 3 nm in addition to well-distributed surface functional groups. The existence of a variety of functional oxygenic groups on the biomass-derived CQD surface, including hydroxyl, alkyl, epoxide, carboxyl, and carbonyl, contributes to their exceptional water-soluble characteristics [25]. Yao et al. reported that diverse functional groups on CQD surface, containing ketonic carbonyl, hydroxy, and carboxyl groups contribute to high quantum yield including enhanced biological capabilities by representing various enzyme catalytic sites [23]. The modification of CQDs allows for the recognition and affinity towards specific analytes to their surface through electrostatic interactions and hydrogen bonds. This modification is used as a platform for sensing applications. Oxidized CQDs have polar functional groups on their surface, such as hydroxyl and carboxyl groups, which enable chemical functionalization along with stable dispersion in water [26]. For selective and biocompatible intracellular Cu^{2+} imaging, Qiang et.al. paired CQDs with an organic molecule using TPEA ([N-(2-aminoethyl)-N, N, N'-tris(pyridin-2-ylmethyl) ethane-1,2-diamine] (AE-TPEA), AE-TPEA as the identification unit. As a result, the CQD-TPEA combination exhibits intense fluorescence at 420 nm excitation, with intensity falling with increasing Cu^{2+} concentration. When it comes to cellular concentrations, this probe has superior selectivity for Cu^{2+} than bare CQDs. It can also detect Cu^{2+} in a linear range of 10^{-6} to 10^{-4} M at pH 4.0–9.0 [27]. Surface functionalization of CQDs enhances PL, QY, water dispersibility, solubility, and biocompatibility significantly, due to which CQDs are effective in energy and environmental applications [26]. Pandey et al. employed CQD-functionalized gold nanorods for doxorubicin administration in a multimodal way that included photothermal treatment, drug delivery, and bioimaging [28].

(b) ***pH sensitivity-*** CQDs are nanoscale carbon particles with unique optical [29] and electronic properties, finding applications in bioimaging [30], sensing [31], drug delivery [32–35], and photocatalysis [36, 37]. Their pH sensitivity stems from surface functional groups like -COOH, -OH, and $-NH_2$, which undergo protonation or deprotonation based on pH, altering their charge and electronic structure. pH-induced changes in surface charge affect electron transfer processes within CQDs, modulating their fluorescence properties. The conjugated π-electron system of CQDs, influenced by pH changes, affects their absorption and emission spectra. CQDs pH sensitivity is governed by the ionization equilibrium of surface groups, used to develop pH-responsive fluorescent probes for intracellular pH imaging and environmental monitoring. Tailoring surface chemistry enhances their sensitivity for specific applications. Liu et al. reported that Carbon dots (CDs) have pH-sensitive hydroxyl, amide, carbonyl, and amine groups on their surface. Protonation and deprotonation activities influence these groups, affecting the optical characteristics of CDs in aqueous solutions. Increasing pH (e.g., from 7 to 9) causes deprotonation, resulting in changes in absorption bands (e.g., 3430 cm^{-1} raised, 3200 cm^{-1} decreased), indicating a stronger photoluminescence (PL) signal. At higher pH, the decrease in H-bonding and deprotonation results in decreased vibrational energy dissipation, which enhances NIR emission [38].

(c) ***Chemical Composition-*** The chemical composition of G-CQDs shapes their applications. CQDs with hydroxyl (-OH) and amino ($-NH_2$) groups offer low cytotoxicity, ideal for cell imaging as well as drug delivery. Surface functional groups like carboxyl (-COOH) enable selective interaction with analytes, aiding in environmental monitoring and healthcare diagnostics. CQDs with suitable bandgap and high quantum yield serve as light-harvesting materials in solar cells, LEDs, and photodetectors. Functionalized CQDs act as catalyst in photocatalysis, electrocatalysis, and heterogeneous catalysis due to their surface chemistry. CQDs with high surface area and conductivity find use in supercapacitors [39] and lithium-ion batteries [40]. CQDs participate in pollutant adsorption and photocatalytic degradation, aiding in environmental cleanup. In essence, the chemical composition of green-synthesized CQDs dictates their properties, enabling diverse applications in bioimaging, sensing [27, 41] energy, catalysis, and environmental remediation.

(d) ***Doping-*** Doping is the introduction of heteroatoms into the constituents of a CQD [42 43, 44]. It is a frequent procedure to improve the emissions of CDs by injecting electrons in the conduction band as well as an upward shift in the Fermi level [45, 46] Introducing various heteroatoms into carbon structure has a substantial impact on nanodot characteristics [47].

   (i) ***N-doping***- Nitrogen is a common dopant because it has carbon-like atomic size and five valence electrons. N atom function as n-type impurities in C-dots, contributing additional electron that cause the Fermi level to rise and the optical characteristics to shift [45, 48 49, 50]. The experimental results revealed that the emission wavelength of C-dots might be easily controlled by varying the N-doping concentration, and the scientists stated that N-doping governs the optical features of the CQD in preference to size. N-doping increases particle luminescence while shifting excitation and emission to longer wavelength. N-doped CQDs have a wavelength range from 420–500 nm, with the highest peak emission at 440 nm. It works as an effective probe for detecting Fe^{3+} and tetracycline antibiotics, with high sensitivity. N-CQDs are a fundamental and an efficient sensor for ultra-sensitive Fe^{3+} detection due to the linear association among fluorescence intensity along with the Fe^{3+} concentration [51]. Heteroatom insertion affects chemical and biological interaction, allowing for selective metal-ion binding [52], antibacterial activity [53], and use as specialized analytical sensors [54–56].

   (ii) ***Phosphorus doping-*** Phosphorus, from group 15 of the periodic table, behaves as an n-type impurity in C-dots. Compared to N, P atoms are bigger than C atoms. Thus, it produce substitutional defects in the C-cluster as well as act as an n-type donor, changing electrical and optical characteristics [44]. Several types of P-doped C-dots were created utilizing P-containing precursors, such as PBr_3 [44], PPh_3 [57], NaH_2PO_4 [58], H_3PO_4 [59], etc.

   (iii) ***Boron doping-*** One valence electron less than C is found in B (i.e., three). Thus, introducing a B-atom in a C-cluster generates p-type carriers within C-dots, modifying their electrical structures as well as optical properties [57]. B-doped C-dots have strong blue fluorescence with a determined emission peak of 368 nm, while non-doped C-dots have weak green emission with a maximum of 440 nm. The high emission blue shift of B-doped C-dots in comparison to non-doped C-dots possibly be recognized to the strong electron-withdrawing nature of B, which is a recognized electron-deficient Lewis acid [43, 60].

(iv) ***Sulphur doping-*** In recent scenario, S-doped C-dots have garnered much interest, because S atoms preserve energy or emissive trap states for photoexcited electron capture, which alters the electronic structure of C-dots [61].

(v) ***Nitrogen/sulfur co-doped C-dots-*** N,S-doped C-dots have been widely researched, because N has an atomic radius near to that of carbon, while S has an electronegativity comparable to that of C [62]. Doping with N and S increases the fluorescence intensity of CQDs [63, 64]. Yu et al. synthesized N,S-doped C-dots using a one-step hydrothermal process employing citric acid (CA) as a C precursor and L-cysteine as a N and S source [65]. The added S-atoms improve the influence of N-atoms on the characteristics of doped C nanomaterials via a cooperative effect. The S-atoms put into the N, S-doped C-dots were competent to eradicate the O-states (surface states generated by the presence of oxygen in C-dots) while enhancing the N-states. As a result, the N, S-doped C-dots demonstrated a great fluorescence quantum yield (73%) and excitation-independent emission. The highest excitation and emission wavelengths of the aqueous N-S-doped C-dot solution is 345 and 415 nm, respectively. Wang et al. demonstrated that adding N or S atoms to graphite lattices might minimize the band gap of the nanoscale graphene structure while co-doping with N and S reduce the band gap more effectively [66].

(e) ***Redox properties-*** G-CQDs exhibit notable redox properties, crucial for electrochemical and catalytic applications. These properties are shaped by factors like surface functional groups, size, and electronic structure. CQDs feature surface groups like -OH, -COOH, $-NH_2$ and $-CONH_2$, influencing redox reactions as electron donors or acceptors acting as reactive sites [53]. Quantum confinement effects [26, 67] and abundant surface sites facilitate electron transfer, enhancing CQDs' utility in diverse redox reactions. CQDs serve as effective electrocatalysts/co-catalysts [68, 69], amplifying electron transfer rates and enabling applications in sensing [27, 41] energy storage, and electrocatalysis [70 71 72]. Incorporation of dopants (e.g., N, S) alters CQDs' electronic states, modifying their redox behavior and band structure [73, 26, 74]. CQDs act as redox mediators [75], expediting electron transfer [76] between electrodes and electrolytes, suitable for energy storage [77, 78], sensing [79, 56, 80, 81], and electrochemical devices. The redox properties of G-CQDs, shaped by surface chemistry and structural characteristics, underpin their versatility in various electrochemical and catalytic contexts.

(f) ***Sensitivity to environmental conditions-*** The sensitivity of G-CQDs to environmental conditions is crucial for their applications. pH and temperature [82] variations influence CQDs' stability and fluorescence properties [83], critical for in vivo imaging, [84] and diagnostics. Environmental factors like pH, temperature, and ions affect CQDs-based sensor sensitivity, and selectivity [85], optimizing applications in monitoring, food safety, and diagnostics. CQDs' sensitivity to light, humidity, and oxygen impacts their efficiency and stability in photovoltaic, and optoelectronic devices [86]. Temperature, pressure, and reaction media composition influence CQDs' catalytic activity, and stability [87], crucial for achieving high efficiency, and selectivity. CQDs sensitivity to electrolyte composition, temperature, and cyclic condition affects their performance in batteries, and supercapacitors [88], influencing energy storage capacity, and stability. Understanding CQDs' sensitivity to parameters like light intensity and pH is vital for designing efficient photocatalytic [89–91] systems for environmental cleanup. Basically, understanding and controlling environmental conditions are essential for optimizing the performance and applications of G-CQDs.

(g) ***Reactivity-*** The reactivity of G-CQDs is pivotal for their applications. CQDs act as catalysts in diverse reactions, hence accelerate reaction kinetics through surface functional groups and defect sites. CQDs selectively interact with target analytes, enhancing sensitivity in environmental monitoring, healthcare diagnostics, and food safety. CQDs efficiently absorb light and facilitate charge transport, improving performance in solar cells, and optoelectronic devices. CQDs aid in drug delivery, imaging, and therapy through surface functionalization as well as production of reactive oxygen species. CQDs exhibit reactivity in electrochemical processes, enhancing capacitance and stability in batteries, and supercapacitors [40]. CQDs degrade pollutants and adsorb contaminants, leveraging their photocatalytic activity and high surface area. Essentially, the reactivity of G-CQDs enables their versatility across catalysis, sensing [27, 41], energy, biomedical, and environmental applications, driving advancements in various fields.

(h) ***Chelating properties-*** G-CQDs form stable compounds with metal ions via coordination chemistry. Although CQDs are not naturally chelating agents, they possibly modified with functional groups such as carboxyl, hydroxyl, amino groups, and conjugated π-domains that allow metal ion binding [92, 93]. These alterations improve their chelating characteristics, making them suitable for metal ion detection, sensing, and environmental remediation. Because of their vacant d-orbitals, transition metals tend to chelate quickly when they come in association with an electron donor group. CQDs have oxygen-containing functional groups that act as electron donors and combine through ions to form non-fluorescent complexes [94]. As a result, CQDs are effective sensors for some transition metals. Fe^{3+} is the most prevalent trace element in both biological and atmospheric system. It is the most common heavy metal contaminant detected in freshwater bodies [95]. Zhao et al. synthesized CQDs from pinewood and used them to detect Fe^{3+} ions [96]. A. Kundu et al. synthesized fluorescent CQDs from rice husk (RH-CQDs) via hydrothermal and develop a versatile sensor platform capable of detecting multiple analytes, including metal ions (Fe^{3+}) and fluoroquinolones, with high selectivity and sensitivity. The carboxyl group of RH-CQDs and the amine group of the piperazine ring form a hydrogen bond, which account for the great selectivity and sensitivity of diverse analytes to fluoroquinolones [97]. Green synthesis methods yield CQDs with specific chelating functions, making them useful in chemistry, material science, and environmental applications.

(i) ***Biological interactions-*** G-CQDs interact with biological systems via cellular uptake [98, 99], imaging, medication administration, and biosensors. G-CQDs are biocompatible, cross biological barriers, and feature fluorescence for sensitive imaging. With ample surface area, they are ideal for drug delivery and function as biosensors. Their characteristics and surface alteration determine interaction, crucial for safe biological applications. CQDs enable high-resolution imaging of biological structures due to their biocompatibility and cellular uptake [98, 99] capability. Gao et al. stated the effective intracellular delivery of drugs/dyes facilitated by fluorine-doped carbon dots (FCDs) represented in Fig. 4.3. Functionalized CQDs penetrate cellular barriers for targeted drug delivery and detect biomarkers with high specificity, enhancing biosensing. They combine diagnostic and therapeutic functions for personalized medicine, aid bioanalytical studies, and promote cell activities in tissue engineering. CQDs generate reactive oxygen species under light, selectively targeting cancer cells in photodynamic therapy. G-CQDs biological interaction features allow for their diverse application in biomedicine, [100] and biotechnology, with promising advances in brain cancer cell diagnostics (Min Zheng et al., 2015), therapies, and regenerative medicine.

2.2 Physical Properties

(a) ***Particle size and morphology-*** When carbon particles are smaller than 10 nm, quantum size effects take place, as a result, there are unique HOMO–LUMO gaps and discrete energy level distributions attained after electron confinement [101]. In addition to increasing the surface area, this size reduction exposes more surface atoms [102]. Zheng et al. detected blue photoluminescence with a strong intensity in small graphitic carbon dots (about 2 nm) in size [103]. The CQDs had consistent size, a narrow distribution, and 88% quantum yield. Their emission wavelength varied according to composition and surface condition, making it possible to synthesize high-yield, affordable, adjustable luminous solid-phase homogenous CQDs [104]. Particle size directly impacts the material's emissive properties due to quantum confinement. Efforts focus on regulating nanodot size through particle separation or template utilization. Surface states significantly affect the luminescence properties of carbon dots [105–107].

Particle size as well as morphology of the G-CQDs significantly impact the application of G-CQDs across various fields like biomedical imaging [108], drug delivery, sensing, catalysis, energy storage, and environmental remediation [52]. Control over these factors is crucial for optimizing CQDs for specific uses. Small sizes enhance tissue penetration and cellular uptake [98, 99] while uniform morphologies ensure consistent performance. Controlled morphologies enable efficient cellular uptake [98, 99] along with targeted drug delivery [109, 110]. Furthermore, particle size as well as morphology influence sensor sensitivity [111], optical properties, catalytic activity, electrochemical performance, and pollutant adsorption efficiency.

(b) ***Optical properties- (i) Photoluminescence -*** CQDs derived from natural sources have higher luminescence, biocompatibility, and solubility than those derived from synthetic materials from various sources [1]. CQDs have more clearly specified emission peaks along with larger excitation spectra. As a result, the identical CQDs produce variable fluorescence by altering their size and excitation wavelength [112]. CQDs produced using citric acid monohydrate and diethylene glycol BIS ether were exceptionally luminous, and their synthesis was simple, eco-friendly, and scalable [112]. CQDs are gaining popularity in the biological field because of their tiny size, stable photoluminescence [2], better surface grafting, biocompatibility, along with high water solubility [1, 113]. Despite being more cytotoxic and having lower photoluminescence (PL) than negatively charged CQDs, positively charged CQDs have a higher labeling efficiency just because of their larger absorption capacity [3]. CQDs show promise as light-harvesting nanomaterials for photovoltaic applications due to their wide absorption spectra along with high absorption coefficient [114]. Mirtchev et al. synthesized CQDs possessing a broad absorption spectrum intended for use as solar cell sensitizers [115]. CQDs display up conversion photoluminescence, absorbing multiple low-energy photons likewise emitting a higher-energy photon [116]. Converting low-energy irradiation to higher energy enhances light absorption, improving photovoltaic energy generation [117].

(c) ***Surface area and porosity-*** G-CQDs' surface area and porosity are vital for diverse applications, optimizing these characteristics is essential for expanding their effectiveness. Higher surface area and porosity enhance their catalytic activity by providing more active sites and improving mass transfer. Increased surface area and porosity contribute to higher sensitivity and selectivity in sensing [27, 41] applications. They enhance the adsorption of target analytes onto CQDs surfaces, improving detection limits and response times. It improve the performance of CQDs in energy storage devices [118] such as batteries and supercapacitors [40]. They allow for more electrolyte infiltration and increase the electrode/electrolyte interface, directing to higher energy storage capacity and faster charge–discharge rates. These facilitate the adsorption of pollutants onto CQDs' surfaces, enabling efficient removal from water or air. They also enhance the efficiency of photocatalytic degradation processes by offering more active sites for pollutant adsorption and photocatalysis. The surface area and porosity of CQDs influence their drug loading capacity, and cellular uptake [98, 99] efficiency in drug delivery [109] applications. Higher surface area allows for greater drug loading, while porosity affects drug release kinetics and cellular interactions [119]. It also impacts the light absorption and emission properties of CQDs in optoelectronic devices. They influence the density of electronic states, which affect the efficiency of light absorption, emission, and charge transport. It is easy to use them for targeted drug delivery because of their high surface area to volume ratio [120].

(d) ***Surface charge-*** CQDs that are green-synthesized are used in diverse fields due to their surface charge. The surface charge of carbon dots is important in determining their bactericidal action. This activity is linked to their capability to compromise cell membrane integrity [121]. Bing et al. examined CQDs with differing surface charges trigger the death of bacteria [122]. + vely charged CQDs exhibit higher cytotoxicity and lower photoluminescence than -vely charged CQDs. However, they have significant labeling efficiency because of their higher absorption capacity. + vely charged CQDs disturb the differentiation of stem cells into adipocytes and osteocytes. Despite their weaker charge, + vely charged CQDs have high biocompatibility and labeling efficiency in human stem cells [123]. + vely and -vely charged dots exhibited bactericidal effects on E. coli, whereas uncharged dots showed no activity. Bacterial growth is mostly inhibited by the formation of reactive oxygen species (ROS), notably hydroxyl radicals, in response to charged dots [124]. The electron-donating nature of amides/amines on N-CQDs generates oxygen radicals. N-type doped CQDs exhibit enhanced electron transfer processes coupled with oxygen radical development due to additional free electron incorporation in CQDs [125]. N-CQDs have increased activity compared to S-CQDs, as +vely charged surface charge (protonation) of N-CQDs improves electrostatic interactions with -vely charged lipid membrane components, increasing antibacterial activity. Quaternary ammonium compounds, effective antibacterial agents, disrupt bacterial cell membranes and inhibit bacterial viability [126]. It impacts electrostatic adsorption, colloidal stability, and interactions with biological molecules. Surface charge influences cellular absorption, and drug delivery systems' targeting effectiveness in biomedical applications. For CQDs to be optimized for a given application and to ensure their stability and efficacy, surface charge must be understood and controlled.

(e) ***Crystallinity-*** The crystallinity of G-CQDs is pivotal across various applications. Highly crystalline CQDs offer improved stability, optical properties [116, 127], and electronic conductivity [128], making them valuable for optoelectronics [129, 130], biosensing [131], and catalysis [132]. Crystallinity influences charge transfer efficiency [133, 134], affecting energy-related applications, while also impacting structural integrity and biocompatibility in biomedical uses like bioimaging [135] and drug delivery. Understanding and optimizing crystallinity are crucial for tailoring CQDs to specific applications, enhancing their functionality. The crystallinity of the core dictates quantum confinement in CQDs. CQDs with a crystalline core exhibit quantum confinement, while those with an amorphous core do not. Additionally, CQDs demonstrate excellent electronic as well as optical properties due to edge effects along with quantum confinement [2]. Their prominent dielectric constant along with extinction coefficient facilitates the dissociation of electrons as well as holes, leading them to promising energy technologies. Furthermore, their low cytotoxicity renders them safe for biomedical applications [26].

(f) ***Thermal stability***- G-CQDs require high thermal stability for a variety of applications. It enables them to perform at high temperature in optoelectronic devices, catalytic reactions, energy storage systems, and biological applications such as drug administration and bioimaging [136]. Overall, thermal stability [137] ensures the long-term viability and effectiveness of G-CQDs across a broad spectrum of benefits.

(g) ***Dispersion and solubility-*** Since CQDs have a lot of carboxyl and carbonyl groups covering their surface, they are very biocompatible and water soluble [1]. CQDs excellent water solubility has led to their application as biosensor carriers (Xin Ting Zheng et al., 2015). CQDs from chia seeds had an average size of 3 nm along with well-distributed surface functional groups. Zeta potential analysis showed a -26.6 mV surface charge for C-dots synthesized at neutral pH and room temperature. This high negative zeta potential improves colloidal stability, preventing aggregation [25]. The surface functional groups have a considerable effect on our CQDs total charge. Travlou et al. reported the measured ζ potential values were -6.47 $\pm$ 0.67 mV for N-CQDs and -47.18 mV for S-CQDs. Despite being negative, suggesting the existence of hydroxyl along with carboxylate functional groups, S-CQDs include sulfonic groups on their surfaces [53]. This sample has a significantly higher negative ζ potential, resulting in increased system stability.

(h) ***Electrochemical properties-*** The electrochemical properties of G-CQDs utilized in various applications, including, G-CQDs can be used as electrochemical sensors for detecting various analytes, such as heavy metals [138–140], biomolecules [141], and pollutants [142–144], ascribed to their high surface area along with excellent electron transfer properties. G-CQDs can be incorporated into electrodes for supercapacitors [145] as well as batteries, while their high surface area as well as conductivity boost charge storage as well as transfer capability. G-CQDs can serve as efficient catalysts for various electrochemical reactions, such as oxygen reduction reaction (ORR) and hydrogen evolution reaction (HER), as consequence of their high catalytic activity and stability. G-CQDs are used in electrochemical cells for energy conversion and electrochemical energy storage [78], such as dye-sensitized solar cells [146, 147] including fuel cells [148], where their electrochemical properties enhance device performance. For each of these uses, the electrochemical properties of G-CQDs, such as their conductivity, surface area together with electrochemical stability, play a crucial role in regulating their performance and efficiency.

2.3 Physicochemical Properties-

(a) ***Surface Area-*** G-CQDs broad surface area and strong electron mobility are essential to their range of uses [26]. High surface area enhances catalytic efficiency, sensitivity in biosensing, and charge storage capacity in energy devices [26]. It also aids in pollutant adsorption for environmental cleanup and facilitates functionalization in drug delivery [149]. Surface area influences light-harvesting efficiency and cell interactions in tissue engineering. Surface area greatly influences the effectiveness of G-CQDs across catalysis, sensing [27, 41], energy storage, environmental cleanup, biomedicine, optoelectronics, and tissue engineering.

(b) ***Dispersion-*** The dispersion properties of G-CQDs significantly influence their utilization. Well-dispersed CQDs enhance mechanical, thermal, and electrical properties. Minimal aggregation ensures accurate bioimaging in biological media. Increased surface area improves sensitivity and selectivity. Uniform dispersion enhances catalytic activity and selectivity. Improved kinetics boost performance in batteries [26] and supercapacitors [4, 150]. Efficient charge transfer along with light emission, enhance device performance. Enhanced adsorption and photocatalytic activity aid in pollutant removal. Achieving optimal dispersion ensures the effectiveness of CQDs across nanocomposite materials, biomedical imaging, sensing [27, 41], catalysis, energy storage, optoelectronics, and environmental remediation applications.

(c) ***Magnetic properties-*** The magnetic properties of G-CQDs are utilized in various applications. The magnetic properties of G-CQDs offer unique opportunities for applications in biomedical imaging, drug delivery [151], hyperthermia therapy, biological separation, environmental remediation, and sensing [27, 41]. These qualities enable precise control and manipulation of CQDs for an extensive extent of empirical administration across different fields. Magnetic CQDs serve as contrast agents in magnetic resonance imaging (MRI) [110] due to their paramagnetic or superparamagnetic properties. They enhance the imaging contrast, enabling better visualization of biological tissues and structures. Magnetic CQDs employed as carriers for targeted drug delivery [151]. By functionalizing CQDs with magnetic nanoparticles, they are guided to specific sites within the body using external magnetic fields, enhancing drug delivery efficiency and reducing off-target effects. Magnetic CQDs [110] are employed in environmental remediation to remove pollutants and in cancer therapy to target tumors because they are capable to induce to heat in an alternating magnetic field, a phenomenon known as magnetic hyperthermia. They are also used in magnetically separated and purified biological molecules, cells, and nanoparticles, as well as in sensors that detect magnetic fields or nanoparticles. Their magnetic properties enable sensitive detection of magnetic signals, which are used in various sensing [27, 41] applications, including biosensors as well as magnetic resonance-based detectors.

(d) ***Electrical conductivity-*** G-CQDs' electrical conductivity finds use in electronics, energy storage, catalysis [90], sensing [152], optoelectronics [153], biomedical devices, and flexible electronics, expanding their role in emerging technologies. They enhance charge transport efficiency and device functionality in electronic devices such as transistors, diodes, and sensors. They are also utilized in energy storage devices [118] such as supercapacitors [4, 154] and batteries, improving charge kinetics and cyclic stability [26, 40]. Additionally, electrically conductive CQDs serve as catalysts in various reactions, enable sensitive detection in sensors, enhance performance in optoelectronic devices [153], facilitate biosensing and bioelectronics, and are essential components in flexible and wearable electronics, broadening their applications in emerging technologies.

(e) ***Density-***The density of G-CQDs can influence their performance in various applications. Higher density indicates a higher concentration of CQDs, which can impact their optical properties, such as fluorescence intensity and stability. In drug delivery, a higher density of G-CQDs could potentially lead to a higher drug-loading capacity. However, the specific impact of density on G-CQDs' performance would depend on the application and the desired characteristics of the CQDs for that particular use.

(f) ***Aggregation behavior-*** The aggregation of G-CQDs affects their effectiveness in applications. Techniques like surface modification or dispersion are crucial to maximize their potential. Aggregation can reduce imaging quality and biocompatibility in biological environments, affecting suitability as contrast agents. It also hinders the effective delivery of therapeutic payloads, impacting drug release kinetics [119], and target-specific delivery, thereby compromising the efficacy of drug delivery systems [33]. Aggregation of CQDs alter their optical as well as electronic properties, affecting sensitivity [111] and selectivity in sensing applications. Aggregation-induced changes in fluorescence emission or conductivity cause inaccurate detection. Aggregation reduces surface area and accessibility of active sites on CQDs, lowering catalytic efficiency and hindering reaction kinetics. Poor dispersion within electrode materials due to aggregation leads to decreased electrochemical performance, reduced charge storage capacity, and compromised cycling stability in energy storage devices [118, 155]. Aggregation-induced quenching of fluorescence or charge carrier recombination impairs CQDs' performance in optoelectronic devices like LEDs and solar cells, reducing device efficiency and response time. Aggregation limits the surface area available for adsorption or photocatalytic degradation of pollutants, reducing the efficiency of CQDs in environmental remediation processes.

(g) ***Mechanical properties-*** While not the main focus, the mechanical properties of G-CQDs are crucial in applications like nanocomposite materials, flexible electronics, sensors, and biomedical devices. Optimizing these properties maximizes the effectiveness of CQDs in various applications. Incorporating CQDs into polymer matrices potentially enhances mechanical characteristics including tensile strength, elasticity, as well as toughness, reinforcing the material and improving its mechanical performance. In flexible and wearable electronics, the mechanical flexibility of CQDs is advantageous, enabling the development of bendable and stretchable electronic devices. Robust mechanical properties of CQDs are essential in sensing applications to withstand environmental conditions and maintain sensing capabilities over time. In biomedical devices and implants, CQDs with suitable mechanical properties ensure compatibility with biological tissues and organs, avoiding adverse reactions or degradation within the body.

(h) ***Solubility-*** The solubility of G-CQDs is vital for their effectiveness in diverse applications. It determines their applicability across various fields, emphasizing the importance of ensuring appropriate solubility properties for optimizing performance. Solubility in aqueous solution is essential for biomedical applications such as drug delivery [33], bioimaging [156], and therapeutics. Water-soluble CQDs are preferred for efficient delivery and interaction with biological system while minimizing cytotoxicity [157]. Solubility affects the dispersion of CQDs in sensing matrices and biological fluids, influencing the sensitivity [111] and selectivity [85] of sensors. Soluble CQDs are key for homogeneous sensing, catalytic activity, and stability in catalytic reactions. They ensure efficient interaction with reactants, enhancing catalytic efficiency. In optoelectronic devices like solar cells and LEDs, solubility is crucial for uniform deposition and improved device performance. Soluble CQDs also enable dispersion in environmental matrices for pollutant detection and remediation, and in electrolytes and electrode materials for energy storage devices [155]. Well-dispersed CQDs enhance ion transport and charge storage capacity, improving the overall performance of batteries and supercapacitors [4].

2.4 *Biological Properties-*

(a) ***Biocompatibility-*** CQDs require high biocompatibility for several biological applications, including bioimaging [31]. CQDs are stable and biocompatible, making them useful for biosensors and bioimaging. They have been utilized to identify a variety of biological macromolecules, containing DNA, microRNA, trypsin, thrombin, hyaluronidase, and alkaline phosphatase. Zhang et al. reported a real-time test for monitoring tyrosinase (TYR) activity and inhibitor screening. It uses dopamine-functionalized carbon quantum dots (Dopa-CQDs) as fluorescent probes, which are CQDs with a high yellowish-green fluorescence coupled with dopamine to increase sensitivity. Functionalized C-dots with quinoline derivatives show exceptional biocompatibility for Zn^{2+} bioanalysis and biomedical detection [41]. Surface functionalization converts precursor carbon nanoparticles into soluble carbon dots, resulting in a solution-like dispersion with high biocompatibility [157]. The biocompatibility of G-CQDs is essential for their biomedical applications like bioimaging, drug delivery [33], biosensing, and therapeutics. Optimizing biocompatibility is crucial for safety and efficacy. Challenges remain in assessing long-term effects and obtaining regulatory approval. Despite hurdles, G-CQDs offer significant promise for healthcare and biotechnology.

(b) ***Cytotoxicity***- Recent studies show that CQDs at low concentrations stimulate cell development and are only slightly harmful to human cells. Higher CQD concentrations, however, cause more cytotoxicity. Research indicates that when the concentration of CQD increases human cell survival rates decline [1]. Ray et al. with HepG2 cells, the survival rate was over 90% at CQD concentrations below 0.5 mg/ml but dropped to 75% at 1.0 mg/ml (Yang, Wang, et al., 2009). Similarly, MCF-7 cell survival rates were close to 100% at CQD concentrations below 20 μg/ml but decreased to 40% at 200 μg/ml. N-doped CQDs had different toxicities on L-929 and MCF-7 cells, with an endurance rate of 90% for L-929 cells as well as 81% for MCF-7 cells at a concentration of 50 μL/ml [158]. Gd-doped CQDs combined with Adriamycin showed excessive cancer cell toxicity (EC50 = 2.5 mg/ml) compared to Adriamycin alone (EC50 = 5 mg/ml), suggesting potential for further study of their combined effects with other biological macromolecules [159]. On account of their prospects for bioimaging and nanoscale size, C-dots are naturally hazardous. Various research groups have done toxicity tests, and while there are few reports at this time, C-dots appear to be low toxicity [160, 161] (Yang, Wang, et al. 2009). CQDs are gaining attention owing to their biocompatibility along with low cytotoxicity, as they lack toxic heavy metals present in distinct fluorescent materials like quantum dots along with certain organic dyes [93]. L.Chai reported Dopamine functionalized CQDs cytotoxicity on HeLa cells was examined using the MTT test. HeLa cells were treated with varying Dopa-CQD doses for 24 h at 37 °C in 96-well plates. After the treatment, MTT solution was added as well as incubated for 4 h. Formazan was dissolved in DMSO afterwards the absorbance at 490 nm was determined. The cell viability rate (VR) was computed as $(A/A_0) \times 100\%$, where A is the absorbance of the treated group and A_0 is the absorbance of the untreated group, with 100% cell survival rate [162]. The cytotoxicity of G-CQDs is crucial in biomedical and biotechnological applications. Addressing cytotoxic effects through biocompatibility assessments and targeted modifications is essential to ensure their safe and effective use. Despite limitations, CQDs hold promise in bioimaging, drug delivery, biosensing, and therapeutics, offering innovative solutions in healthcare and biomedicine [33].

(c) ***Fluorescence***- From UV–visible spectra of CQDs informed by M. Sathish et al. and Y. Deng et al. with the sharp peak at 290 nm, which is associated with the $\pi - \pi^*$ transition of the aromatic C = C bond which is further validated by the FTIR spectra of the CQDs are consistent [163]. The FTIR spectrum identifies various functional groups in the C-dots. Peaks at 1250 cm^{-1} and 1629 cm^{-1} represent the existence of C-O and C = O groups, individually. Peaks at 3424 cm^{-1} signify O–H groups, while 1500 cm^{-1} suggests a phenolic group. Additionally, C-N and C = N bands at 1012 cm^{-1} and 1549 cm^{-1} are observed. Understanding these functional groups provides insights into how they influence the absorption and photoluminescence characteristics of C-dots [52]. The optical properties of CQD were investigated using the steady-state fluorescence spectroscopy method. CQDs that were achieved expanded broad-spectrum area, and their emission included a visible wavelength range that prolonged upto the near-infrared region. This enables them to operate as an enhanced light-capturing mode for nanophotonic purposes such as bioimaging. CQDs are beneficial for fluorescence bioimaging because they generally emit photoluminescence emissions (Xin Ting Zheng et al. 2015) [131]. In certain instances, solvent-dependent fluorescence emission behavior can also be linked to solvation dynamics. S. Mishra reported the photoluminescence spectra of CQDs showed a significant emission peak at 467 nm including an excitation wavelength of 415.8 nm. The size of the CQDs affects their optical characteristics, which may also affect the quantity and type of sp^2 sites that are present in the CQDs [84]. Because of their special optical qualities, low toxicity, and functionalization potential, CQDs have become highly useful nanomaterials for bioimaging. They present a viable foundation for the creation of safer and more efficient imaging agents for a variety of biomedical uses. Among the key properties of carbon-based quantum dots, quantum yield (QY) is crucial when taking into account the numerous applications in imaging and sensing, particularly in the biological domain [164]. CQDs demonstrate potential for imaging tumors. CQDs imposing possibility as bioimaging agents due to their inherent qualities essentially photoluminescence properties [165]. A few of the preferred characteristics of the specific bioimaging agents for CQDs might be enumerated as the comparatively extended fluorescence lifetime, wavelength-dependent excitation and emission, stability and solubility in water, in addition to enhanced biocompatibility [166]. To effectively target specific receptor at the tumor location, an imaging agent is necessary for cancer imaging applications. In order to create the ideal theranostics agent one that act as a multifunctional platform combining targeting and imaging capabilities [167]. It is still difficult to develop targeted imaging mediators by attaching marker molecules to the surface of carbon dots.

(d) ***Antibacterial Properties-*** There have been reports of CQDs targeting both Gram + ve along with Gram -ve bacteria. The mechanism for bacterial targeting was recognised through electrostatic contact linking the cationic residues on the C-dots' surface and the anionic microbial membrane [168]. In antibacterial testing, CQDs could be a valuable alternative to traditional antibiotic medications. Many functional groups found in CQDs may be responsible for the antibacterial activity. These groups may obstruct cellular enzyme operations and prevent cellular growth. The bacterial cell wall was readily contacted by the massive π-conjugated CQD approach via electron transfer [169, 170]. According to the literature, there have been several theories regarding the antibacterial mechanism of CQDs. These theories include light irradiation, ROS, and electrostatic interactions. The production of ROS has significant antibacterial activity [171, 172]. FTIR and XPS analyses validate the existence of nitrogen groups and hydroxyl radicals. CQDs having N elements that have + tive charges connect them to -vely charged germs. CQDs pierce through cell membranes, furthermore, eventually cause microorganisms to die. Numerous investigations have documented CQDs containing nitrogen that have guaranteed antibacterial action against both Gram + ve and Gram − ve bacterium. According to Yadav et al. [173], CNQDs have the ability to interact with bacteria such as E. coli and Staphylococcus aureus while also producing superoxide and hydroxyl radicals. N-doped CQDs with a targeted antibacterial action against B. subtilis including E. coli have been produced by Travlou et al. [53]. It's interesting to note that, when compared to other CQDs, the CQDs in this study exhibit greater inhibition against Bacillus cereus, Staphylococcus aureus, Pseudomonas aeruginosa, Vibrio cholera, and Escherichia coli [174–176]. As a result, medicinal uses for the produced CQDs are possible.

(e) ***Biodegradability-*** The biodegradability of G-CQDs is an important consideration for their utilization across various applications, especially in biomedical and environmental fields. Ensuring that CQDs are biodegradable is crucial to minimize their long-term environmental impact and potential health risks. Biodegradable CQDs are safely metabolized and eliminated from biological systems or degraded in the environment without causing harm. In biomedical applications, biodegradable CQDs are desirable for drug delivery systems to avoid accumulation and potential toxicity. In environmental applications, biodegradable CQDs are used for pollutant remediation without leaving harmful residues. Therefore, the biodegradability of G-CQDs enhances their suitability for diverse applications while minimizing ecological and health concerns.

3 Future and Challenges of Green Synthesized Carbon Quantum Dots (G-CQDs)

The future of G-CQDs offers vast potential for advancement and sustainability. However, addressing challenges like biocompatibility, performance optimization, regulatory compliance, and interdisciplinary collaboration is essential for maximizing their benefits across diverse applications. G-CQDs offer eco-friendly alternatives using renewable resources and green chemistry principles. Future efforts aim to refine green synthesis techniques, minimize waste, and reduce environmental impact. G-CQDs hold promise in biomedical fields like bioimaging and drug delivery due to their biocompatibility, yet trials continue for estimation of long-term biocompatibility as well as toxicity. They similarly establish prospective in optoelectronic devices and energy storage technologies, but optimizing performance and scalability is crucial. G-CQDs exhibit catalytic and photocatalytic properties for environmental applications, but improving efficiency, and recyclability is needed. Integrating G-CQDs into functional materials requires addressing compatibility and stability challenges. Regulatory frameworks and collaboration across disciplines are essential for safe and responsible use and advancement.

Acknowledgements This work received no specific grant from public, commercial, or not-for-profit funding agencies. The authors, D.S. and S.S. and R.V would like to express their gratitude to Banaras Hindu University (BHU), Varanasi, for providing the necessary facilities to conduct the research work. Author J. S. would like to acknowledge BHU for providing a seed grant and BRIDGE grant under the MoE Govt India, Institute of Eminence (IoE) scheme 6031 & 6031A respectively.

References

1. Sun, Y., Zhang, M., Bhandari, B., & Yang, C. (2022). Recent development of carbon quantum dots: biological toxicity, antibacterial properties and application in foods. *Food Reviews International, 38*(7), 1513–1532. https://doi.org/10.1080/87559129.2020.1818255
2. Kim, A., Dash, J. K., Kumar, P., & Patel, R. (2022). Carbon-based quantum dots for photovoltaic devices: A review. *ACS Applied Electronic Materials, 4*(1), 27–58. https://doi.org/10.1021/acsaelm.1c00783
3. Majood, M., Garg, P., Chaurasia, R., Agarwal, A., Mohanty, S., & Mukherjee, M. (2022). Carbon quantum dots for stem cell imaging and deciding the fate of stem cell differentiation. *ACS Omega, 7*(33), 28685–28693. https://doi.org/10.1021/acsomega.2c03285
4. Xu, J., Xue, Y., Cao, J., Wang, G., Li, Y., Wang, W., & Chen, Z. (2016). Carbon quantum dots/nickel oxide (CQDs/NiO) nanorods with high capacitance for supercapacitors. *RSC Advances, 6*(7), 5541–5546. https://doi.org/10.1039/C5RA24192H
5. Hu., Yuefang, Zhang, L., Li, X., Liu, R., Lin, L., & Zhao, S. (2017). Green preparation of S and N co-doped carbon dots from water chestnut and onion as well as their use as an off–on fluorescent probe for the quantification and imaging of coenzyme A. *ACS Sustainable Chemistry & Engineering, 5*(6), 4992–5000. https://doi.org/10.1021/acssuschemeng.7b00393
6. Gang, W., Guo, Q., Chen, D., Liu, Z., Zheng, X., Xu, A., Yang, S., & Ding, G. (2018). Facile and highly effective synthesis of controllable lattice sulfur-doped graphene quantum dots via hydrothermal treatment of Durian. *ACS Applied Materials & Interfaces, 10*(6), 5750–5759. https://doi.org/10.1021/acsami.7b16002
7. Prasannan, A., & Imae, T. (2013). One-pot synthesis of fluorescent carbon dots from orange waste peels. *Industrial & Engineering Chemistry Research, 52*(44), 15673–15678. https://doi.org/10.1021/ie402421s
8. Shen, J., Shang, S., Chen, X., Wang, D., & Cai, Y. (2017). Facile synthesis of fluorescence carbon dots from sweet potato for Fe3+ sensing and cell imaging. *Materials Science and Engineering: C, 76*, 856–864. https://doi.org/10.1016/j.msec.2017.03.178
9. Mahat, N. A., & Shamsudin, S. A. (2020). Transformation of oil palm biomass to optical carbon quantum dots by carbonisation-activation and low temperature hydrothermal processes. *Diamond and Related Materials, 102*, 107660. https://doi.org/10.1016/j.diamond.2019.107660
10. Chaudhary, S., Kumar, S., Kaur, B., & Mehta, S. K. (2016). Potential prospects for carbon dots as a fluorescence sensing probe for metal ions. *RSC Advances, 6*(93), 90526–90536. https://doi.org/10.1039/C6RA15691F
11. Yuan, M., Zhong, R., Gao, H., Li, W., Yun, X., Liu, J., Zhao, X., Zhao, G., & Zhang, F. (2015). One-step, green, and economic synthesis of water-soluble photoluminescent carbon dots by hydrothermal treatment of wheat straw, and their bio-applications in labeling, imaging, and sensing. *Applied Surface Science, 355*, 1136–1144. https://doi.org/10.1016/j.apsusc.2015.07.095
12. Chai, X., He, H., Fan, H., Kang, X., & Song, X. (2019). A hydrothermal-carbonization process for simultaneously production of sugars, graphene quantum dots, and porous carbon from sugarcane bagasse. *Bioresource Technology, 282*, 142–147. https://doi.org/10.1016/j.biortech.2019.02.126
13. Sahu, S., Behera, B., Maiti, T. K., & Mohapatra, S. (2012). Simple one-step synthesis of highly luminescent carbon dots from orange juice: Application as excellent bio-imaging agents. *Chemical Communications, 48*(70), 8835. https://doi.org/10.1039/c2cc33796g
14. Zhou, J., Sheng, Z., Han, H., Zou, M., & Li, C. (2012). Facile synthesis of fluorescent carbon dots using watermelon peel as a carbon source. *Materials Letters, 66*(1), 222–224. https://doi.org/10.1016/j.matlet.2011.08.081
15. Ma, X., Dong, Y., Sun, H., & Chen, N. (2017). Highly fluorescent carbon dots from peanut shells as potential probes for copper ion: The optimization and analysis of the synthetic process. *Materials Today Chemistry, 5*, 1–10. https://doi.org/10.1016/j.mtchem.2017.04.004

16. Jayaweera, S., Yin, K., & Ng, W. J. (2019). Nitrogen-doped durian shell derived carbon dots for inner filter effect mediated sensing of tetracycline and fluorescent ink. *Journal of Fluorescence, 29*(1), 221–229. https://doi.org/10.1007/s10895-018-2331-3
17. Jiao, X.-Y., Li, L., Qin, S., Zhang, Y., Huang, K., & Xu, L. (2019). The synthesis of fluorescent carbon dots from mango peel and their multiple applications. *Colloids and Surfaces A: Physicochemical and Engineering Aspects, 577*, 306–314. https://doi.org/10.1016/j.colsurfa.2019.05.073
18. Sahoo, N. K., Jana, G. C., Aktara, M. N., Das, S., Nayim, S., Patra, A., Bhattacharjee, P., Bhadra, K., & Hossain, M. (2020). Carbon dots derived from lychee waste: Application for Fe3+ ions sensing in real water and multicolor cell imaging of skin melanoma cells. *Materials Science and Engineering: C, 108*, 110429. https://doi.org/10.1016/j.msec.2019.110429
19. Liu, Y., Zhao, Y., & Zhang, Y. (2014). One-step green synthesized fluorescent carbon nanodots from bamboo leaves for copper(II) ion detection. *Sensors and Actuators B: Chemical, 196*, 647–652. https://doi.org/10.1016/j.snb.2014.02.053
20. Thota, S. P., Thota, S. M., Srimadh Bhagavatham, S., Sai Manoj, K., Sai Muthukumar, V. S., Venketesh, S., Vadlani, P. V., & Belliraj, S. K. (2018). Facile one-pot hydrothermal synthesis of stable and biocompatible fluorescent carbon dots from lemon grass herb. *IET Nanobiotechnology, 12*(2), 127–132. https://doi.org/10.1049/iet-nbt.2017.0038
21. Liang, Z., Kang, M., Payne, G. F., Wang, X., & Sun, R. (2016). Probing energy and electron transfer mechanisms in fluorescence quenching of biomass carbon quantum dots. *ACS Applied Materials & Interfaces, 8*(27), 17478–17488. https://doi.org/10.1021/acsami.6b04826
22. Kundu, A., Basu, S., & Maity, B. (2023). Upcycling waste: citrus Limon peel-derived carbon quantum dots for sensitive detection of tetracycline in the nanomolar range. *ACS Omega, 8*(39), 36449–36459. https://doi.org/10.1021/acsomega.3c05424
23. Yao, L., Zhao, M.-M., Luo, Q.-W., Zhang, Y.-C., Liu, T.-T., Yang, Z., Liao, M., Tu, P., & Zeng, K.-W. (2022). Carbon quantum dots-based nanozyme from coffee induces cancer cell ferroptosis to activate antitumor immunity. *ACS Nano, 16*(6), 9228–9239. https://doi.org/10.1021/acsnano.2c01619
24. Liu, Y., Zhu, C., Gao, Y., Yang, L., Xu, J., Zhang, X., Lu, C., Wang, Y., & Zhu, Y. (2020). Biomass-derived nitrogen self-doped carbon dots via a simple one-pot method: Physicochemical, structural, and luminescence properties. *Applied Surface Science, 510*, 145437. https://doi.org/10.1016/j.apsusc.2020.145437
25. Jones, S. S., Sahatiya, P., & Badhulika, S. (2017). One step, high yield synthesis of amphiphilic carbon quantum dots derived from chia seeds: A solvatochromic study. *New Journal of Chemistry, 41*(21), 13130–13139. https://doi.org/10.1039/C7NJ03513F
26. Rasal, A. S., Yadav, S., Yadav, A., Kashale, A. A., Manjunatha, S. T., Altaee, A., & Chang, J.-Y. (2021). Carbon quantum dots for energy applications: A review. *ACS Applied Nano Materials, 4*(7), 6515–6541. https://doi.org/10.1021/acsanm.1c01372
27. Qu, Q., Zhu, A., Shao, X., Shi, G., & Tian, Y. (2012). Development of a carbon quantum dots-based fluorescent Cu2+ probe suitable for living cell imaging. *Chemical Communications, 48*(44), 5473. https://doi.org/10.1039/c2cc31000g
28. Pandey, S., Thakur, M., Mewada, A., Anjarlekar, D., Mishra, N., & Sharon, M. (2013). Carbon dots functionalized gold nanorod mediated delivery of doxorubicin: Tri-functional nano-worms for drug delivery, photothermal therapy and bioimaging. *Journal of Materials Chemistry B, 1*(38), 4972. https://doi.org/10.1039/c3tb20761g
29. Yang, Z., Xu, T., Li, H., She, M., Chen, J., Wang, Z., Zhang, S., & Li, J. (2023). Zero-dimensional carbon nanomaterials for fluorescent sensing and imaging. *Chemical Reviews, 123*(18), 11047–11136. https://doi.org/10.1021/acs.chemrev.3c00186
30. Korepanov, V. I., Hamaguchi, H., Osawa, E., Ermolenkov, V., Lednev, I. K., Etzold, B. J. M., Levinson, O., Zousman, B., Epperla, C. P., & Chang, H.-C. (2017). Carbon structure in nanodiamonds elucidated from Raman spectroscopy. *Carbon, 121*, 322–329. https://doi.org/10.1016/j.carbon.2017.06.012
31. Li, M., Chen, T., Gooding, J. J., & Liu, J. (2019). Review of carbon and graphene quantum dots for sensing. *ACS Sensors, 4*(7), 1732–1748. https://doi.org/10.1021/acssensors.9b00514

32. Dutta, S. D., Hexiu, J., Kim, J., Sarkar, S., Mondal, J., An, J. M., Lee, Y., Moniruzzaman, M., & Lim, K.-T. (2022). Two-photon excitable membrane targeting polyphenolic carbon dots for long-term imaging and pH-responsive chemotherapeutic drug delivery for synergistic tumor therapy. *Biomaterials Science, 10*(7), 1680–1696. https://doi.org/10.1039/D1BM01832A
33. Han, C., Zhang, X., Wang, F., Yu, Q., Chen, F., Shen, D., Yang, Z., Wang, T., Jiang, M., Deng, T., & Yu, C. (2021). Duplex metal co-doped carbon quantum dots-based drug delivery system with intelligent adjustable size as adjuvant for synergistic cancer therapy. *Carbon, 183*, 789–808. https://doi.org/10.1016/j.carbon.2021.07.063
34. Long, W., Ouyang, H., Wan, W., Yan, W., Zhou, C., Huang, H., Liu, M., Zhang, X., Feng, Y., & Wei, Y. (2020). "Two in one": Simultaneous functionalization and DOX loading for fabrication of nanodiamond-based pH responsive drug delivery system. *Materials Science and Engineering: C, 108*, 110413. https://doi.org/10.1016/j.msec.2019.110413
35. Panwar, N., Soehartono, A. M., Chan, K. K., Zeng, S., Xu, G., Qu, J., Coquet, P., Yong, K.-T., & Chen, X. (2019). Nanocarbons for biology and medicine: sensing, imaging, and drug delivery. *Chemical Reviews, 119*(16), 9559–9656. https://doi.org/10.1021/acs.chemrev.9b00099
36. Ma, C., Wang, Y., Han, N., Zhang, R., Liu, H., Sun, X., & Xing, L. (2024). Carbon dot-based artificial light-harvesting systems with sequential energy transfer and white light emission for photocatalysis. *Chinese Chemical Letters, 35*(4), 108632. https://doi.org/10.1016/j.cclet.2023.108632
37. Yu, H., Shi, R., Zhao, Y., Waterhouse, G. I. N., Wu, L., Tung, C., & Zhang, T. (2016). Smart utilization of carbon dots in semiconductor photocatalysis. *Advanced Materials, 28*(43), 9454–9477. https://doi.org/10.1002/adma.201602581
38. Liu, E., Liang, T., Ushakova, E. V., Wang, B., Zhang, B., Zhou, H., Xing, G., Wang, C., Tang, Z., Qu, S., & Rogach, A. L. (2021). Enhanced near-infrared emission from carbon dots by surface deprotonation. *The Journal of Physical Chemistry Letters, 12*(1), 604–611. https://doi.org/10.1021/acs.jpclett.0c03383
39. Sahoo, S., Satpati, A. K., Sahoo, P. K., & Naik, P. D. (2018). Incorporation of carbon quantum dots for improvement of supercapacitor performance of Nickel Sulfide. *ACS Omega, 3*(12), 17936–17946. https://doi.org/10.1021/acsomega.8b01238
40. Prasath, A., Athika, M., Duraisamy, E., Selva Sharma, A., Sankar Devi, V., & Elumalai, P. (2019). Carbon quantum dot-anchored bismuth oxide composites as potential electrode for lithium-ion battery and supercapacitor applications. *ACS Omega, 4*(3), 4943–4954. https://doi.org/10.1021/acsomega.8b03490
41. Zhang, Z., Shi, Y., Pan, Y., Cheng, X., Zhang, L., Chen, J., Li, M.-J., & Yi, C. (2014). Quinoline derivative-functionalized carbon dots as a fluorescent nanosensor for sensing and intracellular imaging of Zn 2+. *J. Mater. Chem. B, 2*(31), 5020–5027. https://doi.org/10.1039/C4TB00677A
42. Hu., Ruoxin, Li, L., & Jin, W. J. (2017). Controlling speciation of nitrogen in nitrogen-doped carbon dots by ferric ion catalysis for enhancing fluorescence. *Carbon, 111*, 133–141. https://doi.org/10.1016/j.carbon.2016.09.038
43. Shan, X., Chai, L., Ma, J., Qian, Z., Chen, J., & Feng, H. (2014). B-doped carbon quantum dots as a sensitive fluorescence probe for hydrogen peroxide and glucose detection. *The Analyst, 139*(10), 2322–2325. https://doi.org/10.1039/C3AN02222F
44. Zhou, J., Shan, X., Ma, J., Gu, Y., Qian, Z., Chen, J., & Feng, H. (2014). Facile synthesis of P-doped carbon quantum dots with highly efficient photoluminescence. *RSC Advances, 4*(11), 5465. https://doi.org/10.1039/c3ra45294h
45. Ayala, P., Arenal, R., Loiseau, A., Rubio, A., & Pichler, T. (2010). The physical and chemical properties of heteronanotubes. *Reviews of Modern Physics, 82*(2), 1843–1885. https://doi.org/10.1103/RevModPhys.82.1843
46. Liu, H., Ding, J., Zhang, K., & Ding, L. (2019). Construction of biomass carbon dots based fluorescence sensors and their applications in chemical and biological analysis. *TrAC Trends in Analytical Chemistry, 118*, 315–337. https://doi.org/10.1016/j.trac.2019.05.051

47. Shamsipur, M., Barati, A., & Karami, S. (2017). Long-wavelength, multicolor, and white-light emitting carbon-based dots: Achievements made, challenges remaining, and applications. *Carbon, 124*, 429–472. https://doi.org/10.1016/j.carbon.2017.08.072
48. Ting, Z. X., Ananthanarayanan, A., Luo, K. Q., & Chen, P. (2015). Glowing graphene quantum dots and carbon dots: properties, syntheses, and biological applications. *Small (Weinheim an der Bergstrasse, Germany), 11*(14), 1620–1636. https://doi.org/10.1002/smll.201402648
49. Chandra, S., Laha, D., Pramanik, A., Ray Chowdhuri, A., Karmakar, P., & Sahu, S. K. (2016). Synthesis of highly fluorescent nitrogen and phosphorus doped carbon dots for the detection of Fe 3+ ions in cancer cells. *Luminescence, 31*(1), 81–87. https://doi.org/10.1002/bio.2927
50. Zhang, Y.-Q., Ma, D.-K., Zhuang, Y., Zhang, X., Chen, W., Hong, L.-L., Yan, Q.-X., Yu, K., & Huang, S.-M. (2012). One-pot synthesis of N-doped carbon dots with tunable luminescence properties. *Journal of Materials Chemistry, 22*(33), 16714. https://doi.org/10.1039/c2jm32973e
51. Park, Y., Yoo, J., Lim, B., Kwon, W., & Rhee, S.-W. (2016). Improving the functionality of carbon nanodots: Doping and surface functionalization. *Journal of Materials Chemistry A, 4*(30), 11582–11603. https://doi.org/10.1039/C6TA04813G
52. Zhou, Y., Liu, Y., Li, Y., He, Z., Xu, Q., Chen, Y., Street, J., Guo, H., & Nelles, M. (2018). Multicolor carbon nanodots from food waste and their heavy metal ion detection application. *RSC Advances, 8*(42), 23657–23662. https://doi.org/10.1039/C8RA03272F
53. Travlou, N. A., Giannakoudakis, D. A., Algarra, M., Labella, A. M., Rodríguez-Castellón, E., & Bandosz, T. J. (2018). S- and N-doped carbon quantum dots: Surface chemistry dependent antibacterial activity. *Carbon, 135*, 104–111. https://doi.org/10.1016/j.carbon.2018.04.018
54. Campos, B. B., Contreras-Cáceres, R., Bandosz, T. J., Jiménez-Jiménez, J., Rodríguez-Castellón, E., Esteves da Silva, J. C. G., & Algarra, M. (2016). Carbon dots as fluorescent sensor for detection of explosive nitrocompounds. *Carbon, 106*, 171–178. https://doi.org/10.1016/j.carbon.2016.05.030
55. Huang, H., Weng, Y., Zheng, L., Yao, B., Weng, W., & Lin, X. (2017). Nitrogen-doped carbon quantum dots as fluorescent probe for "off-on" detection of mercury ions, l-cysteine and iodide ions. *Journal of Colloid and Interface Science, 506*, 373–378. https://doi.org/10.1016/j.jcis.2017.07.076
56. Qi, H., Teng, M., Liu, M., Liu, S., Li, J., Yu, H., Teng, C., Huang, Z., Liu, H., Shao, Q., Umar, A., Ding, T., Gao, Q., & Guo, Z. (2019). Biomass-derived nitrogen-doped carbon quantum dots: Highly selective fluorescent probe for detecting Fe3+ ions and tetracyclines. *Journal of Colloid and Interface Science, 539*, 332–341. https://doi.org/10.1016/j.jcis.2018.12.047
57. Han, Y., Tang, D., Yang, Y., Li, C., Kong, W., Huang, H., Liu, Y., & Kang, Z. (2015). Non-metal single/dual doped carbon quantum dots: A general flame synthetic method and electro-catalytic properties. *Nanoscale, 7*(14), 5955–5962. https://doi.org/10.1039/C4NR07116F
58. Sarkar, S., Das, K., Ghosh, M., & Das, P. K. (2015). Amino acid functionalized blue and phosphorous-doped green fluorescent carbon dots as bioimaging probe. *RSC Advances, 5*(81), 65913–65921. https://doi.org/10.1039/C5RA09905F
59. Chandra, S., Das, P., Bag, S., Laha, D., & Pramanik, P. (2011). Synthesis, functionalization and bioimaging applications of highly fluorescent carbon nanoparticles. *Nanoscale, 3*(4), 1533. https://doi.org/10.1039/c0nr00735h
60. Sadhanala, H. K., Khatei, J., & Nanda, K. K. (2014). Facile hydrothermal synthesis of carbon nanoparticles and possible application as white light phosphors and catalysts for the reduction of nitrophenol. *RSC Advances, 4*(22), 11481. https://doi.org/10.1039/c3ra47527a
61. Sun, Y., Shen, C., Wang, J., & Lu, Y. (2015). Facile synthesis of biocompatible N, S-doped carbon dots for cell imaging and ion detecting. *RSC Advances, 5*(21), 16368–16375. https://doi.org/10.1039/C4RA13820A
62. Wang Yong, Zhuang, Q., & Ni, Y. (2015). Facile microwave-assisted solid-phase synthesis of highly fluorescent nitrogen–sulfur-codoped carbon quantum dots for cellular imaging applications. *Chemistry – A European Journal, 21*(37), 13004–13011. https://doi.org/10.1002/chem.201501723.

63. Magdy, G., Al-enna, A. A., Belal, F., El-Domany, R. A., & Abdel-Megied, A. M. (2022). Application of sulfur and nitrogen doped carbon quantum dots as sensitive fluorescent nanosensors for the determination of saxagliptin and gliclazide. *Royal Society Open Science, 9*(6). https://doi.org/10.1098/rsos.220285.
64. Xu, Q., Liu, Y., Gao, C., Wei, J., Zhou, H., Chen, Y., Dong, C., Sreeprasad, T. S., Li, N., & Xia, Z. (2015). Synthesis, mechanistic investigation, and application of photoluminescent sulfur and nitrogen co-doped carbon dots. *Journal of Materials Chemistry C, 3*(38), 9885–9893. https://doi.org/10.1039/C5TC01912E
65. Dong, Y., Pang, H., Bin, Y. H., Guo, C., Shao, J., Chi, Y., Li, C. M., & Yu, T. (2013). Carbon-based dots co-doped with nitrogen and sulfur for high quantum yield and excitation-independent emission. *Angewandte Chemie International Edition, 52*(30), 7800–7804. https://doi.org/10.1002/anie.201301114
66. Hang-Xing, W., Xiao, J., Yang, Z., Tang, H., Zhu, Z.-T., Zhao, M., Liu, Y., Zhang, C., & Zhang, H.-L. (2015). Rational design of nitrogen and sulfur co-doped carbon dots for efficient photoelectrical conversion applications. *Journal of Materials Chemistry A, 3*(21), 11287–11293. https://doi.org/10.1039/C5TA02057C
67. Baker, S. N., & Baker, G. A. (2010). Luminescent Carbon Nanodots: Emergent Nanolights. *Angewandte Chemie International Edition, 49*(38), 6726–6744. https://doi.org/10.1002/anie.200906623
68. Tian, L., Li, Z., Wang, P., Zhai, X., Wang, X., & Li, T. (2021). Carbon quantum dots for advanced electrocatalysis. *Journal of Energy Chemistry, 55*, 279–294. https://doi.org/10.1016/j.jechem.2020.06.057
69. Tian, L., Qiu, G., Shen, Y., Wang, X., Wang, J., Wang, P., Song, M., Li, J., Li, T., Zhuang, W., & Du, X. (2019). Carbon quantum dots modulated NiMoP hollow nanopetals as efficient electrocatalysts for hydrogen evolution. *Industrial & Engineering Chemistry Research, 58*(31), 14098–14105. https://doi.org/10.1021/acs.iecr.9b01899
70. Zhao, C., Zhang, X., Shu, X., Liu, X., Fang, D., Song, Y., & Wang, J. (2018). Er-doped carbon dots broadening light absorption range and accelerating electron transport for enhancing photovoltaic performance of CdS quantum dots sensitized cells. *Optical Materials, 84*, 242–251. https://doi.org/10.1016/j.optmat.2018.07.016
71. Hoang, V. C., Dave, K., & Gomes, V. G. (2019). Carbon quantum dot-based composites for energy storage and electrocatalysis: Mechanism, applications and future prospects. *Nano Energy, 66*, 104093. https://doi.org/10.1016/j.nanoen.2019.104093
72. Ali, M., Riaz, R., Anjum, A. S., Sun, K. C., Li, H., Ahn, S., Jeong, S. H., & Ko, M. J. (2021). Microwave-assisted ultrafast in-situ growth of N-doped carbon quantum dots on multiwalled carbon nanotubes as an efficient electrocatalyst for photovoltaics. *Journal of Colloid and Interface Science, 586*, 349–361. https://doi.org/10.1016/j.jcis.2020.10.098
73. Xiao, W., Feng, Y., Dong, P., & Huang, J. (2019). A mini review on carbon quantum dots: preparation, properties, and electrocatalytic application. *Frontiers in Chemistry, 7*,. https://doi.org/10.3389/fchem.2019.00671
74. Zhang, X., Zhang, Z., Hu, F., Li, D., Zhou, D., Jing, P., Du, F., & Qu, S. (2019). Carbon-dots-derived 3d highly nitrogen-doped porous carbon framework for high-performance lithium ion storage. *ACS Sustainable Chemistry & Engineering, 7*(11), 9848–9856. https://doi.org/10.1021/acssuschemeng.9b00407
75. Liang, W., Bunker, C. E., & Sun, Y.-P. (2020). Carbon dots: zero-dimensional carbon allotrope with unique photoinduced redox characteristics. *ACS Omega, 5*(2), 965–971. https://doi.org/10.1021/acsomega.9b03669
76. Liang, W., Li, W., Wu, B., Li, Z., Wang, S., Liu, Y., Pan, D., & Wu, M. (2016). Facile synthesis of fluorescent graphene quantum dots from coffee grounds for bioimaging and sensing. *Chemical Engineering Journal, 300*, 75–82. https://doi.org/10.1016/j.cej.2016.04.123
77. Kumar, Y. R., Deshmukh, K., Sadasivuni, K. K., & Pasha, S. K. K. (2020). Graphene quantum dot based materials for sensing, bio-imaging and energy storage applications: A review. *RSC Advances, 10*(40), 23861–23898. https://doi.org/10.1039/D0RA03938A

78. Zhai, Y., Zhang, B., Shi, R., Zhang, S., Liu, Y., Wang, B., Zhang, K., Waterhouse, G. I. N., Zhang, T., & Lu, S. (2022). Carbon dots as new building blocks for electrochemical energy storage and electrocatalysis. *Advanced Energy Materials, 12*(6). https://doi.org/10.1002/aenm.202103426.
79. Boyang, W., Li, J., Tang, Z., Yang, B., & Lu, S. (2019). Near-infrared emissive carbon dots with 33.96% emission in aqueous solution for cellular sensing and light-emitting diodes. *Science Bulletin, 64*(17), 1285–1292. https://doi.org/10.1016/j.scib.2019.07.021
80. Dong, Y., Wang, R., Li, H., Shao, J., Chi, Y., Lin, X., & Chen, G. (2012). Polyamine-functionalized carbon quantum dots for chemical sensing. *Carbon, 50*(8), 2810–2815. https://doi.org/10.1016/j.carbon.2012.02.046
81. Wei, X.-M., Xu, Y., Li, Y.-H., Yin, X.-B., & He, X.-W. (2014). Ultrafast synthesis of nitrogen-doped carbon dots via neutralization heat for bioimaging and sensing applications. *RSC Advances, 4*(84), 44504–44508. https://doi.org/10.1039/C4RA08523J
82. Yang, G., Wan, X., Liu, Y., Li, R., Su, Y., Zeng, X., & Tang, J. (2016). Luminescent Poly(vinyl alcohol)/carbon quantum dots composites with tunable water-induced shape memory behavior in different pH and temperature environments. *ACS Applied Materials & Interfaces, 8*(50), 34744–34754. https://doi.org/10.1021/acsami.6b11476
83. Kasibabu, B. S. B., D'souza, S. L., Jha, S., & Kailasa, S. K. (2015). Imaging of bacterial and fungal cells using fluorescent carbon dots prepared from Carica papaya juice. *Journal of Fluorescence, 25*(4), 803–810. https://doi.org/10.1007/s10895-015-1595-0
84. Mishra, S., Das, K., Chatterjee, S., Sahoo, P., Kundu, S., Pal, M., Bhaumik, A., & Ghosh, C. K. (2023). Facile and green synthesis of novel fluorescent carbon quantum dots and their silver heterostructure: An in vitro anticancer activity and imaging on colorectal carcinoma. *ACS Omega, 8*(5), 4566–4577. https://doi.org/10.1021/acsomega.2c04964
85. Wei, S., Li, L., Du, X., & Li, Y. (2019). OFF–ON nanodiamond drug platform for targeted cancer imaging and therapy. *Journal of Materials Chemistry B, 7*(21), 3390–3402. https://doi.org/10.1039/C9TB00447E
86. Choi, H., Ko, S.-J., Choi, Y., Joo, P., Kim, T., Lee, B. R., Jung, J.-W., Choi, H. J., Cha, M., Jeong, J.-R., Hwang, I.-W., Song, M. H., Kim, B.-S., & Kim, J. Y. (2013). Versatile surface plasmon resonance of carbon-dot-supported silver nanoparticles in polymer optoelectronic devices. *Nature Photonics, 7*(9), 732–738. https://doi.org/10.1038/nphoton.2013.181
87. Marković, Z. M., Labudová, M., Danko, M., Matijašević, D., Mičušík, M., Nádaždy, V., Kováčová, M., Kleinová, A., Špitalský, Z., Pavlović, V., Milivojević, D. D., Medić, M., & Todorović Marković, B. M. (2020). Highly efficient antioxidant F- and Cl-doped carbon quantum dots for bioimaging. *ACS Sustainable Chemistry & Engineering, 8*(43), 16327–16338. https://doi.org/10.1021/acssuschemeng.0c06260
88. Hoang, V. C., Nguyen, L. H., & Gomes, V. G. (2019). High efficiency supercapacitor derived from biomass based carbon dots and reduced graphene oxide composite. *Journal of Electroanalytical Chemistry, 832*, 87–96. https://doi.org/10.1016/j.jelechem.2018.10.050
89. Han, M., Zhu, S., Lu, S., Song, Y., Feng, T., Tao, S., Liu, J., & Yang, B. (2018). *Nano Today, 19*, 201–218. https://doi.org/10.1016/j.nantod.2018.02.008.
90. Li, H., He, X., Kang, Z., Huang, H., Liu, Y., Liu, J., Lian, S., Tsang, C. H. A., Yang, X., & Lee, S. (2010). Water-soluble fluorescent carbon quantum dots and photocatalyst design. *Angewandte Chemie International Edition, 49*(26), 4430–4434. https://doi.org/10.1002/anie.200906154
91. Wang, R., Lu, K.-Q., Tang, Z.-R., & Xu, Y.-J. (2017). Recent progress in carbon quantum dots: Synthesis, properties and applications in photocatalysis. *Journal of Materials Chemistry A, 5*(8), 3717–3734. https://doi.org/10.1039/C6TA08660H
92. Sohal, N., Bhatia, S. K., Basu, S., & Maity, B. (2021). Nanomolar level detection of metal ions by improving the monodispersity and stability of nitrogen-doped graphene quantum dots. *New Journal of Chemistry, 45*(42), 19941–19949. https://doi.org/10.1039/D1NJ04551B
93. Molaei, M. J. (2019). Carbon quantum dots and their biomedical and therapeutic applications: A review. *RSC Advances, 9*(12), 6460–6481. https://doi.org/10.1039/C8RA08088G

94. Weilin, W., Wang, Z., Liu, J., Peng, Y., Yu, X., Weixing, W., Zhang, Z., & Sun, L. (2018). One-pot facile synthesis of graphene quantum dots from rice husks for Fe 3+ sensing. *Industrial & Engineering Chemistry Research, 57*(28), 9144–9150. https://doi.org/10.1021/acs.iecr.8b00913
95. Zhu, L., Shen, D., Liu, Q., Wu, C., & Gu, S. (2021). Sustainable synthesis of bright green fluorescent carbon quantum dots from lignin for highly sensitive detection of Fe3+ ions. *Applied Surface Science, 565*, 150526. https://doi.org/10.1016/j.apsusc.2021.150526
96. Zhao, S., Song, X., Chai, X., Zhao, P., He, H., & Liu, Z. (2020). Green production of fluorescent carbon quantum dots based on pine wood and its application in the detection of Fe3+. *Journal of Cleaner Production, 263*, 121561. https://doi.org/10.1016/j.jclepro.2020.121561
97. Kundu, A., Maity, B., & Basu, S. (2022). Rice husk-derived carbon quantum dots-based dual-mode nanoprobe for selective and sensitive detection of Fe 3+ and fluoroquinolones. *ACS Biomaterials Science & Engineering, 8*(11), 4764–4776. https://doi.org/10.1021/acsbiomaterials.2c00798
98. Gao, P., Liu, S., Su, Y., Zheng, M., & Xie, Z. (2020). Fluorine-doped carbon dots with intrinsic nucleus-targeting ability for drug and dye delivery. *Bioconjugate Chemistry, 31*(3), 646–655. https://doi.org/10.1021/acs.bioconjchem.9b00801
99. Giordani, S., Bartelmess, J., Frasconi, M., Biondi, I., Cheung, S., Grossi, M., Wu, D., Echegoyen, L., & O'Shea, D. F. (2014). NIR fluorescence labelled carbon nano-onions: Synthesis, analysis and cellular imaging. *J. Mater. Chem. B, 2*(42), 7459–7463. https://doi.org/10.1039/C4TB01087F
100. Park, S. Y., Lee, H. U., Park, E. S., Lee, S. C., Lee, J.-W., Jeong, S. W., Kim, C. H., Lee, Y.-C., Huh, Y. S., & Lee, J. (2014). Photoluminescent green carbon nanodots from food-waste-derived sources: large-scale synthesis, properties, and biomedical applications. *ACS Applied Materials & Interfaces, 6*(5), 3365–3370. https://doi.org/10.1021/am500159p
101. Yang, X.-F., Wang, A., Qiao, B., Li, J., Liu, J., & Zhang, T. (2013). Single-atom catalysts: a new frontier in heterogeneous catalysis. *Accounts of Chemical Research, 46*(8), 1740–1748. https://doi.org/10.1021/ar300361m
102. Ishida, T., Murayama, T., Taketoshi, A., & Haruta, M. (2020). Importance of size and contact structure of gold nanoparticles for the genesis of unique catalytic processes. *Chemical Reviews, 120*(2), 464–525. https://doi.org/10.1021/acs.chemrev.9b00551
103. Zheng, L., Chi, Y., Dong, Y., Lin, J., & Wang, B. (2009). Electrochemiluminescence of water-soluble carbon nanocrystals released electrochemically from graphite. *Journal of the American Chemical Society, 131*(13), 4564–4565. https://doi.org/10.1021/ja809073f
104. Niu, X., Zheng, W., Song, T., Huang, Z., Yang, C., Zhang, L., Li, W., & Xiong, H. (2023). Pyrolysis of single carbon sources in SBA-15: A recyclable solid phase synthesis to obtain uniform carbon dots with tunable luminescence. *Chinese Chemical Letters, 34*(2), 107560. https://doi.org/10.1016/j.cclet.2022.05.074
105. Arcudi, F., Đorđević, L., & Prato, M. (2016). Synthesis, separation, and characterization of small and highly fluorescent nitrogen-doped carbon nanodots. *Angewandte Chemie International Edition, 55*(6), 2107–2112. https://doi.org/10.1002/anie.201510158
106. Kwon, W., Lee, G., Do, S., Joo, T., & Rhee, S. (2014). Size-controlled soft-template synthesis of carbon nanodots toward versatile photoactive materials. *Small (Weinheim an der Bergstrasse, Germany), 10*(3), 506–513. https://doi.org/10.1002/smll.201301770
107. Ortega-Liebana, M. C., Chung, N. X., Limpens, R., Gomez, L., Hueso, J. L., Santamaria, J., & Gregorkiewicz, T. (2017). Uniform luminescent carbon nanodots prepared by rapid pyrolysis of organic precursors confined within nanoporous templating structures. *Carbon, 117*, 437–446. https://doi.org/10.1016/j.carbon.2017.03.017
108. Zhijiao, T., Lin, Z., Li, G., & Hu, Y. (2017). Amino nitrogen quantum dots-based nanoprobe for fluorescence detection and imaging of cysteine in biological samples. *Analytical Chemistry, 89*(7), 4238–4245. https://doi.org/10.1021/acs.analchem.7b00284
109. Hailing, Y., Xiufang, L., Lili, W., Baoqiang, L., Kaichen, H., Yongquan, H., Qianqian, Z., Chaoming, M., Xiaoshuai, R., Rui, Z., Hui, L., Pengfei, P., & Hong, S. (2020). Doxorubicin-loaded fluorescent carbon dots with PEI passivation as a drug delivery system for cancer therapy. *Nanoscale, 12*(33), 17222–17237. https://doi.org/10.1039/D0NR01236J

110. Jiang, Q., Liu, L., Li, Q., Cao, Y., Chen, D., Du, Q., Yang, X., Huang, D., Pei, R., Chen, X., & Huang, G. (2021). NIR-laser-triggered gadolinium-doped carbon dots for magnetic resonance imaging, drug delivery and combined photothermal chemotherapy for triple negative breast cancer. *Journal of Nanobiotechnology, 19*(1), 64. https://doi.org/10.1186/s12951-021-00811-w
111. Ahlawat, J., Masoudi Asil, S., Guillama Barroso, G., Nurunnabi, M., & Narayan, M. (2021). Application of carbon nano onions in the biomedical field: Recent advances and challenges. *Biomaterials Science, 9*(3), 626–644. https://doi.org/10.1039/D0BM01476A
112. Singh, I., Arora, R., Dhiman, H., & Pahwa, R. (2018). Carbon quantum dots: Synthesis, characterization and biomedical applications. *The Turkish Journal of Pharmaceutical Sciences, 15*(2), 219–230. https://doi.org/10.4274/tjps.63497.
113. Lin, L., Luo, Y., Tsai, P., Wang, J., & Chen, X. (2018). Metal ions doped carbon quantum dots: Synthesis, physicochemical properties, and their applications. *TrAC Trends in Analytical Chemistry, 103*, 87–101. https://doi.org/10.1016/j.trac.2018.03.015
114. Yan, X., Cui, X., Li, B., & Li, L. (2010). Large, solution-processable graphene quantum dots as light absorbers for photovoltaics. *Nano Letters, 10*(5), 1869–1873. https://doi.org/10.1021/nl101060h
115. Mirtchev, P., Henderson, E. J., Soheilnia, N., Yip, C. M., & Ozin, G. A. (2012). Solution phase synthesis of carbon quantum dots as sensitizers for nanocrystalline TiO 2 solar cells. *Journal of Materials Chemistry, 22*(4), 1265–1269. https://doi.org/10.1039/C1JM14112K
116. Molaei, M. J. (2020). The optical properties and solar energy conversion applications of carbon quantum dots: A review. *Solar Energy, 196*, 549–566. https://doi.org/10.1016/j.solener.2019.12.036
117. Fernando, K. A. S., Sahu, S., Liu, Y., Lewis, W. K., Guliants, E. A., Jafariyan, A., Wang, P., Bunker, C. E., & Sun, Y.-P. (2015). Carbon quantum dots and applications in photocatalytic energy conversion. *ACS Applied Materials & Interfaces, 7*(16), 8363–8376. https://doi.org/10.1021/acsami.5b00448
118. Genc, R., Alas, M. O., Harputlu, E., Repp, S., Kremer, N., Castellano, M., Colak, S. G., Ocakoglu, K., & Erdem, E. (2017). High-capacitance hybrid supercapacitor based on multi-colored fluorescent carbon-dots. *Scientific Reports, 7*(1), 11222. https://doi.org/10.1038/s41598-017-11347-1
119. Peng, Z., Han, X., Li, S., Al-Youbi, A. O., Bashammakh, A. S., El-Shahawi, M. S., & Leblanc, R. M. (2017). Carbon dots: Biomacromolecule interaction, bioimaging and nanomedicine. *Coordination Chemistry Reviews, 343*, 256–277. https://doi.org/10.1016/j.ccr.2017.06.001
120. Dobrovolskaia, M. A., & Mcneil, S. E. (2009). Immunological properties of engineered nanomaterials. In *Nanoscience and Technology*, Co-Published with Macmillan Publishers Ltd, UK. https://doi.org/10.1142/9789814287005_0029.
121. Li, P., Poon, Y. F., Li, W., Zhu, H.-Y., Yeap, S. H., Cao, Y., Qi, X., Zhou, C., Lamrani, M., Beuerman, R. W., Kang, E.-T., Mu, Y., Li, C. M., Chang, M. W., Jan Leong, S. S., & Chan-Park, M. B. (2011). A polycationic antimicrobial and biocompatible hydrogel with microbe membrane suctioning ability. *Nature Materials, 10*(2), 149–156. https://doi.org/10.1038/nmat2915
122. Bing, W., Sun, H., Yan, Z., Ren, J., & Qu, X. (2016). Programmed bacteria death induced by carbon dots with different surface charge. *Small (Weinheim an der Bergstrasse, Germany), 12*(34), 4713–4718. https://doi.org/10.1002/smll.201600294
123. Yan, J., Hou, S., Yu, Y., Qiao, Y., Xiao, T., Mei, Y., Zhang, Z., Wang, B., Huang, C.-C., Lin, C.-H., & Suo, G. (2018). The effect of surface charge on the cytotoxicity and uptake of carbon quantum dots in human umbilical cord derived mesenchymal stem cells. *Colloids and Surfaces B: Biointerfaces, 171*, 241–249. https://doi.org/10.1016/j.colsurfb.2018.07.034
124. Dwyer, D. J., Camacho, D. M., Kohanski, M. A., Callura, J. M., & Collins, J. J. (2012). Antibiotic-induced bacterial cell death exhibits physiological and biochemical hallmarks of apoptosis. *Molecular Cell, 46*(5), 561–572. https://doi.org/10.1016/j.molcel.2012.04.027
125. Barman, M. K., Jana, B., Bhattacharyya, S., & Patra, A. (2014). Photophysical properties of doped carbon dots (N, P, and B) and their influence on electron/hole transfer in carbon

dots-nickel (II) phthalocyanine conjugates. *The Journal of Physical Chemistry C, 118*(34), 20034–20041. https://doi.org/10.1021/jp507080c

126. Datta, K. K. R., Kozák, O., Ranc, V., Havrdová, M., Bourlinos, A. B., Šafářová, K., Holá, K., Tománková, K., Zoppellaro, G., Otyepka, M., & Zbořil, R. (2014). Quaternized carbon dot-modified graphene oxide for selective cell labelling – Controlled nucleus and cytoplasm imaging. *Chemical Communications, 50*(74), 10782. https://doi.org/10.1039/C4CC02637C
127. Kang, S., Jeong, Y. K., Jung, K. H., Son, Y., Kim, W. R., Ryu, J. H., & Kim, K. M. (2020). One-step synthesis of sulfur-incorporated graphene quantum dots using pulsed laser ablation for enhancing optical properties. *Optics Express, 28*(15), 21659. https://doi.org/10.1364/OE.398124
128. Frigione, M., & Lettieri, M. (2020). Recent advances and trends of nanofilled/nanostructured epoxies. *Materials, 13*(15), 3415. https://doi.org/10.3390/ma13153415
129. Ghosh, D., Sarkar, K., Devi, P., Kim, K.-H., & Kumar, P. (2021). Current and future perspectives of carbon and graphene quantum dots: From synthesis to strategy for building optoelectronic and energy devices. *Renewable and Sustainable Energy Reviews, 135*, 110391. https://doi.org/10.1016/j.rser.2020.110391
130. Yuan, T., Meng, T., He, P., Shi, Y., Li, Y., Li, X., Fan, L., & Yang, S. (2019). Carbon quantum dots: An emerging material for optoelectronic applications. *Journal of Materials Chemistry C, 7*(23), 6820–6835. https://doi.org/10.1039/C9TC01730E
131. Naresh, V., & Lee, N. (2021). A review on biosensors and recent development of nanostructured materials-enabled biosensors. *Sensors, 21*(4), 1109. https://doi.org/10.3390/s21041109
132. Yadav, V. K., Malik, P., Khan, A. H., Pandit, P. R., Hasan, M. A., Cabral-Pinto, M. M. S., Islam, S., Suriyaprabha, R., Yadav, K. K., Dinis, P. A., Khan, S. H., & Diniz, L. (2021). Recent advances on properties and utility of nanomaterials generated from industrial and biological activities. *Crystals, 11*(6), 634. https://doi.org/10.3390/cryst11060634
133. Babusenan, A., Pandey, B., Roy, S. C., & Bhattacharyya, J. (2020). Charge transfer mediated photoluminescence enhancement in carbon dots embedded in TiO2 nanotube matrix. *Carbon, 161*, 535–541. https://doi.org/10.1016/j.carbon.2020.01.097
134. Chen, Z., Liu, Y., & Kang, Z. (2022). Diversity and tailorability of photoelectrochemical properties of carbon dots. *Accounts of Chemical Research, 55*(21), 3110–3124. https://doi.org/10.1021/acs.accounts.2c00570
135. Xue, B., Yang, Y., Sun, Y., Fan, J., Li, X., & Zhang, Z. (2019). Photoluminescent lignin hybridized carbon quantum dots composites for bioimaging applications. *International Journal of Biological Macromolecules, 122*, 954–961. https://doi.org/10.1016/j.ijbiomac.2018.11.018
136. Ullah, K. W., Qin, L., Ullah, K. W., Khan, S. U., Hussain, M. M., Ahmed, F., Kamal, S., & Zhou, P. (2023). Fluorescent carbon dots with enhanced thermal stability for multicolor bioimaging in hot conditions. *ACS Applied Nano Materials, 6*(19), 17838–17847. https://doi.org/10.1021/acsanm.3c03131
137. Khan, W. U., Wang, D., & Wang, Y. (2018). Highly green emissive nitrogen-doped carbon dots with excellent thermal stability for bioimaging and solid-state LED. *Inorganic Chemistry, 57*(24), 15229–15239. https://doi.org/10.1021/acs.inorgchem.8b02524
138. Luo, B., Yang, H., Zhou, B., Ahmed, S. M., Zhang, Y., Liu, H., Liu, X., He, Y., & Xia, S. (2020). Facile synthesis of luffa sponge activated carbon fiber based carbon quantum dots with green fluorescence and their application in Cr(VI) determination. *ACS Omega, 5*(10), 5540–5547. https://doi.org/10.1021/acsomega.0c00195
139. Tadesse, A., Hagos, M., RamaDevi, D., Basavaiah, K., & Belachew, N. (2020). Fluorescent-nitrogen-doped carbon quantum dots derived from citrus lemon juice: green synthesis, Mercury(ii) ion sensing, and live cell imaging. *ACS Omega, 5*(8), 3889–3898. https://doi.org/10.1021/acsomega.9b03175
140. Yang, X.-C., Yang, Y.-L., Xu, M.-M., Liang, S.-S., Pu, X.-L., Hu, J.-F., Li, Q.-L., Zhao, J.-T., & Zhang, Z.-J. (2021). Metal-ion-cross-linked nitrogen-doped carbon dot hydrogels for dual-spectral detection and extractable removal of divalent heavy metal ions. *ACS Applied Nano Materials, 4*(12), 13986–13994. https://doi.org/10.1021/acsanm.1c03306

141. Chawre, Y., Satnami, M. L., Kujur, A. B., Ghosh, K. K., Nagwanshi, R., Karbhal, I., Pervez, S., & Deb, M. K. (2023). Förster resonance energy transfer between multicolor emissive N-doped carbon quantum dots and gold nanorods for the detection of H 2 O 2, Glucose, Glutathione, and Acetylcholinesterase. *ACS Applied Nano Materials, 6*(9), 8046–8058. https://doi.org/10.1021/acsanm.3c01518
142. Chen, S., Zhang, S.-Z., & Jiang, H. (2024). Modification of crystal-optimized TiO 2 with biomass-derived carbon quantum dots for highly efficient degradation of favipiravir in water. *ACS ES&T Water, 4*(2), 531–542. https://doi.org/10.1021/acsestwater.3c00595
143. Koç, Ö. K., Üzer, A., & Apak, R. (2022). High quantum yield nitrogen-doped carbon quantum dot-based fluorescent probes for selective sensing of 2,4,6-trinitrotoluene. *ACS Applied Nano Materials, 5*(4), 5868–5881. https://doi.org/10.1021/acsanm.2c00717
144. Saha, A., Das, S., & Devi, P. S. (2022). N-doped fluorescent carbon nanosheets as a label-free platform for sensing bisphenol derivatives. *ACS Applied Nano Materials, 5*(4), 4908–4920. https://doi.org/10.1021/acsanm.1c04467
145. Lv, H., Gao, X., Xu, Q., Liu, H., Wang, Y.-G., & Xia, Y. (2017). Carbon quantum dot-induced MnO 2 nanowire formation and construction of a binder-free flexible membrane with excellent superhydrophilicity and enhanced supercapacitor performance. *ACS Applied Materials & Interfaces, 9*(46), 40394–40403. https://doi.org/10.1021/acsami.7b14761
146. Bora, A., Mohan, K., & Dolui, S. K. (2019). Carbon dots as cosensitizers in Dye-sensitized solar cells and fluorescence chemosensors for 2,4,6-trinitrophenol detection. *Industrial & Engineering Chemistry Research, 58*(51), 22771–22778. https://doi.org/10.1021/acs.iecr.9b05056
147. Qunwei, T., Zhu, W., He, B., & Yang, P. (2017). Rapid conversion from carbohydrates to large-scale carbon quantum dots for all-weather solar cells. *ACS Nano, 11*(2), 1540–1547. https://doi.org/10.1021/acsnano.6b06867
148. Saipanya, S., Waenkaew, P., Maturost, S., Pongpichayakul, N., Promsawan, N., Kuimalee, S., Namsar, O., Income, K., Kuntalue, B., Themsirimongkon, S., & Jakmunee, J. (2022). Catalyst composites of palladium and n-doped carbon quantum dots-decorated silica and reduced graphene oxide for enhancement of direct formic acid fuel cells. *ACS Omega, 7*(21), 17741–17755. https://doi.org/10.1021/acsomega.2c00906
149. Kundu, S., Ghosh, M., & Sarkar, N. (2021). State of the art and perspectives on the biofunctionalization of fluorescent metal nanoclusters and carbon quantum dots for targeted imaging and drug delivery. *Langmuir, 37*(31), 9281–9301. https://doi.org/10.1021/acs.langmuir.1c00732
150. Permatasari, F. A., Irham, M. A., Bisri, S. Z., & Iskandar, F. (2021). Carbon-based quantum dots for supercapacitors: recent advances and future challenges. *Nanomaterials, 11*(1), 91. https://doi.org/10.3390/nano11010091
151. Xu, Y., Wang, C., Wu, T., Ran, G., & Song, Q. (2022). Template-free synthesis of porous fluorescent carbon nanomaterials with gluten for intracellular imaging and drug delivery. *ACS Applied Materials & Interfaces, 14*(18), 21310–21318. https://doi.org/10.1021/acsami.2c00941
152. Pan, J., Zheng, Z., Yang, J., Wu, Y., Lu, F., Chen, Y., & Gao, W. (2017). A novel and sensitive fluorescence sensor for glutathione detection by controlling the surface passivation degree of carbon quantum dots. *Talanta, 166*, 1–7. https://doi.org/10.1016/j.talanta.2017.01.033
153. Li, X., Rui, M., Song, J., Shen, Z., & Zeng, H. (2015). Carbon and graphene quantum dots for optoelectronic and energy devices: A review. *Advanced Functional Materials, 25*(31), 4929–4947. https://doi.org/10.1002/adfm.201501250
154. Prasath, A., Athika, M., Duraisamy, E., Sharma, A. S., & Elumalai, P. (2018). Carbon-quantum-dot-derived nanostructured MnO 2 and its symmetrical supercapacitor performances. *Chemistry Select, 3*(30), 8713–8723. https://doi.org/10.1002/slct.201801950
155. Chen, G., Wu, S., Hui, L., Zhao, Y., Ye, J., Tan, Z., Zeng, W., Tao, Z., Yang, L., & Zhu, Y. (2016). Assembling carbon quantum dots to a layered carbon for high-density supercapacitor electrodes. *Scientific Reports, 6*(1), 19028. https://doi.org/10.1038/srep19028

156. Skaltsas, T., Goulielmaki, M., Pintzas, A., Pispas, S., & Tagmatarchis, N. (2017). Carbon quantum dots/block copolymer ensembles for metal-ion sensing and bioimaging. *Journal of Materials Chemistry B, 5*(27), 5397–5402. https://doi.org/10.1039/C7TB01352C
157. Luo, P. G., Sahu, S., Yang, S.-T., Sonkar, S. K., Wang, J., Wang, H., LeCroy, G. E., Cao, L., & Sun, Y.-P. (2013). Carbon "quantum" dots for optical bioimaging. *Journal of Materials Chemistry B, 1*(16), 2116. https://doi.org/10.1039/c3tb00018d
158. Yao, Y.-Y., Gedda, G., Girma, W. M., Yen, C.-L., Ling, Y.-C., & Chang, J.-Y. (2017). Magnetofluorescent carbon dots derived from crab shell for targeted dual-modality bioimaging and drug delivery. *ACS Applied Materials & Interfaces, 9*(16), 13887–13899. https://doi.org/10.1021/acsami.7b01599
159. Kang, Y.-F., Li, Y.-H., Fang, Y.-W., Xu, Y., Wei, X.-M., & Yin, X.-B. (2015). Carbon quantum dots for zebrafish fluorescence imaging. *Scientific Reports, 5*(1), 11835. https://doi.org/10.1038/srep11835
160. Ray, S. C., Saha, A., Jana, N. R., & Sarkar, R. (2009). Fluorescent carbon nanoparticles: synthesis, characterization, and bioimaging application. *The Journal of Physical Chemistry C, 113*(43), 18546–18551. https://doi.org/10.1021/jp905912n
161. Yang, S.-T., Cao, L., Luo, P. G., Lu, F., Wang, X., Wang, H., Meziani, M. J., Liu, Y., Qi, G., & Sun, Y.-P. (2009). Carbon dots for optical imaging in vivo. *Journal of the American Chemical Society, 131*(32), 11308–11309. https://doi.org/10.1021/ja904843x
162. Chai, L., Zhou, J., Feng, H., Tang, C., Huang, Y., & Qian, Z. (2015). Functionalized carbon quantum dots with dopamine for tyrosinase activity monitoring and inhibitor screening. in vitro and intracellular investigation. *ACS Applied Materials & Interfaces, 7*(42), 23564–23574. https://doi.org/10.1021/acsami.5b06711
163. Deng, Y., Chen, M., Chen, G., Zou, W., Zhao, Y., Zhang, H., & Zhao, Q. (2021). Visible–ultraviolet upconversion carbon quantum dots for enhancement of the photocatalytic activity of titanium dioxide. *ACS Omega, 6*(6), 4247–4254. https://doi.org/10.1021/acsomega.0c05182
164. Zhou, Y., Desserre, A., Sharma, S. K., Li, S., Marksberry, M. H., Chusuei, C. C., Blackwelder, P. L., & Leblanc, R. M. (2017). Gel-like carbon dots: characterization and their potential applications. *Chem Phys Chem, 18*(8), 890–897. https://doi.org/10.1002/cphc.201700038
165. Hola, K., Zhang, Y., Wang, Y., Giannelis, E. P., Zboril, R., & Rogach, A. L. (2014). Carbon dots—Emerging light emitters for bioimaging, cancer therapy and optoelectronics. *Nano Today, 9*(5), 590–603. https://doi.org/10.1016/j.nantod.2014.09.004
166. Ding, C., Zhu, A., & Tian, Y. (2014). Functional surface engineering of C-dots for fluorescent biosensing and in vivo bioimaging. *Accounts of Chemical Research, 47*(1), 20–30. https://doi.org/10.1021/ar400023s
167. Jia, X., Han, Y., Pei, M., Zhao, X., Tian, K., Zhou, T., & Liu, P. (2016). Multi-functionalized hyaluronic acid nanogels crosslinked with carbon dots as dual receptor-mediated targeting tumor theranostics. *Carbohydrate Polymers, 152*, 391–397. https://doi.org/10.1016/j.carbpol.2016.06.109
168. Otis, G., Bhattacharya, S., Malka, O., Kolusheva, S., Bolel, P., Porgador, A., & Jelinek, R. (2019). Selective labeling and growth inhibition of pseudomonas aeruginosa by aminoguanidine carbon dots. *ACS Infectious Diseases, 5*(2), 292–302. https://doi.org/10.1021/acsinfecdis.8b00270
169. Rajendiran, K., Zhao, Z., Pei, D.-S., & Fu, A. (2019). Antimicrobial activity and mechanism of functionalized quantum dots. *Polymers, 11*(10), 1670. https://doi.org/10.3390/polym11101670
170. Thakur, M., Pandey, S., Mewada, A., Patil, V., Khade, M., Goshi, E., & Sharon, M. (2014). Antibiotic conjugated fluorescent carbon dots as a theranostic agent for controlled drug release bioimaging, and enhanced antimicrobial activity. *Journal of Drug Delivery, 2014*, 1–9. https://doi.org/10.1155/2014/282193
171. Kannan, K., Radhika, D., Sadasivuni, K. K., Reddy, K. R., & Raghu, A. V. (2020). Nanostructured metal oxides and its hybrids for photocatalytic and biomedical applications. *Advances in Colloid and Interface Science, 281*, 102178. https://doi.org/10.1016/j.cis.2020.102178

172. Lakshmanan, A., Surendran, P., Manivannan, N., Sathish, M., Balalakshmi, C., Suganthy, N., Rameshkumar, P., Kaviyarasu, K., & Ramalingam, G. (2021). Superficial preparation of biocompatible carbon quantum dots for antimicrobial applications. *Materials Today: Proceedings, 36*, 171–174. https://doi.org/10.1016/j.matpr.2020.02.694
173. Yadav, P., Nishanthi, S. T., Purohit, B., Shanavas, A., & Kailasam, K. (2019). Metal-free visible light photocatalytic carbon nitride quantum dots as efficient antibacterial agents: An insight study. *Carbon, 152*, 587–597. https://doi.org/10.1016/j.carbon.2019.06.045
174. Li, P., Han, F., Cao, W., Zhang, G., Li, J., Zhou, J., Gong, X., Turnbull, G., Shu, W., Xia, L., Fang, B., Xing, X., & Li, B. (2020). Carbon quantum dots derived from lysine and arginine simultaneously scavenge bacteria and promote tissue repair. *Applied Materials Today, 19*, 100601. https://doi.org/10.1016/j.apmt.2020.100601
175. Shahshahanipour, M., Rezaei, B., Ensafi, A. A., & Etemadifar, Z. (2019). An ancient plant for the synthesis of a novel carbon dot and its applications as an antibacterial agent and probe for sensing of an anti-cancer drug. *Materials Science and Engineering: C, 98*, 826–833. https://doi.org/10.1016/j.msec.2019.01.041
176. Singh, S., Nigam, P., Pednekar, A., Mukherjee, S., & Mishra, A. (2020). Carbon quantum dots functionalized agarose gel matrix for in solution detection of nonylphenol. *Environmental Technology, 41*(3), 322–328. https://doi.org/10.1080/09593330.2018.1498133
177. Molaei, M. J. (2019). A review on nanostructured carbon quantum dots and their applications in biotechnology, sensors, and chemiluminescence. *Talanta, 196*, 456–478. https://doi.org/10.1016/j.talanta.2018.12.042
178. Long, W., Bi, Y., Hou, J., Li, H., Xu, Y., Wang, B., Ding, H., & Ding, L. (2016). Facile, green and clean one-step synthesis of carbon dots from wool: Application as a sensor for glyphosate detection based on the inner filter effect. *Talanta, 160*, 268–275. https://doi.org/10.1016/j.talanta.2016.07.020
179. Xu., Manman, Huang, Q., Sun, R., & Wang, X. (2016). Simultaneously obtaining fluorescent carbon dots and porous active carbon for supercapacitors from biomass. *RSC Advances, 6*(91), 88674–88682. https://doi.org/10.1039/C6RA18725K
180. Yang, S.-T., Wang, X., Wang, H., Lu, F., Luo, P. G., Cao, L., Meziani, M. J., Liu, J.-H., Liu, Y., Chen, M., Huang, Y., & Sun, Y.-P. (2009). Carbon dots as nontoxic and high-performance fluorescence imaging agents. *The Journal of Physical Chemistry C, 113*(42), 18110–18114. https://doi.org/10.1021/jp9085969
181. Min, Z., Ruan, S., Liu, S., Sun, T., Qu, D., Zhao, H., Xie, Z., Gao, H., Jing, X., & Sun, Z. (2015). Self-targeting fluorescent carbon dots for diagnosis of brain cancer cells. *ACS Nano, 9*(11), 11455–11461. https://doi.org/10.1021/acsnano.5b05575

Optical and Biomedical Features of Green Carbon Quantum Dots

V. Arul, R. Suresh, P. Chandrasekaran, and K. Radhakrishnan

Abstract The synthesis of carbon quantum dots (CQDs) has witnessed a paradigm shift towards green methodologies, aligning with the global push for sustainable nanomaterials. This book chapter provides a comprehensive exploration of the various properties exhibited by CQDs synthesized through environmentally friendly approaches. By analyzing recent research findings, we unravel the diverse facets of these nanoscale carbonaceous materials, emphasizing optical tunability, surface functionalization, biocompatibility, and multifaceted applications. The discourse commences with an overview of the burgeoning field, contextualizing the significance of green synthesis in the realm of CQDs. Through a meticulous synthesis of relevant literature, the chapter delves into the optical properties of green-synthesized CQDs, elucidating the mechanisms behind their tunability and potential applications in sensors and imaging. The discussion extends to the surface engineering of CQDs, exploring how green synthesis techniques impact their functional groups and surface chemistry, influencing interactions in biological systems. Furthermore, we explore the biocompatibility of green-synthesized CQDs, addressing their suitability for biomedical applications. The chapter culminates in an examination of the diverse applications of these CQDs, ranging from bio-sensing to drug delivery, showcasing their versatility in various fields. This abstract encapsulates the essence of the book chapter, providing a snapshot of the in-depth analysis of green-synthesized CQDs'

V. Arul (✉)
Department of Chemistry, Sri Eshwar College of Engineering (Autonomous), Coimbatore 641202, India
e-mail: kvarulchem6@gmail.com

R. Suresh
Department of Physics, Sri Eshwar College of Engineering (Autonomous), Coimbatore 641202, India

P. Chandrasekaran
Department of Chemical Engineering, National Chung Hsing University, Taichung 402, Taiwan

K. Radhakrishnan
Department of Chemistry, Centre for Material Chemistry, Karpagam Academy of Higher Education, Coimbatore, Tamil Nadu 641 021, India

V. Kumar et al. (eds.), *Green Carbon Quantum Dots*, Engineering Materials,
https://doi.org/10.1007/978-981-97-6203-3_4

properties. The exploration not only contributes to the understanding of these nanomaterials but also underscores the imperative shift towards sustainable and impactful methodologies in their synthesis.

1 Introduction

Carbon quantum dots, as nanoscale carbon-based materials, have attracted substantial interest due to their unique properties, including size-dependent photoluminescence, biocompatibility, and a high surface area. These properties render them suitable for numerous applications, ranging from optoelectronic devices to biosensors and drug delivery systems [1]. While various methods have been employed for CQD synthesis, the green synthesis approach, which uses sustainable and eco-friendly precursors, has garnered particular attention. This chapter explores the properties of CQDs synthesized through green methods, highlighting their significance in modern materials science [2].

Quantum dots (QDs), a diverse assembly of minute fluorescent nanoparticles, represent a novel class of materials with remarkable potential across various fields, particularly in sensing, medical imaging, and diagnostics. QDs, essentially minuscule particles concentrated at a single point, adhere to the laws of quantum theory. Within these artificial atoms, electrically charged particles exhibit well-defined energy levels, akin to those of individual atoms. Typically grown from semiconductor materials, QDs boast diameters ranging from one to several hundred nanometers, housing a few hundred to several thousand atoms. Unlike conventional semiconductor crystals, they exhibit atomic-level behavior, earning them the moniker "artificial atoms". Synthesized through methods like ion implementation, molecular beam epitaxy and X-ray lithography, QDs have become prominent for their precision and applications in bio-analytical labelling [3].

The biomedical community has keenly embraced QDs due to their distinctive properties. These include real-time tissue imaging capabilities, diagnostics, drug delivery, and deployment as single-molecule probes. Their use in therapeutic drug delivery systems is particularly advantageous, given their eco-friendly composition and reduced toxicity compared to traditional materials like polymer or silica spheres [4] (Fig. 1).

The optical properties of Quantum Dots (QDs) are intricately linked to their size, showcasing a remarkable connection between size variations and color emission in crystals. This phenomenon underscores the substantial impact of energy levels within the crystal lattice. Size-Dependent Color Emission of QDs with diverse sizes into a crystal induces a noteworthy shift in color emission. This variation is prominently observed in the fluorescence spectrum, which displays a range of distinct colors. Red light is associated with lower energy levels, while blue light corresponds to higher energy levels. This correlation emphasizes the pivotal role of QD size in determining the observed colors. The band gap energy, defined as the distinction between the excited and resting states of a QD, serves as a crucial factor governing

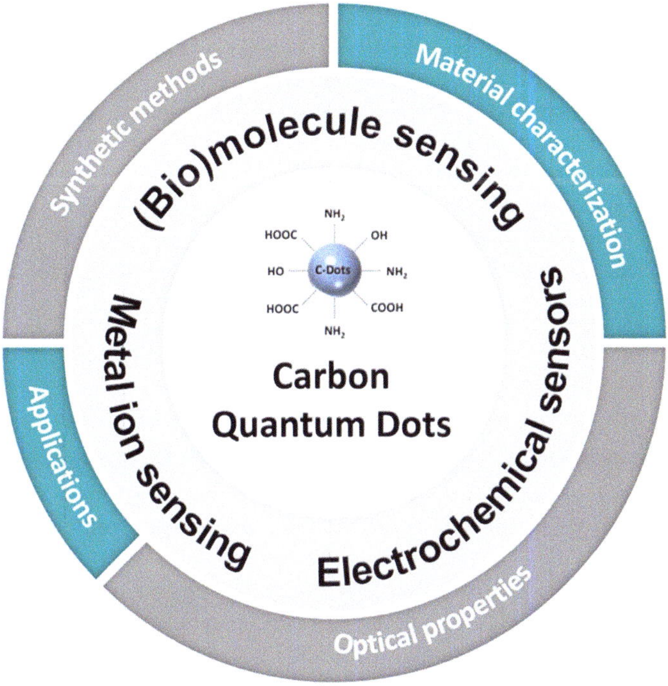

Fig. 1 Potential applications of CQDs [5]

the frequency and, consequently, the color of emitted light. The band gap energy directly influences the transition between energy states and, consequently, the optical characteristics of the QD. Intriguingly, the size of Quantum Dots plays a decisive role in shaping the emitted light's color. Smaller dots exhibit a propensity to emit higher-energy, bluer light, whereas larger dots tend to emit lower-energy, redder light. This size-color relationship is fundamental in understanding and manipulating the optical characteristics of QDs. the size of Quantum Dots is a key determinant of their optical characteristics, influencing the color emitted by crystals. The relationship between QD size and color emission provides valuable insights into the intricate interplay of energy levels within the crystal lattice [6].

The structural and surface properties of QDs play a pivotal role in their applications. Hydrophobic ligands render QDs soluble in organic solvents, while modifications can enhance water solubility and biocompatibility. Various coatings, such as amphiphilic polymers and bi-functional molecules, diversify their applications. Targeted ligands like oligonucleotides, peptides, or antibodies further customize QDs for specific biological interactions.

Despite their myriad advantages, concerns about the toxicity of conventional QDs, particularly those containing cadmium, prompt the exploration of safer alternatives. Carbon-based quantum dots (CQDs) have emerged as compelling substitutes, offering wide excitation spectra, narrow emission spectra, and enhanced photostability. The transition to non-metallic QDs aims to ensure safety in biological applications, circumventing potential harm to human cells.

This comprehensive book chapter underscores the tunable properties of QDs, especially in the context of cancer therapy. While recognizing the potential toxicity of traditional QDs, the spotlight turns to CQDs as safer alternatives. Recent publications highlight the application of CQDs in cancer therapeutics, from bio-imaging to biosensors and bio-molecular/drug delivery. The review explores the historical context, current state, and future prospects of CQDs in biomedical settings, offering insights into potential limitations and avenues for future research.

2 Carbon Quantum Dots: A Green Synthesis Protocol

The synthesizing of carbon dots using green synthesis is a significant topic that is in perfect harmony with sustainability and is acquiring increasing relevance in applications that are used on a daily basis. This debate dives into fundamental techniques, with a particular focus on the polymerization and carbonization of minute organic precursors such as ammonium citrate, Citric acid, ethylene glycol, phenylenediamine, graphite and carbon nanotubes are some instances that are particularly noteworthy. In the realm of ultra-small fluorescent carbon dots, there are two primary methodologies. A primary focus is being placed on the investigation of several techniques for the production of extremely minute fluorescent carbon dots, with the colloidal synthetic approach garnering the most attention. It has been demonstrated that this technology has the capability of producing huge numbers of carbon dots that are precisely controlled. The synthesis of graphene quantum dots (GQDs) from minute aromatic molecules is a notable example. This synthesis makes use of stepwise solution chemistry to produce well-defined structures that are uniform in size. In order to enhance structural control, the formation of a three-dimensional cage surrounding graphene and the reduction of intermolecular π–π^* attraction are both achieved through the formation of covalent bonds between trialkyl-substituted phenyl moieties. Because of their one-of-a-kind characteristics, the colloidal quantum dots that were generated are perfect for use as model systems in the investigation of fundamental processes in complex carbon materials. The fact that these dots have a regulated structure and surface characteristics makes them extremely adaptable and gives them the potential to be used in a variety of industrial applications. Providing environmentally friendly solutions for the manufacture of carbon dots is a fundamental subfield of green chemistry, which plays a pivotal role in this endeavor. Its relevance is highlighted by the multiple advantages it offers, which include safety, environmental friendliness, the avoidance of dangerous chemicals, and functioning under normal conditions [7].

In the manufacturing of carbon dots, environmentally friendly chemical methods make use of a wide variety of natural resources. These resources include chicken eggs and animals, as well as a wide variety of plant species and waste products such as paper and frying oil. Carbon dots are produced by a variety of processes, which exemplifies the adaptability and effectiveness of their manufacturing process. Among the most important methods are: Pyrolysis is a process that involves the breaking down of organic precursors through thermal decomposition at high temperatures. The process of carbonization involves the utilization of controlled heating in order to result in the transformation of organic components into carbon dots. Utilizing microwave radiation to facilitate quick and regulated polymerization reactions is what is meant by the term "microwave-assisted polymerization". The hydrothermal and solvothermal processes include the utilization of high-pressure and high-temperature conditions for the purpose of synthesis. The process of carbonization involves the utilization of controlled heating in order to induce the transformation of organic components into carbon dots [8]. All of the different methods of synthesis have their own unique benefits, which contribute to the increased diversity of carbon dot creation. Carbon dots that have been synthesized using environmentally friendly methods have a controlled structure, surface capabilities, and a sustainable nature, all of which contribute to their prospective uses in a variety of disciplines. The fundamental techniques and methodologies that were presented not only shed light on the environmentally friendly synthesis of carbon dots, but also highlighted the significance of sustainability in the manufacturing process of these dots. These ecologically benign procedures are paving the way for versatile and efficient carbon dot synthesis, which will stimulate applications across a wide range of areas as the field continues to advance.

2.1 Overview of Green Synthesis Methods:

2.1.1 Hydrothermal/Solvothermal Process

Carbon quantum dots (CQDs) have several potential uses in many different industries, including imaging, sensing, and catalysis, which has led to a lot of interest in their production. The hydrothermal and solvothermal carbonisation techniques are two of the most eco-friendly and flexible technologies used to fabricate CQDs. This response provides an overview of these strategies and their applications, highlighting the key features and advantages of the hydrothermal method in CQD synthesis. Hydrothermal carbonization stands out as a notably cost-effective and environmentally friendly procedure, characterized by its one-step method. In this process, a hydrothermal reactor operating at elevated temperatures is employed [9]. The organic precursor solution is sealed inside the reactor, and the procedure utilizes a diverse range of reagents, showcasing its adaptability. Among these reagents are acids such as citric and ascorbic acid, animal products including bovine serum, egg albumin, and cow milk, as well as plant-derived components from the diet. The extensive variety of components contributes significantly to the adaptability of the process.

The resulting carbon quantum dots (CQDs) exhibit remarkable fluorescence, enhancing their overall flexibility. Within the hydrothermal process, precursor molecules undergo heating and pressurization inside a stainless steel autoclave lined with Teflon, with the entire process occurring in water. This combination of factors makes hydrothermal carbonization a versatile and efficient method for synthesizing CQDs with potential applications across various fields. The precursor molecules encompass a diverse range, including glucose, proteins, amino acids, polymers and various natural products. The hydrothermal method presents numerous advantages. Key properties: Homogeneity: Hydrothermally derived CQDs are highly homogeneous. Water Solubility: The CQDs are water-soluble, contributing to their applicability in aqueous environments.

Monodispersity: The particles are mono-dispersed, ensuring uniformity in size. Photostability: Hydrothermal CQDs demonstrate robust photostability. Salt Tolerance: They exhibit tolerance to high ionic salt conditions, specifically observed in 2 M NaCl. Controlled Particle Size: The hydrothermal method allows for controlled particle size during synthesis. Elevated Quantum Yield (QY): Hydrothermally synthesized CQDs achieve a high QY without the need for surface passivation (Fig. 2).

Both hydrothermal and solvothermal methods contribute to the growing body of research on CQD synthesis. While solvothermal approaches use ammonia, alcohol, and other solvents as alternatives to water, the hydrothermal method stands out for its simplicity, single-step operation, non-toxicity, cost-effectiveness, and eco-friendliness. The hydrothermal method's advantages position it as a leading choice in the evolving landscape of CQD fabrication, providing a promising avenue for tailored properties to meet specific application requirements.

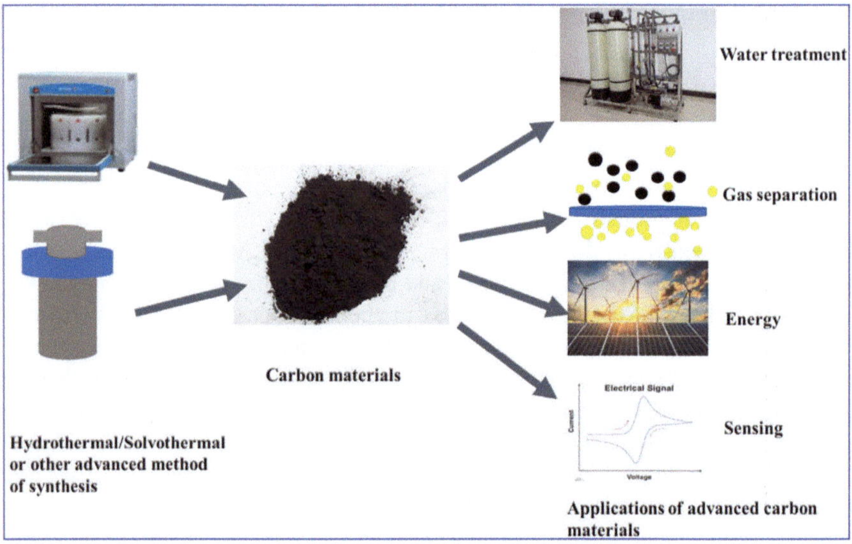

Fig. 2 Hydrothermal/solvothermal synthesis of CQDs [10]

2.1.2 Microwave-Assisted Heating

Microwave-assisted heating has emerged as a cost-effective and time-efficient method for synthesizing carbon quantum dots (CQDs). This approach, surpassing electrochemical and hydrothermal methods, utilizes microwaves to achieve targeted contact with carbon-based materials, allowing for efficient and neighborhood-specific heating. The simplified carbonization processes result in the production of diverse nanostructure morphologies.

Numerous environmentally friendly synthesis investigations using microwave-assisted techniques have been conducted. Notably, the controlled synthetic method involving branching polyethylene mine and citric acid induces changes in internal structure, showcasing the adaptability of carbon dots through straightforward fabrication processes. This method also highlights the catalytic capabilities and photoluminescence of carbon dots.

A major application of microwave-assisted heating involves producing spherical Graphene Quantum Dots (GQDs) from cow's milk. The photoluminescence properties of these GQDs are influenced by both ionic strength and the duration of heating. Remarkably, the combination of these factors demonstrates strong cytotoxicity effects against various cancer cells while maintaining biocompatibility with the L929 cell line. This underscores the potential of GQDs in biomedical applications [10].

Investigations into safe production methods have led to the microwave-assisted synthesis of sulfur- and oxygen-coated green brilliant graphitic carbon nitride quantum dots. The treatment of citric acid and thiourea under microwave conditions yields luminous compounds with excellent biocompatibility, resistance to interference in high ionic strength solutions, and a high quantum yield. This method holds promise for applications in imaging and sensing due to its favorable properties.

Recent research highlights an eco-friendly approach to synthesizing CQDs using Vaccinium Meridionale Swartz extract and microwave-assisted carbonization. This method enables the rapid generation of large quantities of CQDs with high mass fraction concentrations in under five minutes. Notably, these CQDs exhibit excellent heat resistance up to 300 degrees Celsius, opening up possibilities for high-temperature applications [11].

The synthesis of Carbon Quantum Dots has evolved with the introduction of microwave-assisted techniques and eco-friendly approaches. These methods offer unique advantages, from producing cytotoxic GQDs for cancer cell applications to creating versatile CQDs with excellent heat resistance. The multi-functionality of CQDs positions them as promising candidates for a wide range of scientific and technological applications.

2.1.3 Pyrolysis

Pyrolysis, involving the thermal degradation of carbon precursors at elevated temperatures, stands out as the most popular and preferred approach for the manufacture of carbon quantum dots (CQDs). This method offers numerous advantages, making it a highly favored choice in CQD synthesis. Pyrolysis provides several key advantages: Quicker Operations: The process is characterized by faster reaction times, contributing to efficiency in CQD production. Solvent-Free Methods: Pyrolysis eliminates the need for solvents, enhancing environmental friendliness and simplicity. Wide Tolerance for Precursors: The method accommodates a broad range of precursors, showcasing versatility in material selection. Shorter Reaction Times: Pyrolysis ensures efficient and rapid synthesis, reducing the overall time required for CQD formation. Cost-Effectiveness: The economical nature of pyrolysis makes it a cost-effective choice for large-scale CQD production. Scalability: The scalability of pyrolysis adds to its appeal, making it suitable for various production scales.

Green Synthesis of CQDs: Examples and Insights, Eleusine Coracana Pyrolysis: An example highlighting green synthesis involves the pyrolysis of E. coracana, resulting in the production of carbon quantum dots with a size of around 6 nm. Notably, the fluorescence of these CQDs was significantly suppressed by Cu^{2+} compared to other metal ions. This quenching phenomenon was attributed to the preferential adsorption of Cu^{2+} on the CQD surface through the π-bonding of aromatic C = C. This contrasted with other divalent metals, which favored σ-bonding with CQDs. Mono-dispersed carbon quantum dots (CQDs) were synthesized through a one-step thermal breakdown technique using fennel seeds from the plant Foeniculum vulgare. Notably, these CQDs exhibited elevated photostability and colloidal stability, demonstrating resilience to pH variations. Remarkably, the particles displayed fluorescence without the need for an extra surface passivation step. Their excitation-independent emission and enduring photoluminescence activity persisted consistently throughout their lifetime [12] (Fig. 3).

The pyrolysis process provides crucial insights into the controlled synthesis of carbon quantum dots (CQDs). Factors such as reaction time and temperature play pivotal roles in influencing the characteristics of the synthesized CQDs. Understanding these parameters contributes to tailoring CQDs with specific properties, advancing the controlled synthesis of this versatile carbon-based material. Pyrolysis stands as a preferred method for the green synthesis of carbon quantum dots, offering a suite of advantages that contribute to its popularity. The examples presented underscore the versatility of pyrolysis in producing CQDs with diverse properties, paving the way for advancements in applications ranging from sensing to imaging [14].

2.2 *Biomass Waste in Synthesizing CQDs*

Biomass waste, including food waste and agricultural leftovers, has emerged as a promising and sustainable source for Carbon Quantum Dot (CQD) synthesis in recent years, receiving substantial attention. The vast spectrum of biomass waste types, with high carbon content (up to 40–50%) and ample oxygen and nitrogen, presents it as a successful precursor for CQD valorization. lemon peel, Watermelon peel, orange peel, orange juice, fennel seeds, and coffee grounds have all been shown to be

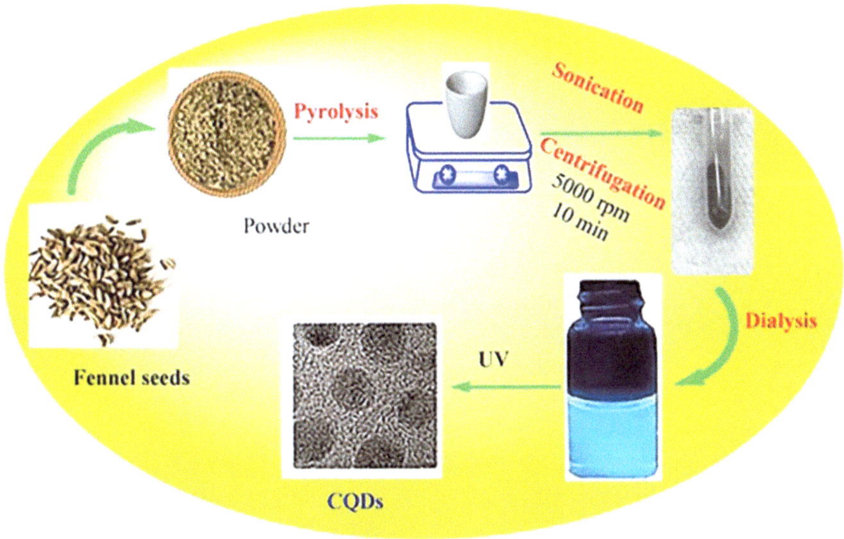

Fig. 3 Synthesis of CQDs by pyrolysis technique [13]

useful for "green" CQD synthesis without the requirement for post-treatment. These CQDs perform admirably in bio-sensing, cellular imaging, and chemical sensing applications, indicating the effectiveness of utilising biomass waste.

Bottom-up CQD generation using biomass waste uses biomass waste as the beginning material for the carbon source, which then goes through the appropriate procedures to make CQDs. Walnut shell is an especially excellent biomass waste in CQD synthesis, with improved photostability and photoluminescence capabilities when compared to typical metal-based Quantum Dots (QDs). Because of its shown lack of toxicity, it is a safe alternative for future bio-imaging applications. Chitin, another biomass waste, is successfully converted into CQDs by bottom-up methodologies. Chitin is converted into chitosan gel, which is known for its long-term stability without degradation, which extends the shelf life of the source material for CQD preparation. This demonstrates the adaptability of biomass waste in contributing to long-term CQD synthesis (Fig. 4).

Bottom-up processes, including hydrothermal carbonization, pyrolysis, and microwave treatment, leverage diverse biomass waste products for the synthesis of carbon quantum dots (CQDs). Each biomass waste product has various benefits and drawbacks, emphasising the significance of careful analysis when deciding which biomass waste to use. Yield/quantum yield, photostability, photoluminescence characteristics, and consistent CQD size must all be carefully considered. Watermelon peel and orange peel, for example, produce CQDs with great water solubility and good luminescence characteristics but have poor yields/quantum yields. Walnut shells, on the other hand, exhibit excellent photostability, consistent CQD size, and high yield, making them an excellent choice for bottom-up CQD synthesis. The use of biomass

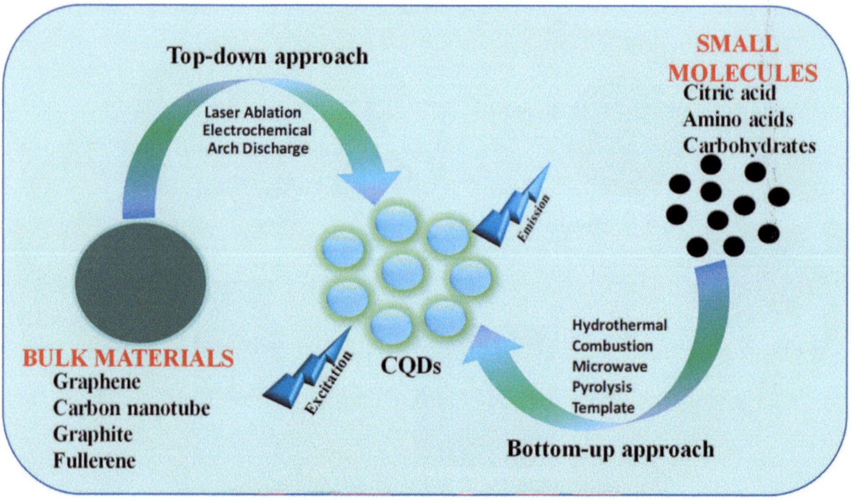

Fig. 4 The typical approaches for the synthesis of CQDs [15]

waste in CQD synthesis is a sustainable and efficient method that contributes to the creation of green and diverse carbon-based products. Researchers can tailor CQDs with desirable qualities by carefully selecting and optimising biomass waste sources in various bottom-up techniques, supporting improvements in diverse applications [16].

2.3 *Advantages of Green Synthesis Methods*

Discussion on the environmental benefits and sustainability of green synthesis methods, emphasizing reduced chemical waste and energy consumption.

- Environmental Sustainability: Reduced Carbon Footprint: The application of green synthesis techniques not only decreases the reliance on hazardous chemicals but also substantially lowers the energy demand in comparison to traditional methods, thereby reducing the overall carbon footprint of CQD production.
- Biocompatible Materials: The inclusion of natural precursors, such as biomass waste or plant extracts, not only ensures the biocompatibility of the resulting CQDs but also aligns with the growing demand for eco-friendly and non-toxic materials in biomedical applications.
- Avoidance of Toxic Substances: By minimizing the use of toxic reagents, green synthesis enhances the safety profile of CQDs, making them more suitable for various biological and medical applications.

- Cost-Effectiveness, Utilization of Renewable Resources: The utilization of renewable resources, such as plant extracts, in green synthesis contributes to cost-effectiveness by reducing dependency on expensive precursor materials.
- Minimized Processing Steps: The simplified synthetic procedures in green methods not only reduce costs associated with complex processing but also make CQD production more economically viable.
- Diverse Source Materials-Biomass Waste Utilization: Green synthesis efficiently transforms various biomass wastes, including food waste and agricultural residues, into valuable CQDs, promoting the efficient utilization of these materials and waste valorization strategies.
- Energy Efficiency-Mild Reaction Conditions: Green synthesis often operates under mild reaction conditions, minimizing energy consumption and making the process more energy-efficient compared to high-temperature processes involved in traditional synthesis methods.
- Tailored Functionalities-Versatile Precursors: The flexibility in choosing diverse natural precursors in green synthesis allows for the tailoring of CQD properties, enabling the development of functionalized CQDs with specific characteristics suited for various applications.
- Wide Range of Applications-Biomedical Applications: The biocompatible and non-toxic nature of CQDs synthesized through green methods makes them particularly suitable for applications in bio-imaging, bio-sensing, and drug delivery, contributing to advancements in biomedical research.
- Regulatory -Compliance with Green Chemistry Principles: Green synthesis aligns with the principles of green chemistry, meeting regulatory requirements and addressing concerns about the environmental impact of chemical processes, enhancing the overall compliance of CQD production.
- Community and Stakeholder Acceptance-Positive Perception: The sustainable and eco-friendly nature of green synthesis fosters a positive perception among communities, stakeholders, and researchers, leading to increased acceptance and support for these methods in CQD production.

3 Chemical Structures of C-dots

C-dots have been conceptualized as entities that are defined by a carbogenic core that is ornamented with surface functional groups as a result of the examination into the chemical structures of C-dots, which has led to several theoretical models. Elemental examination, in particular the use of X-ray photoelectron spectroscopy (XPS), exhibits a constant illumination of elevated oxygen levels, which is indicative of either an oxidized core or the presence of surface functional groups (Fig. 5).

Surface Functional Groups: An exhaustive investigation into the surface chemistry of carbon dots (C-dots) has revealed that these dots have distinctive compositions that are the result of a variety of fabrication techniques. When compared to the molecules of the original soot, the pure C-dots that are produced as a result of

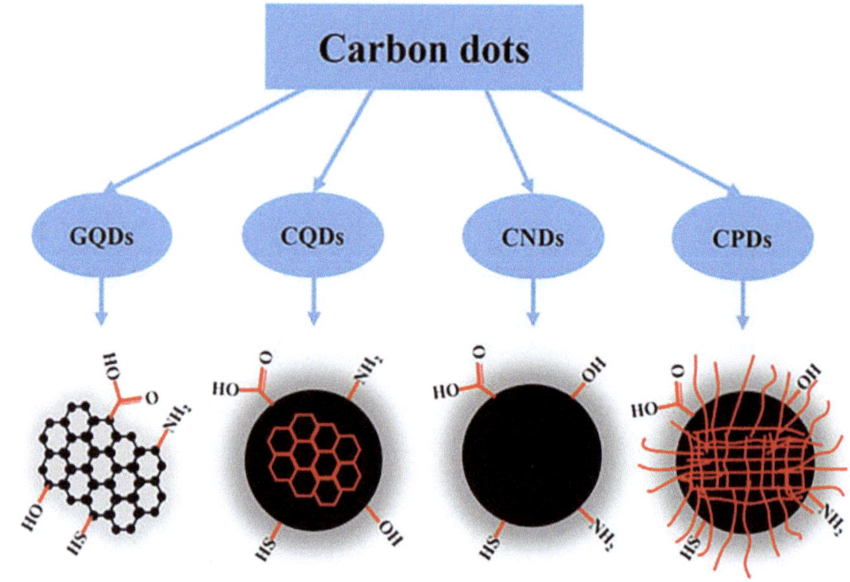

Fig. 5 Architectural depiction of diverse Carbon Quantum Dots (CQDs) exhibiting enriched functional groups and polymer chains adorning the surface, encapsulating a central carbon core [17]

the wet oxidation of candle soot are distinguished by a much higher concentration of oxygen. The incorporation of carbonyl groups into the structure of the C-dots is responsible, at the very least in part, for the increased oxygen concentration that has been observed. The surface chemistry of carbon dots (C-dots) has successfully identified carboxylic functional groups on their surface, employing Fourier transform infrared (FT-IR) spectroscopy as a pivotal instrument in this exploratory endeavor. The effective application of FT-IR spectroscopy played a crucial role in achieving this significant milestone.

For instance, C-dots produced through methods such as laser ablation/passivation and the purification of single-walled carbon nanotubes (SWNTs) exhibit distinct FT-IR signatures. These signatures unequivocally confirm the presence of carboxyl functional groups on the surface of the C-dots, and they are clearly observable and recognizable in the FT-IR spectrum. To further validate the chemical structure of C-dots with non-carboxyl functionalities, rigorous research efforts have been undertaken, utilizing both FT-IR and nuclear magnetic resonance (NMR) analyses. These analyses contribute to a comprehensive understanding of the intricate surface chemistry of C-dots, marking significant advances. These findings, when taken as a whole, contribute to the ongoing evolution of our understanding of the complex surface chemistry of C-dots, which represents tremendous leaps in the past few decades.

Carbogenic Core Structures: Despite the fact that there has been progress in identifying surface functional groups, the precise nature of the carbogenic core continues to be a topic of discussion. The difficulty derives from the inherent heterogeneity in C-dots that is brought about by the various synthesis techniques. To be more specific, the assignment of lattice fringes of X-ray diffraction planes presents difficulties in reaching findings that are free of ambiguity. This is especially true when comparing diamond-like carbon to graphitic carbon.

3.1 *Structural Classifications for the C-dot Core:*

The synthesis of C-dots through electrochemical etching of carbon fibers has unveiled intriguing structural aspects. The graphite structure of C-dots, characterized by a lattice spacing corresponding to the (002) facet of graphite, signifies a breakthrough. The X-ray photoelectron spectroscopy (XPS) analysis, specifically the C1s spectra, revealed the presence of graphitic sp2 carbons, indicated by a peak at 284.5 electron volts. This discovery lays the foundation for subsequent investigations. Influence of Gallate Esters on C-dot Size and Surface Chemistry has been explored by employing gallate esters with varying alkyl chain lengths (lauryl, propyl, methyl). This experiment aimed to regulate C-dot size and surface chemistry. Raman spectroscopy demonstrated distinct intensity ratios of the D and G bands (ID/IG), with lauryl exhibiting the highest prominence. The alkyl chain length was identified as a factor affecting graphitization of the carbogenic core, providing crucial insights. Laser Ablation/Passivation Process for Diamond-Like C-dots a notable advancement involves the pursuit of a diamond-like structure in C-dots through laser ablation/passivation. Two types of C-dots, passivated by polymer (PEG200N) and untreated, were investigated. Transmission electron microscopy (TEM) revealed uniformly dispersed C-dots with a 2-nm size. Selected area electron diffraction (SAED) patterns exhibited diamond-like structures, evidenced by ring radius square ratios corresponding to diamond planes. Formation of Graphite Oxide Structure via Thermal Breakdown The graphite oxide structure of C-dots, formed through the thermal breakdown of ammonium citrate salts, adds a distinct dimension. X-ray diffraction (XRD) patterns unveiled reflections indicating the presence of disordered carbon and securely packed phenyls in the polymeric corona. Comparative analysis with other carbogenic networks emphasized similarities with lignite, coal, and humic compounds. One-Step Microwave-Assisted Pyrolysis of Polyethylenimine (PEI) Functionalized C-dots Exploring amorphous carbon structure in C-dots synthesized by one-step microwave-assisted pyrolysis of PEI revealed significant diffraction peaks at $2q = 18.3°$ in XRD patterns, confirming amorphous nature. Raman spectra indicated a faint G band at 1557 cm^{-1}. The findings emphasized consistency in properties among C-dots produced by the pyrolysis method. Paracrystalline Carbon in C-dots has introduced the term “paracrystalline carbon” for C-dots, describing a structure with graphitic clusters within an amorphous carbon matrix.

HR-TEM images revealed submerged graphitic (sp^2) clusters in an oxygenic matrix. The distinctive Raman spectrum with discrete D and G bands supported the presence of numerous sp^2 clusters, setting paracrystalline carbon apart from amorphous carbon. Nuanced Characterizations and the Dynamic Link Collectively, these diverse studies underscore the complex nature of C-dot formations, emphasizing the need for nuanced characterizations. The intricate interplay between graphitic clusters and amorphous carbon becomes evident through these investigations, offering valuable insights into the dynamic structure of C-dots [18].

3.2 Electronic Structures of C-dots

Molecular Orbital Theory and Electronic Transitions in C-dots; The electronic transitions observed in C-dots, primarily composed of carbon (C), oxygen (O), and nitrogen (N), exhibit similarities to organic molecules. This section delves into the intricate details of these transitions, emphasizing their connection to the chemical structures of C-dots.

Molecular Orbital Theory: Foundation for Understanding Transitions. The comprehensive description of electronic transitions in C-dots is rooted in molecular orbital theory (MO). MO theory categorizes electrons into nonbonding, antibonding, and bonding MOs, each possessing distinct energy attributes. The refinement of orbital classification in organic molecules, considering symmetry of rotation around the inter nuclear axis, introduces five types of MO: s, p, n (nonbonding), s*, and p*. This classification becomes crucial in understanding the electronic states and their diverse energy levels.

Classification of Orbitals and Symmetry; Organic compounds play a pivotal role in refining orbital classification based on symmetry. The five types of MOs (s, p, n, s*, p*) contribute to various electronic states, showcasing different energy levels. The symmetry of rotation around the internuclear axis becomes a key factor in this classification.

UV–Vis Spectroscopy: Unravelling Transitions; Electronic transitions in C-dots occur when light absorption lifts an electron to a higher energy state. UV–VIS spectroscopy becomes a powerful tool for detecting transitions such as s–s*, s-p*, p-p*, n-p*, and n-s*. Each transition corresponds to specific absorption ranges, and the molar absorptivities highlight the significance of these transitions, with n-p* transitions showing lower values than p-p* transitions.

Relationship to Chemical Structures; C-dots manifested the transitions like s–s*, s-p*, p-p*, n-p*, and n-s* due to their unique chemical structures. The p-states, associated with aromatic sp2-hybridized carbons, exhibit a decreasing energy gap with an increasing number of aromatic carbons, mirroring p-conjugation seen in organic molecules.

Influence of Functional Groups; The electron lone pairs present in functional groups, such as thiols, carbonyls, amides, and amines, assume a pivotal role in contributing to the formation of n-states in carbon dots (C-dots). Specifically, when these functional groups are attached to aromatic sp2-hybridized carbons, a notable transition occurs from the n- p* states associated with the functional groups of the aromatic carbons. The intricate interplay between p-n states is intricately influenced by the interaction of overlapping orbitals and the electron-withdrawing or donating capabilities inherent in the functional groups. This dynamic interaction serves as a significant determinant in shaping the electronic properties of C-dots. The resulting electronic structure modulation holds promising implications for the tunability and optimization of C-dots for various applications, ranging from sensors to biomedical imaging.

Significance for Photophysical Behavior; Understanding these electronic transitions becomes critical in unravelling the photophysical behavior of C-dots, dictating optical features like light absorption and luminescence. This comprehension serves as a foundation for further research, exploring the diverse applications of C-dots across various scientific domains. The parallels between electronic transitions in C-dots and organic molecules underscore their unique photophysical activity, opening avenues for potential applications. The interplay between molecular orbitals and chemical structures highlights the importance of this knowledge in harnessing the capabilities of C-dots.

4 Structural Properties

It is essential to investigate the size and size distribution of green-synthesized carbon quantum dots (CQDs) in order to gain an understanding of the influence that the choice of precursor and the conditions of synthesis have on these attributes. Transmission electron microscopy (TEM) tests reveal that green-synthesized CDs often have a diameter that falls somewhere between 2 and 12 nm. According to the findings of a size distribution analysis, the vast majority of these CDs are in the smaller end of the size range that is being discussed. Consequently, this highlights the importance of customizing the size of these semiconductor-like dots in order to achieve photon confinement through band gap modulation technology. Crystallographic information can be derived from the lattice order and spacing of CDs that have been synthesized from black pepper. High-resolution transmission electron micrographs of CDs offer this information. This thorough characterisation contributes to our improved comprehension of the structural characteristics of green-synthesized carbon quantum dots (CQDs). However, amorphous CDs do not possess a periodic crystal lattice, which makes it difficult to examine them using transmission electron microscopy (TEM). They are unable to diffract electrons and frequently display an X-ray diffraction pattern that is characterized by a broad reflection, known as an amorphous halo. This pattern typically reaches its peak in the range of $2\theta = 23.4$–$24.6°$. [7] It is essential to make this distinction between crystalline and amorphous CDs in order to achieve satisfactory characterisation.

The similarity in composition between formvar-coated TEM grids and CDs presents difficulties for transmission electron microscopy (TEM) imaging of CDs, particularly amorphous CDs. A careful deposition on the transmission electron microscopy grid is required in order to prevent aggregation, which could result in sizing artifacts. Successfully overcoming these hurdles will result in measurements that are accurate and trustworthy. The emergence of atomic force microscopy (AFM) as an alternative for size measurements enables the collection of data regarding the length, width, and height of compact discs (CDs). This technique avoids the contrast problems that are typically encountered when electron microscopy is performed on carbon nanomaterials. When combined, atomic force microscopy (AFM) and transmission electron microscopy (TEM) provide a comprehensive method for acquiring information on both size and topography. The measurements of dynamic light scattering play an important part in the context of biological applications since they provide a hydrodynamic radius. There is a direct connection between this parameter and the diffusivity of CDs in their respective biological contexts. It is possible that CDs, regardless of their size, have the capability of forming quasi-spherical morphologies, which may include hollow architectures. This ability positions CDs as prospective drug delivery agents. The complete analysis of size and size distribution in green-synthesized CQDs, taking into consideration a variety of characterization approaches, helps to the advancement of our understanding of these nanomaterials and promotes the prospective uses of these nanomaterials in a variety of disciplines.

5 Properties of Carbon Quantum Dots (CQDs)

Carbon Quantum Dots (CQDs) exhibit diverse optical properties, and this section explores key aspects such as absorption properties, fluorescence mechanisms, upconversion photoluminescence, quantum confinement effect, defective state emission, and the role of surface functionalization.

5.1 Absorption Properties

5.1.1 Well-Defined Absorption Bands

CQDs demonstrate specific absorbance characteristics with well-defined absorption bands in the visible region at approximately 280 nm and 350 nm, accompanied by a UV tail. Recent studies, attribute the 280 nm absorption to a π-π^* transition of a $C = C$ bond and the 350 nm absorption to an n-π^* transition of the $C = O$ bond. Heterogeneity in absorption band placement is linked to factors like raw precursor and synthesis methods.

5.1.2 Tailoring Absorption Properties

Doping with heteroatoms (B, N, F, P, S) and surface modification offer a sophisticated approach to alter CQDs' absorption properties. This strategic modulation enables precise tuning of band gap, electrical structure, and optical properties, broadening potential applications. Surface passivation, utilizing substances like thiols and spiropyrans, not only protects CQDs but enhances optical activity, extending fluorescence into the near-infrared range. Specialized passivation further extends absorbance to longer wavelengths (350–550 nm).

5.1.3 Photocatalytic and Photovoltaic Applications

CQDs exhibit significant absorption coefficients and broad spectra, making them appealing for photocatalytic and photovoltaic applications. The absorption spectrum typically peaks in the UV region (230–320 nm), enhancing photon harvesting, especially in the short-wavelength region. The interplay between heteroatom doping, surface modification, and inherent absorption features forms a comprehensive toolset for modifying CQDs' absorption properties.

5.2 *Fluorescence Mechanisms*

Fluorescence in CQDs is a complex phenomenon driven by distinct mechanisms originating from specific structural features. Understanding these fluorescence mechanisms is pivotal for harnessing the optical properties of CQDs in various applications [19].

5.2.1 Band Gap Transitions in Conjugated π-Domains

Fluorescence initiation in CQDs is intricately linked to band gap transitions occurring within conjugated π-domains. This process unfolds when the infinite π-network of a graphene sheet transforms into a finite structure. The creation of sp2 islands, achieved through the reduction of graphene oxide, is a critical step in facilitating fluorescence emission. Maintaining a single-layer graphene structure with non-connected π-domains is essential to prevent quenching between Islands. The dimensions of these sp2 islands significantly influence the band gap energy of CQDs. The recombination of electron–hole pairs within these sp2 domains results in fluorescence emission. Elucidating the fluorescence mechanism associated with band gap transitions provides valuable insights for fine-tuning the optical properties of CQDs for specific applications.

5.2.2 Surface Flaws and Surface Passivation

Fluorescence emission in Carbon Quantum Dots (CQDs) is a multifaceted phenomenon influenced by intricate processes, particularly the formation of surface flaws during oxidation and subsequent passivation. A nuanced understanding of these mechanisms is crucial for tailoring and optimizing the optical characteristics of CQDs for diverse applications. Surface flaws, a product of the oxidation process, play a pivotal role in the fluorescence emission of CQDs. These flaws act as energy traps, contributing to the fluorescence phenomenon. The degree of oxidation directly influences the number of surface flaws and, consequently, impacts the energy levels associated with fluorescence. This insight underscores the significance of controlling the oxidation process to modulate the fluorescence properties of CQDs. Surface passivation emerges as a strategic approach to manipulate and augment the fluorescence characteristics of CQDs. Substances like thiols and spiropyrans, when employed for passivation, not only contribute to enhancing the stability of CQDs by forming protective layers but also intensify the fluorescence emission. The passivation process mitigates the impact of surface flaws, promoting a more controlled and robust fluorescence response.

5.3 Upconversion Photoluminescence

Upconversion photoluminescence in Carbon Quantum Dots (CQDs) unfolds as a captivating process characterized by the sequential absorption of low-energy photons, culminating in the emission of high-energy luminescence. This phenomenon bears distinctive advantages, particularly in the realm of in-vivo monitoring, owing to its attributes of high spatial resolution, low photon toxicity, and deep tissue penetration. The essence of upconversion photoluminescence lies in the stepwise absorption of low-energy photons, a unique feature of CQDs. This process, intricately orchestrated within the quantum confines of these nanoscale structures, leads to the generation of luminescence with higher energy levels. Understanding the dynamics of this successive photon absorption provides a foundation for harnessing the full potential of upconversion in CQDs. The distinctive attributes of upconversion photoluminescence position CQDs as promising candidates for in-vivo monitoring applications. The high spatial resolution ensures precise imaging, while the low photon toxicity safeguards the biological milieu. Additionally, the capability for deep tissue penetration enhances the applicability of CQDs in biological studies, diagnostics, and therapeutic monitoring. A noteworthy characteristic supporting the evidence of upconversion in CQDs is the quadratic relationship between power and excitation laser intensity. This observation signifies two-photon absorption, a phenomenon that further enhances the efficiency of the upconversion process. The quadratic relationship serves as a distinctive fingerprint of the underlying mechanisms driving upconversion photoluminescence in CQDs. The profound understanding of upconversion processes in CQDs holds paramount importance for their applications in biological domains. From bio-imaging to therapeutic interventions, the ability to exploit upconversion photoluminescence opens avenues for groundbreaking advancements. Researchers and practitioners can leverage this knowledge to develop innovative strategies for non-invasive monitoring and targeted treatments.

In essence, the exploration of upconversion photoluminescence in CQDs not only unravels the fascinating physics governing their optical behavior but also paves the way for transformative applications in the biological landscape. This comprehension serves as a cornerstone for scientists and researchers venturing into the dynamic intersection of nanotechnology and biomedicine.

5.4 *Quantum Confinement Effect*

Carbon Quantum Dots (CQDs), with their roots embedded in quantum-sized graphite shards, unveil a mesmerizing interplay between luminescence and cluster size. This intricate relationship shapes the optical properties of CQDs and contributes to their potent emission capabilities. In this section, we delve into the quantum origins of CQDs, exploring the fascinating connection between luminescence and the size of these nanoscale clusters.

At the heart of CQDs lies their origin from quantum-sized graphite shards. These shards, characterized by their nanoscale dimensions, serve as the foundational building blocks for the powerful emission exhibited by CQDs. Understanding the quantum essence of these graphite shards provides insights into the unique optical behavior displayed by CQDs.

The journey into the quantum origins of CQDs involves meticulous purification processes and theoretical calculations. Researchers, through systematic purification techniques, isolate CQDs of varying sizes. The subsequent theoretical calculations shed light on the energy landscape of these quantum-sized clusters. Notably, as the fragment size increases, there is a discernible reduction in the energy gap a phenomenon in harmony with the principles of visible light emission [20] (Fig. 6).

Contrary to conventional expectations, the intense photoluminescence observed in CQDs is not solely attributed to the carbon–oxygen surface. Instead, it finds its origins in the quantum-sized graphite structure inherent to these nanomaterials. The interplay of quantum dimensions and luminescence unveils a captivating synergy, where the size of graphite clusters becomes a determining factor in the optical prowess of CQDs. The correlation between luminescence and cluster size, substantiated by theoretical frameworks and experimental validations, marks a significant stride in the understanding of CQDs. Bridging the realms of theory and experimentation, researchers unravel the nuances of quantum phenomena governing the emission characteristics of CQDs. This synthesis of knowledge opens avenues for tailoring CQDs with specific luminescent properties through controlled manipulation of their quantum dimensions. The quantum origins of CQDs underscore their intricate relationship with luminescence, where the quantum-sized graphite structure emerges as a key player in shaping their optical behavior. This exploration not only enriches our theoretical understanding but also charts a course for precision engineering of CQDs for diverse applications in sensing, imaging, and beyond.

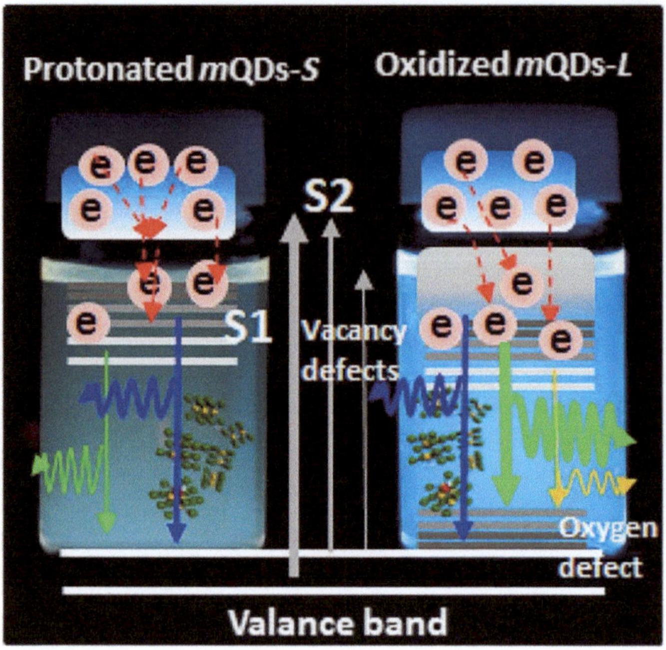

Fig. 6 Schematic representation for quantum confinement effect in CQDs [21]

5.5 Defective State Emission

Luminescence in CQDs correlates with carboxylate ligands on their surface. Green luminescence derives from unique edge states and C = O functional groups. Surface passivation techniques influence the interaction between emission centres and traps, governing optical properties. These insights contribute to a nuanced understanding of the photophysical behavior of CQDs, laying the groundwork for diverse applications in scientific domains.

5.6 Surface Functionalization

Surface functionalization proves to be a versatile method for tailoring CQDs' properties, impacting electronic structures, mechanical properties, and optical characteristics.

5.6.1 Passivation Techniques

Surface passivation techniques, such as employing thin layers of passivating materials like thiols and spiropyrans, enhance photostability and quantum yields. Diverse passivation methods, including ionic liquid (IL) passivation and inorganic shell passivation, offer tailored quantum yields and biocompatibility.

5.6.2 Unconventional Enhancement Effects

Unconventional enhancement effects, exemplified by Ag^{+} ions enhancing photoluminescence characteristics, provide innovative avenues. The conversion of Ag^{+} ions to Ag0 clusters on C-dot surfaces increases fluorescence, showcasing unique enhancement mechanisms.

5.6.3 Hybrid Passivation Approaches

Hybrid passivation involving surface doping with inorganic salts and functionalization with PEG1500N molecules yields C-dots with high fluorescence emission quantum yields. The introduction of new emitting surface states through surface functionalization alters excitation and emission characteristics.

5.6.4 Red-Shifted Emission by Surface Oxidation

Electrochemically etching carbon fibers induces red-shifted emission in C-dots due to surface oxidation. The degree of oxidation influences the formation of emissive surface states responsible for photoluminescence. Surface functionalization emerges as a powerful tool for addressing defect sites, modifying optical characteristics, and expanding the applications of C-dots. The versatility of this approach is evident in its impact on mechanical properties and dispersibility, offering new possibilities for C-dot utilization.

6 Solid-State Architectures and Mechanical Properties

Surface functionalization stands as a pivotal technique, not only addressing solubility challenges but also unlocking the potential for tailored solid-state designs, thereby expanding the applications of carbon dots (C-dots). The synthesis of highly luminous C-dots, for instance, highlights the utilization of an organosilane coordinating solvent, yielding an impressive quantum yield of approximately 47%. Surface functionalization plays a crucial role not only in providing dispersibility to C-dots but also in enabling the creation of solid-state designs, expanding the potential applications of C-dots. By utilizing an organosilane coordinating solvent, the synthesis of highly luminous C-dots showcased an impressive quantum yield of approximately 47%. The incorporation of surface methoxysilyl groups further facilitated the direct fabrication of a hybrid fluorescent film or monolith through simple heating, eliminating the need for additional polymers or inorganic substances.

In-depth exploration of the morphology of surface-functionalized C-dots led to the development of a flexible, free-standing, and transparent ionogel. This innovative material was achieved by combining carboxyl-functionalized ionic liquid with organosilane-functionalized C-dots. Notably, the ionogel demonstrated thickness-dependent emission tunability at a single excitation wavelength, highlighting its potential for optoelectronic applications [22].

Beyond addressing solubility challenges, surface functionalization emerges as a critical technique for customizing the mechanical properties of C-dots. The ability to control dispersibility and craft diverse solid-state structures opens up new possibilities for applications spanning solution-based technologies to advanced optoelectronic materials. This section underscores the transformative impact of surface functionalization in shaping the mechanical properties and solid-state designs of C-dots, contributing to their versatility and applicability across various technological domains.

7 Doping Strategies: Tailoring the Electronic Landscape of C-Dots

Doping, the strategic introduction of atomic impurities into carbon dots (C-dots), stands out as a powerful method to finely tune their intrinsic properties, particularly the electronic and optical characteristics. This section explores various doping elements, their impacts, and the emerging field of co-doping for expanded possibilities.

7.1 Elemental Dopants for Shaping C-Dot

Nitrogen Doping: Shaping the Electronic Landscape: Nitrogen, a quintessential dopant in C-dots, brings its five valence electrons and comparable atomic size to the intricate realm of carbon modification. As n-type impurities, nitrogen atoms inject excess electrons, orchestrating an upward shift in the Fermi level. This orchestrated dance leads to profound alterations in optical properties, making nitrogen doping a cornerstone in tailoring C-dot characteristics. The versatility of nitrogen-containing precursors further accentuates the strategic role of nitrogen in this intricate modulation [23].

Phosphorus Doping: Navigating Group 15 Dynamics: Phosphorus, positioned in group 15 of the periodic table, assumes the mantle of an n-type impurity in the realm of C-dots. Its larger stature than carbon allows it to intricately form substitutional defects within carbon clusters, thereby becoming a potent modulator of electronic and optical properties. The utilization of diverse phosphorus-containing precursors, ranging from phosphorous tribromide to phosphoric acid, exemplifies the expansive landscape of possibilities in the phosphorus doping saga.

Boron Doping: Unveiling P-Type Transformations: Boron, with its one less valence electron than carbon, emerges as a transformative force introducing p-type carriers upon assimilation into C-dots. This incorporation sets in motion a cascade of changes in electronic structures and optical properties, offering a distinctive pathway for modulation. The dance of boron within C-dots not only alters their inherent characteristics but opens up a unique avenue for tailoring their behavior in diverse applications.

Sulfur Doping: Illuminating Trap States and Beyond: Sulfur-doped C-dots step into the limelight, heralded by sulfur atoms providing pivotal energy or emissive trap states for the capture of photo excited electrons. The synthesis of these sulfur-infused counterparts, facilitated by precursors like sulfuric acid and sodium hydrosulfide, showcases the expanding repertoire of strategies in crafting C-dots with enhanced functionalities. The allure of sulfur doping lies in its potential to elevate the performance of C-dots, making them luminous contenders in the landscape of nanomaterials.

The C-dot modulation with nitrogen, phosphorus, boron, and sulfur emerge as virtuoso dopants, each adding a unique note to the composition. These doping strategies not only alter the electronic and optical properties but also pave the way for a myriad of applications. As we delve deeper into the intricacies of these elemental dopants, the canvas of possibilities for tailored C-dots unfolds, promising advancements in the ever-evolving realm of nanomaterial science.

7.2 Co-Doping: Expanding Possibilities

Single-element doping to co-doping strategies to enhance the capabilities of luminous materials, aiming to expand the emission spectrum and create new possibilities in optical applications. Co-doping, involving the simultaneous introduction of two or more atoms, provides increased control over light absorption and emission behaviors.

One notable co-doping approach involves the combination of nitrogen and sulfur, referred to as N,S-doped C-dots. The synthesis of these C-dots is achieved through hydrothermal treatment, utilizing citric acid as a carbon precursor and L-cysteine as a source of nitrogen and sulfur. The inclusion of sulfur atoms resulted in a high fluorescence quantum yield (73%) and excitation-independent emission, enhancing the nitrogen-induced characteristics of the C-dots. Additionally, the hydrothermal oxidative synthesis process, known as boron/nitrogen co-doping, produces fluorescent boron/nitrogen co-doped C-dots (B,N-doped C-dots) with bright green photoluminescence suitable for catalysis.

Co-doping with phosphorus and nitrogen, known as P,N-doped C-dots, was achieved using a combustion flame technique. These C-dots demonstrated exceptional performance in the oxygen reduction reaction (ORR), indicating their potential in energy-related applications.

Furthermore, co-doping with nitrogen, sulfur, and phosphorus was explored for the environmentally friendly synthesis of water-soluble nitrogen, sulfur, and phosphorus co-doped carbon dots (N,S,P-doped C-dots). This was accomplished through hydrothermal treatment of cucumber juice saturated with nitrogen, sulfur, and phosphorus, allowing for the modification of photoluminescence size and wavelength. These co-doping strategies offer versatile methods for tailoring the properties of C-dots for various applications, including catalysis and energy research.

7.3 Other Atom Doping

Beyond the mentioned elements, Si, Ge, Se, Gd, and Tb have been employed as dopants, showcasing their diverse impacts on C-dot properties. For instance, Se-doped C-dots exhibited yellow luminescence and found applications in bio-imaging. The doping emerges as a versatile tool for tailoring the electronic structure of C-dots, offering a spectrum of possibilities from single-element modulation to synergistic co-doping strategies. These approaches not only expand the optical and electronic landscape of C-dots but also open avenues for diverse applications ranging from sensing to bio-imaging [24].

8 Illuminating C-Dots Analytical Potential Application

8.1 Sensors and Biosensors in Water Treatment

Surface functionalization stands out as a critical technology shaping the characteristics of carbon dots (C-dots), offering tailoring capabilities for diverse sensing applications. The ability to modify surface functional groups renders C-dots adaptable receptors, enabling the detection of various ions and chemicals with high specificity.

To develop highly sensitive and selective detectors for heavy metal ions, surface functionalization of C-dots is a crucial step. For instance, the functionalization of C-dots with NH2-(polyethylene glycol) (PEG200) and N-acetyl-L-cysteine (NAC) was employed to create a biosensor for Hg2 + , demonstrating sensitivity to micro molar concentrations of Hg2 + and Cu^{2+}. Improved selectivity for Cu^{2+} particles was achieved by utilizing C-dots functionalized with branched poly(ethylenimine) (BPEI), showcasing the potential for targeted detection of heavy metal ions.

Luminescent C-dots, modified with [Zr(H2O)2EDTA], exhibited versatility in fluoride ion detection by incorporating b-cyclodextrin and calix[4]arene-25,26,27,28-tetrol. These components served as selective elements for fluoride ion detectors, showcasing the adaptability of C-dot-based sensors beyond heavy metal ion detection and their potential for various applications.

In non-enzymatic glucose sensing, fluorescent C-dots treated with boronic acid demonstrated precise assembly and simultaneous fluorescence quenching in the presence of glucose, enabling measurements in the range of 9–900 mM. Additionally, surface-modified C-dots exhibited excellent performance in detecting a wide variety of substances, including m-phenylenediamine, methyl parathion, nifedipine, dipicolinic acid, water, quercetin, and ascorbic acid. These examples underscore the versatility and wide-ranging applications of C-dot-based sensors in analytical tasks.

Utilizing a unique amidation process to passivate C-dots with cysteamine facilitated the identification of nanoparticles. This strategy enhanced the photoluminescent capabilities of gold nanoparticles, allowing for the development of a photoluminescent probe selectively detecting gold nanoparticles. Furthermore, amine-terminated groups on C-dots achieved selective detection of citrated silver nanoparticles through the amidatorization process.

The adaptability of surface-functionalized C-dots as versatile tools for sensing applications which holds promising potential due to their ease of functionalization, versatility, and ability to detect a wide variety of chemicals. These attributes make them compelling candidates for various analytical tasks, and the future holds intriguing possibilities for C dot based sensors, including applications in biological diagnostics and environmental monitoring.

8.2 *Biomedical and Biotechnological Applications*

Surface functionalization is an important technology that allows carbon dots (C-dots) to have properties that may be adjusted for various sensing uses. Their versatility as receptors for a wide range of ions and chemicals is due to the fact that C-dots can have their surface functional groups changed. Functionalizing C-dots opens up opportunities for the development of highly sensitive and selective detectors, particularly for heavy metal ions. For example, the functionalization of C-dots with N-acetyl-L-cysteine (NAC) and NH_2-(polyethylene glycol) (PEG200) resulted in the successful creation of a Hg^{2+} biosensor, demonstrating sensitivity to micro molar amounts of Hg^{2+} and Cu^{2+}. Further functionalization with branched poly(ethylenimine) (BPEI) improved selectivity for Cu^{2+}. In another study, C-dots functionalized with a quinoline derivative were employed to design a Zn^{2+} nanosensor system, showcasing outstanding selectivity and rapid response in detecting Zn^{2+}. Fluoride ion detection saw advancements through the use of luminescent C-dots modified with [Zr(H2O)2EDTA] and functionalized C-dots incorporating b-cyclodextrin and calix[4]arene-25,26,27,28-tetrol. This innovative technique enhanced sensitivity and precision in fluoride ion detection.

These discoveries expand the possibilities for the application of functionalized C-dots in the development of sophisticated sensors for various analytical and environmental purposes. C-dot-based sensors demonstrate the capability to detect a broad range of targets, including heavy metal ions. For instance, luminous C-dots treated with boronic acid were developed for non-enzymatic glucose sensing, achieving measurements in the range of 9–900 mM. Surface modification techniques successfully isolated various substances, such as water, nifedipine, methyl parathion, dipicolinic acid, ascorbic acid, m-phenylenediamine, and quercetin. The unique amidation process for passivated C-dots with cysteamine facilitated the identification of nanoparticles. Additionally, functionalized C-dots through amidation were employed for the selective detection of citrated silver nanoparticles.

In particular, surface-functionalized C-dots have enormous potential as a sensing tool. Some examples of such applications include the identification of nanoparticles, focused chemical analysis, and the detection of heavy metal ions. From biomedical diagnostics to environmental monitoring, C-dots are a great fit for a variety of analytical problems due to their adaptability and ease of functionalization.

8.3 Cancer Therapy and Drug Delivery

See Fig. 7.

The photoluminescence, biocompatibility, and non-toxic nature of carbon quantum dots (CQDs) position them as promising candidates for serving as versatile

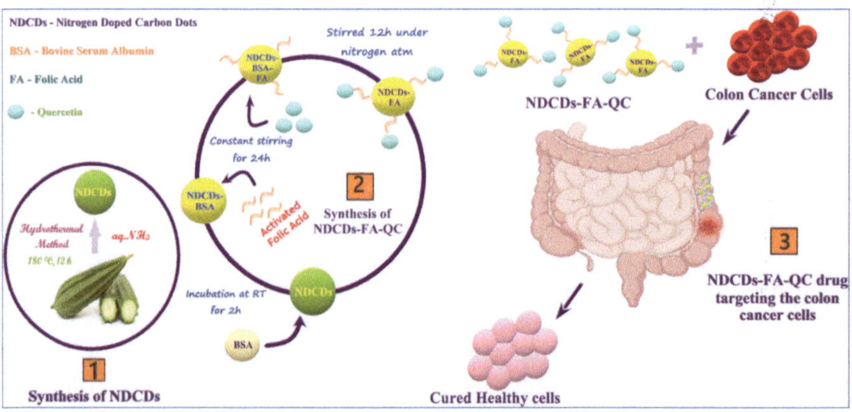

Fig. 7 Targeted drug delivery for cancer therapy using CQDs [25]

carriers for gene and drug delivery. Their small size and large surface area allow for rapid absorption by cells, facilitating effective medication delivery without compromising efficacy. An illustrative example involves the development of a composite, specifically arginine-glycine-aspartic acid-GQDs, loaded with the anticancer drug doxorubicin. This composite demonstrated potent cytotoxic effects when applied to U251 glioma cells, showcasing its utility for targeted imaging and drug delivery.

The interaction between doxorubicin and GQDs, facilitated by the formation of hydrogen bonds, influenced pH-dependent regulated release, enhancing therapeutic efficacy significantly. In another study, researchers explored the potential of a smart stimuli-response drug delivery system using CQDs coated with alginate beads and a crude garlic extract. This system exhibited pH-dependent controlled release of medication in response to pathogens present in the target environment. The result was an increased therapeutic efficacy of drug delivery, highlighting the potential of CQDs in developing advanced and responsive drug delivery platforms. Additionally, GQD-coupled folic acid showed efficient cellular absorption, medication release, and significant cytotoxicity against HeLa cells with minimal effect on non-target cells.

Among other advancements, a PEGylated nanographene oxide has shown great promise in the treatment of colon cancer when administered with the insoluble aromatic drug SN38. By forming an assembly conjugated with PEI-pDNA, CQDs linked to Au nanoparticles demonstrated the critical transfection efficiency. The assembly was able to deliver pDNA into the nuclei of cells by entering cells that contained cytoplasmic QDs. Also it is explored for potential usage in treating pancreatic cancer were biodegradable charged polyester vectors including GQDs [26]. These vectors were doxorubicin nanocarriers and interfering ribonucleic acid nanocarriers; they were also physiologically stable, efficiently down-regulated K-ras, and caused photothermal cell death when exposed to laser light. The ability of the positively charged CQDs to efficiently transfer genes with minimum toxicity when attached to plasmid DNA was demonstrated, highlighting their potential for usage in clinical settings.

8.4 Imaging and Bio-Imaging

Live cell bio-imaging has recently emerged as an indispensable tool for investigating biological mechanisms, providing insights into cellular processes' functions and dynamics. Bio-imaging allows the direct visualization of biological processes in real-time, aiding in understanding the three-dimensional structure of specimens without the need for physical interaction. Considering the optical properties of carbon-based quantum dots is crucial for their practical application. Quantum dots derived from carbon exhibit bright fluorescence, low cytotoxicity, and good biocompatibility, making them ideal for Bio-imaging. Graphene quantum dots (GQDs) and carbon quantum dots (CQDs) have significantly advanced Bio-imaging, enabling the imaging of cells from various cell lines, including those derived from Escherichia

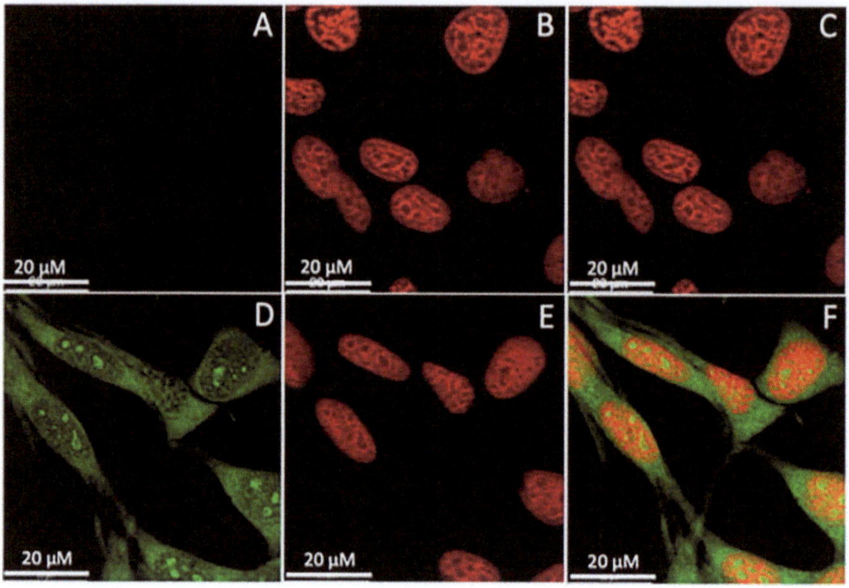

Fig. 8 Bio-imaging of Cancer cell using CQDs [28]

coli (E. coli), HepG2, A549, HeLa, Ehrlich ascites carcinoma and fibroblasts from the National Institutes of Health (3T3). The remarkable qualities of carbon quantum dots, such as low toxicity, eco-friendliness, and emission of visible to near-infrared (NIR) light, have propelled their application in Bio-imaging. For instance, injecting these molecules into mice facilitated optical imaging using CQDs in conjunction with wheat straw. Despite these advancements, challenges persist in Bio-imaging applications involving CQDs modified with a Cu(II) complex and GQDs with varied emission wavelengths, warranting further research [27] (Fig. 8).

The challenges persist in the development of graphene quantum dots (GQDs), prompting an urgent need for further exploration. The recent revelation that the quantum yield of GQDs is lower than that of conventional semiconductor quantum dots underscores the pressing necessity for their improvement.

GQDs with robust emission in the red-near infrared spectrum are envisioned as potential nano-probes for bio-imaging applications. Leveraging their optical and radioactive properties, GQDs hold promise in the realms of imaging and therapy, addressing prevalent issues in industrial and environmental sectors. Notably, GQD-coupled antibodies or peptides exhibit potential in cancer imaging, yet a deeper understanding of their role in imaging cancer targets necessitates additional research.

The challenges associated with GQDs extend to the need for excitation of a substantial number of photons, addressing brain gene treatment, exploring novel applications for neurobehavior, and overcoming the blood–brain barrier. Additionally, a comprehensive examination into the cytotoxic effects of GQDs with varying sizes, surface coatings, and morphologies is imperative.

The development of GQDs on a large scale, supplemented with polyethyleneimine or (3-carboxyl) phenyl bromide phosphine, showcases an environmentally friendly and cost-effective approach. The resulting composites exhibit selectivity for imaging mitochondria or cell nuclei, possess notable optical characteristics, and demonstrate minimal cytotoxicity. This underscores the indispensability of GQDs for bio-imaging applications, whether conducted in vitro or in vivo.

8.5 *Anti-Microbial Activity*

The interaction of carbon quantum dots (CQDs) with various viruses has demonstrated their potential in inhibiting infection spread. Specifically, functionalized CQDs with amino groups or boronic acid have shown effectiveness against herpes simplex virus type 1. Utilizing a hydrothermal carbonization technique with 4-aminophenyl boronic acid hydrochloride successfully produced CQDs capable of efficiently suppressing the herpes simplex virus type 1.5.

Phenylboronic acid-based CQDs, produced in a similar manner, exhibited no antiviral activity at the applied concentration, highlighting the specificity of antiviral actions. Furthermore, these CQDs hold promise in combating pathogenic human illnesses, including corona viruses, through mechanisms involving the inhibition of human coronavirus-229E (HCoV-229E) entry.

Beyond antiviral applications, CQDs showcase potent antibacterial properties against various bacteria, including Staphylococcus aureus, E. coli, and Pseudomonas aeruginosa. They contribute significantly to imaging microbial structures, aiding in gram-type determination, vitality evaluation, and bio-film imaging. The layers of CQDs serve as carriers for traditional disinfection agents, simplifying investigations of bacterial cells.

The specificity of CQDs in bacterial interaction is highlighted by their selective connections with Gram-negative bacteria. The mechanism involves adsorption on the bacterial surface, leading to increased fluorescence emission, breakdown of the cell wall, and bacterial inactivation. The simplicity, speed, and ease of execution distinguish this differentiation method.

This article addresses two key aspects: the technology of manufacturing carbon dots and their applications in selectively imaging and eradicating Gram-positive bacteria. Glycerol, an economical carbon source, enhances water dispersibility in the final CQD product. The organosilane molecule Si-QAC, acting as both a carbon supply and surface passivation agent, exhibits potent antibacterial activity due to electrostatic and hydrophobic interactions with bacterial surfaces. The quaternary ammonium group and long hydrocarbon chain in Si-QAC contribute significantly to its antibacterial efficacy.

9 Conclusion

In the exploration of "Various Properties of Green-Synthesized Carbon Quantum Dots", our journey has unfolded a compelling narrative of innovation and sustainability. Green synthesis has emerged as a powerful avenue for the fabrication of carbon quantum dots (CQDs), offering not only enhanced properties but also a sustainable and eco-friendly approach. As we conclude this chapter, several key themes and insights underscore the multifaceted nature of green-synthesized CQDs.

Eco-Friendly Synthesis: A Sustainable Odyssey: The utilization of green synthesis routes has not only enriched the properties of CQDs but has also positioned them as champions of sustainability. From biomass precursors to plant extracts, the chapters of green synthesis narrate a story of environmental consciousness. This approach not only aligns with contemporary ecological concerns but also ensures the scalability and accessibility of CQDs.

Photophysical Marvels: The optical properties of green-synthesized CQDs have captivated our attention, revealing a kaleidoscope of possibilities. Their tunable fluorescence, influenced by synthesis parameters and precursor choices, offers a palette for applications in Bio-imaging, sensors, and optoelectronics. The green synthesis approach adds a unique layer to their photophysical prowess, often resulting in enhanced quantum yields and stabilities.

Tailored Surface Chemistry: Green synthesis has not only crafted CQDs with remarkable optical properties but has also endowed them with tailored surface chemistry. This facet opens avenues for functionalization and customization, making them versatile tools in diverse applications. Surface functionalization, achieved through green methodologies, enhances their compatibility with biological systems, paving the way for bio-applications and drug delivery.

Multifunctional Applications: Green-synthesized CQDs, with their rich repertoire of properties, find applications across diverse domains. From sensing applications, where they exhibit prowess in detecting ions and molecules, to energy-related fields, green CQDs prove their mettle. The intersection of their green origin and multifunctional applications exemplifies a harmonious synergy between innovation and environmental responsibility.

Future Frontiers: As we draw the curtains on this chapter, the future of green-synthesized CQDs appears both promising and expansive. The ongoing exploration of synthesis methodologies, coupled with a deepening understanding of their properties, sets the stage for breakthroughs in various scientific and technological domains. The fusion of green principles with quantum dot research opens doors to a sustainable future where advanced materials coexist with environmental consciousness. In this concluding chapter, we find ourselves at the intersection of innovation, sustainability, and diverse applications. The properties of green-synthesized CQDs echo a harmonious blend of scientific advancement and environmental responsibility. As researchers continue to delve into the uncharted territories of green synthesis and CQD applications, the chapters yet to be written hold the promise of new revelations and transformative discoveries. The journey of green-synthesized CQDs is a testament to the synergy between nature-inspired methodologies and cutting-edge materials science, forging a path towards a greener and technologically enriched future.

References

1. Nekoueian, K., Amiri, M., Sillanpää, M., Marken, F., Boukherroub, R., & Szunerits, S. (2019). Carbon-based quantum particles: An electroanalytical and biomedical perspective. *Chemical Society Reviews, 48*(15), 4281–4316.
2. Azam, N., Najabat Ali, M., & Javaid Khan, T. (2021). Carbon quantum dots for biomedical applications: Review and analysis. *Frontiers in Materials, 8*, 700403.
3. Alaghmandfard, A., Sedighi, O., Rezaei, N. T., Abedini, A. A., Khachatourian, A. M., Toprak, M. S., & Seifalian, A. (2021). Recent advances in the modification of carbon-based quantum dots for biomedical applications. *Materials Science and Engineering: C, 120*, 111756.
4. Wegner, K. D., & Hildebrandt, N. (2015). Quantum dots: Bright and versatile in vitro and in vivo fluorescence imaging biosensors. *Chemical Society Reviews, 44*(14), 4792–4834.
5. Safranko, S., Goman, D., Stankovic, A., Medvidović-Kosanovic, M., Moslavac, T., Jerkovic, I., & Jokić, S. (2021). An Overview of the recent developments in carbon quantum dots—Promising nanomaterials for metal ion detection and (Bio) molecule sensing. *Chemosensors, 9*(6), 138.
6. Sarkar, S.K., 2023. Quantum Dots as Optical Materials: Small Wonders and Endless Frontiers. In *Handbook of Materials Science, Volume 1: Optical Materials* (pp. 545–596). Singapore: Springer Nature Singapore.
7. Han, Y., Liccardo, L., Moretti, E., Zhao, H., & Vomiero, A. (2022). Synthesis, optical properties and applications of red/near-infrared carbon dots. *Journal of Materials Chemistry C*.
8. Kumar, D. S., Kumar, B. J., & Mahesh, H. M. (2018). Quantum nanostructures (QDs): an overview. *Synthesis of inorganic nanomaterials*, pp. 59–88.
9. Tohamy, H. A. S., El-Sakhawy, M., & Kamel, S. (2023). Microwave-assisted synthesis of amphoteric fluorescence carbon quantum dots and their chromium adsorption from aqueous solution. *Scientific Reports, 13*(1), 11306.
10. Ndlwana, L., Raleie, N., Dimpe, K. M., Ogutu, H. F., Oseghe, E. O., Motsa, M. M., Msagati, T. A., & Mamba, B. B. (2021). Sustainable hydrothermal and solvothermal synthesis of advanced carbon materials in multidimensional applications: A review. *Materials, 14*(17), 5094.
11. Nasilowski, M., Mahler, B., Lhuillier, E., Ithurria, S., & Dubertret, B. (2016). Two-dimensional colloidal nanocrystals. *Chemical reviews, 116*(18), 10934–10982.
12. Molaei, M. J. (2019). Carbon quantum dots and their biomedical and therapeutic applications: A review. *RSC Advances, 9*(12), 6460–6481.
13. El-Shabasy, R. M., Farouk Elsadek, M., Mohamed Ahmed, B., FawzyFarahat, M., Mosleh, K. N., & Taher, M. M. (2021). Recent developments in carbon quantum dots: Properties, fabrication techniques, and bio-applications. *Processes, 9*(2), 388.
14. Wu, M., Zhan, J., Geng, B., He, P., Wu, K., Wang, L., Xu, G., Li, Z., Yin, L., & Pan, D. (2017). Scalable synthesis of organic-soluble carbon quantum dots: Superior optical properties in solvents, solids, and LEDs. *Nanoscale, 9*(35), 13195–13202.
15. Yadav, P. K., Chandra, S., Kumar, V., Kumar, D., & Hasan, S. H. (2023). Carbon quantum dots: Synthesis, structure, properties, and catalytic applications for organic synthesis. *Catalysts, 13*(2), 422.
16. Yuan, F., Wang, Z., Li, X., Li, Y., Tan, Z. A., Fan, L., & Yang, S. (2017). Bright multicolor bandgap fluorescent carbon quantum dots for electroluminescent light-emitting diodes. *Advanced Materials, 29*(3), 1604436.
17. Li, G., Xu, J., & Xu, K. (2023). Physiological functions of carbon dots and their applications in agriculture: A review. *Nanomaterials, 13*(19), 2684.

18. Nguyen, K. G., Baragau, I. A., Gromicova, R., Nicolaev, A., Thomson, S. A., Rennie, A., Power, N. P., Sajjad, M. T., & Kellici, S. (2022). Investigating the effect of N-doping on carbon quantum dots structure, optical properties and metal ion screening. *Scientific Reports, 12*(1), 13806.
19. Wang, C., Yang, M., Shi, H., Yao, Z., Liu, E., Hu, X., Guo, P., Xue, W., & Fan, J. (2022). Carbon quantum dots prepared by pyrolysis: Investigation of the luminescence mechanism and application as fluorescent probes. *Dyes and Pigments, 204*, 110431.
20. Fatima, I., Rahdar, A., Sargazi, S., Barani, M., Hassanisaadi, M., & Thakur, V. K. (2021). Quantum dots: Synthesis, antibody conjugation, and HER2-receptor targeting for breast cancer therapy. *Journal of Functional Biomaterials, 12*(4), 75.
21. Kim, B. H., Yang, J. Y., Park, K. H., Lee, D., & Song, S. H. (2023). Competitive effects of oxidation and quantum confinement on modulation of the photophysical properties of metallic-phase tungsten dichalcogenide quantum dots. *Nanomaterials, 13*(14), 2075.
22. Mishra, R. K., Chianella, I., Goel, S., & Nezhad, H. Y. Carbon quantum dots and polymer nanocomposites: Synthesis, properties, and versatile applications.
23. Arul, V., Radhakrishnan, K., Sampathkumar, N., Vinoth Kumar, J., Abirami, N., & Inbaraj, B. S. (2023). Detoxification of toxic organic dye by heteroatom-doped fluorescent carbon dots prepared by green hydrothermal method using Garcinia mangostana extract. *Agronomy, 13*(1), 205.
24. Li, Q., Noffke, B. W., Liu, Y., & Li, L. S. (2015). Understanding fundamental processes in carbon materials with well-defined colloidal graphene quantum dots. *Current Opinion in Colloid & Interface Science, 20*(5–6), 346–353.
25. Chandrasekaran, P., Sivaraman, G., Rasala, S., Sethuraman, M. G., Kotla, N. G., & Rochev, Y. (2022). Quercetin conjugated fluorescent nitrogen-doped carbon dots for targeted cancer therapy application. *Soft Matter, 18*(30), 5645–5653.
26. Thakur, M., Mewada, A., Pandey, S., Bhori, M., Singh, K., Sharon, M., & Sharon, M. (2016). Milk-derived multi-fluorescent graphene quantum dot-based cancer theranostic system. *Materials Science and Engineering: C, 67*, 468–477.
27. Guo, L., Li, L., Liu, M., Wan, Q., Tian, J., Huang, Q., Wen, Y., Liang, S., Zhang, X., & Wei, Y. (2018). Bottom-up preparation of nitrogen doped carbon quantum dots with green emission under microwave-assisted hydrothermal treatment and their biological imaging. *Materials Science and Engineering: C, 84*, 60–66.
28. Şenel, B., Demir, N., Büyükköroğlu, G., & Yıldız, M. (2019). Graphene quantum dots: Synthesis, characterization, cell viability, genotoxicity for biomedical applications. *Saudi Pharmaceutical Journal, 27*(6), 846–858.
29. Arul, V., Chandrasekaran, P., Sivaraman, G., & Sethuraman, M. G. (2023). Biogenic preparation of undoped and heteroatoms doped carbon dots: Effect of heteroatoms doping in fluorescence, catalytic ability and multicolour in-vitro bio-imaging applications-a comparative study. *Materials Research Bulletin, 162*, 112204.
30. Yang, Z., Xu, T., Li, H., She, M., Chen, J., Wang, Z., Zhang, S., & Li, J. (2023). Zero-dimensional carbon nanomaterials for fluorescent sensing and imaging. *Chemical Reviews, 123*(18), 11047–11136.
31. Pourmadadi, M., Rahmani, E., Rajabzadeh-Khosroshahi, M., Samadi, A., Behzadmehr, R., Rahdar, A., & Ferreira, L. F. R. (2023). Properties and application of carbon quantum dots (CQDs) in biosensors for disease detection: a comprehensive review. *Journal of Drug Delivery Science and Technology*, 104156.

Green Synthesis of Carbon Quantum Dots and Their Environmental Application for the Detection of Heavy Metal Ions

Sutha Rahupathy, Monisha Sivanandhan, and Amutha Parasuraman

Abstract The increased spotlight on Sustainable Chemistry has progressively drawn attention to the greener synthesis of Carbon Dots. Among the endless production pathways, the evolution of green synthetic methods is uplifted by the abundance availability of natural precursors. This greener approach is attempted in the production of Carbon Dots (CDs) that flourished in the past two decades. Carbon Dots are crowned in the carbon family for their astonishing features. Owing to their tuneable optical properties, large surface area to volume ratio, and effective photon harvesting in the short-wavelength region, they serve as excellent choices for heavy metal detection. Industrial operations over the centuries tremendously increased the heavy metal ion concentration. The most prominent commercial heavy metals disrupting biological functions are mercury, chromium, cadmium, lead, and arsenic. Despite the high consciousness of heavy metal risks, the frequency of poisoning continues, demanding proactive and efficient management. Hence, designing viable, selective, and dependable sensors beneficial over conventional techniques that request complex instrumentation and sample preparation is vital. The emergence of fluorescent Carbon Dots with exceptional photoluminescent properties and simplicity of functionalization makes them a valuable tool for innocuous metal detection by fluorescence quenching. Certain suggested mechanisms for metal ion sensing by CDs include Complexation, Inner Filter Effect (IFE), Light induced Electron Transfer (LET), ion binding, and aggregation. Among these, the most widely recognized is the static fluorescence inhibition, which implies the development of a non-nuclear interaction between the CDs and the metal cation, thereby enabling the detection of metals. This chapter presents an outlook on the role of green synthesized CDs in sensing heavy metal ions shedding light on future studies.

Keywords Carbon dots · Optical properties · Metal ion sensing · Green precursors · Fluorescence quenching

---

S. Rahupathy · M. Sivanandhan · A. Parasuraman (✉)
Department of Chemistry, PSGR Krishnammal College for Women, Coimbatore 641004, India
e-mail: amuchem@gmail.com

V. Kumar et al. (eds.), *Green Carbon Quantum Dots*, Engineering Materials,
https://doi.org/10.1007/978-981-97-6203-3_5

1 Introduction

In the wake of scientific enthusiasm, experts are fast approaching greener research strategies to gain control over the pathetic irony that environmental studies are frequently contributing to further environmental issues due to the chemicals employed during the research [1]. To patch up and restore the collapsed ecosystem, Green Chemistry operates at a molecular level by adopting chemical innovation to reach the ecological and economical goals concurrently [2]. Several breakthroughs are being made by the Green Revolution, which creates a significant impact globally on the challenging issues they are addressing. Due to this, the field of Green Chemistry has spread the waves of confidence to foresee potential risks and preserve a safe environment for the succeeding generations. This thriving green age operates in tandem with the existing carbon era to bring about progress in scientific innovations. When it comes to the nuts and bolts of constructing basic human necessities, Carbon Chemistry takes the lead as it is plentiful in the atmosphere. The central role of elemental carbon in science and technology is known far and wide and need not be emphasized. This role is driven by carbon's capability to develop bonds to itself and almost every element, in practically limitless variations [3]. The collaborative ideas of Carbon Chemistry with Green Chemistry ignited the spark for the escalation of 'Green Carbon Dots'. They are one of the currently evolving disciplines with immense potential for assisting both science and society [4]. Carbon dots (CDs) are a novel subclass of carbon nanomaterials with characteristic diameters of size less than 10 nm, featuring a broad range of physiochemical properties [5]. The breathtaking surge of interest in these microscopic particles has fueled application-related research in the fields of material science, chemistry, and biology [6].

1.1 History of Carbon Dots

For generations, carbon nanoparticles have been embodied as fullerenes, graphene, Carbon Nanotubes, Carbon Nanofiber, Carbon Black, Ultrafine Carbon Aggregates (UCA), and so on [7], [8]. It became apparent that carbon nanoparticles and their complexes make headway as advantageous tools for elevating the molecular makeup and properties of materials [9]. Since the turn of this century, manually engineered nanoparticles have evolved under multiple titles. By a twist of fate in 2004, Xiaoyou Xu and his fellow researchers became the first to stumble upon Carbon Dots like nanoparticles. The researchers encountered minute segments of Single-Walled Carbon Nanotubes (SWCNTs) and were captivated by their photoluminescent emission properties, during electrophoretic purification of SWCNTs derived from arc-discharge carbon soot that were previously chemically oxidized with concentrated HNO_3 and extracted with basic water [10]. In the next couple of years, Sun's crew assembled luminous carbon nanoparticles (less than 10 nm) through laser ablation of graphite by subsequently performing chemical oxidation and surface passivation that

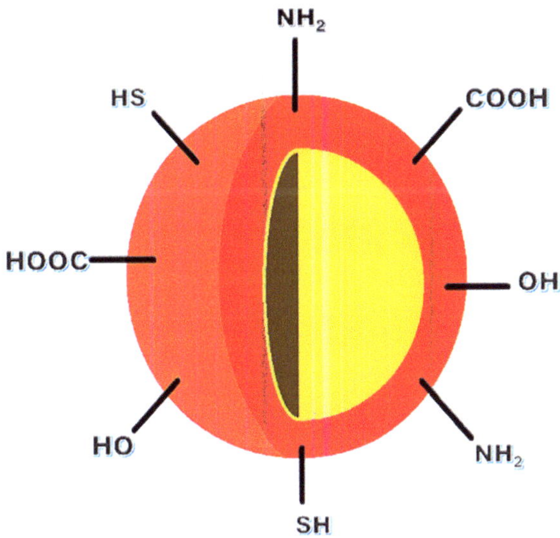

Fig. 1 Carbon dots with surface functional groups

produced bright and excitation-dependent PL emission. They identified it as water-stable fluorescent carbon nanoparticles with an optimistic future and entitled them as 'Carbon Dots'[11]. Within a year, Sun's crew disclosed yet another innovation of hydrophilic CDs passivated with poly-propionyl ethylenimine-co-ethylenimine (PPEI-EI). This piece of work highlighted their potential for cell imaging in human breast cancer MCF-7 cells, most likely by endocytosis, using the two-photon luminescence microscopy of the CDs [12]. This novel discovery was indeed an important milestone in the recognition of CDs. Developments in the last decade, about CDs were witnessed to have intriguing possibilities in metal sensing, membrane separation, pollutant adsorption, photocatalytic degradation, and antimicrobial activities. Even after substantial research on CDs, accurately forecasting their physiochemical properties continues to be a challenging task, partly because of their composite chemical structure. Despite plenty of challenges and constraints along the way, CDs are anticipated to unlock even more appealing possibilities for nano-green technologies [13] (Fig. 1).

1.2 *An Alternative to Semiconductor Quantum Dots*

Traditional Quantum Dots (TQDs) technology emerged in the 1970s to address the global energy deficit. TQDs can be visualized as a link between discrete energy levels of tiny particles and crystalline bulk semiconductors, typically engineered as an inorganic crystalline core, encased with different ligands on their surface, as a colloidal suspension. They are characterized as semiconducting, quasi-crystalline, three-dimensional nanostructures with confinement of the holes in the valence band

(VB), the excitons (electron–hole pairs), and the electrons in the conduction band (CB) [14]. They are often fabricated from II-VI (CdS, CdSe, and CdTe), IV-VI (PbS and PbSe), and III-V (InP and InAs) group semiconductors. However, little attention has been given to TQDs since their synthesis demands extremely specialized equipment but ends up producing a very low yield. Furthermore, the presence of potentially dangerous heavy metals in TQDs has its own disadvantages in biological and related biomedical studies [15]. The coordination interaction between the surface capping ligands and core metal atoms in TQDs is in contrast with the CDs. The surface functional groups in CDs comprise mainly an oxygen-containing moiety that is bonded covalently to the carbon core via C–C or other C-X bonds (X = O, N, S, etc.), which is optimal for water solubility and appropriate functioning [16], [17]. The positive effects of Quantum Dots and carbon compounds are put together in one dot which is considerably better than TQDs, thereby enabling great flexibility in the manipulation of physiochemical and structural properties. CDs have aroused the interest of scientists as they are privileged to compensate for the primary disadvantages of TQDs.

2 Green Synthesis of Carbon Dots

To procure CDs with desirable qualities, three key issues must be addressed during CD synthesis: carbonaceous aggregation, uniformity and size control, and surface qualities. Two primary approaches that have been documented for the synthesis of CDs are Top-down and Bottom-up strategies. Among them, hydrothermal carbonization is the predominantly chosen method to design CDs from green resources due to its low cost, benign nature, and ecologically beneficial procedure. Still, the major hurdles for CDs are large-scale manufacturing and the selection of an ideal precursor that is carbon-intensive, affordable, sustainable, and conveniently available for an extended time duration [18] (Fig. 2).

2.1 Types of Green Precursors

CDs crafted from natural precursors have captured the attention of researchers due to their admirable characteristics and sustainable synthetic procedures [19]. Since CDs serve primarily in biomedicine applications such as drug administration, biosensing, bioimaging, and gene transfer, the transition from chemical precursors to "green precursors" seems more favourable [20]. Habitually, a dilemma arises in opting for the appropriate precursor for the desired specific application. The concept of employing eco-friendly sources addresses diverse substances that fall under the rubrics of plant biomass, animal products, food items, and waste materials. These substances are perfectly green and biodegradable carbon assets since they mostly comprise proteins, starch, lipids, and nucleic acids. Incorporating these low-value raw materials into

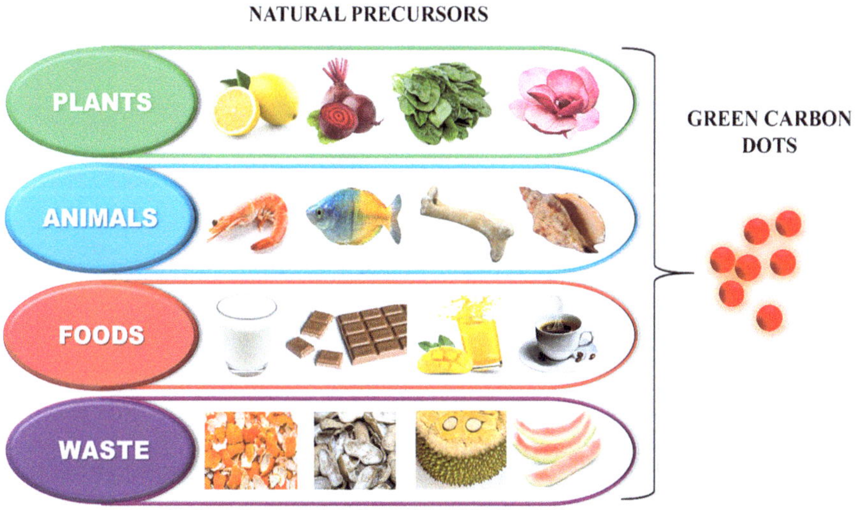

Fig. 2 Types of Green precursors employed for CDs synthesis

functional materials makes room for heteroatom doping (N, S, and P) and also yields numerous advantages [21].

2.1.1 Plant Biomass

Different plant parts such as the seeds, root, stem, leaf, bud, flower, fruit, gum, resin, and bulbs are efficiently implemented as inexpensive reservoirs of C-rich precursors to create CDs with a range of physicochemical properties. Citric acid, amino acids, ascorbic acid, carbohydrates, and proteins contribute to the majority of carbon sources. Other elements, such as cellulose, hemicellulose, lignin, starch, and polysaccharides, which are widespread heteropolymers in plant parts also naturally speed up surface passivation, post-modification, and doping in CDs [22]. Doping with superior auxochromes and the presence of electron-rich N, S, and P functional groups are responsible for the increased Quantum Yield (QY) and efficiency. The properties and usage of CDs are governed by the type of plant precursors and the technique of synthesis employed. A few of the plant-derived CDs are featured in Table 1.

2.1.2 Animal Products

Animal-based products are feasible alternatives to chemical doping agents since they are high in protein, saturated fat, and carbohydrates, as well as micronutrients like zinc, calcium, vitamin B_{12}, and iron. In broad terms, milk and its products

Table 1 Quantum Yield and synthetic method of different plant precursors

Plant precursors	Synthetic method	QY (%)	Particle Size(nm)	Applications	References
Curaua fibers	Hydrothermal	31	2.4	Detecting Fe(III) ions	[23]
Dwarf banana + aq. ammonia	Hydrothermal	23	4	Detecting Fe(III) ions	[24]
Catheranthus roseus	Hydrothermal	28.2	5	Detecting Al(III) and Fe(III) ions	[25]
Nerium Oleander L. Petals	Hydrothermal	3.5	5 ~ 6	Detecting Co(II) ions	[26]
Mesquite leaves	Pyrolysis	5	5.8	Detecting Hg(II) ions	[27]
Papaya Pulp + EDTA	Pyrolysis	23.7	7	Detecting Cr(III) and Cr(VI) in water	[28]
Gum tragacanth + EDA	Hydrothermal	66.74	2.5	Detecting Au(III) ions	[29]
Kelp + EDA	Microwave	23.5	3.7	Detecting Co(II) ions	[30]
Muskmelon	Acid oxidation	26.9	4.3	Detecting Hg(II) ions and cellular imaging	[31]
Dunaliella salina	Hydrothermal	8	4.7	Intracellular detection of Hg(II) and Cr(VI) ions	[32]
Lemon Juice	Hydrothermal	2.4	-	Detecting Hg(II) ions	[33]
Bamboo leaves	Solvothermal	4.7	3 ~ 7	Detecting Pb(II) and Hg(II) ions	[34]
Jackfruit seeds + Phosphoric acid	Microwave	17.91	5	Detecting Au(III) ions	[35]
Soybeans	Ultrasonic	16.7	2.4	Detecting Fe(III) ion	[30]
Quince fruit	Microwave	8.55	4.9	Detecting As(III) ions	[36]
Tulsi leaves	Hydrothermal	3.06	5	Detecting Cr(VI) ions	[37]
Pineapple Peel	Hydrothermal	42.0	3	Detection of Hg(II) ions in water	[38]
Pear juice	Hydrothermal	10.8	2–3	Detecting Al(III) ions	[39]
Syringa oblata Lindl	Hydrothermal	12.4	1 ~ 5	Detecting Fe(III) and H^+ ions	[40]
Phyllanthus acidus + aq. ammonia	Hydrothermal	14	4.5–5.5	Detecting Fe(III) ions	[41]
Hongcaitai	Hydrothermal	21	1.9	Detecting ClO^- and Hg(II)ions	[42]
Sweet potato	Hydrothermal	8.64	3.39	Detecting Fe(III) ions	[43]
Whey	Pyrolysis	11.4	4	Detecting Sn(II) ions	[44]

comprise an emulsion or colloid of butterfat globules holding carbs and proteins [45]. On the other hand, marine-based precursors prove to be an accessible bio-resource for CD synthesis at the quantum scale to increase their effectiveness. The benefits of synthesized animal CDs pertain to their sustainability, exceptional fluorescent nature, and biocompatibility [46]. As a result, animal products could potentially act as initial supplies for synthesizing CDs. The table delivers a handful of CDs that are synthesized from animal products (Table 2).

2.1.3 Food Products

Food products are utilized as sources of carbon for CDs synthesis because they are widely accessible and produce a significant amount of QY when subjected to thermal treatment. Small molecules including citric acid, ascorbic acid, polyethylene glycol, glucose, etc. are present in these precursors [65]. Due to their superior fluorescence performance, CDs made from food can realize their utility in analysis and sensing. Compared to non-biomass carbon dots, they have more functional groups on the surface which can interact with metal ions and non-metal molecules to alter their fluorescence characteristics. Below is a list of certain CDs originating from foodstuffs and their corresponding applications (Table 3).

2.1.4 Waste Materials

Thc production and consumption framcwork known as thc "circular cconomy" trics to reuse and recycle current resources and products as much as possible, in contrast to the conventional linear economy model, which is based on the simple flow of materials from resources to goods and directly into waste. This circular economy model seeks to reduce waste by reusing, restoring, and recycling both raw materials and finished goods. Additionally, it seeks to substitute primary materials, like fossil carbon sources with biomass wastes. Waste materials could be extracted from a variety of sources, including plants, animal waste, agricultural wastes, food scraps, by-products from industrial operations, and human activity waste. Although it is mostly made up of the elements carbon, oxygen, and hydrogen, it also contains trace amounts of other heteroatoms like phosphorus, sulphur, nitrogen and certain toxic metals [80]. The innate presence of heteroatoms in waste-derived precursors eliminates the necessity for additional processing to incorporate external heteroatoms [22]. The table hands out some of the CDs that have been designed using discarded materials (Table 4).

Table 2 Quantum Yield and synthetic method of CDs from different animal precursors

Animal precursors	Synthetic methods	QY (%)	Particle size(nm)	Applications	References
Camel milk	Hydrothermal	24.6	3 ~ 15	Detecting Mn(VII) ions	[47]
Wool	Pyrolysis	8	2 ~ 6	Detecting Fe(III) and Cr(VI) ions	[48]
Gelatin	Hydrothermal	22.7	0.5 ~ 5	Detecting Fe(III) ions	[49]
Shark cartilage	Hydrothermal	20.46	19.6	Fluorescent imaging in zebrafish	[46]
Fish scales	Hydrothermal	6.9	5 ~ 10	Detecting Fe(III) ions	[50]
Casein	Microwave	18.7	1.64	Bioimaging	[45]
Milk	Hydrothermal	38	7	Detecting Sn(II) ions	[51]
Collagen	Hydrothermal	15	1.25	Cell imaging in 3D scaffolds	[52]
Silk Fibroin	Microwave	15	5.4 ± 0.9	Bioimaging, sensing, and drug delivery	[53]
Pigeon	Pyrolysis	33.5	3.8 ~ 4.2	Detecting Fe(III) and Hg(II) ions	[54]
Prawn shell	Hydrothermal	9	4	Detecting Cu(II) ions	[55]
Serum albumin	Microwave	14	2.4~5	Detecting Ag(I) ions	[56]
Silkworm chrysalis	Microwave	46	19	Bio-imaging	[57]
Pigskin	Hydrothermal	24.1	5.6	Detecting Co(II) ions	[58]
Dried shrimps	Hydrothermal	-	6	Biosensors and bio-imaging	[59]
Bee pollens	Hydrothermal	6.1–12.8	1 ~ 2	Bioimaging and catalysis	[60]
Goose feathers	Microwave-hydrothermal	17.1	21.5	Detecting Fe(III) ions	[61]
Egg white	Hydrothermal	64	2.1	Detecting Fe(III) ions	[60]

(continued)

Table 2 (continued)

Animal precursors	Synthetic methods	QY (%)	Particle size(nm)	Applications	References
Honey	Hydrothermal	19.8	2	Detecting Fe(III) ion in real samples	[62]
Human hair	Hydrothermal	10.75	4.56	Detecting Hg(II) ions	[63]
Eggshell membrane	Microwave- Hydrothermal	14	5	Detecting Cu(II) ions	[64]

Table 3 Quantum Yield and synthetic method of CDs from food products

Food products	Synthetic methods	QY (%)	Particle size (nm)	Applications	References
Tofu	Hydrothermal	64	2.9	Detecting Co(II) ions and EDTA	[66]
Pearl Millet	Pyrolysis	52	4 ~ 5	Detecting Pb(II) ions	[67]
Cat feedstocks	Hydrothermal	28	2 ~ 5	Detecting Fe(III) ions	[68]
Black soya beans	Pyrolysis	39	5.5	Detecting Fe(III) ions	[69]
Finger millet ragi	Thermal	~	3 ~ 8	Detecting Cu(II) ions	[70]
Groundnut powder	Hydrothermal carbonization	17.6	2.5	Detecting Cr(VI) ions and in vitro bioimaging	[71]
Watermelon juice	Hydrothermal	10.6	3 ~ 7	Detecting Fe(III) ions and cysteine	[72]
Table sugar	Microwave	2.5	3.5	Visible detection of Pb(II) ions	[73]
Caffeine	Pyrolysis	69	13	Detecting Ag(I) ions	[74]
Yogurt	Hydrothermal	1.5	4.7	Detection of formic acid vapour and metal ions	[75]
Mangosteen pulp	Calcination	-	5	Detecting Fe(III) ions	[76]
Papaya powder	Hydrothermal	18.98	3.4	Detecting Fe(III) ions	[77]
Chocolate	Thermal	-	6.41	Detecting Fe(III) ions	[56]
Soft drink + Bread	Hydrothermal	0.26	4 ~ 20	LED application	[65]
Naked oats	Pyrolysis	1.2	8.64 ± 0.84 nm	Detecting Al(III) ions	[78]
Cornflour	Hydrothermal	7.7	3.5	Cell imaging and Detecting Cu(II) ions	[79]

Table 4 Synthetic methods of CDs from waste materials with their QY and applications

Waste materials	Synthetic methods	QY (%)	Particle size (nm)	Applications	References
Jackfruit peel	Hydrothermal	13.04	6.4	In vitro cytotoxicity studies	[81]
Palm kernel shell	Microwave	44	6–7	Detecting Cu(II) ions	[82]
Waste tea	Hydrothermal	7.1	10	Detecting Fe(III) ions,$CrO4^{2-}$,ascorbic acid and L-cysteine	[83]
Wheat straw	Hydrothermal	13	3.5	Cellular Imaging and In Vivo Bioimaging	[84]
Expired milk	Hydrothermal	8.64	2	Bioimaging and Detecting Fe(III) ions	[51]
Waste chimney oil	H_2SO_4 treatment	7.5	1–4	Detecting Fe(III) ions, fluorescent marker ink and light emitting polymer film	[85]
Rice husk	Pyrolysis	15	3–6	Bioimaging	[77]
Peanut shell	Pyrolysis	9.91	0.42.4	Bioimaging	[86]
Cow manure	Hydrothermal	0.65	2–7	Cellular imaging	[87]
Waste paper	Hydrothermal	10.8	4.5	Biolabelling	[79]
Sago waste	Thermal pyrolysis	-	6–17	Detection of Cu(II) and Pb(II) ions	[88]
Watermelon peel	Pyrolysis	7.1	2	Bioimaging	[89]
Coffee ground	Pyrolysis	3.8	5	Bioimaging	[90]

2.2 *Synthetic Approaches*

2.2.1 Top-Down Approach

The top-down method typically deals with the disintegration of macrostructures into multiple nanostructures through various means. This strategy for CD synthesis signifies the transformation of carbon-based compounds into nanoparticle-sized carbon materials by adopting laser, arc discharge, electrochemical, or ultrasonication procedures. Nanostructures encompassing activated carbon, nanodiamonds, carbon nanotubes, graphite, carbon soot, and graphite oxide are reshaped into CDs using this top-down strategy [91]. However, they require harsh experimental conditions, laborious production stages, prolonged reaction times, and utilizes expensive equipment, which severely restricts their practical applicability. Furthermore, CDs prepared utilizing top-down techniques with no further purification or surface modification have weak luminescence and poor QY [92].

(i) Laser-ablation

Laser ablation is the thermal or adiabatic process of eliminating atoms from a solid by exposing it to a pulsed laser beam or powerful continuous wave (CW) which is a straightforward, affordable, and trouble-free technique [93]. This method vaporizes the target surface immersed in a fluid medium by using a high-intensity pulsed laser which develops flumes packed with atomic or ionic particles. As a result of the flume's reaction with the liquid medium, CDs are generated. The target material is altered by evaporation at low laser flux and into plasma at high laser flux. The disintegrating components are discovered to have nanoscale dimensions. Despite its appealing advantages, laser ablation has been witnessed to have a diminished QY and poor size control [94].

(ii) Arc discharge

The arc discharge process accounts for the fission of gas molecules to generate plasma by applying direct or alternating current. The resulting plasma (4000–6000 K temperature) would eventually sublime the bulk carbon precursors to carbon vapours which will undergo a phase transition and condense into liquid at a lower temperature [95]. Precursor temperature and arc stability, which govern the interaction at the anode border, may be used to regulate the morphologies and productivity of CDs [96]. Even though, this method is a popular approach for generating flawless nanomaterials with sustainable properties, the yield of CDs produced remains inadequate, further comprising a range of complicated components. As a result, purifying CDs produced by arc discharge remains challenging [97].

(iii) Electrochemical/chemical oxidation

The electrochemical approach brings forth ultrapure CDs from larger molecular materials such as carbon nanotubes, graphite, and carbon fiber using an electrolytic process where larger molecules are utilized as electrodes in the presence of appropriate electrolytes. The number of electrochemical cycles modified the structural breakdown of bulk carbon. On subjecting to a thousand cycles of electrochemical treatment, severe deformity was induced in Multiwalled Carbon Nanotubes (MWCNTs), whereas, at the end of 100 cycles MWCNTs were entangled with coiled properties. As a result, the extent of electrochemical treatment cycles affected the surface reaction and active area of CDs [98]. This method cuts down the synthesis time, expensive chemicals, and strong acids and diminishes the need for further surface modification [99]. However, electrochemical carbonization has a poor QY of CDs.

(iv) Ultrasonication Method

Ultrasound can generate oscillating low and high-pressure waves in liquid media, resulting in the formation and explosion of tiny vacuum bubbles. These cavitations induce dissipation, severe hydrodynamic shear stresses, and impinging liquid jets at high speeds. While the bubbles expand, compress, and collapse, ultrasonic waves create extreme pressure and energy instantaneously, crushing the carbon–carbon bonds. By blending all these qualities, ultrasound has a knack for smashing the bulk carbon-based supplies into Carbon Dots. This technique has been advocated because of its unmatched benefits, especially due to green sustainability, cheap, uniform and rapid permeability [100].

2.2.2 Bottom-Up Approach

On the other hand, the bottom-up approach manifests CDs by supplying energy in the form of thermal or microwave irradiation to escalate the chosen chemical precursors to "quantum-sized" particles. Thermal decomposition, hydrothermal methods, and microwave irradiation are glimpses of physical or chemical treatments that aid in the self-assembly of small molecules. Citrates, sugars, and organic acids are commonly employed as precursors in this strategy. To generate CDs, the source molecules are ionized, dissociated, evaporated, or sublimated, followed by the building up of resulting units. Bottom-up techniques provide significant benefits since the product's shape and size distribution can be accurately regulated for surface passivation [22].

(i) Thermal Decomposition

Thermal decomposition is an elevated temperature, endothermic, kinetic process that promotes degradation by dint of dehydration and carbonization of chemical bonds at the molecular level. The onset temperature synchronized will usually rely on the sensitivity and conditions of the experimental setup executed. At elevated temperatures and controlled pressures, CDs nucleate from precursor molecules. This technique offers an affordable, rapid, facile, solvent-free pathway by accommodating a broad spectrum of precursors [101].

(ii) Hydrothermal/Solvothermal Method

The most exhaustively studied method for the manufacture of CDs is the hydrothermal/solvothermal method that deals with agglutination of different molecules under liquid phase at high temperature ($100 \sim 1000°C$) and pressure (1 MPa ~ 1 GPa) in an autoclave condition. The yield is determined by the solubility of the precursor in hot solution which encourages the crystal morphology of CDs. Rapid supersaturation is advantageous to the production of crystal nuclei, which affects the dimensions of CDs. The solvothermal method varies from the hydrothermal approach such that the solvents are organic like benzene, dimethylformamide (DMF), or dimethyl sulfoxide (DMSO) instead of water-based. Figuring out an appropriate solvent for CDs formation and structural management is critical. This route provides an inexpensive, environmentally beneficial, and secure way of integrating CDs using raw materials which obviates the need for hefty chemicals and multistep passivation [102]. Yet the low-temperature hydrothermal/solvothermal method calls for supplemental metal ions or templates to yield distinct forms of carbon materials for the upsurge in applications.

(iii) Microwave Irradiation

Microwave-assisted synthesis is notable for its instantaneous heating which spikes the reaction rate. Carbon has been touted as a potent microwave absorber as it produces heat when exposed to microwave radiation [103]. The acute penetrative power of microwaves causes interaction between the charged particles [104]. This phenomenon weakens the chemical bonds by lowering activation energy and thus alters thermodynamic functions. Using this approach, CDs could precisely spring from carbon precursors in a fleeting time with superior QY. Coupling hydrothermal and microwave techniques upshot the microwave hydrothermal procedure in which the chemical reaction increases twofold.

3 Analytical Techniques Employed in Carbon Dots Characterization

The diverse techniques of characterization are invoked to explore the framework (structure and dimension), molecular geometry, elemental formulation, crystallographic parameters, particle sizing, and granular dispersal of individual CDs processed using specific methods. These techniques essentially comprise microscopic, spectroscopic, and diffraction studies.

3.1 Microscopic Studies

Microscopic techniques offer evaluation and clarification on the intricate details of CDs nanostructure, surface morphology, chemical composition, and interfacial properties.

(i) Atomic Force Microscopy (AFM)

AFM is one of the high-resolution topographical tools used to investigate the mechanical characteristics of nanoscale CDs. It captures deformation or impression to elucidate the firmness or softness of CDs along with their receptivity to external stimuli by retrieving two-dimensional and three-dimensional data. A topographical map ensures the measurement of transverse dimensions, along with surface morphology displaying texture, surface flaws, and the existence of agglomeration or clusters that are vital for quality evaluation [105].

(ii) Transmission Electron Microscopy (TEM)

TEM is an advanced microscopy tool that entraps high resolution images of CDs to unveil the particle size and distribution precisely. The image details lattice fringes, homogeneity and graphitic/crystalline nature. When a beam of focused electrons strikes a particular area of thin material, it generates a pattern designated as

Selected Area Electron Diffraction (SAED) which showcases crystallographic data, especially lattice arrangement and crystal orientation, about the atomic structure of the material [106]. TEM can also be coupled with Energy-Dispersive X-ray Spectroscopy (EDS) to analyze the chemical composition and recognize any impurities or functionalization present.

(iii) Scanning Electron Microscopy (SEM)

With the use of potent SEM imaging technology, it is possible to determine whether the CDs occur as individual entities or as aggregated particles. Additionally, it offers details on surface morphology, size distribution, elemental analysis through Energy-Dispersive X-ray Spectroscopy (EDS), and chemical composition mapping, which illustrates the spatial distribution of elements in CDs [80].

3.2 Spectroscopic Studies

Spectroscopic studies provide essential data regarding the absorption and emission properties possessed by CDs. Yet another perspective of spectroscopy is to acknowledge the interactions between protons, electrons, and ions in the CDs.

(i) Ultraviolet–Visible (UV–Vis) and Photoluminescence (PL) Spectroscopy

The optical absorption and optical fluorescence of CDs are measured by UV–Vis and PL spectroscopy respectively, as a function of wavelength. The $C = C$ bonds and $C = O$ surface groups in CDs give rise to UV absorption peaks that pave the way for the study of linear optical absorption activity. PL spectroscopy examines the electronic structure and properties of CDs corresponding to the intensity of peak light emitted at the specified wavelength. The fluorescence emission reveals the excitation behaviour of CDs concerning wavelength [105]. UV–vis and PL spectroscopic results are used together to determine the Quantum Yield of CDs.

(ii) Fourier Transform Infrared (FT-IR) and X-Ray Photoelectron Spectroscopy (XPS)

FTIR and XPS are significant spectroscopic methods that scrutinize the surface properties of CDs. FTIR spectrum characterizes functional groups in CDs namely hydroxyl, carboxyl, carboxylic acid, ether, epoxy, amine or amide functionalities. While XPS evaluates the elemental makeup, electronic state of the elements and density of the electronic states that exist within the CD sample [107].

(iii) Dynamic Light Scattering (DLS) Spectroscopy and Zeta-Potential (ZP) measurement

DLS (also known as photon correlation spectroscopy) and ZP measurements (called electrokinetic potential) are effective techniques to measure the hydrodynamic radius and potential difference of the outermost layer respectively for electrophoretically migrating CDs in solution, [108]. Both the measurements are based on colloidal dispersion properties of molecules (Brownian movement) and light scattering. As a consequence, these techniques are applicable only for transparent diluted CD solutions with defined ionic strength and pH (Fig. 3).

(iv) Raman Spectroscopy

Raman spectroscopy is persistently employed to arrive at the carbon state of individual CDs by Raman scattering. Raman spectra of CDs regularly exhibit two significant first-order bands namely the Disorder (D) band and Graphitic (G) band. The D band shows out-of-plane vibrations corresponding to dangling carbon bonds in disordered graphite, while the G band is attributed to the in-plane vibration of carbon atoms that make up the graphitic structures [109]. The purity of CDs is presumed by the ratio of D/G band intensity. CDs that are crystalline have a lower D/G ratio than CDs that are amorphous (Fig. 4).

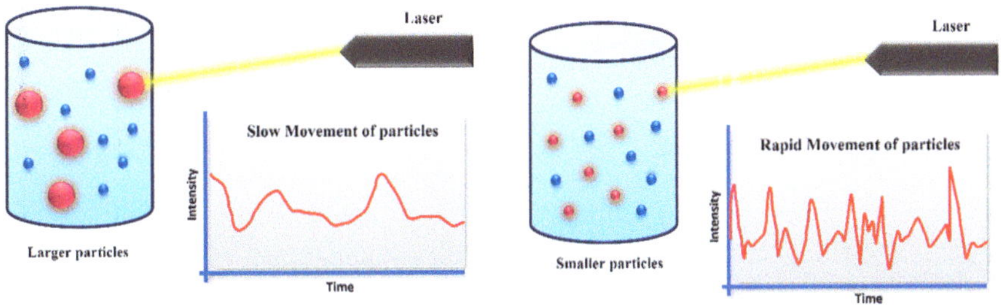

Fig. 3 Dynamic light scattering technique to analyze particle size

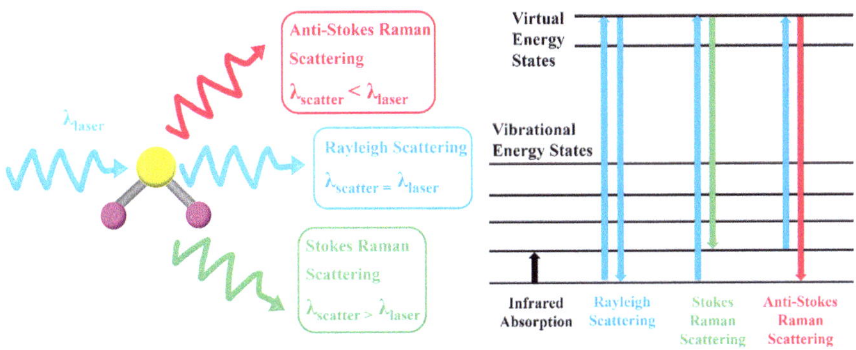

Fig. 4 Principle of raman spectroscopy

3.3 Diffraction Studies

The diffraction technique aids in promptly evaluating the crystal type, phase purity, and particle size of CDs based on their diffraction pattern. This is one of the widely used non-destructive ways of measuring superficial residual stress on crystalline materials.

(i) Powder X-ray Diffraction (XRD)

The analytical tool XRD reveals the phase-type and crystal nature of CDs by analyzing the X-rays dispersed from long-range ordered crystalline materials. As a general rule, diffraction of X-rays is determined by the interatomic spacing in crystallographic structures, hence the beam that is scattered from randomly coordinated nano materials reveals the type of crystal arrangement in CDs [109].

4 Properties of Carbon Dots

4.1 Optical Properties

(i) Absorption

In the visible and ultraviolet (UV) regions of the electromagnetic spectrum, CDs feature photon-absorbing properties [110].

(ii) Phosphorescence

It is the emission of light observed when CDs capture photons from an external light source and subsequently radiate these photons to a higher wavelength [111] (Fig. 5).

(iii) Chemiluminescence

The process of producing electromagnetic radiation in the form of light emission by consuming energy released from a chemical reaction is termed Chemiluminescence [112] (Fig. 6).

(iv) Electrochemical luminescence

Electrochemical luminescence is a form of light emission that uses energy from controlled redox reactions that apply electrical potential to electrodes immersed in a solution (or) the electrochemical generation of electrons and holes in CDs to produce excited states, which eventually relaxes to their ground state upon emission [112] (Fig. 7).

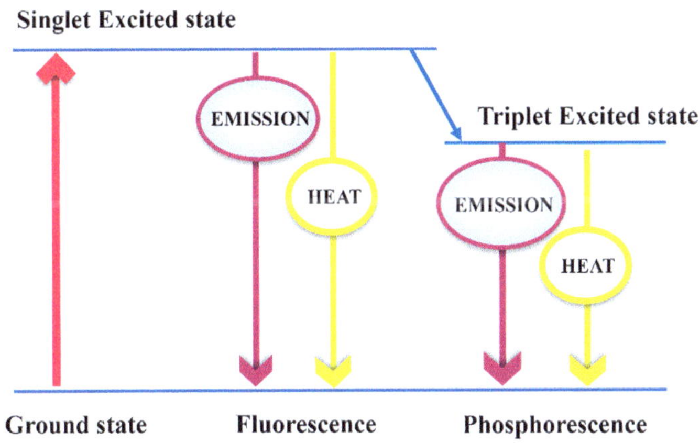

Fig. 5 Mechanism of phosphorescence

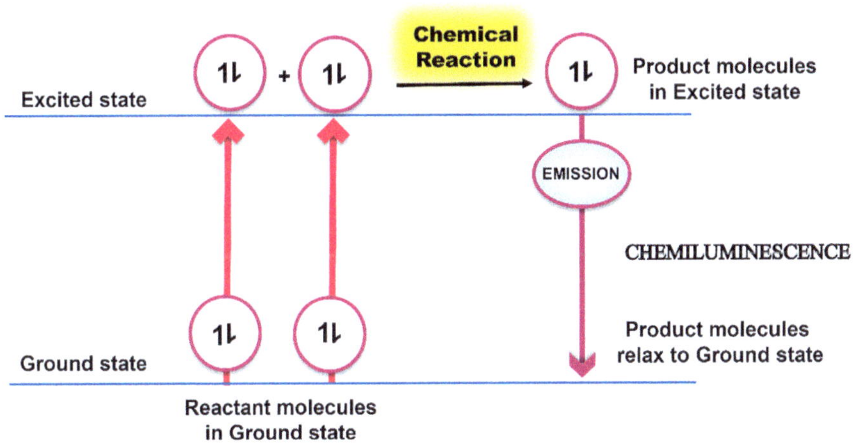

Fig. 6 Mechanism of chemiluminescence

(v) Up-conversion Luminescence

Up-Conversion Luminescence refers to anti-stokes light emission in the course of transforming lower-energy photons from the infrared region into higher-energy photons in the visible light region through non-linear optical interactions [113] (Fig. 8).

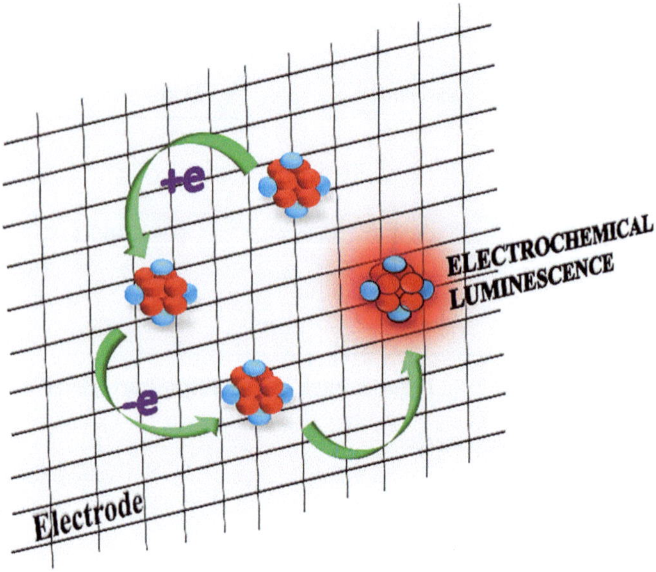

Fig. 7 Mechanism of electrochemiluminescence

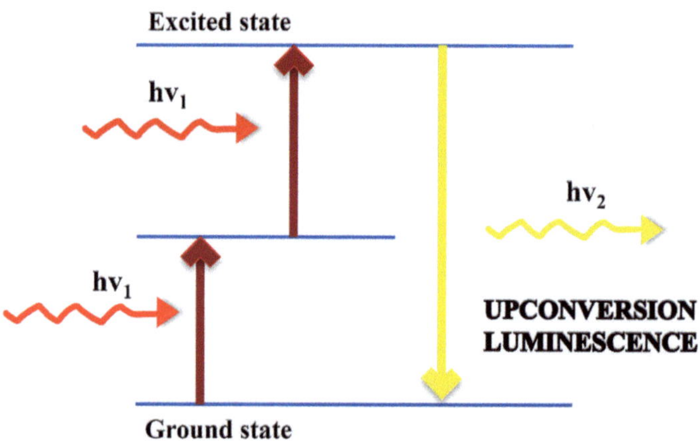

Fig. 8 Mechanism of up-conversion luminescence

4.2 Physical Properties

(i) Photostability

The ability to resist optical property loss or degradation when subjected to light, preventing unfavourable changes in structure or luminescent properties for reliable applications is termed photostability [62].

(ii) Crystallinity

Crystallinity defines the degree of long-range ordered, repeating atomic structures in the crystal lattice of CDs [114].

(iii) Quantum Yield

It is the ratio of photons absorbed relative to photons emitted by CDs (or) the efficiency of CDs to convert the fraction of absorbed photons into luminescence [115].

4.3 Biological Properties

(i) Biocompatibility

It is the ability to interact reliably with biological systems in a well-tolerated manner rather than producing harmful or undesirable effects [110].

(ii) Cytotoxicity

The potential to induce damage or toxicity to cells and tissues of biological systems which greatly influences the suitability for biomedical applications is termed cytotoxicity [110].

5 Photoluminescent Mechanism of Carbon Dots

Before customizing optical properties, in-depth knowledge of the causes of CD luminescence phenomena must be acquired. CDs are fabricated using numerous precursors, pretreatment, and synthetic methods, resulting in variable optical qualities [115]. However, no exact definition of the CD structure is in harmony with existing photoluminescence mechanisms due to the diverse and complicated nature of CDs. Certain hypotheses like Molecular Fluorescence, Surface State Emission, and Quantum Confinement Effect have been offered so far to throw some light on the origin of CD fluorescence.

5.1 Molecular Fluorescence

The molecular state is the wave function produced by all of the electrons in a molecule. This molecular state is the photoluminescence (PL) core developed exclusively by an organic fluorophore attached to the exterior or interior of the carbon framework and could radiate PL effectively [116]. This type of PL mechanism is widespread in CDs produced by bottom-up methods. Since this approach provides extremely reactive conditions, it is realistic to anticipate other competing reactions leading to small fluorescent molecules. In general, during CD synthesis, tiny molecules or perhaps oligomeric luminophores would be formed. Consequently, these luminophores could adhere to the outer layer of CDs or remain lodged within them, rendering emission properties to CDs [117]. With a rise in reaction temperature, the dimension of the carbon core grows while the superficial fluorescent groups shrink, thereby decreasing the molecular fluorescence [115].

5.2 *Surface State Emission*

Surface states in CDs are a synergistic combination of carbon cores and linked functional groups. The band gap energy of CDs corresponding to the degree of π-electron system in carbon cores couples with these functional groups, thereby modifying their electronic structures and exhibiting recombination. On the flip side, these surface functional groups possess plentiful structural configurations that intercalate more electronic structures with different energy levels, leading to additional recombination probabilities of electrons and holes trapped by surface states. Surface defects of CDs can also interact with the environment, altering their electronic structures and causing visible optical changes. Furthermore, these defects are qualified to serve as a capture site for excitons. The energy from the recombination of captured excitons gives rise to a red shift in surface state emission when excited by light. The ratio of oxygen on the CD surface is intimately connected to their red-shifted emissions. This phenomenon could be supported by the fact that an increase in surface oxidation corresponds to a decrease in band gap energy between the HOMO and LUMO states [34]. The extent of emission by CDs and its properties are influenced by particle diameter, heteroatom doping, functional groups, surface imperfections, surface chemical composition, valence atoms, as well as charge transfer on surface adsorbents [118] (Fig. 9).

5.3 *Quantum Confinement Effect*

The Quantum Confinement Effect (QCE) is the spatial confinement of electron–hole pairs from one or more dimensions to zero dimensions leading to the formation of nanosized particles. CDs smaller than the defined “Bohr exciton radius” in general display this effect. The kind of confinement is determined by the Bohr radius; for instance, weak confinement is determined for particles with a radius of 3–10 nm and strong confinement is estimated for particles with a radius of less than 3 nm. Electron–hole pairs become spatially confined when particle size reduces and the wavefunction of electrons gets restricted, leading to quantized energy levels [119]. As the enclosure space decreases, the energy levels become more discrete, and the bandgap widens, which is a fundamental feature of nanostructures. CDs with large

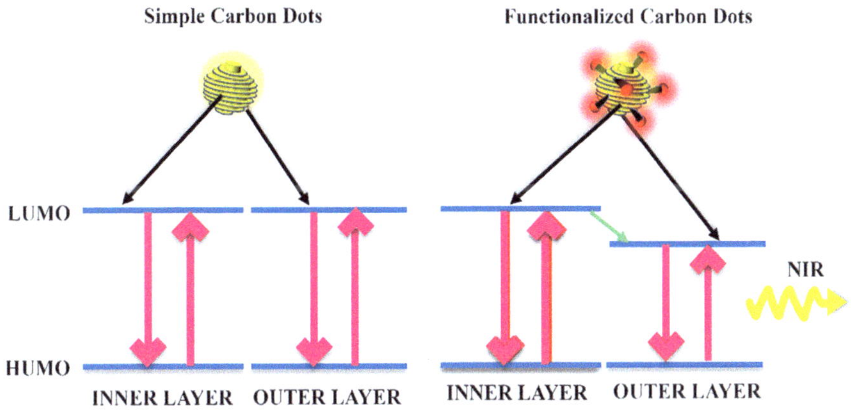

Fig. 9 Fluorescence of CDs based on surface functional groups

band gap energies due to QCE tend to show emission at higher energy and shorter wavelengths that pose wider applications in optoelectronic devices (Fig. 10).

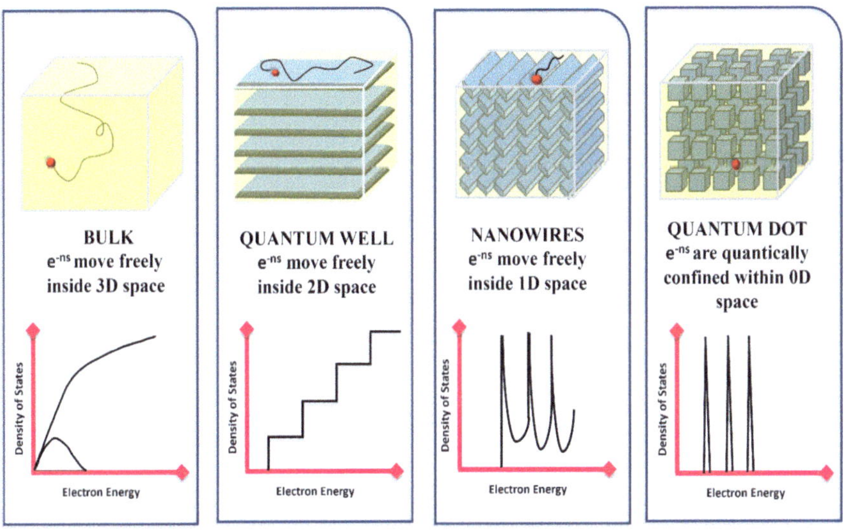

Fig. 10 Spatial confinement of electrons from 3D to zero dimension leading to nanosized particles

6 Quenching Mechanisms Involved in Sensing Application of Carbon Dots

Grasping the PL mechanism in CDs and embracing it in futuristic applications rely on the study of quenching behaviours. This phenomenon demands molecular interaction between the fluorochrome and the quenching factor, provoking momentary effects in the fluorescence properties of CDs. The quenching may arise due to excited-state response, chemical rearrangements, ground-state complex configuration, molecular collision, energy/electron transfer, and emission group annihilation. Conventional quenching mechanisms addressed are Static, Dynamic, FRET, PET and IFE mechanisms.

6.1 Static Quenching Mechanism

Static Quenching (SQ) of CD fluorescence takes place when the quenching agent adheres to the CDs to create a non-fluorescent ground-state complex. Unmodified CDs create a non-fluorescent pair with tiny molecules or metal atoms. The fluorescence lifetimes of CD remain unaltered during this mechanism. The absorption spectra may shift due to the emergence of ground-state complexes. The fluorescence may rise with increasing temperature, due to the dissociation of the loosely bound quenching agent, thereby diminishing the quenching effect. Due to the simplicity of validating the SQ mechanism, it is often used to figure out analytes. It is important to highlight that there are no observable changes in fluorescence lifespan or the distribution of particle size, indicating the dominance of the SQ mechanism in that particular CD [120] (Fig. 11).

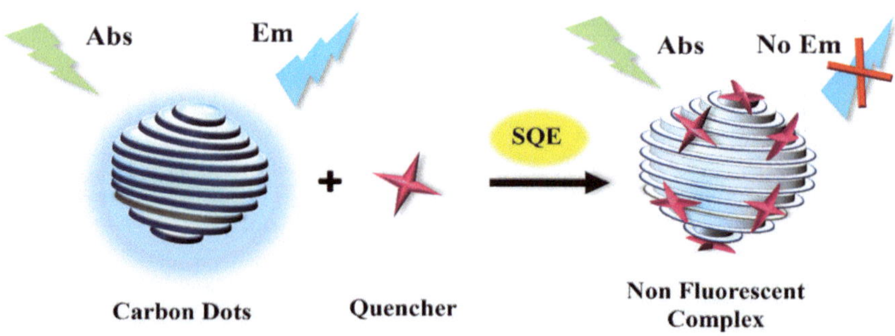

Fig. 11 Static quenching mechanism

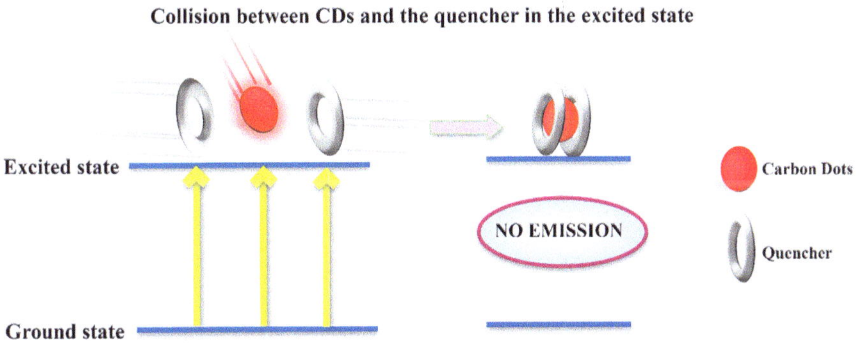

Fig. 12 Dynamic quenching mechanism

6.2 Dynamic Quenching (DQ) Mechanism

During collision between the quencher and CDs in the excited state, energy or charge carried by the CDs is transferred to the quenching molecule. This enables the CDs to revert to the ground state without emitting photons. This kind of fluorescence quenching is designated as Dynamic Quenching (DQ). No ground-state complexes are produced in DQ, unlike in SQ, but its efficiency is limited to the longevity of CDs and concentration of quencher. Furthermore, the collision rate becomes enormous at elevating temperatures enhancing the quenching efficiency. Owing to the simple model of CDs functioning on the DQ mechanism, they are employed in the construction of fluorescence sensors for detecting a wide range of analytes [120] (Fig. 12).

6.3 Forster Resonance Energy Transfer (FRET) Mechanism

FRET was previously known as Fluorescence Resonance Energy Transfer. FRET involves the long-range dipole–dipole interactive exchange of nonradiative energy from an excited donor to the acceptor. In 1948, Theodor Forster presented an equation for estimating the transfer capacity of electronic excitation energy across an energy donor (D) and acceptor (A). Particular standards pertain to fluorescent probes that employ the FRET process in CDs. The distance (R) from the donor to the acceptor should lie within the range of 10 Å to 100 Å (Fig. 13a). The donor's emission spectrum and the acceptor's absorption spectrum should be substantially overlapped (referred to as J) so that the energy can be easily transferred from the donor to the acceptor. On the contrary, the donor's absorption spectrum must be completely distinct from the acceptor's emission spectrum (denoted by R). In general, the greater the J value and the shorter the distance (R), the stronger would be the FRET effect (Fig. 13b). There have been several reports of CD-based FRET systems where CDs serve as donors.

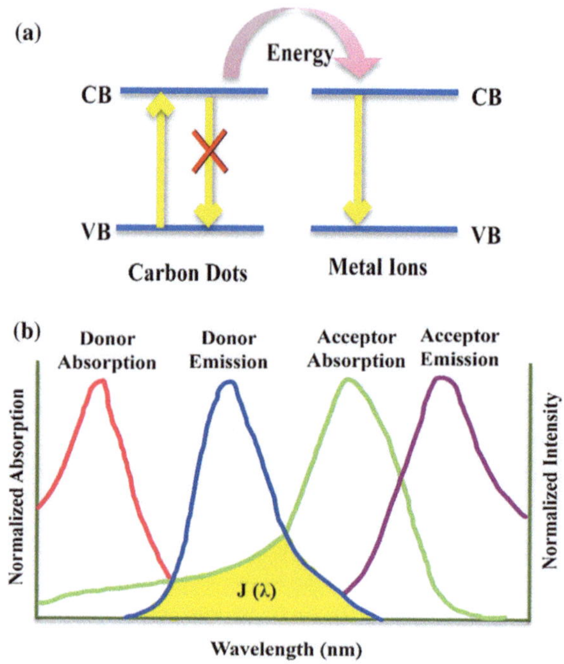

Fig. 13 FRET mechanism

Due to the significant overlap of the CD emission spectrum on the absorption spectrum of targets, FRET can happen from CDs to targets, when they are within a short accessible range. This either subsides the CD fluorescence or shifts the CD emission wavelength to blue or red. Since the majority of target absorption spectra do not appear in the visible region, they cannot be employed directly as FRET acceptors. Quenching agents are thus involved in acting as acceptors from CDs [121].

6.4 Photoinduced Electron Transfer (PET) Mechanism

PET is a quenching mechanism involving electron transfer across an electron donor (D) and an electron acceptor (A). PET occurs in two different forms. (i) Electrons held in CDs can excite from the highest occupied molecular orbital (HOMO) to the lowest unoccupied molecular orbital (LUMO) under the influence of excitation. On the other hand, the transfer of electrons from the acceptor to the CDs is also possible when the acceptor has higher HOMO energy than the HOMO of fluorescent CDs. This phenomenon is regarded as a PET mechanism (Fig. 14). The variation in CD fluorescence produced by a PET may be employed as a sensor for acceptor or target detection. (ii) The second type is the d-PET mechanism in which the LUMO of CDs contributes electrons to the LUMO of the acceptor, causing fluorescence quenching. But still, when the acceptor bonds to the target molecule, the

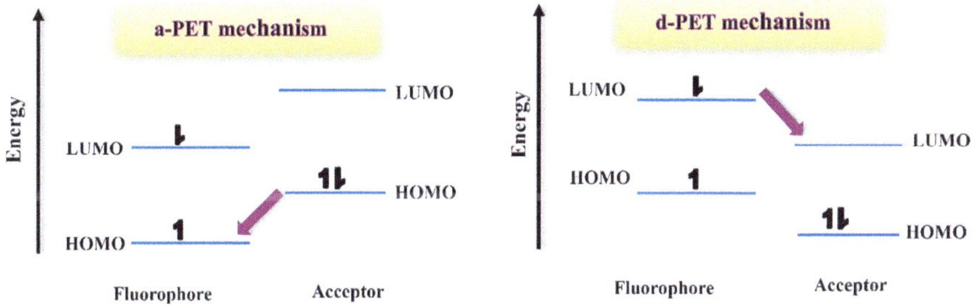

Fig. 14 Frontier orbital energy diagram on PET mechanism **a** a-PET **b** d-PET mechanisms

PET process is constrained, and the CD acquires its fluorescence [122]. Typical PET sensors have been extensively used for sensing metal ions and reactive small molecules using individual atoms as electron donors (O, S, N, Se, Te, etc.).

6.5 Inner Filter Effect (IFE)

IFE is a non-irradiation energy conversion process in which excitation or emission light is absorbed by the absorber in a detecting system. Due to light absorption, IFE may reduce the excitation intensity peak at the site of observation or minimize the visibly perceived fluorescence. The absorption of excitation and/or emission radiation by solution lowers fluorescence intensity, resulting in a nonlinear relation between fluorescence intensity and CD concentration. At times, attenuations are referred to as primary or secondary IFE based on the type of radiation absorbed. The absorption of excitation light by chromophores (CDs) in a solution or matrix is known as the primary inner filter effect (pIFE). On the contrary, the secondary inner filter effect (sIFE) involves the absorption of emission radiation by the same chromophores. To design an IFE-based metal ion sensor, one must choose an analyte-independent fluorescent and an analyte-sensitive absorber. It is important to note that absorption peaks for ground state compounds should not be noticed in IFE, so species employed should be chemically inert to each other. CDs are capable of acting both as a fluorescent and an absorber in IFE-based sensors because of their sustained fluorescence and numerous surface functional groups [123] (Fig. 15).

7 Various Metal Sensing by Green Carbon Dots

Certain metal ions are capable of quenching fluorescence by binding with specific structures or functional groups on the surface of CDs. Electron transfer between metal ions and CDs will result in fluorescence quenching. Several green CDs are employed in the sensitive and specific detection of metal ions using colorimetric or fluorometric

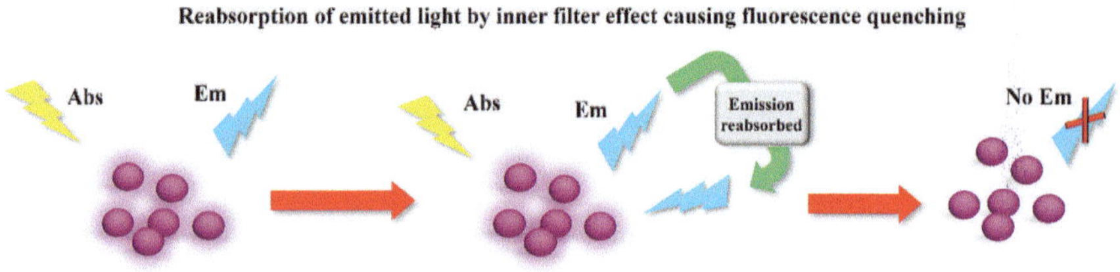

Fig. 15 Inner filter effect mechanism

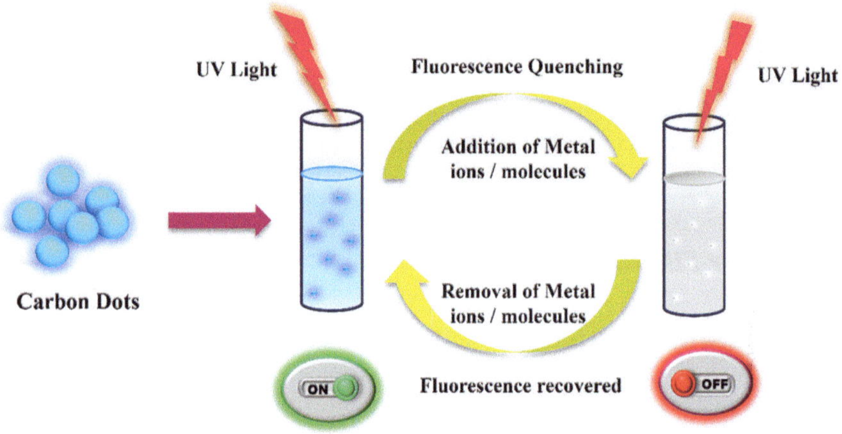

Fig. 16 Metal ion sensing by green carbon dots

examinations. It is feasible to detect a particular metal ion through CD sensors which are designed to have fluorescent turn-on and turn-off modes. Meanwhile, this technology successfully resolves the challenges of laborious procedures and the expensive cost of conventional detection methods [123] (Fig. 16).

7.1 Sources and Toxic Effects of Heavy Metal Pollutants

The remarkable spike in heavy metal usage has resulted in a surge of metallic elements in both the upland and underwater environments. Metal pollution has emerged as an effect of anthropogenic activity, mainly due to metal-based industries. Agricultural application of heavy metals, like herbicides, insecticides, chemical fertilizers, and other chemicals, serves as a secondary cause of heavy metal pollution. Natural sources of heavy metal contamination include volcanic eruptions, corrosion of metals, soil erosion, etc. Metals are mostly stable and non-biodegradable. When heavy metals enter our systems through ingestion or inhalation, they can bioaccumulate in our bodies. Bioaccumulation has its own biological and physiological complications leading to serious health ailments. Metal ions can be detoxified in the human body by insulating the active element with a protein coating or accumulating them in intracellular granules in an insoluble state for excretion. However, these processes are tedious and complicated. As a result, metallic poisoning is considered to be highly hazardous [124].

7.2 *Lead (Pb^{2+}) Sensor*

One of the dangerous p-metal that is known to cause detrimental illness in humans is lead (Pb) [125]. Despite having three oxidation states Pb(0), Pb(II), and Pb(IV), Pb(II) state is the most commonly found form in the environment. Pb (II) concentrations in the blood of more than 5 mmol/L can cause mental impairment, memory loss, anemia, migraine attacks, and even death. The proposed limit of lead intake by the Environmental Protection Agency (EPA) in drinking water is fewer than 15 ppb. Wee et al. [126] pioneered the investigation of Pb(II) detection through the electron transfer mechanism, utilizing CDs derived from protein-based bovine serum albumin in their study. Conventional Stern–Volmer relation was used to determine the fluorescence quenching behaviour. By the recombination of excited electrons to the valence band from the conduction band, the Pb (II) ions efficiently quench the fluorescence of CDs once they occur closer to the surface. Similar quenching mechanisms were spotted in CDs from chocolates [56] and industrial Sago waste [88]. From the flavonoid extracts of Ginkgo biloba leaves Xing [127] synthesized carbon dots in their research endeavour [127]. Among the many metal ions subjected, only Pb^{2+} exhibited particular affinity to flavonoid glycosides on the CD surface. Enhanced visual perception of fluorescence quenching was achieved by doping flavonoid CDs with agarose-hydrogel. Rajkumar Bandi et al. [128] produced nitrogen-doped CDs using lantana camara berries and provided a comprehensive insight into the quenching mechanism of photoluminescence (involving both dynamic and static approaches) for Pb (II) detection. This study suggested a linear relationship between the quenching constant and temperature in both static and dynamic quenching mechanisms which was evidenced by the Stern–Volmer plots. Furthermore, V. A. Ansi et al. created CDs from table sugar for effective Pb (II) detection [73]. The specific detection of Pb (II) was unaffected by the presence of metal ions mixture. Electron transfer reactions within the excited state could be the cause of fluorescence quenching. The ultrasensitive and selective Pb^{2+} analysis was made possible with the use of CDs derived from Potato-Dextrose Agar (PDA). The reduction in zeta potential verified the complex formation between Pb(II) and CDs causing fluorescence quenching [129]. To examine Pb(II) ions in river water, Liu et al. crafted multi-emission fluorescent CDs by employing bamboo leaf extracts [34]. They subsequently demonstrated the use of these CDs in environmental monitoring. Jing et al. incorporated ample functional groups onto the CD surface, [130] by fabricating CDs using hydrochar, which has a surplus of conjugated double bonds, carbonyl groups, hydroxyl groups, and quinoid groups. These CDs exhibited potential use in Pb^{2+} detection.

They increased fluorescence efficiency by altering the oxidation conditions of CDs to vary their surface state.

7.3 Mercury (Hg^{2+}) Sensor

One of the undecomposable and poisonous ions is Mercury(II) [131]. The buildup of metallic elements within the body through the food cycle causes harm to the vital organs present [132]. World Health Organization recommended that drinking water with more than 1 mg/L of mercury can be lethal [34]. Therefore, detecting mercury metal is vital for preserving the ecosystem and preventing health risks. Lu et al. [133] presented CDs out of Pomelo peel waste that were used in the Hg(II) sensing in liquid samples. Introducing a potent Hg(II) chelator, the Hg(II) establishes an HgS bond on the exterior of CDs, enabling the CDs' fluorescence to be quenched. This revealed that the electron transfer process was the probable mechanism accountable for quenching the fluorescence in this study. Hong Huang et al. [131] created N-CDs with strawberry juice to detect Hg(II) levels in samples. Since the fluorescence dropped as the concentration of Hg(II) increased, the research concluded that a dynamic/ultrafast electron transfer mechanism was responsible for the fluorescence quenching. Pineapple peel-based CDs [38] were capable of sensing Hg(II) with desirable reliability in various aqueous samples. According to Vandarkuzhali et al., the dynamic quenching process showed that adding Hg(II) to CDs reduced their lifespan. In contrast, to detect Hg(II) in aqueous and biological samples, Desai et al. [31] synthesized multicolour fluorescent CDs from muskmelon. After adding Hg(II) to CDs, they noticed no change in the lifespan, indicating a static quenching process. Lemon juice-derived CDs, synthesized by thermal decomposition were examined in an aqueous solution with neutral reactants. At a basic pH environment, CDs showed fluorescence quenching with Hg(II) ions [33].

Highland barley-derived N-CDs containing surface functionalities exhibited significant chelating capacity with Hg(II) ions [134] which caused the fluorescence quenching. Using fairy fruit (*Jinhua bergamot*), Jing Yu et al. [135] produced luminescent CDs that were water-soluble. Using cucumber juice, Wang et al. [92] synthesized N/S/P co-doped fluorescent CDs that were water-soluble. Ye et al. [54] used egg white, egg yellow, and pigeon feathers to fabricate CDs. They all emphasized that non-radiative recombination arising between the surface functional groups and Hg(II) could be the cause of fluorescence quenching.

7.4 Gold (Au^{3+}) Sensor

Two existing oxidation states of gold are Au(I) and Au(III). Compared to metallic gold, Au (III) could potentially be a threat to human health. The ecological discharge of effluents containing gold traces poses a risk to human well-being, causing impairment in the liver, kidneys, and nervous system, as well as impeding the growth of ecosystem components. Identification of Au(III) in the environment as well as in biological systems is therefore recommended [136].

Liao et al. [137] reported hydrothermal carbonization of water-soluble nitrogen-doped CDs (NCDs) from ethylene diamine and natural peach gum polysaccharide to detect Au(III) ions. It was the first study where N-CDs exhibited highly sensitive and specific fluorescence quenching action towards Au(III) ions within a minute. The CDs exhibited no interference from other organic molecules and elements present in the sample. Rahmani et al. [29] utilized gum tragacanth ethylenediamine to produce N-CDs as a fluorescent sensor for Au(III) ions selectively. The synergistic effect of FRET (overlap of CDs emission spectra with metal ion absorption spectra) and electron transfer accounts for the fluorescence quenching mechanism in these investigations. This impact may be caused by functional groups on the outermost layer of CD surface which includes oxygen and nitrogen. These groups can interact with Au(III) to form complexes, which in turn causes a non-radiative electron shift from C-dots to Au(III) ions. The addition of an electron donor atom like nitrogen to CDs, efficiently encourages the synergistic interaction between Au(III) and oxygen-based functional groups on the N-CD edge. Hence this non-radiative electron transfer process and subsequent interaction induce the N-CD/Au(III) system to fluoresce less. Raji et al. [35] interestingly stated the reduction of Au(III) to polydispersed gold nanoparticles during Au(III) sensing using N-CDs produced from Jackfruit seeds, which is facilitated by the reductive exchange of electrons from N-CDs to Au(III) ions.

7.5 Copper (Cu^{2+}) Sensor

Cu (II) is an essential part of the protein Complex (IV) found in the inner mitochondrial membrane. However, a brief period of exposure to elevated Cu(II) levels might upset the gastrointestinal tract, while prolonged exposure causes significant harm to the kidneys and liver. The permissible level of copper in drinking water is fixed to be 1.3 ppm by the U.S. Environmental Protection Agency [138]. Copper creates a diverse spectrum of compounds in its + 1(cuprous) and + 2 (cupric) oxidation states.

Tan et al. [88] synthesized CDs from starch effluent which was used in the selective sensing of Cu (II) ions in aqueous conditions. Cu (II) ions exhibited exceptional absorption affinity towards the CDs in comparison to other transition metal ions. This high affinity of metal ions to CDs inhibits the initial electronic conversion process within the CDs, thereby disrupting the non-radiative pathway resulting in effective quenching. Ma et al. [139] synthesized luminescent CDs out of peanut shells by an eco-friendly pyrolysis process which did not involve any chemicals. Their application is expanded to Cu(II) sensing capability by serving as a fluorescence sensor. The chelation of cupric ions with CDs through oxygen and nitrogen functional groups on their surfaces may be responsible for this specific quenching action of Cu (II). Das et al. [138] thermally blended L-arginine with lemon extract into CDs for accurate sensing of Cu(II) ions. The amino clusters on the CD surface trap the Cu(II) ions and create a cupric ammine complex through the inner filter effect, which extinguishes the fluorescence of CDs, thereby detecting their presence. X. Zhu et al. [140] produced CDs using natural kelp and polyethyleneimine (PEI), and further fabricated CD composites by incorporating fluorescein isothiocyanate (FITC) with CDs. Subsequently, these CDs were capable of detecting Cu(II) ions due to the presence of PEI-containing amine groups which triggered fluorescence quenching by the Inner Filter Effect. Iodide was introduced to the CD composites' solution to reduce the minor hindrance caused by Hg (II) ions. Using photoluminescent polymer nanodots (PPNDs) derived from grass [141], Cu(II) ions in water samples were detected by fluorescence quenching. PPNDs quenching was due to the energy transfer occurring between Cu(II) ions and N, O functional groups bound to it. The N and O containing CDs showed enhanced chelating action and stronger thermodynamic attraction towards Cu(II) ions than other metals. Zhang et al. [142] prepared CDs by the process of brewing Pu-erh tea. The acquired CDs along with o-phenylenediamine were used to develop a sensing system that simultaneously used colorimetric and fluorometric sensing for Cu(II) ions. The detection process depends upon the combined effect of Cu(II) interaction and the fluorescent resonance energy transfer from CDs to the oxidized product of o-phenylenediamine.

7.6 Iron (Fe^{3+}) Sensor

Iron occurs in two crucial oxidation states, ferric (III) and ferrous (II), important for life-sustaining reactions. Fe(III) is the most hazardous of the two oxidation states due to its insoluble nature and tendency to form toxic radicals. The primary cause of Parkinson's disease, cytotoxicity, metabolic abnormalities, etc. is due to excess of Fe(III) accumulation in the human body. The ability of free iron in the body to take up electrons from the near surroundings might affect the biological systems [143]. Thus, it is recommended to examine for Fe(III) ions in the surroundings and in biological systems.

Abhay Sachdev and his group [144] showcased the pH-sensitive optical response of fluorescent CDs that were produced by the hydrothermal process using coriander leaves. This response successfully sensed Fe(III) without the use of additional surface-modifying agents. The reduction in quenching efficiency was observed under both acidic and basic conditions as surface carboxylic groups underwent protonation. Conversely, in the pH range of 7–9, a rise in quenching efficiency was noted as surface carboxylic groups were deprotonated.

Xiaoming Yang et al. [62], stated the use of honey-mediated CDs produced solvothermally to identify Fe(III) ions. This research work optimized various reaction parameters like reaction time, pH, temperature, and buffer solution to enhance the sensitivity of Fe(III) detection. The existence of Fe(III) ions enlarged the

CDs' size, as validated by HR-TEM images, illustrating that CD-Fe(III) coordination triggered aggregations leading to the fluorescence quenching of CDs. In the pursuit of producing CDs to detect Fe(III) ions, numerous precursors like sugarcane molasses [145], mangosteen pulp [76], *Syringa oblata Lindl* [40], and sweet potatoes [43] were employed. These works highlighted the transfer of electrons between surface groups of CDs and Fe(III), causing aggregation and consequently photoluminescence quenching. The non-radiative recombination of electrons and holes, resulting from dynamic quenching, was observed by assessing the fluorescence lifespan of highly luminous CDs synthesized from onion waste [146]. Yellow, green, and blue-fluorescent CDs were achieved from tomato (*Solanum Lycopersicum*) by employing phosphoric acid and sulphuric acid as oxidative agents. Enhanced Fe (III) sensing ability was exhibited by blue-CDs. [147]. Fe(III) was also detected using Lycii Fructus derived CDs [148] in live HeLa (Henrietta Lacks) cells, and Yellow River water samples. Nitrogen-doped CDs [149] synthesized from rice residue and glycine displayed sensitivity towards Fe (III) ions in aqueous samples. The fluorescence reduction with Fe(III) ions was ascribed to the specific binding between Fe(III) ions and surface hydroxyl groups on N-CDs. The fluorescence reduction in NCDs produced from peels of dwarf bananas, under the influence of Fe(III), was elucidated by the reactivity of surface functionalities inherent to the CDs [24]. From residual kiwi fruit peel [150], separate CDs were made, with and without ammonium hydroxide (NH_4OH). Due to the greater attraction of CDs for Fe(III) ions, the complex formation is believed to be the cause of the reduced fluorescence intensities in CDs. N-CDs fabricated using betel leaves implied a significant sensitivity and selectivity to Fe(III), along with blue fluorescence. The surface atoms on N-CDs and ferric ion bind strongly, quenching the fluorescence [151].

7.7 Chromium (Cr^{6+}) Sensor

The toxic and carcinogenic heavy metal, Cr(VI), is extensively used in industrial, domestic, agricultural, medical, and technological realms, posing a threat to health [152]. In the environment, both Cr(VI) and Cr(III) oxidation states prevail, with Cr(VI) causing prolonged adverse effects owing to its considerable solubility and mobility. According to the World Health Organization (WHO), the acceptable level of Cr(VI) in drinking water is 0.05 mg/L. Pooja D. et al. [28] created EDTA-modified CDs using Papaya waste to detect chromium ions. N and O functional groups on the CDs' surface are responsible for the observed fluorescence quenching, facilitating detection without the need for pre-reduction or oxidation steps. CDs sourced from Tulsi leaves [37] were effective in detecting Cr(VI) ions in contaminated water from households and industries using the mechanism of the Inner Filter Effect. By introducing ascorbic acid, the effect of CDs can be reverted and repurposed. As reported by Tyagi et al. [153], the water-soluble CDs produced from lemon peels showed good photostability and were capable of identifying Cr(VI) in drinking water. Non-radiative electron–hole pair recombination and subsequent fluorescence quenching in CDs is a consequence of surface functional groups, reduced redox potentials, and lowered d-d transition states in Cr(VI). Groundnut-derived N-doped CDs have demonstrated notable selectivity for Cr(VI) ions and also exhibited the capability for Cr(VI) reduction when exposed to glutathione and humic acid. Singh et al. [32] demonstrated an eco-friendly approach (employing algal biomass of *Dunaliella salina*) for the one-pot hydrothermal synthesis of nitrogen, and phosphorus dual-doped CDs without resorting to any chemical or passivating substances. They interestingly documented an absorbance and fluorescence emission spectral overlap, along with a shift in the lifetime value. A dual-mode mechanism comprising dynamic quenching and the Inner Filter Effect is involved in the detection of Cr(VI).

7.8 Silver (Ag^+) Sensors

Silver ions Ag(I) are acknowledged as a pervasive pollutant, finding widespread usage in various sectors like electronics, photography, and pharmaceuticals. Approximately 2500 tonnes of Ag(I) are released into the environment annually with the escalating use of silver. On that account, Ag(I) exposure to the human body can deactivate sulfhydryl enzymes or proteins, impacting their roles in physiological processes. Therefore, the precise identification of Ag(I) holds considerable importance in addressing environmental pollution and ensuring human health [154].

Ayaz Ahmad et al.[155] discovered that CDs derived from pectin (present in citrus peels) could identify silver ions by generating yellow-coloured silver nanoparticles. The calculated lower reduction potential of CDs compared to that of silver ions supported sensing through an electron transfer mechanism. Cyclic voltammetry analysis affirmed the capability of CDs to sense Ag(I) ions and reduce them to Ag(0), resulting in silver nanoparticle formation. Water-soluble fluorescent CDs were synthesized by Murugesan et al. [156] through a hydrothermal process utilizing coconut sap as the precursor. The surface characterization indicated the existence of hydroxyl, carbonyl, and carboxylic acid groups, and their effectiveness in detecting Ag(I) ions was validated. Dang et al. [74] presented a pioneering one-step synthesis method for the production of N and S-doped fluorescent CDs. The procedure entailed the thermal heating of a solid blend of caffeine, ammonium persulfate, and urea. The resultant CDs exhibited the potential to serve as a fluorescent detector for Ag(I) ions sensing, thereby showcasing both sensitivity and selectivity. Lu et al. [157] implemented an eco-friendly technique to produce S, N doped CDs, employing bean pods and onions as precursors. The synthesized S N/CDs experienced quenching selectively in the presence of Ag(I) ions. The proposed quenching mechanism has pertained to aggregation-induced quenching that resulted from the coagulation process.

7.9 *Cadmium (Cd^{2+}) Sensors*

Cadmium is a commonly encountered heavy metal pollutant due to its extensive usage in dyeing, electroplating, agriculture, and semiconductor industries, resulting in abundant environmental distribution. Exposure to cadmium is associated with a spectrum of health complications, including renal dysfunction and the onset of cancer [158]. The World Health Organization (WHO) accredits a regulated limit of 3 ppb for Cd(II) ion concentration in drinking water. The presence of other interfering transition metal ions, such as Zn(II), presented a notable challenge in Cd(II) sensing. These ions share similar chemical reactivity and are grouped in the periodic table. This poses a challenge in distinguishing whether the fluorescence quenching is caused by Cd(II) or Zn(II) ions [159]. Hence there exists a critical necessity for synthesizing a fluorophore capable of differentiating Cd(II) ions from Zn(II) ions with enhanced specificity and sensibility. Gu et al. [160] established a simple route to craft S and N-doped CDs from scallions (spring onions) as the carbon source. These synthesized CDs demonstrated considerable utility in the sensing of Cd(II) ions. The modest decrease in the fluorescent lifetime served as evidence, indicating a static quenching mechanism in the interaction between Cd(II) ions and CDs. The result implicated that Cd(II) ions had a stronger affinity and faster chelating kinetics towards heteroatom doped on CDs. Pandey et al. [159] documented the synthesis of innovative fluorescent CDs derived from *Murraya koenigii* leaves (curry leaves). The preparation involved a single-step hydrothermal process devoid of toxic chemicals. The CDs exhibited efficacy in selective detection of Cd (II) ions. The sensing mechanism was a charge transfer process from ligand to metal, substantiated through analytical techniques encompassing UV–visible spectroscopy, lifetime decay, and zeta potential studies.

7.9.1 Miscellaneous Ion Sensors

(i) Aluminium Sensor

Despite the diverse applications of aluminium compounds, an elevated Al(III) ion level is linked to Alzheimer's and Parkinson's diseases, emphasizing the importance of environmental monitoring for safety. Bhamore et al. [39] synthesized fluorescent CDs using pear fruit as a carbon source. The CDs featured multiple organic groups on their surfaces due to the presence of various phytochemicals in pear fruit. Through a Chelation-Enhanced Fluorescence (CHEF) mechanism, the synthesized CDs demonstrated high selectivity for detecting Al(III) ions. Notably, the fluorescent intensity of the CDs was substantially enhanced with the addition of Al(III) ions, elucidating the selective and strong interactions of rich electron donor groups with Al(III) ions.

(ii) Arsenic Sensor

Increased arsenic contamination from urbanization and agriculture poses health risks. As(III) was specified as a human carcinogen by the International Agency for

Research into Cancer (IARC). Resultantly practical sensors are required for the effective detection of As(III). Radhakrishnan and panneerselvam [161] developed fluorescent CDs from prickly pear cacti wherein the surface was passivated with glutathione (GSH) to create a sensor probe. Experimental results revealed the specific affinity of As(III) ions towards the acid groups on the CDs over other metal ions. The elevated sensitivity of CDs towards As(III) ions was attributed to fluorescence quenching through a turn-off mechanism, influenced by factors such as functional groups, electron-transfer mechanisms, and inner filter effect.

(iii) Cobalt Sensor

The exceptional physicochemical characteristics of cobalt find applications across various fields. However, its extensive use raises environmental concerns with potential implications for human health. The fluorescence probes overcome traditional drawbacks in sensing Co(II) ions. C. Zhao et al. developed CDs by using natural kelp as a carbon source along with ethylenediamine serving as a nitrogen dopant [30]. The fluorescent CDs exhibited selective quenching in the presence of Co(II). The considerable spectral overlap between the UV absorption spectrum of Co(II) and the emission spectrum of CDs indicated that the possible mechanism governing fluorescence quenching could be either the Inner Filter Effect (IFE) or Fluorescence Resonance Energy Transfer (FRET). However, examination of the fluorescence lifetime decay curves in the presence and absence of Co(II) remains unchanged confirming the inner filter effect.

(iv) Palladium Sensor

The focus on Pd(II) detection arises from its potential adverse health effects, causing severe eye and skin irritations through complex formation with biomolecules. Sensors that use fluorescence have been employed for detecting Pd(II) in recent times. Hashemi et al. [162] achieved the production of fluorescent CDs utilizing Red Beetroot. The introduction of Pd(II) led to the quenching of CD fluorescence, elucidated by a static quenching mechanism due to the formation of a ground-state complex between CDs and Pd(II) ions.

(v) Zinc Sensor

Despite its low toxicity, zinc intake in excess can be detrimental to health, highlighting the necessity to monitor its presence in dietary consumption. Jayan et al. [163] successfully prepared CDs from tender coconut water, showcasing their application in the identification of Zn(II) ions in water. The suggested mechanism for the Zn(II) sensing action of CDs involved an on–off mechanism. Table 5 lists the metal ion sensing by various green synthesized carbon dots.

Table 5 Metal ion detection using green synthesized carbon dots

Precursor	Synthetic method	Quantum yield (%)	Metal ion detected	Reference
Pectin from citrus peel	Solvothermal	-	Ag (I)	[155]
Caffeine	Pyrolysis	69	Ag (I)	[74]
Bean pod and Onion	Hydrothermal	5.55	Ag (I)	[157]
Coconut Sap	Hydrothermal	-	Ag (I)	[156]
Pyrus pyrifolia fruit	Hydrothermal	10.8	Al (III)	[39]
Prickly Pear Cactus	Hydrothermal	12.7	As (III)	[161]
Quince fruit	Microwave irradiation	8.55	As (III)	[36]
Gum tragacanth	Hydrothermal	66.74	Au (III)	[29]
Waste expanded Polystyrene	Solvothermal	20	Au (III)	[136]
Jackfruit seeds	Microwave irradiation	17.19	Au (III)	[35]
Spring onion	Microwave irradiation	18.6	Cd (II)	[160]
Curry leaves	Hydrothermal	5.4	Cd (II)	[159]
Natural Kelp	Microwave irradiation	-	Co (II)	[30]
Dried rose petals and Ripe banana peel	Hydrothermal	28 and 27	Cr (VI) and Fe (III)	[164]
Tulsi leaves	Hydrothermal	3.06	Cr (VI)	[37]
Papaya Waste	Pyrolysis	23.7	Cr (III) and Cr (VI)	[28]
Dunaliella salina	Hydrothermal	8	Cr (VI) and Hg (II)	[32]
Lemon peel waste	Hydrothermal	14	Cr (VI)	[153]
Shallot	Hydrothermal	32.34	Cr (VI)	[165]
Denatured milk	Hydrothermal	-	Cr (VI)	[166]
Pineapple	Hydrothermal	10.06	Cr (VI)	[20]
Poria cocos polysaccharide	Hydrothermal	4.82	Cr (VI)	[167]
Spirulina algae	Hydrothermal	32	Cu (II)	[168]
Lemon juice and L-arginine	Thermolysis	7.7	Cu (II)	[138]
Prawn shell	Thermolysis	9	Cu (II)	[55]
Grass	Hydrothermal	4.2	Cu (II)	[141]
Peanut shells	Pyrolysis	10.58	Cu (II)	[139]
Sago waste	Pyrolysis	-	Cu (II) and Pb (II)	[88]

(continued)

Table 5 (continued)

Precursor	Synthetic method	Quantum yield (%)	Metal ion detected	Reference
Pu-erh tea	Brewing	-	Cu (II)	[142]
Natural Kelp	Hydrothermal	12.3	Cu (II)	[140]
Spinach leaves	Hydrothermal	0.68	Fe (III)	[169]
Coriander leaves	Hydrothermal	6.48	Fe (III)	[144]
Lemon juice	Hydrothermal	38	Fe (III)	[170]
Bombyx mori silk	Hydrothermal	61.1	Fe (III)	[171]
Rice residue	Hydrothermal	23.48	Fe (III)	[149]
Sesame	Microwave irradiation	8.02	Fe (III)	[172]
Osmanthus frangrans	Hydrothermal	18.53	Fe (III)	[17]
Dwarf Banana	Hydrothermal	23	Fe (III)	[41]
Kiwi Fruit Peel	Hydrothermal	19	Fe (III)	[150]
Onion waste	Hydrothermal	28	Fe (III)	[128]
Syringa oblata Lindl	Hydrothermal	6.5	Fe (III)	[40]
Sugarcane Molasses	Hydrothermal	5.8	Fe (III)	[145]
Tomato	Hydrothermal	12.7	Fe (III)	[147]
Betel leaves	Hydrothermal	4.21	Fe (III)	[151]
Coriander leaves	Hydrothermal	6.48	Fe (III)	[144]
Sweet Potato	Hydrothermal	8.64	Fe (III)	[43]
Mango leaves	Pyrolysis	18.2	Fe (III)	[143]
Lycii Fructus	Hydrothermal	17.2	Fe (III)	[148]
Honey	Hydrothermal	19.8	Fe (III)	[62]
Mangosteen pulp	Pyrolysis	-	Fe (III)	[76]
Chinese yam	Hydrothermal	9.3	Hg (II)	[173]
Coconut milk	Thermal pyrolysis	-	Hg (II)	[174]
Corn bract	Solvothermal	6.9	Hg (II)	[175]
Flour	Microwave irradiation	5.4	Hg (II)	[176]
Lotus root	Microwave irradiation	19	Hg (II)	[132]
Muskmelon Fruit	Sonication	14.3	Hg (II)	[31]
Lemon Juice	Thermal decomposition	0.07	Hg (II)	[33]
Lotus root	Microwave irradiation	19	Hg (II)	[132]
Strawberry juice	Hydrothermal	6.3	Hg (II)	[131]

(continued)

Table 5 (continued)

Precursor	Synthetic method	Quantum yield (%)	Metal ion detected	Reference
Bamboo leaves	solvothermal treatment	4.7	Hg (II) and Pb (II)	[34]
Pomelo peel	Hydrothermal	6.9	Hg (II)	[133]
Pineapple Peel	Hydrothermal	0.42	Hg (II)	[38]
Cucumber Juice	Hydrothermal	-	Hg (II)	[177]
Highland Barley	Hydrothermal	14.4	Hg (II)	[134]
Pigeon feathers	Pyrolysis	33.5	Hg (II)	[54]
Jinhua bergamo	Hydrothermal	50.78	Hg (II)	[135]
Jamun or Black plum	Hydrothermal	6.1	Mn (II)	[178]
Common Meadow-grass	Hydrothermal	7	Mn (II) and Fe (III)	[179]
Table sugar	Microwave irradiation	2.5	Pb (II)	[73]
Lantana camara Berries	Hydrothermal	33.15	Pb (II)	[128]
Potato Dextrose Agar	Microwave irradiation	9	Pb (II)	[129]
Biomass resources	Hydrothermal	22.6	Pb (II)	[130]
Chocolate	Hydrothermal	-	Pb (II)	[56]
Ginkgo biloba leaves	Hydrothermal	16.1	Pb (II)	[127]
Red Beetroot	Hydrothermal	27.6	Pd (II)	[162]
Coconut water	Hydrothermal	13.4	Zn (II)	[163]

8 Summary and Outlook

An overall view of the achievements of plant-mediated green CDs is presented, highlighting their significant progress in optically sensing environmental contaminants, with a specific focus on heavy metal ions. Despite advancements, certain challenges persist, requiring attention to the prospective scalability and practical application of these CDs as economically viable sensing probes in day-to-day life. Besides figuring out ways to make green CDs stable and effective, it is crucial to ensure a narrow bandwidth of fluorescence for specific sensing applications. In comparison to Traditional Quantum Dots synthesized from chemical precursors, the green Carbon Dots exhibit lower levels of fluorescence intensity and quantum yield. It is important to explore sustainable source materials including recycled waste, residuals, and biomass to upgrade the production of inherently modified CDs with elevated quantum yield. Though the operational mechanism of ecofriendly produced CDs is identified, their preferential binding to specific metal ions depending on the chosen precursor, continues to be a puzzle [180]. Further thorough investigations are needed to formulate selective sensing probes that display fluorescence emission within the UV–visible region. It is anticipated that upcoming investigations on CDs captivate widespread attention, particularly in the fields of food, agriculture, and environmental pollutant sensing, owing to their inherent qualities of biocompatibility, simplicity, and affordability.

References

1. Keith, L. H., Gron, L. U., Young, J. L., Safety, C., Way, C., College, H., & Avenue, W. (2007). Green Analytical Methodologies.
2. Anastas, P., & Eghbali, N. (2010). Green chemistry : Principles and practice 301–312. https://doi.org/10.1039/b918763b.
3. Hirsch, A. (2010). The era of carbon allotropes. *Nat Publ Gr, 9*, 868–871. https://doi.org/10.1038/nmat2885
4. Feng, Z., Adolfsson, K. H., Xu, Y., Fang, H., Hakkarainen, M., & Wu, M. (2021). Carbon dot / polymer nanocomposites : From green synthesis to energy, environmental and biomedical applications. *Sustainable Materials and Technologies, 29*, e00304. https://doi.org/10.1016/j.susmat.2021.e00304
5. Yao, B., Huang, H., Liu, Y., & Kang, Z. (2019). Carbon dots : A small conundrum. *Trends in Cognitive Sciences, 1*, 235–246. https://doi.org/10.1016/j.trechm.2019.02.003
6. Stark, W. J., Stoessel, P. R., Wohlleben, W., & Hafner, A. (2015). Chem soc rev industrial applications of nanoparticles. *Chemical Society Reviews.* https://doi.org/10.1039/C4CS00362D
7. Kempi, W., Szymon, Ł., Kempi, M., & Markowski, D. (2014). Experimental techniques for the characterization of carbon nanoparticles – A brief overview, pp. 1760–1766. https://doi.org/10.3762/bjnano.5.186.
8. Piotrovskiy, L. B., Kokorina, A. A., Prikhozhdenko, E. S., Sukhorukov, G. B., Sapelkin, & A. V. (2017). Related content Luminescent carbon nanoparticles : synthesis , methods of investigation , applications Luminescent carbon nanoparticles : synthesis , methods of investigation , applications. https://doi.org/10.1070/RCR4751.
9. Shpilevsky, E., Penyazkov, O., Filatov, S., Shilagardi, G., Tuvshintur, P. T. D. (2018). *Modification of Materials By Carbon Nanoparticles, 271,* 70–75. https://doi.org/10.4028/www.scientific.net/SSP.271.70.
10. Xu, X., Ray, R., Gu, Y., Ploehn, H. J., Gearheart, L., Raker, K., & Scrivens, W. A. (2004). Electrophoretic Analysis and Purification of Fluorescent Single-Walled Carbon Nanotube Fragments, pp. 12736–12737.
11. Sun, Y., Zhou, B., Lin, Y., Wang, W., Fernando, K. A. S., Pathak, P., Meziani, M. J., Harruff, B. A., Wang, X., Wang, H., Luo, P. G., Yang, H., Kose, M. E., Chen, B., Veca, L. M., Xie, S., & Carolina, S. (2006). Quantum-Sized Carbon Dots for Bright and Colorful Photoluminescence, pp. 7756–7757. https://doi.org/10.1021/ja062677d.
12. Cao, L., Wang, X., Meziani, M. J., Lu, F., Wang, H., Luo, P. G., Lin, Y., Harruff, B. A., Veca, L. M., Murray, D., Xie, S., & Sun, Y. (2007). Carbon Dots for Multiphoton Bioimaging, pp. 11318–11319.
13. Long, C., Jiang, Z., Shangguan, J., Qing, T., Zhang, P., & Feng, B. (2021). Applications of carbon dots in environmental pollution control : A review. *Chemical Engineering Journal, 406*, 126848. https://doi.org/10.1016/j.cej.2020.126848
14. Brus, L. E., & Brus, L. E. (1983). A simple model for the ionization potential , electron affinity , and aqueous redox potentials of small semiconductor crystallites A simple model for the ionization potential , electron affinity , and aqueous redox potentials of small semiconductor crysta, p. 5566. https://doi.org/10.1063/1.445676.

15. Xu, G., Zeng, S., Zhang, B., Swihart, M. T., Yong, K., & Prasad, P. N. (2016). *New Generation Cadmium-Free Quantum Dots for Biophotonics and Nanomedicine*. https://doi.org/10.1021/acs.chemrev.6b00290
16. Kelarakis, A. (2015). Current opinion in colloid & interface science graphene quantum dots : In the crossroad of graphene, quantum dots and carbogenic nanoparticles. *Current Opinion in Colloid & Interface Science, 20*, 354–361. https://doi.org/10.1016/j.cocis.2015.11.001
17. Wang, M., Wan, Y., Zhang, K., Fu, Q., Wang, L., Zeng, J., Xia, Z., & Gao, D. (2019). Green synthesis of carbon dots using the flowers of Osmanthus fragrans (Thunb.) Lour. as precursors: application in Fe 3+ and ascorbic acid determination and cell imaging. Anal Bioanal Chem. https://doi.org/10.1007/s00216-019-01712-6.
18. Zhu, S., Meng, Q., Wang, L., Zhang, J., Song, Y., Jin, H., Zhang, K., Sun, H., Wang, H., & Yang, B. (2013). Highly Photoluminescent Carbon Dots for Multicolor Patterning , Sensors , and Bioimaging ** Angewandte, pp. 1–6. https://doi.org/10.1002/anie.201300519.
19. Yan, Z., Zhang, Z., & Chen, J. (2016). Biomass-based carbon dots: Synthesis and application in imatinib determination. *Sensors Actuators, B Chem, 225*, 469–473. https://doi.org/10.1016/j.snb.2015.10.107
20. Sharma, S., Umar, A., Mehta, S. K., & Kansal, S. K. (2017). *Fluorescent spongy carbon nanoglobules derived from pineapple juice: A potential sensing probe for specific and selective detection of chromium (VI) ions*. Elsevier. https://doi.org/10.1016/j.ceramint.2017.02.127
21. Kumari, A., Kumar, A., Sahu, S. K., & Kumar, S. (2017). Synthesis of green fluorescent carbon quantum dots using waste polyolefins residue for Cu 2 + ion sensing and live cell imaging. *Sensors Actuators B Chem*. https://doi.org/10.1016/j.snb.2017.07.075
22. Meng, W., Bai, X., Wang, B., Liu, Z., Lu, S., & Yang, B. (2019). Biomass-derived carbon dots and their applications. *Energy Environ Mater*. https://doi.org/10.1002/eem2.12038
23. Raja, S., Buhl, E. M., Dreschers, S., Schalla, C., Zenke, M., Sechi, A., & Mattoso, L. H. C. (2021). Curauá-derived carbon dots: Fluorescent probes for effective Fe(III) ion detection, cellular labeling and bioimaging. *Materials Science and Engineering C, 129*, 112409. https://doi.org/10.1016/j.msec.2021.112409
24. Atchudan, R., Edison, T. N. J. I., Perumal, S., Muthuchamy, N., & Lee, Y. R. (2020). Hydrophilic nitrogen-doped carbon dots from biowaste using dwarf banana peel for environmental and biological applications. *Fuel, 275*, 117821. https://doi.org/10.1016/j.fuel.2020.117821
25. Arumugham, T., Alagumuthu, M., Amimodu, R. G., Munusamy, S., & Iyer, S. K. (2020). A sustainable synthesis of green carbon quantum dot (CQD) from Catharanthus roseus (white flowering plant) leaves and investigation of its dual fluorescence responsive behavior in multi-ion detection and biological applications. *Sustainable Materials and Technologies, 23*, e00138. https://doi.org/10.1016/j.susmat.2019.e00138
26. Dutta, A., Rooj, B., Mondal, T., Mukherjee, D., & Mandal, U. (2020). Detection of Co2+ via fluorescence resonance energy transfer between synthesized nitrogen-doped carbon quantum dots and Rhodamine 6G. *Journal of the Iranian Chemical Society, 17*, 1695–1704. https://doi.org/10.1007/s13738-020-01891-5
27. Pourreza, N., & Ghomi, M. (2019). Green synthesized carbon quantum dots from Prosopis juliflora leaves as a dual off-on fluorescence probe for sensing mercury (II) and chemet drug. *Materials Science and Engineering C, 98*, 887–896. https://doi.org/10.1016/j.msec.2018.12.141
28. Pooja, D., Singh, L., Thakur, A., & Kumar, P. (2019). Green synthesis of glowing carbon dots from Carica papaya waste pulp and their application as a label-freechemo probe for chromium detection in water. *Sensors Actuators, B Chem, 283*, 363–372. https://doi.org/10.1016/j.snb.2018.12.027
29. Rahmani, Z., & Ghaemy, M. (2019). One-step hydrothermal-assisted synthesis of highly fluorescent N-doped carbon dots from gum tragacanth: Luminescent stability and sensitive probe for Au3+ ions. *Opt Mater (Amst), 97*, 109356. https://doi.org/10.1016/j.optmat.2019.109356

30. Zhao, W. B., Liu, K. K., Song, S. Y., Zhou, R., & Shan, C. X. (2019). Fluorescent nano-biomass dots: Ultrasonic-assisted extraction and their application as nanoprobe for FE3+ detection. *Nanoscale Research Letters, 14*,. https://doi.org/10.1186/s11671-019-2950-x
31. Desai, M. L., Jha, S., Basu, H., Singhal, R. K., Park, T. J., & Kailasa, S. K. (2019). Acid oxidation of muskmelon fruit for the fabrication of carbon dots with specific emission colors for recognition of Hg2+ ions and cell imaging. *ACS Omega, 4*, 19332–19340. https://doi.org/10.1021/acsomega.9b02730
32. Singh, A. K., Singh, V. K., Singh, M., Singh, P., Khadim, S. R., Singh, U., Koch, B., Hasan, S. H., & Asthana, R. K. (2019). One pot hydrothermal synthesis of fluorescent NP-carbon dots derived from Dunaliella salina biomass and its application in on-off sensing of Hg (II), Cr (VI) and live cell imaging. *Journal of Photochemistry and Photobiology, A: Chemistry, 376*, 63–72. https://doi.org/10.1016/j.jphotochem.2019.02.023
33. Gharat, P. M., Pal, H., & Dutta Choudhury, S. (2019). Photophysics and luminescence quenching of carbon dots derived from lemon juice and glycerol. *Spectrochim Acta - Part A Mol Biomol Spectrosc, 209*, 14–21. https://doi.org/10.1016/j.saa.2018.10.029
34. Liu, M. L., Chen, B. B., Li, C. M., & Huang, C. Z. (2019). Carbon dots: Synthesis, formation mechanism, fluorescence origin and sensing applications. *Green Chemistry, 21*, 449–471. https://doi.org/10.1039/c8gc02736f
35. Raji, K., Ramanan, V., & Ramamurthy, P. (2019). Facile and green synthesis of highly fluorescent nitrogen-doped carbon dots from jackfruit seeds and its applications towards the fluorimetric detection of Au3+ ions in aqueous medium and in: In vitro multicolor cell imaging. *New J Chem 43*, 11710–11719. https://doi.org/10.1039/c9nj02590a.
36. Ramezani, Z., Qorbanpour, M., & Rahbar, N. (2018). Green synthesis of carbon quantum dots using quince fruit (Cydonia oblonga) powder as carbon precursor: Application in cell imaging and As3+ determination. *Colloids Surfaces A Physicochem Eng Asp, 549*, 58–66. https://doi.org/10.1016/j.colsurfa.2018.04.006
37. Bhatt, S., Bhatt, M., Kumar, A., Vyas, G., Gajaria, T., & Paul, P. (2018). Green route for synthesis of multifunctional fluorescent carbon dots from Tulsi leaves and its application as Cr(VI) sensors, bio-imaging and patterning agents. *Colloids Surfaces B Biointerfaces, 167*, 126–133. https://doi.org/10.1016/j.colsurfb.2018.04.008
38. Vandarkuzhali, S. A. A., Natarajan, S., Jeyabalan, S., Sivaraman, G., Singaravadivel, S., Muthusubramanian, S., & Viswanathan, B. (2018). Pineapple peel-derived carbon dots: applications as sensor, molecular keypad lock, and memory device. *ACS Omega, 3*, 12584–12592. https://doi.org/10.1021/acsomega.8b01146
39. Bhamore, J. R., Jha, S., Singhal, R. K., Park, T. J., & Kailasa, S. K. (2018). Facile green synthesis of carbon dots from Pyrus pyrifolia fruit for assaying of Al3+ ion via chelation enhanced fluorescence mechanism. *Journal of Molecular Liquids, 264*, 9–16. https://doi.org/10.1016/j.molliq.2018.05.041
40. Diao, H., Li, T., Zhang, R., Kang, Y., Liu, W., Cui, Y., Wei, S., Wang, N., Li, L., Wang, H., Niu, W., & Sun, T. (2018). Facile and green synthesis of fluorescent carbon dots with tunable emission for sensors and cells imaging. *Spectrochim Acta - Part A Mol Biomol Spectrosc, 200*, 226–234. https://doi.org/10.1016/j.saa.2018.04.029
41. Atchudan, R., Edison, T. N. J. I., Aseer, K. R., Perumal, S., Karthik, N., & Lee, Y. R. (2018). Highly fluorescent nitrogen-doped carbon dots derived from Phyllanthus acidus utilized as a fluorescent probe for label-free selective detection of Fe3+ ions, live cell imaging and fluorescent ink. *Biosensors & Bioelectronics, 99*, 303–311. https://doi.org/10.1016/j.bios.2017.07.076
42. Li, L. S., Jiao, X. Y., Zhang, Y., Cheng, C., Huang, K., & Xu, L. (2018). Green synthesis of fluorescent carbon dots from Hongcaitai for selective detection of hypochlorite and mercuric ions and cell imaging. *Sensors Actuators, B Chem, 263*, 426–435. https://doi.org/10.1016/j.snb.2018.02.141
43. Shen, J., Shang, S., Chen, X., Wang, D., & Cai, Y. (2017). Facile synthesis of fluorescence carbon dots from sweet potato for Fe3 + sensing and cell imaging. *Materials Science and Engineering C, 76*, 856–864. https://doi.org/10.1016/j.msec.2017.03.178

44. Devi, P., Kaur, G., Thakur, A., Kaur, N., Grewal, A., & Kumar, P. (2017). Waste derivitized blue luminescent carbon quantum dots for selenite sensing in water. *Talanta, 170*, 49–55. https://doi.org/10.1016/j.talanta.2017.03.069
45. Bajpai, S. K., D'Souza, A., & Suhail, B. (2019). Blue light-emitting carbon dots (CDs) from a milk protein and their interaction with Spinacia oleracea leaf cells. *Int Nano Lett, 9*, 203–212. https://doi.org/10.1007/s40089-019-0271-9
46. Kim, K. W., Choi, T. Y., Kwon, Y. M., & Kim, J. Y. H. (2020). Simple synthesis of photoluminescent carbon dots from a marine polysaccharide found in shark cartilage. *Electronic Journal of Biotechnology, 47*, 36–42. https://doi.org/10.1016/j.ejbt.2020.07.003
47. Kumar, R., Vincy, A., Rani, K., Jain, N., Singh, S., Agarwal, A., & Vankayala, R. (2023). Facile synthesis of multifunctional carbon dots derived from camel milk for Mn 7+ sensing and antiamyloid and anticancer activities. *ACS Omega*. https://doi.org/10.1021/acsomega.3c05485
48. Song, Y., Qi, N., Li, K., Cheng, D., Wang, D., & Li, Y. (2022). Green fluorescent nanomaterials for rapid detection of chromium and iron ions: Wool keratinbased carbon quantum dots. *RSC Advances, 12*, 8108–8118. https://doi.org/10.1039/d2ra00529h
49. Latief, U., ul Islam, S., Khan, Z.M.S.H., Khan, M.S., 2021. A facile green synthesis of functionalized carbon quantum dots as fluorescent probes for a highly selective and sensitive detection of Fe3+ ions. Spectrochim Acta - Part A Mol Biomol Spectrosc 262, 120132. https://doi.org/10.1016/j.saa.2021.120132
50. Zhang, Y., Gao, Z., Yang, X., Chang, J., Liu, Z., & Jiang, K. (2019). Fish-scale-derived carbon dots as efficient fluorescent nanoprobes for detection of ferric ions. *RSC Advances, 9*, 940–949. https://doi.org/10.1039/C8RA09471C
51. Su, R., Wang, D., Liu, M., Yan, J., Wang, J. X., Zhan, Q., Pu, Y., Foster, N. R., & Chen, J. F. (2018). Subgram-scale synthesis of biomass waste-derived fluorescent carbon dots in subcritical water for bioimaging, sensing, and solid-state patterning. *ACS Omega, 3*, 13211–13218. https://doi.org/10.1021/acsomega.8b01919
52. Dehghani, A., Ardekani, S. M., Hassan, M., & Gomes, V. G. (2018). Collagen derived carbon quantum dots for cell imaging in 3D scaffolds via two-photon spectroscopy. *Carbon N Y, 131*, 238–245. https://doi.org/10.1016/j.carbon.2018.02.006
53. Ko, N. R., Nafiujjaman, M., Cherukula, K., Lee, S. J., Hong, S. J., Lim, H. N., Park, C. H., Park, I. K., Lee, Y. K., & Kwon, I. K. (2018). Microwave-assisted synthesis of biocompatible silk fibroin-based carbon quantum dots. *Particle & Particle Systems Characterization, 35*, 1–8. https://doi.org/10.1002/ppsc.201700300
54. Ye, Q., Yan, F., Luo, Y., Wang, Y., Zhou, X., & Chen, L. (2017). Formation of N, S-codoped fluorescent carbon dots from biomass and their application for the selective detection of mercury and iron ion. *Spectrochim Acta - Part A Mol Biomol Spectrosc, 173*, 854–862. https://doi.org/10.1016/j.saa.2016.10.039
55. Gedda, G., Lee, C. Y., Lin, Y. C., & Wu, H. F. (2016). Green synthesis of carbon dots from prawn shells for highly selective and sensitive detection of copper ions. *Sensors Actuators, B Chem, 224*, 396–403. https://doi.org/10.1016/j.snb.2015.09.065
56. Liu, X., Li, T., Hou, Y., Wu, Q., Yi, J., & Zhang, G. (2016). Microwave synthesis of carbon dots with multi-response using denatured proteins as carbon source. *RSC Advances, 6*, 11711–11718. https://doi.org/10.1039/c5ra23081k
57. Feng, J., Wang, W. J., Hai, X., Yu, Y. L., & Wang, J. H. (2016). Green preparation of nitrogen-doped carbon dots derived from silkworm chrysalis for cell imaging. *J Mater Chem B, 4*, 387–393. https://doi.org/10.1039/c5tb01999k
58. Wen, X., Shi, L., Wen, G., Li, Y., Dong, C., Yang, J., & Shuang, S. (2016). Green and facile synthesis of nitrogen-doped carbon nanodots for multicolor cellular imaging and Co2+ sensing in living cells. *Sensors Actuators, B Chem, 235*, 179–187. https://doi.org/10.1016/j.snb.2016.05.066
59. D'Souza, S. L., Deshmukh, B., Bhamore, J. R., Rawat, K. A., Lenka, N., & Kailasa, S. K. (2016). Synthesis of fluorescent nitrogen-doped carbon dots from dried shrimps for cell imaging and boldine drug delivery system. *RSC Advances, 6*, 12169–12179. https://doi.org/10.1039/c5ra24621k

60. Zhang, Z., Sun, W., & Wu, P. (2015). Highly photoluminescent carbon dots derived from egg white: facile and green synthesis, photoluminescence properties, and multiple applications. *ACS Sustain Chem Eng, 3*, 1412–1418. https://doi.org/10.1021/acssuschemeng.5b00156
61. Liu, R., Zhang, J., Gao, M., Li, Z., Chen, J., Wu, D., & Liu, P. (2015). A facile microwave-hydrothermal approach towards highly photoluminescent carbon dots from goose feathers. *RSC Advances, 5*, 4428–4433. https://doi.org/10.1039/c4ra12077a
62. Yang, X., Zhuo, Y., Zhu, S., Luo, Y., Feng, Y., & Dou, Y. (2014). Novel and green synthesis of high-fluorescent carbon dots originated from honey for sensing and imaging. *Biosensors & Bioelectronics, 60*, 292–298. https://doi.org/10.1016/j.bios.2014.04.046
63. Sun, D., Ban, R., Zhang, P. H., Wu, G. H., Zhang, J. R., & Zhu, J. J. (2013). Hair fiber as a precursor for synthesizing of sulfur- and nitrogen-co-doped carbon dots with tunable luminescence properties. *Carbon N Y, 64*, 424–434. https://doi.org/10.1016/j.carbon.2013.07.095
64. Wang, Q., Liu, X., Zhang, L., & Lv, Y. (2012). Microwave-assisted synthesis of carbon nanodots through an eggshell membrane and their fluorescent application. *The Analyst, 137*, 5392–5397. https://doi.org/10.1039/c2an36059d
65. Sarswat, P. K., & Free, M. L. (2015). Light emitting diodes based on carbon dots derived from food, beverage, and combustion wastes. *Physical Chemistry Chemical Physics: PCCP, 17*, 27642–27652. https://doi.org/10.1039/c5cp04782j
66. Wen, X., Wen, G., Li, W., Zhao, Z., Duan, X., Yan, W., Trant, J. F., & Li, Y. (2021). Carbon dots for specific "off-on" sensing of Co2+ and EDTA for in vivo bioimaging. *Materials Science and Engineering C, 123*, 112022. https://doi.org/10.1016/j.msec.2021.112022
67. Chauhan, P., Chaudhary, S., & Kumar, R. (2021). Biogenic approach for fabricating biocompatible carbon dots and their application in colorimetric and fluorometric sensing of lead ion. *Journal of Cleaner Production, 279*, 123639. https://doi.org/10.1016/j.jclepro.2020.123639
68. Ahn, J., Song, Y., Kwon, J. E., Lee, S. H., Park, K. S., Kim, S., Woo, J., & Kim, H. (2019). Food waste-driven N-doped carbon dots: Applications for Fe3+ sensing and cell imaging. *Materials Science and Engineering C, 102*, 106–112. https://doi.org/10.1016/j.msec.2019.04.019
69. Jia, J., Lin, B., Gao, Y., Jiao, Y., Li, L., Dong, C., & Shuang, S. (2019). Highly luminescent N-doped carbon dots from black soya beans for free radical scavenging, Fe3+ sensing and cellular imaging. *Spectrochim Acta - Part A Mol Biomol Spectrosc, 211*, 363–372. https://doi.org/10.1016/j.saa.2018.12.034
70. Murugan, N., Prakash, M., Jayakumar, M., Sundaramurthy, A., & Sundramoorthy, A. K. (2019). Green synthesis of fluorescent carbon quantum dots from Eleusine coracana and their application as a fluorescence 'turn-off' sensor probe for selective detection of Cu 2+. *Applied Surface Science, 476*, 468–480. https://doi.org/10.1016/j.apsusc.2019.01.090
71. Roshni, V., Misra, S., Santra, M. K., & Ottoor, D. (2019). One pot green synthesis of C-dots from groundnuts and its application as Cr(VI) sensor and in vitro bioimaging agent. *Journal of Photochemistry and Photobiology, A: Chemistry, 373*, 28–36. https://doi.org/10.1016/j.jphotochem.2018.12.028
72. Lu, M., Duan, Y., Song, Y., Tan, J., & Zhou, L. (2018). Green preparation of versatile nitrogen-doped carbon quantum dots from watermelon juice for cell imaging, detection of Fe3+ ions and cysteine, and optical thermometry. *Journal of Molecular Liquids, 269*, 766–774. https://doi.org/10.1016/j.molliq.2018.08.101
73. Ansi, V. A., & Renuka, N. K. (2018). Table sugar derived Carbon dot – a naked eye sensor for toxic Pb2+ ions. *Sensors Actuators, B Chem, 264*, 67–75. https://doi.org/10.1016/j.snb.2018.02.167
74. Dang, D. K., Sundaram, C., Ngo, Y. L. T., Chung, J. S., Kim, E. J., & Hur, S. H. (2018). One pot solid-state synthesis of highly fluorescent N and S co-doped carbon dots and its use as fluorescent probe for Ag+ detection in aqueous solution. *Sensors Actuators, B Chem, 255*, 3284–3291. https://doi.org/10.1016/j.snb.2017.09.155
75. Moonrinta, S., Kwon, B., In, I., Kladsomboon, S., Sajomsang, W., & Paoprasert, P. (2018). Highly biocompatible yogurt-derived carbon dots as multipurpose sensors for detection of

formic acid vapor and metal ions. *Opt Mater (Amst), 81*, 93–101. https://doi.org/10.1016/j.optmat.2018.05.021

76. Yang, R., Guo, X., Jia, L., Zhang, Y., Zhao, Z., & Lonshakov, F. (2017). Green preparation of carbon dots with mangosteen pulp for the selective detection of Fe 3+ ions and cell imaging. *Applied Surface Science, 423*, 426–432. https://doi.org/10.1016/j.apsusc.2017.05.252
77. Wang, Z., Yu, J., Zhang, X., Li, N., Liu, B., Li, Y., Wang, Y., Wang, W., Li, Y., Zhang, L., Dissanayake, S., Suib, S. L., & Sun, L. (2016). Large-scale and controllable synthesis of graphene quantum dots from rice husk biomass: A comprehensive utilization strategy. *ACS Applied Materials & Interfaces, 8*, 1434–1439. https://doi.org/10.1021/acsami.5b10660
78. Shi, L., Li, X., Li, Y., Wen, X., Li, J., Choi, M. M. F., Dong, C., & Shuang, S. (2015). Naked oats-derived dual-emission carbon nanodots for ratiometric sensing and cellular imaging. *Sensors Actuators, B Chem, 210*, 533–541. https://doi.org/10.1016/j.snb.2014.12.097
79. Wei, J., Zhang, X., Sheng, Y., Shen, J., Huang, P., Guo, S., Pan, J., & Feng, B. (2014). Dual functional carbon dots derived from cornflour via a simple one-pot hydrothermal route. *Materials Letters, 123*, 107–111. https://doi.org/10.1016/j.matlet.2014.02.090
80. Zhuo, C., Alves, J. O., Tenorio, J. A. S., & Levendis, Y. A. (2012). Synthesis of carbon nanomaterials through up-cycling agricultural and municipal solid wastes. *Industrial and Engineering Chemistry Research, 51*, 2922–2930. https://doi.org/10.1021/ie202711h
81. Paul, A., & Kurian, M. (2021). Facile synthesis of nitrogen doped carbon dots from waste biomass: Potential optical and biomedical applications. *Clean Eng Technol, 3*, 100103. https://doi.org/10.1016/j.clet.2021.100103
82. Ang, W. L., Boon Mee, C. A. L., Sambudi, N. S., Mohammad, A. W., Leo, C. P., Mahmoudi, E., Ba-Abbad, M., & Benamor, A. (2020). Microwave-assisted conversion of palm kernel shell biomass waste to photoluminescent carbon dots. *Science and Reports, 10*, 1–15. https://doi.org/10.1038/s41598-020-78322-1
83. Chen, K., Qing, W., Hu, W., Lu, M., Wang, Y., & Liu, X. (2019). On-off-on fluorescent carbon dots from waste tea: Their properties, antioxidant and selective detection of CrO 42−, Fe 3+, ascorbic acid and L-cysteine in real samples. *Spectrochim Acta - Part A Mol Biomol Spectrosc, 213*, 228–234. https://doi.org/10.1016/j.saa.2019.01.066
84. Huang, C., Dong, H., Su, Y., Wu, Y., Narron, R., & Yong, Q. (2019). Synthesis of carbon quantum dot nanoparticles derived from byproducts in bio-refinery process for cell imaging and in vivo bioimaging. *Nanomaterials, 9*,. https://doi.org/10.3390/nano9030387
85. Das, P., Ganguly, S., Maity, P. P., Bose, M., Mondal, S., Dhara, S., Das, A. K., Banerjee, S., & Das, N. C. (2018). Waste chimney oil to nanolights: A low cost chemosensor for tracer metal detection in practical field and its polymer composite for multidimensional activity. *Journal of Photochemistry and Photobiology, B: Biology, 180*, 56–67. https://doi.org/10.1016/j.jphotobiol.2018.01.019
86. Xue, M., Zhan, Z., Zou, M., Zhang, L., & Zhao, S. (2016). Green synthesis of stable and biocompatible fluorescent carbon dots from peanut shells for multicolor living cell imaging. *New Journal of Chemistry, 40*, 1698–1703. https://doi.org/10.1039/c5nj02181b
87. D'Angelis Do E. S. Barbosa, C., Corrêa, J. R., Medeiros, G. A., Barreto, G., Magalhães, K. G., De Oliveira, A.L., Spencer, J., Rodrigues, M. O., & Neto, B. A. D. (2015). Carbon dots (C-dots) from cow manure with impressive subcellular selectivity tuned by simple chemical modification. *Chem - A Eur J, 21*, 5055–5060. https://doi.org/10.1002/chem.201406330.
88. Tan, X. W., Romainor, A. N. B., Chin, S. F., & Ng, S. M. (2014). Carbon dots production via pyrolysis of sago waste as potential probe for metal ions sensing. *Journal of Analytical and Applied Pyrolysis, 105*, 157–165. https://doi.org/10.1016/j.jaap.2013.11.001
89. Zhou, J., Sheng, Z., Han, H., Zou, M., & Li, C. (2012). Facile synthesis of fluorescent carbon dots using watermelon peel as a carbon source. *Materials Letters, 66*, 222–224. https://doi.org/10.1016/j.matlet.2011.08.081
90. Hsu, P. C., Shih, Z. Y., Lee, C. H., & Chang, H. T. (2012). Synthesis and analytical applications of photoluminescent carbon nanodots. *Green Chemistry, 14*, 917–920. https://doi.org/10.1039/c2gc16451e

91. Lim, C. S., Hola, K., Ambrosi, A., Zboril, R., & Pumera, M. (2015). Graphene and carbon quantum dots electrochemistry. *Electrochemistry Communications, 52*, 75–79. https://doi.org/10.1016/j.elecom.2015.01.023
92. Wang, Y., & Hu, A. (2014). Carbon quantum dots: Synthesis, properties and applications. *J Mater Chem C, 2*, 6921–6939. https://doi.org/10.1039/c4tc00988f
93. Shahidi, S., Rashidian, M., & Dorranian, D. (2018). Preparation of antibacterial textile using laser ablation method. *Optics & Laser Technology, 99*, 145–153. https://doi.org/10.1016/j.optlastec.2017.08.025
94. Namdari, P., Negahdari, B., & Eatemadi, A. (2017). Synthesis, properties and biomedical applications of carbon-based quantum dots: An updated review. *Biomedicine & Pharmacotherapy, 87*, 209–222. https://doi.org/10.1016/j.biopha.2016.12.108
95. Arora, N., & Sharma, N. N. (2014). Arc discharge synthesis of carbon nanotubes: Comprehensive review. *Diamond and Related Materials, 50*, 135–150. https://doi.org/10.1016/j.diamond.2014.10.001
96. Liang, F., Tanaka, M., Choi, S., & Watanabe, T. (2017). Formation of different arc-anode attachment modes and their effect on temperature fluctuation for carbon nanomaterial production in DC arc discharge. *Carbon N Y, 117*, 100–111. https://doi.org/10.1016/j.carbon.2017.02.084
97. Zuo, J., Jiang, T., Zhao, X., Xiong, X., Xiao, S., & Zhu, Z. (2015). Preparation and application of fluorescent carbon dots. *Journal of Nanomaterials, 2015*,. https://doi.org/10.1155/2015/787862
98. Zhou, J., Booker, C., Li, R., Zhou, X., Sham, T. K., Sun, X., & Ding, Z. (2007). An electrochemical avenue to blue luminescent nanocrystals from multiwalled carbon nanotubes (MWCNTs). *Journal of the American Chemical Society, 129*, 744–745. https://doi.org/10.1021/ja0669070
99. Hou, Y., Lu, Q., Deng, J., Li, H., & Zhang, Y. (2015). One-pot electrochemical synthesis of functionalized fluorescent carbon dots and their selective sensing for mercury ion. *Analytica Chimica Acta, 866*, 69–74. https://doi.org/10.1016/j.aca.2015.01.039
100. Lin, X., Xiong, M., Zhang, J., He, C., Ma, X., Zhang, H., Kuang, Y., Yang, M., & Huang, Q. (2021). Carbon dots based on natural resources: Synthesis and applications in sensors. *Microchemical Journal, 160*, 105604. https://doi.org/10.1016/j.microc.2020.105604
101. Sharma, A., & Das, J. (2019). Small molecules derived carbon dots: Synthesis and applications in sensing, catalysis, imaging, and biomedicine. *J Nanobiotechnology, 17*, 1–24. https://doi.org/10.1186/s12951-019-0525-8
102. Khan, Z. M. S. H., Rahman, R. S., Shumaila, Islam, S., & Zulfequar, M. (2019). Hydrothermal treatment of red lentils for the synthesis of fluorescent carbon quantum dots and its application for sensing Fe3+. *Opt Mater (Amst), 91*, 386–395. https://doi.org/10.1016/j.optmat.2019.03.054.
103. Menéndez, J. A., Arenillas, A., Fidalgo, B., Fernández, Y., Zubizarreta, L., Calvo, E. G., & Bermúdez, J. M. (2010). Microwave heating processes involving carbon materials. *Fuel Processing Technology, 91*, 1–8. https://doi.org/10.1016/j.fuproc.2009.08.021
104. Quan, X., Zhang, Y., Chen, S., Zhao, Y., & Yang, F. (2007). Generation of hydroxyl radical in aqueous solution by microwave energy using activated carbon as catalyst and its potential in removal of persistent organic substances. *Journal of Molecular Catalysis A: Chemical, 263*, 216–222. https://doi.org/10.1016/j.molcata.2006.08.079
105. Mansuriya, B. D., & Altintas, Z. (2021). Carbon dots: Classification, properties, synthesis, characterization, and applications in health care-an updated review (2018–2021). *Nanomaterials, 11*,. https://doi.org/10.3390/nano11102525
106. Sivakumar, M., & Dasgupta, A. (2019). Selected Area Electron Diffraction, a technique for determination of crystallographic texture in nanocrystalline powder particle of Alloy 617 ODS and comparison with precession electron diffraction. *Materials Characterization, 157*, 109883. https://doi.org/10.1016/j.matchar.2019.109883
107. Bokare, A., Nordlund, D., Melendrez, C., Robinson, R., Keles, O., Wolcott, A., & Erogbogbo, F. (2020). Surface functionality and formation mechanisms of carbon and graphene quantum

dots. *Diamond and Related Materials, 110*, 108101. https://doi.org/10.1016/j.diamond.2020.108101

108. Bhattacharjee, S. (2016). DLS and zeta potential - What they are and what they are not? *Journal of Controlled Release, 235*, 337–351. https://doi.org/10.1016/j.jconrel.2016.06.017
109. Hu, Q., Gong, X., Liu, L., & Choi, M. M. F. (2017). *Characterization and Analytical Separation of Fluorescent Carbon Nanodots, 2017*, 30–37.
110. Li, M., Chen, T., Gooding, J. J., & Liu, J. (2019). Review of carbon and graphene quantum dots for sensing. *ACS Sensors, 4*, 1732–1748. https://doi.org/10.1021/acssensors.9b00514
111. Lewis, G. N., & Kasha, M. (1944). Phosphorescence and the triplet state. *Journal of the American Chemical Society, 66*, 2100–2116. https://doi.org/10.1021/ja01240a030
112. Jelinek, R. (2017). Characterization and physical properties of carbon-dots. *Carbon Nanostructures,* 29–46. https://doi.org/10.1007/978-3-319-43911-2_3
113. Yao, J., Huang, C., Liu, C., & Yang, M. (2020). Upconversion luminescence nanomaterials: A versatile platform for imaging, sensing, and therapy. *Talanta, 208*, 120157. https://doi.org/10.1016/j.talanta.2019.120157
114. Khairol Anuar, N. K., Tan, H. L., Lim, Y. P., So'aib, M. S., & Abu Bakar, N. F. (2021). a review on multifunctional carbon-dots synthesized from biomass waste: design/ fabrication, characterization and applications. *Front Energy Res, 9,* 1–22. https://doi.org/10.3389/fenrg.2021.626549.
115. Wang, L., Weng, S., Su, S., & Wang, W. (2023). Progress on the luminescence mechanism and application of carbon quantum dots based on biomass synthesis. *RSC Advances, 13*, 19173–19194. https://doi.org/10.1039/d3ra02519e
116. Zhu, S., Song, Y., Zhao, X., Shao, J., Zhang, J., & Yang, B. (2015). The photoluminescence mechanism in carbon dots (graphene quantum dots, carbon nanodots, and polymer dots): Current state and future perspective. *Nano Research, 8*, 355–381. https://doi.org/10.1007/s12274-014-0644-3
117. Zhi, B., Yao, X. X., Cui, Y., Orr, G., & Haynes, C. L. (2019). Synthesis, applications and potential photoluminescence mechanism of spectrally tunable carbon dots. *Nanoscale, 11*, 20411–20428. https://doi.org/10.1039/c9nr05028k
118. Ding, H., Li, X. H., Chen, X. B., Wei, J. S., Li, X. B., & Xiong, H. M. (2020). Surface states of carbon dots and their influences on luminescence. *Journal of Applied Physics, 127*,. https://doi.org/10.1063/1.5143819
119. Ilaiyaraja, N., Fathima, S. J., & Khanum, F. (2018). *Quantum dots: A novel fluorescent probe for bioimaging and drug delivery applications.* Elsevier Inc. https://doi.org/10.1016/B978-0-12-813661-4.00012-2
120. Hu, J., Sun, Y., Aryee, A. A., Qu, L., Zhang, K., & Li, Z. (2022). Mechanisms for carbon dots-based chemosensing, biosensing, and bioimaging: A review. *Analytica Chimica Acta, 1209*, 338885. https://doi.org/10.1016/j.aca.2021.338885
121. Yuan, L., Lin, W., Zheng, K., & Zhu, S. (2013). FRET-based small-molecule fluorescent probes: Rational design and bioimaging applications. *Accounts of Chemical Research, 46*, 1462–1473. https://doi.org/10.1021/ar300273v
122. Sun, W., Li, M., Fan, J., & Peng, X. (2019). Activity-based sensing and theranostic probes based on photoinduced electron transfer. *Accounts of Chemical Research, 52*, 2818–2831. https://doi.org/10.1021/acs.accounts.9b00340
123. Chen, S., Yu, Y. L., & Wang, J. H. (2018). Inner filter effect-based fluorescent sensing systems: A review. *Analytica Chimica Acta, 999*, 13–26. https://doi.org/10.1016/j.aca.2017.10.026
124. Mahurpawar, M. (2015). Effects of heavy metals on human healtheffects of heavy metals on human health. *Int J Res –Granthaalayah, 3,* 1–7. https://doi.org/10.29121/granthaalayah.v3.i9se.2015.3282.
125. Tchounwou, P. B., Yedjou, C. G., Patlolla, A. K., & Sutton, D. J. (2012). Molecular, clinical and environmental toxicicology Volume 3: Environmental toxicology. *Molecular, Clinical and Environmental Toxicology.* https://doi.org/10.1007/978-3-7643-8340-4
126. Wee, S. S., Ng, Y. H., & Ng, S. M. (2013). Synthesis of fluorescent carbon dots via simple acid hydrolysis of bovine serum albumin and its potential as sensitive sensing probe for lead (II) ions. *Talanta, 116*, 71–76. https://doi.org/10.1016/j.talanta.2013.04.081

127. Xu, J., Jie, X., Xie, F., Yang, H., Wei, W., & Xia, Z. (2018). Flavonoid moiety-incorporated carbon dots for ultrasensitive and highly selective fluorescence detection and removal of Pb2+. *Nano Research, 11*, 3648–3657. https://doi.org/10.1007/s12274-017-1931-6
128. Bandi, R., Dadigala, R., Gangapuram, B. R., & Guttena, V. (2018). Green synthesis of highly fluorescent nitrogen – Doped carbon dots from Lantana camara berries for effective detection of lead(II) and bioimaging. *Journal of Photochemistry and Photobiology, B: Biology, 178*, 330–338. https://doi.org/10.1016/j.jphotobiol.2017.11.010
129. Gupta, A., Verma, N. C., Khan, S., Tiwari, S., Chaudhary, A., & Nandi, C. K. (2016). Paper strip based and live cell ultrasensitive lead sensor using carbon dots synthesized from biological media. *Sensors Actuators, B Chem, 232*, 107–114. https://doi.org/10.1016/j.snb.2016.03.110
130. Jing, S., Zhao, Y., Sun, R. C., Zhong, L., & Peng, X. (2019). Facile and high-yield synthesis of carbon quantum dots from biomass-derived carbons at mild condition. *ACS Sustain Chem Eng, 7*, 7833–7843. https://doi.org/10.1021/acssuschemeng.9b00027
131. Huang, H., Lv, J. J., Zhou, D. L., Bao, N., Xu, Y., Wang, A. J., & Feng, J. J. (2013). One-pot green synthesis of nitrogen-doped carbon nanoparticles as fluorescent probes for mercury ions. *RSC Advances, 3*, 21691–21696. https://doi.org/10.1039/c3ra43452d
132. Gu, D., Shang, S., Yu, Q., & Shen, J. (2016). Green synthesis of nitrogen-doped carbon dots from lotus root for Hg(II) ions detection and cell imaging. *Applied Surface Science, 390*, 38–42. https://doi.org/10.1016/j.apsusc.2016.08.012
133. Lu, W., Qin, X., Liu, S., Chang, G., Zhang, Y., Luo, Y., Asiri, A. M., Al-Youbi, A. O., & Sun, X. (2012). Economical, green synthesis of fluorescent carbon nanoparticles and their use as probes for sensitive and selective detection of mercury(II) ions. *Analytical Chemistry, 84*, 5351–5357. https://doi.org/10.1021/ac3007939
134. Xie, Y., Cheng, D., Liu, X., & Han, A. (2019). Green hydrothermal synthesis of N-doped carbon dots from biomass highland barley for the detection of Hg2+. *Sensors (Switzerland), 19*,. https://doi.org/10.3390/s19143169
135. Yu, J., Song, N., Zhang, Y. K., Zhong, S. X., Wang, A. J., & Chen, J. (2015). Green preparation of carbon dots by Jinhua bergamot for sensitive and selective fluorescent detection of Hg2+ and Fe3+. *Sensors Actuators, B Chem, 214*, 29–35. https://doi.org/10.1016/j.snb.2015.03.006
136. Ramanan, V., Siddaiah, B., Raji, K., & Ramamurthy, P. (2018). Green synthesis of multifunctionalized, nitrogen-doped, highly fluorescent carbon dots from waste expanded polystyrene and its application in the fluorimetric detection of Au3+ ions in aqueous media. *ACS Sustain Chem Eng, 6*, 1627–1638. https://doi.org/10.1021/acssuschemeng.7b02852
137. Liao, J., Cheng, Z., & Zhou, L. (2016). Nitrogen-doping enhanced fluorescent carbon dots: Green synthesis and their applications for bioimaging and label-free detection of Au3+ ions. *ACS Sustain Chem Eng, 4*, 3053–3061. https://doi.org/10.1021/acssuschemeng.6b00018
138. Das, P., Ganguly, S., Bose, M., Mondal, S., Das, A. K., Banerjee, S., & Das, N. C. (2017). A simplistic approach to green future with eco-friendly luminescent carbon dots and their application to fluorescent nano-sensor 'turn-off' probe for selective sensing of copper ions. *Materials Science and Engineering C, 75*, 1456–1464. https://doi.org/10.1016/j.msec.2017.03.045
139. Ma, X., Dong, Y., Sun, H., & Chen, N. (2017). Highly fluorescent carbon dots from peanut shells as potential probes for copper ion: The optimization and analysis of the synthetic process. *Mater Today Chem, 5*, 1–10. https://doi.org/10.1016/j.mtchem.2017.04.004
140. Zhu, X., Jin, H., Gao, C., Gui, R., & Wang, Z. (2017). Ratiometric, visual, dual-signal fluorescent sensing and imaging of pH/copper ions in real samples based on carbon dots-fluorescein isothiocyanate composites. *Talanta, 162*, 65–71. https://doi.org/10.1016/j.talanta.2016.10.015
141. Liu, S., Tian, J., Wang, L., Zhang, Y., Qin, X., Luo, Y., Asiri, A. M., Al-Youbi, A. O., & Sun, X. (2012). Hydrothermal treatment of grass: A low-cost, green route to nitrogen-doped, carbon-rich, photoluminescent polymer nanodots as an effective fluorescent sensing platform for label-free detection of Cu(II) ions. *Advanced Materials, 24*, 2037–2041. https://doi.org/10.1002/adma.201200164
142. Zhang, W., Li, N., Chang, Q., Chen, Z., & Hu, S. (2020). Making a cup of carbon dots for ratiometric and colorimetric fluorescent detection of Cu2+ ions. *Colloids Surfaces A Physicochem Eng Asp, 586*, 124233. https://doi.org/10.1016/j.colsurfa.2019.124233

143. Singh, J., Kaur, S., Lee, J., Mehta, A., Kumar, S., Kim, K. H., Basu, S., & Rawat, M. (2020). Highly fluorescent carbon dots derived from Mangifera indica leaves for selective detection of metal ions. *Science of the Total Environment, 720*, 137604. https://doi.org/10.1016/j.scitotenv.2020.137604
144. Sachdev, A., & Gopinath, P. (2015). Green synthesis of multifunctional carbon dots from coriander leaves and their potential application as antioxidants, sensors and bioimaging agents. *The Analyst, 140*, 4260–4269. https://doi.org/10.1039/c5an00454c
145. Huang, G., Chen, X., Wang, C., Zheng, H., Huang, Z., Chen, D., & Xie, H. (2017). Photoluminescent carbon dots derived from sugarcane molasses: Synthesis, properties, and applications. *RSC Advances, 7*, 47840–47847. https://doi.org/10.1039/c7ra09002a
146. Bandi, R., Gangapuram, B. R., Dadigala, R., Eslavath, R., Singh, S. S., & Guttena, V. (2016). Facile and green synthesis of fluorescent carbon dots from onion waste and their potential applications as sensor and multicolour imaging agents. *RSC Advances, 6*, 28633–28639. https://doi.org/10.1039/c6ra01669c
147. Kailasa, S. K., Ha, S., Baek, S. H., Phan, L. M. T., Kim, S., Kwak, K., & Park, T. J. (2019). Tuning of carbon dots emission color for sensing of Fe 3+ ion and bioimaging applications. *Materials Science and Engineering C, 98*, 834–842. https://doi.org/10.1016/j.msec.2019.01.002
148. Sun, X., He, J., Yang, S., Zheng, M., Wang, Y., Ma, S., & Zheng, H. (2017). Green synthesis of carbon dots originated from Lycii Fructus for effective fluorescent sensing of ferric ion and multicolor cell imaging. *Journal of Photochemistry and Photobiology, B: Biology, 175*, 219–225. https://doi.org/10.1016/j.jphotobiol.2017.08.035
149. Qi, H., Teng, M., Liu, M., Liu, S., Li, J., Yu, H., Teng, C., Huang, Z., Liu, H., Shao, Q., Umar, A., Ding, T., Gao, Q., & Guo, Z. (2019). Biomass-derived nitrogen-doped carbon quantum dots: Highly selective fluorescent probe for detecting Fe 3+ ions and tetracyclines. *Journal of Colloid and Interface Science, 539*, 332–341. https://doi.org/10.1016/j.jcis.2018.12.047
150. Atchudan, R., Edison, T. N. J. I., Perumal, S., Vinodh, R., Sundramoorthy, A. K., Babu, R. S., & Lee, Y. R. (2021). Leftover kiwi fruit peel-derived carbon dots as a highly selective fluorescent sensor for detection of ferric ion†. *Chemosensors, 9*, 1–15. https://doi.org/10.3390/chemosensors9070166
151. Kalanidhi, K., & Nagaraaj, P. (2021). Facile and Green synthesis of fluorescent N-doped carbon dots from betel leaves for sensitive detection of Picric acid and Iron ion. *Journal of Photochemistry and Photobiology, A: Chemistry, 418*, 113369. https://doi.org/10.1016/j.jphotochem.2021.113369
152. Li, F. M., Liu, J. M., Wang, X. X., Lin, L. P., Cai, W. L., Lin, X., Zeng, Y. N., Li, Z. M., & Lin, S. Q. (2011). Non-aggregation based label free colorimetric sensor for the detection of Cr (VI) based on selective etching of gold nanorods. *Sensors Actuators, B Chem, 155*, 817–822. https://doi.org/10.1016/j.snb.2011.01.054
153. Tyagi, A., Tripathi, K. M., Singh, N., Choudhary, S., & Gupta, R. K. (2016). Green synthesis of carbon quantum dots from lemon peel waste: Applications in sensing and photocatalysis. *RSC Advances, 6*, 72423–72432. https://doi.org/10.1039/c6ra10488f
154. Ma, Y., Lv, W., Chen, Y., Na, M., Liu, J., Han, Y., Ma, S., & Chen, X. (2019). Facile preparation of orange-emissive carbon dots for the highly selective detection of silver ions. *New Journal of Chemistry, 43*, 5070–5076. https://doi.org/10.1039/C8NJ06109B
155. Ayaz Ahmed, K. B., Kumar, P. S., & Veerappan, A. (2016). A facile method to prepare fluorescent carbon dots and their application in selective colorimetric sensing of silver ion through the formation of silver nanoparticles. *Journal of Luminescence, 177*, 228–234. https://doi.org/10.1016/j.jlumin.2016.04.053
156. Murugesan, P., Moses, J. A., & Anandharamakrishnan, C. (2020). One step synthesis of fluorescent carbon dots from neera for the detection of silver ions. *Spectroscopy Letters, 53*, 407–415. https://doi.org/10.1080/00387010.2020.1764589
157. Lu, H., Li, C., Wang, H., Wang, X., & Xu, S. (2019). Biomass-derived sulfur, nitrogen co-doped carbon dots for colorimetric and fluorescent dual mode detection of silver (I) and cell imaging. *ACS Omega, 4*, 21500–21508. https://doi.org/10.1021/acsomega.9b03198

158. Bai, H., Tu, Z., Liu, Y., Tai, Q., Guo, Z., & Liu, S. (2020). Dual-emission carbon dots-stabilized copper nanoclusters for ratiometric and visual detection of Cr2O72- ions and Cd2+ ions. *Journal of Hazardous Materials, 386*, 121654. https://doi.org/10.1016/j.jhazmat.2019.121654
159. Pandey, S. C., Kumar, A., & Sahu, S. K. (2020). Single step green synthesis of carbon dots from Murraya Koenigii leaves; A unique turn-off fluorescent contrivance for selective sensing of Cd (II) ion. *Journal of Photochemistry and Photobiology, A: Chemistry, 400*, 112620. https://doi.org/10.1016/j.jphotochem.2020.112620
160. Gu, D., Hong, L., Zhang, L., Liu, H., & Shang, S. (2018). Nitrogen and sulfur co-doped highly luminescent carbon dots for sensitive detection of Cd (II) ions and living cell imaging applications. *Journal of Photochemistry and Photobiology, B: Biology, 186*, 144–151. https://doi.org/10.1016/j.jphotobiol.2018.07.012
161. Radhakrishnan, K., & Panneerselvam, P. (2018). Green synthesis of surface-passivated carbon dots from the prickly pear cactus as a fluorescent probe for the dual detection of arsenic(iii) and hypochlorite ions from drinking water. *RSC Advances, 8*, 30455–30467. https://doi.org/10.1039/c8ra05861j
162. Hashemi, N., & Mousazadeh, M. H. (2021). Green synthesis of photoluminescent carbon dots derived from red beetroot as a selective probe for Pd2+ detection. *Journal of Photochemistry and Photobiology, A: Chemistry, 421*, 113534. https://doi.org/10.1016/j.jphotochem.2021.113534
163. Jayan, S. S., Jayan, J. S., Sneha, B., & Abha, K. (2021). Materials Today : Proceedings Facile synthesis of carbon dots using tender coconut water for the fluorescence detection of heavy metal ions. *Mater Today Proc, 43*, 3821–3825. https://doi.org/10.1016/j.matpr.2020.11.417
164. Das, M., Thakkar, H., Patel, D., & Thakore, S. (2021). Repurposing the domestic organic waste into green emissive carbon dots and carbonized adsorbent: A sustainable zero waste process for metal sensing and dye sequestration. *Journal of Environmental Chemical Engineering, 9*, 106312. https://doi.org/10.1016/j.jece.2021.106312
165. Sakaew, C., Sricharoen, P., Limchoowong, N., Nuengmatcha, P., Kukusamude, C., Kongsri, S., & Chanthai, S. (2020). Green and facile synthesis of water-soluble carbon dots from ethanolic shallot extract for chromium ion sensing in milk, fruit juices, and wastewater samples. *RSC Advances, 10*, 20638–20645. https://doi.org/10.1039/d0ra03101a
166. Athika, M., Prasath, A., Duraisamy, E., Sankar Devi, V., Selva Sharma, A., & Elumalai, P. (2019). Carbon-quantum dots derived from denatured milk for efficient chromium-ion sensing and supercapacitor applications. *Materials Letters, 241*, 156–159. https://doi.org/10.1016/j.matlet.2019.01.064
167. Huang, Q., Bao, Q., Wu, C., Hu, M., Chen, Y., Wang, L., & Chen, W. (2022). Carbon dots derived from Poria cocos polysaccharide as an effective "on-off" fluorescence sensor for chromium (VI) detection. *J Pharm Anal, 12*, 104–112. https://doi.org/10.1016/j.jpha.2021.04.004
168. Emami, E., & Mousazadeh, M. H. (2021). Green synthesis of carbon dots for ultrasensitive detection of Cu2+ and oxalate with turn on-off-on pattern in aqueous medium and its application in cellular imaging. *Journal of Photochemistry and Photobiology, A: Chemistry, 418*, 113443. https://doi.org/10.1016/j.jphotochem.2021.113443
169. Ran, Y., Wang, S., Yin, Q., Wen, A., Peng, X., Long, Y., & Chen, S. (2020). Green synthesis of fluorescent carbon dots using chloroplast dispersions as precursors and application for Fe3+ ion sensing. *Luminescence, 35*, 870–876. https://doi.org/10.1002/bio.3794
170. Mondal, T. K., Gupta, A., Shaw, B. K., Mondal, S., Ghorai, U. K., & Saha, S. K. (2016). Highly luminescent N-doped carbon quantum dots from lemon juice with porphyrin-like structures surrounded by graphitic network for sensing applications. *RSC Advances, 6*, 59927–59934. https://doi.org/10.1039/c6ra12148a
171. Liu, H., Zhang, Y., Liu, J. H., Hou, P., Zhou, J., & Huang, C. Z. (2017). Preparation of nitrogen-doped carbon dots with high quantum yield from: Bombyx mori silk for Fe(III) ions detection. *RSC Advances, 7*, 50584–50590. https://doi.org/10.1039/c7ra10130a

172. Roshni, V., & Divya, O. (2017). One-step microwave-assisted green synthesis of luminescent N-doped carbon dots from sesame seeds for selective sensing of Fe(III). *Curr Sci, 112,* 385–390. https://doi.org/10.18520/cs/v112/i02/385-390.
173. Li, Z., Ni, Y., & Kokot, S. (2015). A new fluorescent nitrogen-doped carbon dot system modified by the fluorophore-labeled ssDNA for the analysis of 6-mercaptopurine and Hg (II). *Biosensors & Bioelectronics, 74*, 91–97. https://doi.org/10.1016/j.bios.2015.06.014
174. Roshni, V., & Ottoor, D. (2015). Synthesis of carbon nanoparticles using one step green approach and their application as mercuric ion sensor. *Journal of Luminescence, 161*, 117–122. https://doi.org/10.1016/j.jlumin.2014.12.048
175. Zhao, J., Huang, M., Zhang, L., Zou, M., Chen, D., Huang, Y., & Zhao, S. (2017). Unique approach to develop carbon dot-based nanohybrid near-infrared ratiometric fluorescent sensor for the detection of mercury ions. *Analytical Chemistry, 89*, 8044–8049. https://doi.org/10.1021/acs.analchem.7b01443
176. Qin, X., Lu, W., Asiri, A. M., Al-Youbi, A. O., & Sun, X. (2013). Microwave-assisted rapid green synthesis of photoluminescent carbon nanodots from flour and their applications for sensitive and selective detection of mercury(II) ions. *Sensors Actuators, B Chem, 184*, 156–162. https://doi.org/10.1016/j.snb.2013.04.079
177. Wang, C., Sun, D., Zhuo, K., Zhang, H., & Wang, J. (2014). RSC advances application †. *RSC Advances, 4*, 54060–54065. https://doi.org/10.1039/C4RA10885J
178. Bhamore, J. R., Park, T. J., & Kailasa, S. K. (2020). Glutathione-capped Syzygium cumini carbon dot-amalgamated agarose hydrogel film for naked-eye detection of heavy metal ions. *J Anal Sci Technol, 11*,. https://doi.org/10.1186/s40543-020-00208-8
179. Krishnaiah, P., Atchudan, R., Perumal, S., Salama, E. S., Lee, Y. R., & Jeon, B. H. (2022). Utilization of waste biomass of Poa pratensis for green synthesis of n-doped carbon dots and its application in detection of Mn2+ and Fe3+. *Chemosphere, 286*, 131764. https://doi.org/10.1016/j.chemosphere.2021.131764
180. Torres Landa, S. D., Reddy Bogireddy, N. K., Kaur, I., Batra, V., & Agarwal, V. (2022). Heavy metal ion detection using green precursor derived carbon dots. *iScience, 25*, 103816. https://doi.org/10.1016/j.isci.2022.103816.

Peroxidase-Like Activity of Green Synthesized Carbon Quantum Dots

K. Radhakrishnan and Panneerselvam Perumal

Abstract This book chapter delves into the peroxidase-like activity of Carbon Quantum Dots (CQDs) synthesized through environmentally friendly methods. Green synthesis methods, such as hydrothermal and microwave-assisted techniques, are explored for their role in producing CQDs with unique catalytic properties. The investigation emphasizes the significance of sustainable nanomaterials, aligning their catalytic efficiency with eco-friendly practices. CQDs, known for biocompatibility, low toxicity, and tunable luminous properties, exhibit promising peroxidase-like activity. The study addresses the mechanistic insights behind CQDs' catalytic behavior, examining surface functional groups, heteroatom doping, quantum confinement effects, and the role of defects. These factors contribute to the enzyme-mimicking capabilities of CQDs, paving the way for optimized applications in biosensing, environmental remediation, and beyond. The synthesis methods are scrutinized, highlighting recent advances in green approaches using natural precursors. The chapter synthesizes diverse research sources, offering a comprehensive understanding of green-synthesized CQDs' peroxidase-like capabilities. From catalytic mechanisms to applications in cancer therapy and water purification, this chapter encapsulates the transformative potential of green-synthesized CQDs in science and technology.

Keywords Nanozymes · Carbon dots (CDs) · Enzyme-mimicking · Catalytic performance · Biomedical applications

K. Radhakrishnan
Department of Chemistry, Centre for Material Chemistry, Karpagam Academy of Higher Education, Coimbatore, Tamil Nadu 641021, India

P. Perumal (✉)
Department of Chemistry, SRM Institute of Science and Technology, Kattankulathur, Chennai, Tamil Nadu 603203, India
e-mail: panneerp1@srmist.edu.in; panneerchem82@gmail.com

V. Kumar et al. (eds.), *Green Carbon Quantum Dots*, Engineering Materials,
https://doi.org/10.1007/978-981-97-6203-3_6

1 Introduction

Carbon quantum dots (CQDs), characterized by their extraordinary catalytic properties, have garnered significant attention in the nanomaterials domain. This study delves into the peroxidase-like catalytic activity of CQDs, with a specific focus on those synthesized using eco-friendly processes. The introduction establishes a compelling connection between nanotechnology and biomimicry by emphasizing the pivotal role of CQDs' catalytic efficiency in promoting environmentally beneficial activities. This study provides an in-depth analysis of the fundamental challenges associated with CQDs, spanning production processes, intrinsic properties, and diverse applications. Notably, the study underscores the distinct advantages offered by CQDs in comparison to conventional quantum dots. These advantages include high water solubility, biocompatibility, tunable light characteristics, minimal toxicity, and facile manufacturing. A noteworthy aspect of this study is the observation that many CQDs exhibit nanozyme properties, opening up a novel avenue for exploration. Utilizing scientific terms, the study showcases the unique benefits of these nanozyme characteristics over enzymes derived from proteins through a comparative analysis. The discussion elucidates the ways in which CQDs, endowed with nanozyme properties, represent a promising frontier in the realm of catalysis [1].

Anticipating a notable upsurge in the exploration of nanomaterials with applications akin to peroxidase, this study investigates the catalytic properties of carbon quantum dots (CDs). The study is motivated by the widespread interest in peroxidases, which exhibit the ability to produce chromogenic products at low substrate concentrations. Peroxidases play a crucial role in various analytical, industrial, and environmental applications. The involvement of hydrogen peroxide in the oxidation of organic substrates, serving as a mediator between the synthesis and destruction of reactive oxygen species, adds a vital dimension to this process. Carbon quantum dots (CDs) have been extensively studied for their peroxidase-like activity, with a focus on both their production methods and diverse applications. These applications encompass bioimaging, cholesterol sensing, glucose detection, and immunoassays. Various efforts have been made to enhance the peroxidase-like properties of CDs, involving doping with chemicals or combination with other molecules. Despite significant advancements, our comprehension of the intricate mechanisms underlying CD-induced peroxidase-like activity remains limited. In contrast, other nanomaterials such as graphene oxide, metal-doped carbon quantum dots, carbon nanodots, and oxygenated carbon nanotubes derived from candle soots have seen elucidation of their mechanisms through research. However, controversies persist regarding the specific oxygen-containing functional groups responsible for these reactions [2].

The oxygen-based functional groups contribute to CDs' peroxidase-like activity, their specific roles vary widely. Importantly, the precise functions of these oxygen-containing groups in the context of CD's peroxidase-like activity and their connection with photophysical properties remain largely unexplored and warrant further investigation. The multifaceted nature of CDs, with their unique properties and potential applications, particularly in mimicking peroxidase activity, underscores the importance of ongoing research to unravel the intricate mechanisms and fully realize their capabilities [3].

The exploration of the peroxidase-like activity of green-synthesized CQDs extends across an expansive landscape of scientific inquiry. From the synthesis methods that strive to reduce environmental impact to the catalytic mechanisms underpinning their peroxidase-like behavior, the literature encompasses a wealth of insights. Moreover, the applications of CQD peroxidase mimics span across biosensing, cancer therapy, and water purification, showcasing their versatility and potential to address critical challenges in various domains. This chapter will synthesize and distill this wealth of knowledge from diverse research sources, providing readers with a comprehensive understanding of the green-synthesized CQDs' peroxidase-like capabilities and their potential transformative impact on science and technology.

2 Green Synthesis of Carbon Quantum Dots

Recent advances in the synthesis of CQDs have focused on environmentally friendly approaches that minimize the use of hazardous chemicals. Green synthesis methods, such as microwave-assisted, hydrothermal, and pyrolysis techniques, have gained popularity for their simplicity, cost-effectiveness, and reduced environmental impact. Several recent studies have highlighted the synthesis of CQDs using green precursors such as fruit extracts, biomass, and waste materials. These approaches not only yield CQDs with excellent peroxidase-like activity but also contribute to sustainable nanomaterial production [4].

2.1 Top-Down Strategy

To implement the top-down method, larger carbon compounds, such as activated carbon, fullerene, carbon nanotubes, graphene, graphite, and carbon soot, are disassembled into their component parts according to the top-down strategy. This disassembly is achieved through methods such as arc discharge, electrochemical techniques, and laser ablation. While this strategy is particularly effective for carbon structures with sp2 hybridization, it may face challenges related to energy gaps.

2.1.1 Laser Ablation Process

The laser ablation process, initially introduced to the scientific community by Sun and colleagues in 2006, employs high-energy laser pulses to produce carbon quantum dots (CQDs) from a target surface. Recent advancements, has high-lighted the consistent production of small, uniform CQDs with promising applications in cell imaging. The successfully synthesized homogeneous CQDs with high quantum yields (QYs) for bio-imaging using an ultrafast dual-beam pulsed laser ablation technique. Additionally, Buendia and his team have achieved the development of fluorescent CQDs for cell labeling through laser ablation methods. Nevertheless, addressing issues such as size variation, low quantum yields, and non-fluorescent properties in CQDs generated by this process is crucial. Common pre-treatments such as surface passivation and oxidation are employed to enhance fluorescence characteristics and QY.

2.1.2 Electrochemical Approach

The electrochemical oxidation method, was used for CQD synthesis. Typically, multi-wall carbon nanotubes (CNTs) serve as the initial carbon material and are reduced in size using tetra-butyl ammonium perchlorate as an electrolyte. Researchers like Zhao have successfully produced bright carbon nanomaterials through electrochemical oxidation, employing a graphite rod as the working electrode and phosphate as a pH buffer. This approach has been used to create water-soluble CQDs with adjustable luminosity. The synthesis of CQDs through electrochemical oxidation, highlighting its applicability at room temperature and pressure. Hou and colleagues have utilized the electrochemical treatment of sodium citrate and urea in deionized water to create CQDs emitting a brilliant blue light. While this method offers advantages like cost-effectiveness and simplicity, it is less commonly used due to limitations such as a restricted number of small molecule precursors and challenging purification processes.

2.1.3 Arc Discharge Method

The arc discharge method, initially purifying single-wall carbon nanotubes, involves the use of nitric acid as an oxidizing agent. This process enhances the water solubility of CQDs and introduces various functional groups on their surface. CQDs produced through this method exhibit 1.66% quantum yields when stimulated at 366 nm. However, it results in CQDs with varying particle sizes. Electrical flash techniques have been employed to isolate fluorescent nanomaterials from both untreated and nitric acid-oxidized carbon nanostructures. The carbon by-products from arc synthesis to create CQDs with up-conversion fluorescence, CQDs through arc discharge in water. However, this approach may introduce challenging-to-remove pollutants due to the complex composition of the CQDs.

2.2 Bottom-Up Approach

The bottom-up technique involves the creation of larger carbon structures by assembling smaller carbon sources, including polymers, amino acids, carbohydrates, and various organic materials. This approach necessitates the assembly of diverse carbon sources using techniques such as hydrothermal and solvothermal processes, pyrolysis, combustion, and microwave irradiation. The resulting carbon quantum dots (CQDs) exhibit variations in size and structure, influenced by parameters such as the solvent, precursor molecule structures, and reaction conditions. The cost-effectiveness, simplicity, and scalability of the bottom-up approach make it a preferred choice for large-scale production scenarios.

Selection of Precursor Materials: Both chemical and natural sources are viable options for selecting precursor materials in the bottom-up synthesis of CQDs. Chemical precursors encompass a range of substances, including sucrose, glucose, lactic acid, citric acid, glycerol, ascorbic acid, and ethylene glycol. On the other hand, natural sources provide diverse options, such as the latex of the aloe vera plant, the Ficus benghalensis tree, the seeds of the Artocarpus lakoocha tree, rice husks, the peel of the pomelo tree, and the leaves of the Azadirachta indica tree.

Influence of Precursor Selection on CQD Properties: The choice of precursor materials significantly impacts the size, structure, and properties of the synthesized CQDs. This study explores how different precursors, whether chemical or natural, contribute to the characteristics of the resulting CQDs. The discussion includes an examination of the influence of specific precursor molecules on the assembly process and the properties of the final carbon quantum dots (Fig. 1).

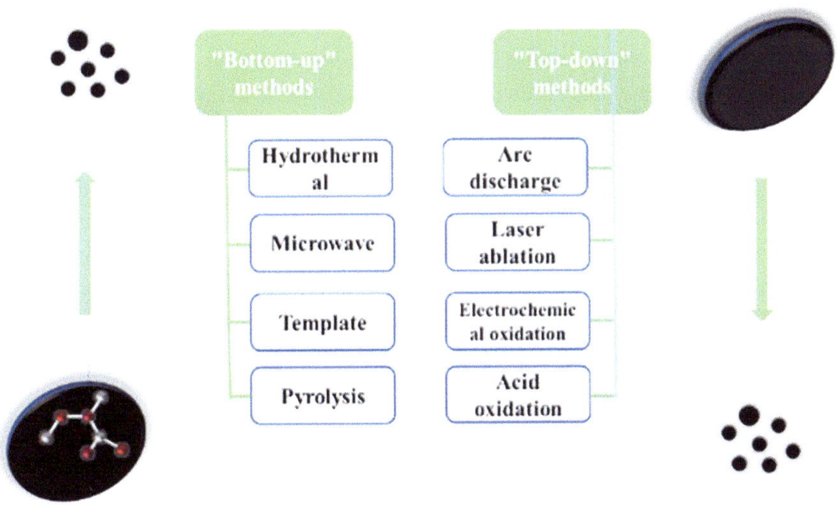

Fig. 1 Synthesis of Carbon Quantum Dots. (Adapted with permission [5], Copyrights 2023 MDPI)

3 Peroxidase-Like Activity of Carbon Quantum Dots

The peroxidase-like activity of CQDs involves the catalysis of peroxidase substrates, such as 3,3',5,5'-tetramethylbenzidine (TMB) and o-phenylenediamine (OPD), in the presence of hydrogen peroxide (H_2O_2). Recent studies have provided a comprehensive understanding of the dual response exhibited by carbon quantum dots (CQDs) in terms of colorimetric and fluorescence sensing. These investigations have successfully elucidated the intricate catalytic mechanisms underlying the peroxidase-like activity of CQDs. The acceleration of peroxidase substrates, specifically hydrogen peroxide (H_2O_2) and 2,2'-Azino-bis(3-ethylbenzothiazoline-6-sulfonic acid) (ABTS) occurs in presence of CDs. The detection of this reaction, characterized by an emission peak wavelength of CQDs, sets the stage for a comprehensive exploration of the CDs' kinetic parameters. Utilizing the Michaelis–Menten model and Lineweaver–Burk plot analysis, to unravel the intricate kinetics of CDs. The exceptional thermostability of CDs, operational even at 90 degrees Celsius, positions them as a formidable alternative to traditional enzymes like horseradish peroxidase (HRP). CDs can catalyze the oxidation of substrates by utilizing the surface functional groups, defects, and heteroatom doping as active sites. The generation of reactive oxygen species (ROS) during the reaction plays a pivotal role in mimicking the peroxidase activity. Researchers have utilized various spectroscopic and electrochemical techniques to probe these mechanisms and optimize CDs for specific applications [6]. Carbon Quantum Dots (CQDs) exhibit notable catalytic properties in the oxidation of diverse substrates, including 3,3,5,5-tetramethylbenzidine (TMB), o-phenylenediamine (OPD), and dopamine hydrochloride (DA) in the presence of hydrogen peroxide (H_2O_2). These catalytic reactions result in the generation of distinct colored products, providing a visual indication of CQDs' peroxidase-like activities. TMB Oxidation: TMB undergoes oxidation in the presence of H2O2, producing a blue-colored product with a maximum absorbance at 652 nm. This reaction mirrors the color change seen in traditional peroxidase-catalyzed reactions, highlighting CQDs' peroxidase-like activity. OPD Oxidation: OPD oxidation, catalyzed by CQDs in the presence of H_2O_2, yields a yellow-colored product with an absorbance peak at 417 nm. This observed absorbance indicates HRP-like activity of CQDs, resembling the colorimetric responses associated with horseradish peroxidase. DA Oxidation: DA undergoes oxidation in the presence of H_2O_2, resulting in an orange-colored product. Time-scan mode measurements, monitoring the absorbance change of DA at 480 nm, provide insights into the catalytic efficiency and kinetics of CQDs in the oxidation of DA, showcasing their potential in various applications.

3.1 Surface Functional Groups:

In the bottom-up technique, the synthesis of larger carbon structures involves the assembly of smaller carbon sources, encompassing amino acids, polymers, carbohydrates, and waste materials. This method necessitates the strategic assembly of these smaller carbon sources, employing various techniques such as hydrothermal and solvothermal processes, combustion, pyrolysis, and microwave irradiation. The resulting size and structure of Carbon Quantum Dots (CQDs) are influenced by parameters such as the solvent used, the structures of precursor molecules, and the specific reaction conditions. The simplicity, cost-effectiveness, and scalability of the bottom-up approach have contributed to its popularity in large-scale production scenarios. This method proves particularly advantageous when dealing with substantial quantities of CQDs. Precursor materials for CQD synthesis can be sourced from both chemical and natural origins. Chemical precursors encompass a variety of substances, including glucose, sucrose, citric acid, lactic acid, ascorbic acid, glycerol, and ethylene glycol. On the other hand, natural sources provide diverse options, such as the latex of the Ficus benghalensis tree, the aloe vera plant, rice husks, the seeds of the Artocarpous lakoocha tree, the leaves of the Azadirachta indica tree, and the peel of the pomelo tree. The careful selection of precursor materials, combined with the versatility of the bottom-up approach, enables the tailored synthesis of CQDs with desired properties, making them suitable for a wide range of applications in various fields [7].

3.2 Heteroatom Doping:

New active sites are introduced into the carbon lattice of CDs through the process of heteroatom doping, which is particularly effective with non-carbon elements such as nitrogen. As a result of the presence of nitrogen atoms, heteroatom-doped CDs have been shown to have improved peroxidase-like activity, as revealed by recent research. In order to make CDs more effective at catalyzing the oxidation of peroxidase substrates, these nitrogen atoms have the ability to function as catalytic centers, which means that they can change the electronic structure of CDs. As a result of the doping of nitrogen atoms, electron-rich and electron-poor sites are introduced into the structure of the CD, which enables effective electron transport throughout the catalytic process. The change of the electronic structure has the potential to improve the CDs' capacity to create reactive oxygen species (ROS) and successfully catalyze processes similar to those of peroxidase [8].

3.3 Quantum Confinement Effects

The small size of CDs leads to quantum confinement effects, which impact their electronic structure and, subsequently, their catalytic properties. Recent research has shown that the confinement of electrons within CDs can influence the redox processes involved in the catalysis of peroxidase substrates, making CDs efficient nanocatalysts. The quantum confinement of electrons alters the energy levels within CDs, affecting their ability to donate or accept electrons during the redox reactions. This quantum confinement effect is particularly crucial in controlling the kinetics of the catalytic process, making CDs effective peroxidase mimics.

3.4 Synergistic Effects

Some studies suggest that the peroxidase-like activity of CDs may result from a combination of factors, including the presence of surface functional groups, heteroatom doping, and quantum confinement effects. These synergistic effects can lead to enhanced catalytic performance. Understanding the interplay between these factors is crucial for optimizing CDs for specific applications. By considering the combined influence of surface functional groups, heteroatom doping, and quantum confinement effects, researchers can tailor CDs for various applications, optimizing their catalytic efficiency to achieve the desired outcomes.

3.5 Oxygen Vacancies and Defects

The presence of oxygen vacancies and defects within the structure of carbon dots (CDs) stands out as a significant factor influencing their peroxidase-like activity. Recent investigations have examined into the intricate role of defects in CDs, particularly oxygen vacancies, and their impact on catalytic behavior. These defects, acting as additional active sites for catalysis, induce alterations in the electronic properties of CDs. The introduction of oxygen vacancies and defects significantly amplifies the opportunities for electron transfer and reactive oxygen species (ROS) generation, thereby providing a robust enhancement to the peroxidase-like activity exhibited by CDs. This section sheds light on the pivotal role of defects in shaping the catalytic prowess of CDs, offering valuable insights into the mechanisms underlying their enhanced functionality.

4 Enzyme-Mimicking Properties of CDs

Carbon dots (CDs), nanoscale carbon-based materials, have garnered significant attention due to their remarkable enzyme-mimicking capabilities. These unique physicochemical qualities, including biocompatibility, water-solubility, fluorescence, and surface functionalization, make CDs versatile entities with applications in sensing, biomedical, optoelectronic, and catalytic fields. CDs exhibit enzyme-mimicking properties in three distinct ways: pure CDs with inherent enzyme-mimicking capabilities, CDs doped with different elements, and hybrid nanozymes incorporating CDs. The peroxidase activity of perfect CDs, attributing it to the presence of oxygen functional groups, especially carboxylic acid groups, on their surface. Notably, CDs surpass natural enzymes in terms of stability, robustness, and ease of production, enhancing their utility in diverse applications [9].

Doping as a Strategy for Enhancement: Doping has emerged as a strategy to enhance the characteristics of CDs. Copper-doped CDs, for instance, exhibited superior catalytic activity compared to pristine CDs, with enhanced electrical properties and surface chemical reactivities. Similarly, N-CDs and iron-doped CDs demonstrated successful applications in colorimetric glucose and hydrogen peroxide detection. Precise modulation of CDs' electronic structure and surface chemical reactivities is crucial for optimizing their physicochemical properties and catalytic activity [10].

Hybrid Nanozymes for Synergistic Effects: Hybrid nanozymes, such as the N-CDs/manganese oxide/ferric oxide hybrid, present a promising avenue for improving catalytic capabilities through the synergy of different elements. Despite challenges like the regular use of reducing agents and complex procedures, these hybrid systems outperform individual components [11].

Influence of Precursors on CD Quality: The choice of precursor materials significantly influences CD quality, affecting both optical properties and enzyme-mimicking activities. The impact of precursor selection on peroxidase-like activity, emphasizing the control of carboxylic groups by these choices. Understanding how precursors shape the visual properties and enzyme-mimicking powers of CDs is crucial for harnessing their full potential. The CDs, with their versatile enzyme-mimicking characteristics, offer a broad spectrum of applications. Their properties are intricately linked to manufacturing methods and precursor selection, emphasizing the need for precise control in their synthesis. This understanding is paramount for optimizing CDs for applications in biosensing and environmental investigation.

5 Peroxidase-Like Activity of Carbon Dots

Peroxidases play a crucial role in the acceleration of oxidation processes, which are responsible for the release of water molecules when peroxides are present. The purpose of this section is to investigate the peroxidase-like activity of carbon dots

(CDs), with a specific emphasis on their capacity to imitate the enzymatic capabilities of horseradish peroxidase (HRP). Catalysis is made easier by natural peroxidases, which frequently include a heme group in their structure. A triphasic oxidation cycle is carried out by HRP isoform C, which is a peroxidase that has been the subject of substantial research. This cycle includes the binding of peroxide to the heme group, the oxidation of iron atoms, and subsequent reduction steps that lead to the renewal of the enzyme [12].

CDs are examples of peroxidases that are built on nanoparticles that participate in Fenton reactions. These processes include the direct reduction of hydrogen peroxide to hydroxyl radicals (OH). The ability of CDs to simulate the activity of peroxidase is achieved through the alteration of their surfaces with hydrophilic groups such as carbonyl, carboxylic, amine, and hydroxyl groups. Particularly noteworthy is the fact that hydrogen bonding makes it easier for the carboxylic group to engage in surface binding affinity with hydrogen peroxide. As can be seen in Fig. 2, an inquiry is currently being carried out to determine whether or not it is possible to recreate the microenvironments where amino acids are found within materials. Taking the example of a histidine-functionalized graphene quantum dot (His-GQD)/hemin complex, for instance, a non-sequential catalytic mechanism is successfully demonstrated. It is important to execute catalytic oxidation of chromogenic substrates such as 2,2′-azino-bis(3-ethylbenzothiazoline-6-sulfonic acid) (ABTS) and 3,3′,5,5′-tetramethylbenzidine (TMB) in the presence of hydrogen peroxide in order to accomplish the evaluation of peroxidase-like activity [13].

As a result of their tiny size and considerable specific surface area, CDs are capable of producing a significant amount of OH during catalytic reactions. When

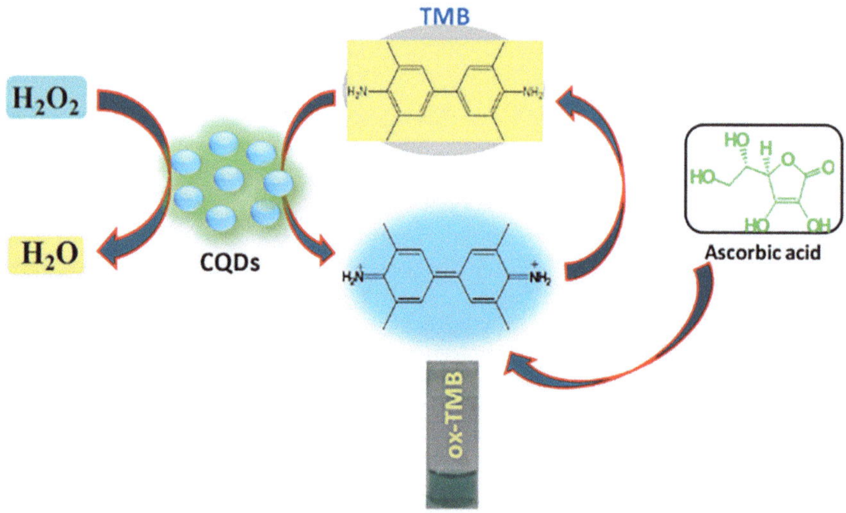

Fig. 2 Demonstration of peroxidase-like activity the CQDs.-assisted oxidation of TMB and H_2O_2. (Adapted with permission [5], Copyrights 2023 MDPI)

it comes to catalytic reactions that are assisted by CDs, electron transfer processes play an extremely important role. Increased radical production can be attributed to the transfer of electrons between substrates and hydrogen peroxide. In terms of kinetic characteristics, pure CDs and doped CDs, such as N and Fe-CDs, have catalytic activity that is comparable to or even surpasses that of HRP. CDs are superior to natural enzymes in terms of their ability to tolerate temperature, pH, and concentration ranges that are more expansive. In a wide variety of detection systems, CD-based nanozymes have demonstrated their potential. These systems include glucose and H2O2 detection, as well as cholesterol, glutathione, uric acid, and oxidative ions such as Fe3 + and Ag + . In addition, CDs have a remarkable catalytic efficiency in the elimination of phenolic compounds, in addition to other benefits like as thermal stability, cost-effectiveness, and reusability for the treatment of wastewater. The purpose of this section is to give a detailed examination of the peroxidase-like activity of CDs, with an emphasis on the various applications of CDs and their superior properties in comparison to natural enzymes [14, 15].

6 Oxidase and Laccase-Like Functionalities of Carbon Dots

In a wide variety of applications, oxidases play a key role since they are essential for catalysing the oxidation of substrates and producing water or hydrogen peroxide when they are present. In this section, the nanozyme properties of carbon dots (CDs) are investigated, with a particular emphasis placed on the oxidase and laccase-like capabilities of these nanozymes. Among the enzymes that are classified as oxidoreductases and cuproproteins, laccases are responsible for catalysing the oxidation of substances that contain ortho- and para-diphenol groups. Research on laccase-mimicking nanozymes, particularly in the context of biosensor applications, has been rather limited, in contrast to research on nanozymes that mimic electron transport chains. Because of the obvious color shift that occurs in their oxidising products, organic substrates that are often utilised, such as TMB, o-phenylenediamine (OPD), and ABTS, are utilised for the purpose of measuring oxidase-mimetic activity.

The CDs are known to be very effective nanozymes that are similar to oxidases. They exhibit oxidase-like activity, which can be related to the fact that they adsorb dissolved oxygen. The transformation of carbon dioxide into oxygen dioxide is accomplished through the process of electron transfer from delocalized electrons in CDs to oxygen. Nevertheless, in conditions of neutral pH, the catalytic activity of CDs that have not been altered in any way tends to gradually decrease. In order to imitate the actions of oxidase and laccase, novel CD-based hybrid materials have been developed. These materials incorporate components such as cerium ions and come equipped with a pH-dependent oxidase-like activity that can be easily controlled. The oxidase-like activity observed in pristine CDs is associated with the adsorption of dissolved oxygen, leading to the conversion of carbon dioxide to oxygen dioxide through electron transfer. However, under neutral pH conditions, the catalytic activity of pristine CDs tends to decline gradually. To address this issue, innovative CD-based hybrid materials that mimic the actions of oxidase and laccase have been developed. Examples of these materials include those containing cerium ions. These hybrids have a pH-dependent oxidase-like activity that can be tuned to specific amounts. Nanozymes based on CDs with metals added to them, such as cerium or manganese, were developed to aid in the absorption of oxygen by the active site of the nanocatalyst. This absorbed oxygen undergoes an electron transfer process, which ultimately culminates in the formation of singlet oxygen (O21). The inclusion of manganese to CDs-based nanozymes correlates with the reduction of dioxygen to oxygen. Due to the presence of manganese ions, this reduction produces hydrogen peroxide, which eventually degrades into hydroxyl (OH) [16].

Enzyme-mimicking research has predominantly focused on peroxidase-like activity, but there is a growing interest in nanozymes with oxidase-like functionality. This approach accelerates oxidation reactions without the need for additional oxidizing agents like hydrogen peroxide, mitigating the risk of analyte degradation. A variety of carbonaceous materials, including graphene and carbon nanospheres, mimic oxidase action, as do metal oxides (PtO2, CeO2), metal–organic frameworks, and metallic compounds (Ag, Au, and Pt). CDs, encompassing diverse types such as GQDs, CQDs, and CNDs, are particularly promising for oxidase-mimicking activity due to their low cytotoxicity, photochemical stability, and intense fluorescence.

Hybrid nanozymes based on CDs, exhibiting activities akin to oxidases and laccases, are instrumental in detecting a wide array of chemicals. For instance, CDs synthesized through hydrothermal polymerization display oxidase-like activity, inhibiting bacterial growth by catalyzing the rapid oxidation of TMB when exposed to ultraviolet light. Hybrid nanozymes combining CDs with materials like cerium (IV) have been designed for ratiometric fluorescence quantitative detection of pharmaceuticals such as alendronate sodium (ALDS). The development of nanozymes capable of mimicking oxidase and laccase functions has garnered significant attention due to their potential applications in diverse sectors, offering advantages in terms of sensitivity, efficiency, and adaptability [17].

7 Catalytic Properties of Carbon Dots Analogous to Catalase Enzyme

The catalase enzyme plays a pivotal role in mitigating oxidative stress induced by reactive oxygen species (ROS) by converting hydrogen peroxide (H_2O_2) into water and oxygen, as indicated by Eq. 6. Carbon dots (CDs) present a promising avenue for creating novel materials that mimic enzymatic functions, owing to their high biocompatibility and economic feasibility. Given the association of ROS and oxidative stress with numerous illnesses, catalase-mimicking systems, capable of neutralizing ROS like H_2O_2 and •OH, are crucial for therapeutic interventions in over 200 diseases, including cardiovascular disease, Alzheimer's disease, and metabolic disorders.

Recent advancements in catalase-mimicking systems have yielded notable examples, with a prime illustration being carbon dots (CDs) pyrolytically doped with copper and zinc (CuZn-CDs). This innovative development signifies a breakthrough in the creation of catalase-mimicking entities, showcasing the application of pyrolytic methods for metal incorporation. CuZn-CDs surpass their pure and single metal-doped counterparts in electron abundance, electron transport capacity, and stability, demonstrating enhanced antioxidant activity. Even under challenging conditions such as high temperatures and low pH levels, CuZn-CDs exhibit substantial antioxidant activity, proving their therapeutic potential for myocardial ischemia in both in vitro and in vivo studies, particularly in situations associated with lower pH, such as anaerobic metabolism and CO_2 retention.

The production of N-doped CDs with catalase-like activity has demonstrated how sensitive these CDs are when it comes to recognising iodide ions (I) in urine. The dual-readout approach is utilised by N-CDs in order to catalyse the H2O2-o-phenylenediamine (OPD) reaction. This method enhances sensitivity and selectivity while simultaneously minimising the impact of external influences. Because of this, the color of the aqueous solution changes from being colorless to yellow, as is generally expected. In addition, graphene oxide quantum dots (GOQDs) that are created through sonication exhibit features similar to those of catalase and antioxidant cells. The neuroprotective potential of these GOQDs has been demonstrated through their ability to mitochondrial damage, reduce oxidative stress, and apoptosis in vitro with PC12 cells that were stimulated with 1-methyl-4-phenyl-pyridinium ion (MPP +) and in vivo with zebrafish after MPP + therapy.

While creating CDs capable of effectively simulating catalase remains a challenge, recent research has made significant strides by constructing hybrid systems incorporating doped CDs. These hybrid systems hold enormous promise for specific therapeutic applications, including the treatment of oxidative stress-related disorders like heart ischemia and the reliable detection of oxygen in urine.

8 Resemblance to Superoxide Dismutase Activity

Superoxide dismutase, commonly referred to as SOD, assumes a critical role in the conversion of superoxide radicals (O2•) into hydrogen peroxide and, ultimately, molecular oxygen. Recent investigations have delved into nanozymes exhibiting superoxide dismutase activity derived from carbon. This exploration employs methodologies including heteroatomic doping and the fabrication of composite nanoparticles housing a substantial quantity of nanozymes.

In human cells, four distinct SOD isoforms can be detected. Copper, zinc, and manganese SOD are located in the cytoplasm, while manganese SOD is found in the mitochondria. Naturally occurring SODs require metallic cofactors for catalytic action. CDs, in general, exhibit the ability to scavenge reactive oxygen species (ROS) and generate hydrogen peroxide. Various catalytic processes have been proposed to elucidate this nanozyme activity, such as a hybrid structure composed of Au–N-CDs, charge and energy transfer pathways, and ROS formation. CDs with multifunctional capabilities mimicking superoxide dismutase (SOD) can eliminate harmful substances like hydrogen peroxide and superoxide radicals [18].

Carbon-based nanozymes have led to the development of a diverse range of nano-antioxidants imitating SOD activity, serving as viable alternatives to traditional SODs. Comparative studies between native SODs and SOD-mimicking carbon compounds reveal superior performance of carbon materials, especially at elevated pH levels. Heteroatomic doping of carbon quantum dots with elements like fluorine and chlorine enhances their antioxidant capabilities. Anti-oxidative and pro-oxidative capabilities are displayed by hybrid nanozymes, which are created by incorporating gold nanoparticles with nitrogen-doped carbon dots. Hybrid nanozymes have made it possible to imitate numerous enzymes, which has expanded the usefulness of nanozymes to levels that are comparable to those of natural enzymes. Nanoparticles that have been engineered or manufactured using a variety of enzyme-mimicking materials offer a wide range of capabilities and provide fresh insights. The ability of SOD-mimicking CDs to scavenge reactive oxygen species (ROS) makes them potentially useful for a wide variety of applications, including the treatment of oxidative stress-related disorders and the development of antibacterial medications. Bioimaging, antibacterial activity, and cytoprotective uses against diseases which are associated with oxidative stress, such as Alzheimer's disease, Parkinson's disease, and cardiac abnormalities, are all areas in which these CDs have shown promise [19].

9 Applications of CQD Peroxidase Mimics

Carbon Quantum Dots (CQDs) have beneficial properties that emphasize their stability and biocompatibility, resulting in efficient reactive oxygen species (ROS) generation within cells. CQDs, whether in pure form or hybrid structures, provide advantages such as degradation resistance, making them suitable for a wide range of applications. CQDzymes, enzyme-like entities based on CQDs, have advantages over natural enzymes, such as longer shelf life, fewer operating limitations, and lower prices. CQDzymes, on the other hand, may have lower catalytic activity than regular enzymes. Their application in very sensitive sensing systems, particularly for environmental monitoring, is emphasized, with an emphasis on their peroxidase-like enzymatic activity. CQDs catalyze the oxidation of hydrogen peroxide, resulting in the formation of hydroxyl radicals that alter the properties of substrates. The text also discusses the use of CQDs to accelerate the H2O2-mediated oxidation of peroxidase substrates, explaining the "ping-pong" catalytic pathway. Because of their fast electron transfer rate and strong substrate affinity, CQDs produced from precursors containing heteroatoms often exhibit increased catalytic activity.

9.1 Carbon Quantum Dots in Biological Sensing

Carbon quantum dots (C-dots) have a highly effective catalytic activity that is akin to peroxidase (POD). C-dots' distinctive color reactions have been proven using effective catalysis of POD substrates such as TMB, OPD, and THB in the presence of hydrogen peroxide (H_2O_2) in a sodium acetate buffer. The catalytic process, which involves an increase in electron density and mobility in C-dots, leads in accelerated electron transfer from C-dots to H_2O_2 and better TMB oxidation. The mechanism, however, is not fully known. Optimal catalytic activity is observed under particular circumstances (35 °C, pH 3.5, and 300 mM hydrogen peroxide). Temperature, pH, and the concentrations of hydrogen peroxide and carbon dots all affect this activity. It is worth noting that C-dots have steady catalytic activity throughout a wide pH range (2 to 12) and temperature range (0 to 9 degrees Celsius), distinguishing them from Horseradish Peroxidase (HRP). According to kinetic studies, this ping-pong mechanism is similar to that of HRP and suggests greater affinities to H_2O_2 and POD substrates as shown in Fig. 3.

Graphene quantum dots (GQDs) have received a lot of attention due to their amazing properties as a result of quantum confinement and edge effects. Strong photoluminescence (PL), great photostability, and low cytotoxicity are among these features. GQDs are created by hydrothermally treating carbon black with nitric acid. These GQDs show excellent catalysis of the TMB + H_2O_2 system, resulting in a

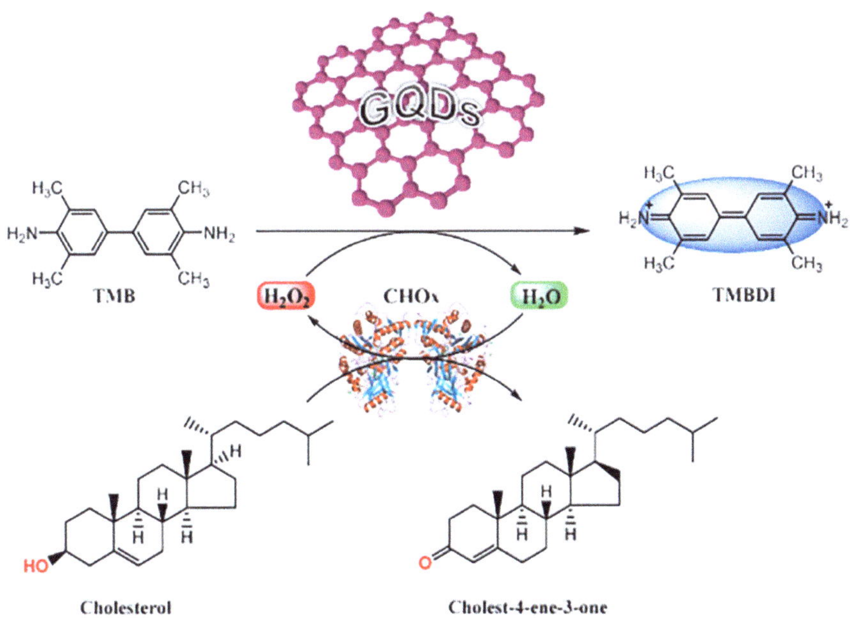

Fig. 3 Biosensing of cholesterol by peroxidase mimics of Cholesterol oxidase (CHOx) and GQDs-catalyst. (Adapted with permission [20], Copyrights 2023 MDPI)

distinctive blue hue when placed in an acetate buffer with a pH of 4.0. Because they have intrinsic POD activity, they have a high sensitivity for sensing H_2O_2, glucose, and GSH. GQD selectivity is highly excellent, especially when used to complex biological materials like cell lysate. It is worth noting that graphene quantum dots (GQDs) are expected to have stronger catalytic activity than large-sized graphene oxide (GO) due to their rapid diffusion rate and efficient interaction with proteins.

10 Future Directions and Challenges

Mechanistic Insights: While substantial progress has been achieved in comprehending the peroxidase-like activity of green-synthesized CQDs, there remains a compelling need for further research to unravel the intricate catalytic mechanisms underlying CQD peroxidase mimics. A deeper understanding of these mechanisms will significantly contribute to the rational design of more efficient catalytic systems. This entails a comprehensive exploration of the interactions between CQDs, peroxidase substrates, and hydrogen peroxide. Researchers should aim to elucidate the specific roles of surface functional groups, defects, heteroatom doping, and quantum confinement effects in the catalytic process. Moreover, the use of advanced analytical techniques, such as in-situ spectroscopy and computational modeling, can provide invaluable insights into the mechanistic intricacies of CQDs' peroxidase-like activity.

Biocompatibility: The biocompatibility of green-synthesized CQDs is a critical concern, especially in the context of biomedical applications. As CQDs gain momentum in fields like drug delivery, bioimaging, and theranostics, a comprehensive assessment of their long-term toxicity and safety profiles is imperative. This entails rigorous in vitro and in vivo studies to evaluate the impact of CQDs on cellular function, tissue compatibility, and overall organism health. Researchers should focus on elucidating the biodistribution, pharmacokinetics, and potential adverse effects of CQDs when administered in different forms, such as nanoparticles or functionalized with bioactive molecules. These studies are essential for ensuring the safe use of CQDs in medical applications and reducing potential risks to patients and the environment.

Scalability: For the practical application of green-synthesized CQDs, the development of scalable and reproducible synthesis methods is paramount. The ability to produce CQDs in large quantities with consistent quality is essential for their commercialization and widespread utilization. Researchers should explore various green precursors, such as waste materials, biomass, or agricultural by-products, to ensure a sustainable and eco-friendly source for CQD synthesis. Optimization of synthesis parameters, such as temperature, reaction time, and precursor composition, is crucial to achieve reproducibility and scalability. Moreover, the development of continuous flow synthesis methods and the integration of automation can streamline the production of CQDs for various applications, ranging from sensors to water purification systems.

Diversity in Nanozymes: Given the current limited repertoire of nanozymes, fostering further diversity in this field is imperative. The next generation of nanozymes should aspire to mimic the roles performed by semi-artificial enzymes.

Challenges in Creating Multifunctional Nanozymes: Crafting multifunctional nanozymes poses challenges. To achieve effective control over the catalytic process, a thorough comprehension of each CD enzyme is necessary. The application of coordination chemistry principles could enhance the rational design of nanozymes.

11 Conclusion

In the exploration of green-synthesized Carbon Quantum Dots (CQDs) and their peroxidase-like activity, a captivating landscape of possibilities has unfolded across nanotechnology, biomedicine, and environmental science. CQDs have emerged as versatile and eco-friendly nanomaterials, showcasing promise in diverse applications. Throughout this chapter, fundamental aspects, green synthesis methods, and the mechanisms driving peroxidase-like activity in CQDs are explored. The role of surface functional groups, defects, heteroatom doping, and quantum confinement effects in catalyzing peroxidase substrates and generating reactive oxygen species became apparent. This knowledge serves as a foundation for designing more efficient catalytic systems. The applications of CQD peroxidase mimics were explored extensively, ranging from biosensing and cancer therapy to water purification. Their high sensitivity, selectivity, and low detection limits position them as ideal candidates for point-of-carc diagnostics and cnvironmcntal monitoring. Thc ability to gcncratc cytotoxic reactive oxygen species opens avenues in targeted cancer treatment, while their efficacy in degrading organic contaminants contributes to efficient water purification systems.

As we navigate the future of CQDs, challenges and opportunities loom large. Deeper mechanistic insights, thorough investigations into biocompatibility, and the development of scalable synthesis methods are crucial. Enhanced understanding of catalytic mechanisms allows fine-tuning for specific applications. Ensuring safety and minimal toxicity in biomedical contexts is vital, and scalable production methods will broaden accessibility. The peroxidase-like activity of green-synthesized CQDs presents a dynamic field with the potential to impact multiple domains, offering innovative solutions. The journey has just begun, and future chapters are anticipated to bring new discoveries, further unlocking the secrets of CQDs and shaping the forefront of scientific inquiry. In this unfolding narrative, CQDs stand as beacons of innovation, promising transformative contributions to science and technology.

References

1. Wei, H., & Wang, E. (2013). Nanomaterials with enzyme-like characteristics (nanozymes):Next-generation artificial enzymes. *Chemical Society Reviews, 42*(14), 6060–6093.
2. Li, R., Zhen, M., Guan, M., Chen, D., Zhang, G., Ge, J., Gong, P., Wang, C., & Shu, C. (2013). A novel glucose colorimetric sensor based on intrinsic peroxidase-like activity of C60-carboxyfullerenes. *Biosensors & Bioelectronics, 47*, 502–507.
3. Chen, K., & Arnold, F. H. (2020). Engineering new catalytic activities in enzymes. *Nature Catalysis, 3*(3), 203–213.
4. Yadav, P. K., Chandra, S., Kumar, V., Kumar, D., & Hasan, S. H. (2023). Carbon quantum dots: synthesis, structure, properties, and catalytic applications for organic synthesis. *Catalysts, 13*(2), 422.
5. Chai, Y., Feng, Y., Zhang, K., & Li, J. (2022). Preparation of fluorescent carbon dots composites and their potential applications in biomedicine and drug delivery—A review. *Pharmaceutics., 14*(11), 2482.
6. Xu, C., Zhao, C., Li, M., Wu, L., Ren, J., & Qu, X. (2014). Artificial evolution of graphene oxide Chemzyme with Enantioselectivity and near-infrared Photothermal effect for cascade Biocatalysis reactions. *Small (Weinheim an der Bergstrasse, Germany), 10*, 1841–1847.
7. Jiang, Z., Li, H., Ai, R., Deng, Y., & He, Y. (2020). Electrostatic-driven coordination interaction enables high specificity of UO_2^{2+} peroxidase mimic for visual colorimetric detection of UO_2^{2+}. *ACS Sustainable Chem Eng, 8*, 11630–11637.
8. Su, C., Wang, B., Li, S., Wie, Y., Wang, Q., & Li, D. (2021). Fabrication of Pd@ZnNi-MOF/GO nanocomposite and its application for H_2O_2 detection and catalytic degradation of methylene blue dyes. *ChemistrySelect, 6*, 8480–8489.
9. Zhu, W., Zhang, J., Jiang, Z., Wang, W., & Liu, X. (2014). High-quality carbon dots: Synthesis, peroxidase-like activity and their application in the detection of H_2O_2, Ag^+ and Fe^{3+}. *RSC Advances, 4*, 17387–17392.
10. Lou, Y., Hao, X., Liao, L., Zhang, K., Chen, S., Li, Z., Ou, J., Qin, A., & Li, Z. (2021). Recent advances of biomass carbon dots on syntheses, characterization, luminescence mechanism, and sensing applications. *Nano Select, 2*, 1117–1145.
11. Liu, M. L., Chen, B. B., Li, C. M., & Huang, C. Z. (2019). Carbon dots: Synthesis, formation mechanism, fluorescence origin and sensing applications. *Green Chemistry, 21*, 449–471.
12. Zheng, C., Ke, W., Yin, T., & An, X. (2016). Intrinsic peroxidase-like activity and the catalytic mechanism of gold@carbon dots nanocomposites. *RSC Advances, 6*, 35280–35286.
13. Berglund, G. I., Carlsson, G. H., Smith, A. T., Szöke, H., Henriksen, A., & Hajdu, J. (2002). The catalytic pathway of horseradish peroxidase at high resolution. *Nature, 417*, 463–468.
14. Veitch, N. C. (2004). Horseradish peroxidase: A modern view of a classic enzyme. *Phytochemistry, 65*, 249–259.
15. Xin, Q., Jia, X., Nawaz, A., Xie, W., Li, L., & Gong, J. R. (2020). Mimicking peroxidase active site microenvironment by functionalized graphene quantum dots. *Nano Research, 13*, 1427–1433.
16. Dan, X., Ruiyi, L., Qinsheng, W., Yongqiang, Y., Guangli, W., & Zaijun, L. (2021). Synthesis of silver nanocrystal with an excellent oxidase-like activity and its application in colorimetric detection of D-penicillamine. *Microchemical Journal, 166*, 106204.
17. Ju, P., Wang, Z., Zhang, Y., Zhai, X., Jiang, F., Sun, C., & Han, X. (2020). Enhanced oxidase-like activity of Ag@Ag2WO4 nanorods for colorimetric detection of Hg^{2+}. *Colloids and Surfaces A: Physicochemical and Engineering Aspects, 603*, 125203.
18. Spagnolo, L., Törö, I., D'Orazio, M., O'Neill, P., Pedersen, J. Z., Carugo, O., Rotilio, G., Battistoni, A., & Djinović-Carugo, K. (2004). Unique features of the sodC-encoded superoxide dismutase from mycobacterium tuberculosis, a fully functional copper-containing enzyme lacking zinc in the active site. *Journal of Biological Chemistry, 279*, 33447–33455.

19. Zhao, L., Ren, X., Zhang, J., Zhang, W., Chen, X., & Meng, X. (2020). Dendritic silica with carbon dots and gold nanoclusters for dual nanozymes. *New Journal of Chemistry, 44*, 1988–1992.
20. Garg, B., & Bisht, T. (2016). Carbon nanodots as peroxidase nanozymes for biosensing. *Molecules, 21*(12), 1653.
21. Sun, H., Zhao, A., Gao, N., Li, K., Ren, J., & Qu, X. (2015). Deciphering a nanocarbon-based artificial peroxidase: Chemical identification of the catalytically active and substrate-binding sites on graphene quantum dots. *Angewandte Chemie International Edition, 54*, 7176–7180.

Sensing Activity of Green Synthesized Carbon Quantum Dots for the Detection of Heavy Metal Ions

K. Radhakrishnan and Panneerselvam Perumal

Abstract Heavy metal ions pose a significant threat to human health and the environment due to their toxicity and persistence. The development of efficient and environmentally friendly sensors for their detection is of utmost importance. In this chapter, we explore the recent literature on the use of green-synthesized carbon quantum dots (CQDs) as sensing platforms for heavy metal ions. We discuss the synthesis methods, sensing mechanisms, and potential applications of these innovative materials, emphasizing their eco-friendly nature.

1 Introduction

In recent years, the exploration of nanomaterials for their diverse applications has witnessed a significant surge. Among these, Carbon Quantum Dots (CQDs) have emerged as promising candidates, particularly in the realm of heavy metal ion detection [1]. This book chapter delves into the intriguing world of "Sensing Activity of Green Synthesized Carbon Quantum Dots for the Detection of Heavy Metal Ions [2]." The synthesis of carbon quantum dots (CQDs) through the utilisation of methods that are both environmentally friendly and sustainable, popularly known as "green synthesis," has achieved a great deal of interest due to the fact that it has the potential to revolutionise sensor technologies [3].

Beginning with the purification of carbon nanotubes, the tiniest members of the carbon family, CQDs were first discovered. Although their precise origins are still being investigated, these tiny carbon nanolights have shown themselves to have extraordinary properties such as a high quantum yield and tunable luminosity. They

K. Radhakrishnan
Department of Chemistry, Centre for Material Chemistry, Karpagam Academy of Higher Education, Coimbatore, Tamil Nadu 641021, India

P. Perumal (✉)
Department of Chemistry, SRM Institute of Science and Technology, Kattankulathur, Chennai, Tamil Nadu 603203, India
e-mail: panneerp1@srmist.edu.in; panneerchem82@gmail.com

V. Kumar et al. (eds.), *Green Carbon Quantum Dots*, Engineering Materials,
https://doi.org/10.1007/978-981-97-6203-3_7

are chosen sensor probes because of their intrinsic diversity, which is reflected in their surface capabilities, environmental compatibility, and non-toxicity [4].

While several counterparts, including indicator dyes and semiconductor quantum dots, have encountered challenges related to complex synthesis, bleaching, or toxicity concerns, CQDs have surged ahead. The advantages of green synthesis methods have been utilised in the production of these nanomaterials. These approaches entail the utilisation of sustainable and environmentally friendly precursors, such as plant extracts, biomass, and waste materials [5, 6]. Recent literature has illuminated various green synthesis techniques, including hydrothermal, microwave-assisted, and pyrolysis methods [7]. These environmentally conscious approaches not only reduce the ecological footprint but also yield CQDs with desirable properties, making them ideal for sensing applications [8].

CQDs have captured the scientific community's interest as versatile sensory platforms, especially in the context of heavy metal ion detection [9]. Their sensing mechanisms are rooted in their unique properties and interactions with the target analytes. Analyte-induced quenching or enhanced fluorescence of CQDs, often mediated through electron or energy transfer mechanisms, is the foundation upon which their sensing prowess rests. Recent research has shed light on a multitude of heavy metal detection applications, highlighting the significant role CQDs can play in safeguarding water resources and human health [10].

The pollution of water resources by heavy metals is a global concern, given their toxicity and non-biodegradable nature, even at trace levels [11]. Monitoring their levels and trends in water sources is of utmost importance, as it affects water safety and human well-being. These inorganic pollutants primarily manifest as cations (e.g., Hg^{2+}, Cd^{2+}, Pb^{2+}) and anions (e.g., AsO_4^{3-}, $Cr_2O_7^{2-}$), with some elements being essential for certain metabolic and cellular functions in trace amounts [12]. However, the consequences of exposure to toxic metals like Hg and As are dire, leading to health issues, cellular dysfunction, oxidative stress, and the generation of reactive oxygen species [13].

Traditional methods for qualitative and quantitative metal measurement, such as inductively atomic absorption spectroscopy, coupled plasma mass spectrometry, and X-ray fluorescence analysis, have proved invaluable but have their drawbacks. In contrast, optical and electrochemical sensing techniques have emerged as attractive alternatives due to their sensitivity, simple technology, and field applicability. However, the usefulness of such field-applicable methods is contingent on their meeting crucial characteristics including sensitivity, selectivity, stability, non-toxicity, reusability, and cost-effectiveness, and producing accurate and reliable results within a genuine matrix.

Metallic nanoparticles, oxides, semiconductor nanostructures, chalcogenides and polymers are only some of the nanomaterials that have been investigated as possible sensors for heavy metals in water over the years [14]. CQDs are a new type of sensor that has greatly increased the variety of technologies that may be used to detect heavy metals [15]. In light of the continued difficulties in heavy metal detection, CQDs present a promising alternative due to their distinctive features and the ease with which their surfaces can be modified. Here, we take a systematic look at the growing importance of CQDs synthesised using environmentally friendly methods for detecting metals in water. The adaptable, cost-effective, and environmentally benign nature of CQDs presents them as possible game-changers in this vital field [16].

Recent developments have shed light on the utility of both unmodified and modified Carbon Quantum Dots (CQDs) in detecting heavy metal ions across diverse water sources, ranging from tap water to groundwater and wastewater [17]. Researchers have meticulously explored the intricacies of these versatile nanomaterials, employing innovative approaches such as surface functionalization, surface passivation, and atomic doping to fine-tune their spectral properties. These strategic interventions have paved the path for cutting-edge sensing solutions that are adept at addressing the ever-expanding challenges associated with heavy metal monitoring in water.

This chapter will provide a comprehensive examination of the advantages and disadvantages associated with the use of CQDs as eco-sensitive sensors for the detection of heavy metals and metalloids in water. Heavy metal detection might undergo a revolutionary change if CQDs were to be combined with emerging technologies that could be used in the field. This would make the process far more approachable, productive, and cost-effective across a broad spectrum of uses. At the conclusion of this chapter, readers will have a comprehensive knowledge of how CQDs are destined to transform the recognition of heavy metal ions and contribute to a future that is both safer and healthier [18].

2 Green Synthesis Approaches for Carbon Quantum Dots

The synthesis of Carbon Quantum Dots (CQDs) has been a subject of extensive research, with a focus on achieving efficiency, cleanliness, and high yield. The synthesis approaches can be broadly categorized into two main routes: the top-down approach and the bottom-up approach [19]. These methods not only reduce the environmental impact but also yield CQDs with desirable properties, making them suitable for sensing applications.

2.1 Top-Down Methodology

Using the top-down technique, bigger carbonaceous materials or structures are split or broken up into tiny bits. This is often performed through the use of processes such as chemical oxidation, discharge, electrochemical oxidation, and ultrasonic treatments. The top-down strategy has been successful, despite the fact that it has a number of limitations. Some of these limitations include the necessity to employ expensive components, the application of demanding reaction conditions, and the imposition of lengthy reaction periods. As a result of these drawbacks, the top-down strategy is less desirable than other strategies [20].

2.2 *Bottom-Up Methodology*

The bottom-up method primarily aims to achieve the transition of more compact carbon structures into CQDs of the necessary size. Many other types of bottom-up synthesis have been investigated by scientists. These include solvothermal synthesis, hydrothermal synthesis, ultrasonic synthesis, thermal decomposition synthesis, pyrolysis synthesis, carbonization synthesis, and microwave synthesis. These approaches are employed, most notably in biomedical applications, because to the fact that they are able to construct CQDs with the qualities that are needed as well as uniform size distributions [21].

Benefits and Drawbacks: For CQD synthesis, a number of top-down techniques have been developed, including thermal oxidation, electrochemical oxidation, laser ablation, and pyrolysis. These techniques do, however, have certain drawbacks, such as low product yields, limited spectrum efficiency, an inability to adjust size, and the use of hazardous chemicals or high temperatures. The need for innovative, simple, and eco-friendly processes that adhere to the ideas of sustainability, renewable resources, and green chemistry is therefore rising. In comparison to bottom-up approaches, top-down methods are less favorable due to their impure results, limited yields, and non-environmentally friendly processes. Higher yields, more control over CQD size, and greener, cleaner synthesis techniques are all provided by bottom-up approaches. Bottom-up approaches are also more cost-effective [22].

2.2.1 Dominant Methods: Hydrothermal/Pyrolysis

The hydrothermal and pyrolysis methods have gained prominence due to their high quantum yields, straightforward synthesis, and environmentally friendly procedures. Additionally, these methods can effectively utilize biomass waste, representing an abundant and clean carbon source. Although bottom-up methods, such as hydrothermal and pyrolysis, may have some downsides, like low quantum yields, their overall benefits overshadow those of the top-down approaches as shown in Fig. 1.

(i) Hydrothermal Synthesis:

Hydrothermal synthesis is a common and effective method for preparing CQDs. In this process, a precursor material is subjected to high-temperature and high-pressure conditions in an aqueous environment. The combination of heat and pressure facilitates the controlled formation of CQDs. Plant extracts, biomass, or waste materials are often used as the carbon source, and the reaction conditions can be fine-tuned

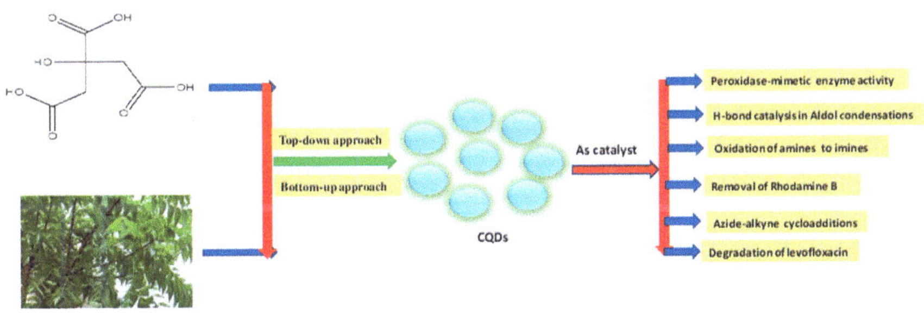

Fig. 1 Synthesis approaches of CQDs classified as top-down approach and the bottom-up. (Adapted with permission [26], Copyrights 2023 MDPI)

to yield CQDs with specific properties. The hydrothermal method is favored for its ability to produce CQDs with high quantum yields and tailored surface functionalities [23].

(ii) Microwave-assisted Synthesis:

Microwave-assisted synthesis is a rapid and efficient technique for generating CQDs. It involves exposing the precursor material to microwave radiation, which accelerates the carbonization process. This method is known for its speed and the capability to produce CQDs with tunable luminescence properties. Plant extracts and biomass can serve as carbon sources, and the microwave conditions can be adjusted to control the size and properties of the CQDs [24].

(iii) Pyrolysis:

Pyrolysis is another method used for the green synthesis of CQDs. During pyrolysis, the precursor material is heated to high temperatures in the absence of oxygen. This results in the decomposition of the precursor into CQDs. Pyrolysis is recognized for its simplicity and ability to yield CQDs with excellent photoluminescent properties. Waste materials, such as agricultural residues or even food waste, can be employed as carbon sources in this process [25].

Each of these green synthesis methods offers distinct advantages and allows for the tailoring of CQD properties. The choice of method depends on factors such as the desired CQD characteristics and the availability of suitable precursors. Green-synthesized CQDs have garnered attention in sensing applications for their eco-friendliness, sustainability, and the ability to produce nanomaterials with exceptional optical and electronic properties, making them valuable tools in various fields, including environmental monitoring and health sciences.

3 Sensing Mechanisms of Carbon Quantum Dots

The sensing of heavy metal ions by CQDs relies on their unique properties and interactions with the target analytes. Recent research has demonstrated several sensing mechanisms, including:

3.1 Fluorescence Quenching

CQDs exhibit strong and tunable fluorescence properties, which can be quenched by the presence of heavy metal ions through surface interactions or energy transfer processes. Recent studies have elucidated the quenching mechanisms and explored their applications in fluorescence-based sensing.

In the realm of fluorescence sensing, Carbon Quantum Dots (CQDs) have left indelible marks due to their non-toxic and hydrophilic nature, aligning seamlessly with environmental considerations. Consequently, there has been an upsurge of interest among researchers in CQDs, driven by the necessity to establish facile synthetic routes and harness renewable carbon sources. CQDs, celebrated for their ability to minimize production costs and enable large-scale synthesis, have emerged as favored fluorescent nanoprobes. Their fluorescence characteristics are subject to modification upon interaction with metal ions, leading to fluorescence quenching. The thoughtful design of CQD-based sensors holds the key to optically discerning specific analytes, notably metal ions. The choice of the carbon source for CQD synthesis remains a pivotal criterion in this context [27].

3.1.1 Optical Sensing Mechanisms of CQDs

CQDs' optical signals are modified after interacting with metal ions, causing a decrease in emission intensity or a shift in the absorption spectrum during optical sensing. This change is instructive because it reveals how the target metal ions are acting. Fluorescent sensors based on metal ions that do not interfere with the CQDs' signal are nevertheless inconvenient. Therefore, it is crucial to build appropriate sensors using effective sensing methods. The following sections will go into detail about the various sensing strategies typically used with CQD-based sensors. The most common of these is called the "Turn On–Off" method, and it involves interacting directly with CQDs and metal ions to dampen the optical emission intensity and therefore turn off the fluorescence of the CQDs. Calibration curves allow for the quantification of this quenching effect [28].

3.1.2 Mechanisms for Suppressing Excitation in CQD-Based Sensors

The "Turn On–Off-On" method employs Carbon Quantum Dots (CQDs) in conjunction with fluorophores and chelating agents. Metal ions quench CQD fluorescence, but as their concentration increases, the connection between CQDs and metal ions can be disrupted, leading to the revival of CQD fluorescence. Analyte analysis is determined by the percentage of recovered CQDs, with the analyte requiring a high affinity for the quencher to release and restore CQD fluorescence.

"Post-functionalization" involves introducing heteroatoms or groups into CQDs to enhance receptor affinity. Specifically, the addition of nitrogen facilitates the conjugation of heavy metal ions like Cu^{2+}, Hg^{2+}, and Fe^{3+}. This post-functionalization technique encompasses various quenching mechanisms crucial for metal ion sensing detection.

- **Dynamic Quenching Effect (DQE)** quencher continually associates and dissociates with the fluorophore, influencing its fluorescence intensity.
- **Static Quenching Effect (SQE)** results in the formation of a stable compound between the quencher and the fluorophore, extinguishing fluorescence.
- **Photo-induced Electron Transfer (PET)** quenches fluorescence by absorbing electrons from the excited fluorophore.
- **Inner Filter Effect (IFE)** occurs when the quencher absorbs transmitted light, diminishing the fluorescence signal and affecting measurement accuracy.
- **Fluorescence Resonance Energy Transfer (FRET)** comprises the transmission of energy from the excited fluorophore to the neighboring quencher, resulting in fluorescence quenching.

This comprehensive method aids in the detection and interpretation of metal ions, showcasing versatility and adaptability in various sensing applications. Despite the extensive body of research on the chemical synthesis of CQDs for metal ion sensing, our primary emphasis is on utilizing environmentally safe and renewable precursors to detect a wide range of metal ions. In the following sections, we will delve into the theory behind metal ion sensing as well as the practical uses of this technique for the detection of heavy metal ions [29].

3.1.3 Static Quenching Effect (SQE) in Green Charged Quantum Dots (CQDs) as a Quenching Mechanism

The non-fluorescent transformation of the surface of green CQDs is achieved through the SQE. This involves the formation of a complex with a quencher, typically metal ions in their ground state. The [CQDs-quencher] complex absorbs light upon activation, hindering emission and returning the system to its ground state. Key characteristics of SQE include: Alterations in the absorption spectra of CQDs due to the formation of a non-fluorescent molecule acting as a quencher for CQDs in the ground state. The fluorescence lifespan remains nearly identical ($t_0/t = 1$) before and after the CQD quenching process Exposure to heat reduces the stability of the [CQDs-quencher] complex, which in turn reduces SQE.

The incorporation of a quencher, such as heavy metal ions, into Carbon Quantum Dots (CQDs) results in a decrease in fluorescence emission intensity. Quantifying this quenching effect is often achieved through the use of a Stern–Volmer plot. By applying the Stern–Volmer equation, the quenching rate constant (kq) can be determined. A linear relationship is observed between the emission intensity in the absence of the metal ion quencher (F_0) and the intensity in the presence of increasing concentrations of the quencher (F), where Q is greater than F. The Stern–Volmer constant is represented by the slope of the Stern–Volmer plot, and the intercept of the plot is equal to one.

3.1.4 Dynamic Quenching Effect (DQE) and Green Charge-Coupled Devices

The Dark Quenching Effect (DQE) is the opposite of the Stern–Volmer Quenching Effect (SQE). It occurs when excited CQDs encounter metal ion quenchers, causing the CQDs to return to their ground state without emitting light. DQE is characterized by the following features as shown in Fig. 2.

(i) Unlike SQE, CQDs do not exhibit changes in their absorption bands during the collision in the excited state.
(ii) Significant alterations in fluorescence lifetime measurements are observed before and after quenching CQDs with metal ions.
(iii) DQE is more stable at higher temperatures.

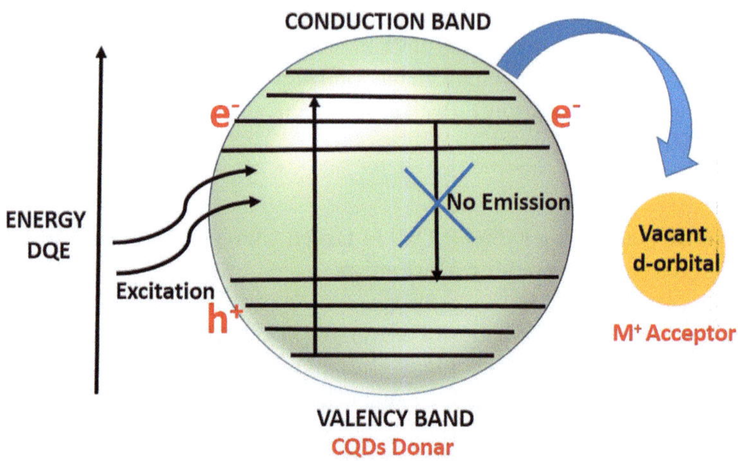

Fig. 2 Systematic Representation of Green Carbon Quantum Dots (CQDs) for Heavy Metal Ion Sensing Using Dynamic Quenching Effect (DQE)

Experimental evidence supports these distinctive characteristics, providing strong support for the underlying mechanism of DQE. Metal ions such as Pb^{2+}, Fe^{3+}, V^{5+}, Cd^{2+}, and Cu^{2+} have been successfully detected using DQEs, with green CQDs being widely employed for this purpose [30].

3.1.5 The Mechanism of Quenching in Green CQDs, Photoinduced Electron Transfer (PET)

Non-emitting electron transfer events between CQDs and metal ion quenchers are orchestrated by the Photoinduced Electron Transfer (PET) mechanism, a version of the electron transfer (ET) mechanism as shown in Fig. 3. Here, CQDs and the metal ion quencher both play a role as electron donors or acceptors, resulting in the formation of anion or cation radicals. Reductive PET, in which CQDs operate as electron acceptors, and oxidative PET, in which CQDs act as electron donors, are two distinct types of PET. By preventing electrons from returning to their ground state after being photo excited, fluorescence quenching occurs in PET when an electron is transported from the donor to the acceptor. For both reductive PET and oxidative PET, the driving force of this electron transfer is the energy gap between the CQDs and the metal ion quencher. Notably, nitrogen doping in CQDs offers them the capacity to rapidly conjugate with heavy metal ions [31].

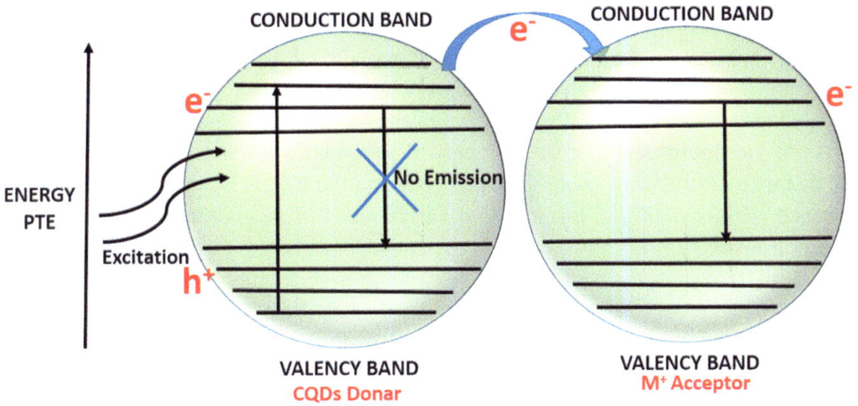

Fig. 3 Scalable Representation of Green Carbon Quantum Dots (CQDs) for Heavy Metal Ion Sensing by Photoinduced Electron Transfer (PET)

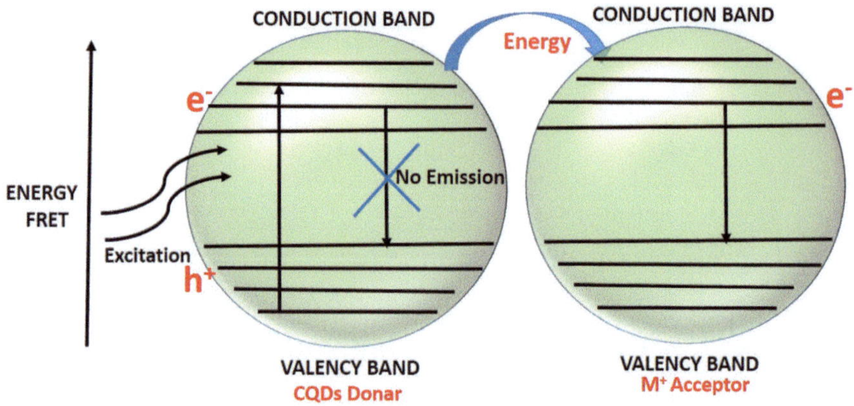

Fig. 4 Scalable Representation of Green Carbon Quantum Dots (CQDs) for Heavy Metal Ion Sensing Utilizing by Photoinduced Electron Transfer (PET)

3.1.6 Fluorescence Resonance Energy Transfer (FRET) in Green CQDs as a Quenching Mechanism

FRET is a technique that involves the transfer of energy between a donor molecule and an acceptor molecule within a range of around 10 nm, mostly through dipole–dipole interactions. This process does not entail the use of radiation. As can be seen in Fig. 4, the molecules that act as donors can cause a variety of different types of FRET to take place. Two fluorophores, one of which has a higher energy than the other, are the components that make up a conventional FRET pair. This transfer of energy leads in a drop in the fluorescence of the donor, while concurrently producing a rise in the fluorescence of the acceptor. It is possible to get an approximation of the distance between the metal ion quencher and the CQDs by utilizing the Förster distance (R0), which is a formula that can be used to assess the efficiency of this energy transfer [32].

3.1.7 Green CQDs Exhibit a Quenching Mechanism Known as the Inner Filter Effect (IFE).

In the development of fluorescence-based sensors, Electronic Energy Transfer (EET) involves the transfer of excitation energy between chromophores without photon emission. This method is utilized in sensors to induce changes in fluorescence intensity. Fluorescence Resonance Energy Transfer (FRET) transfers energy from an excited donor to an acceptor fluorophore through non-radiative coupling, widely used in sensors to study molecular interactions. Intramolecular Charge Transfer (ICT) involves electron transfer within a molecule, causing changes in electronic structure and optical properties. In sensors, ICT induces alterations in fluorescence intensity or wavelength, detecting changes in the sensor environment. These methods, requiring

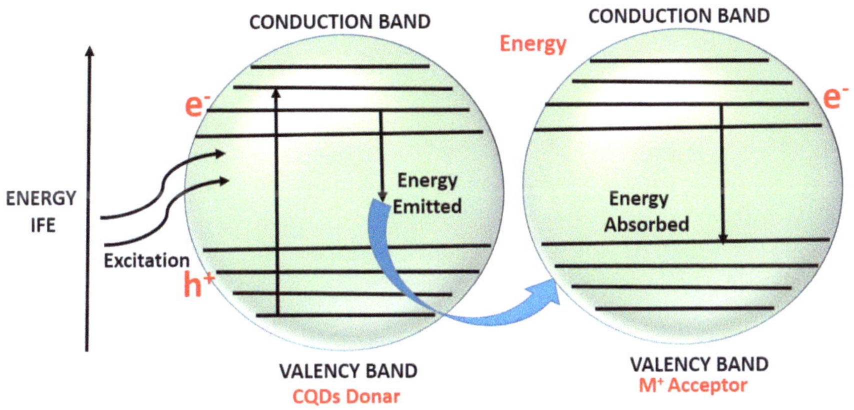

Fig. 5 Intrinsic fluorescence enhancement (IFE) is used to illustrate the systematic representation of green carbon quantum dots (CQDs) for the purpose of heavy metal ion sensing

precise chromophore arrangements, contribute to the design and optimization of complex yet versatile fluorescence-based sensors (see Fig. 5).

An intriguing alternative to these methods is Intra-Fluorescence Emission (IFE), which can be employed in sensors. IFE occurs due to the overlap of the quencher's absorption spectrum with the excitation or emission spectra of the fluorophore, resulting in induced fluorescence. Unlike more conventional quenching effects, IFE does not involve the loss of energy or electron transfer. Instead, IFE is observed when there is a high concentration of fluorophores or other absorbing species in a solution, causing attenuation of the excitation beam and absorption of the emitted light. Importantly, IFE fluorescence intensity decreases exponentially, distinguishing it from typical quenching events.

In-situ Fluorescence Enhancement (IFE) is a fluorescence-based sensor strategy that does not require modification of the fluorophore or involve electron/energy transfer processes like FRET or PET. Furthermore, IFE can occur even when the separation between the emitter and absorber exceeds 10 nm, simplifying sensor design and operation.

By applying a correction to the fluorescence spectra, as shown in Equation:

$$= \frac{I_{corr}}{I_{absd}} = 10(Aex.x1 + Aem.y1)$$

We can confirm that the fluorescence quenching mechanism of CQDs is IFE.

When the excitation and emission wavelengths are considered, the absorbance per centimeter is denoted by the letters Aex and Aem. The optical path length of excitation and emission beams is denoted by the symbols $\times$ 1 and y1, respectively. This correction helps confirm the presence of IFE in the fluorescence quenching mechanism of CQDs, offering a clearer understanding of the sensor's operation.

3.1.8 Quenching Mechanism: Combined Mechanisms in Green CQDs

In some instances, fluorescence quenching of CQDs emerges from the synergy of two or more principles, a testament to the adaptability of CQDs in metal ion sensing. These combined mechanisms, such as Surface Quenching Effect (SQE) and Intra-Fluorescence Emission (IFE), highlight the multifaceted nature of CQDs in this domain. Notably, green CQDs derived from flax straw have proven instrumental in sensing Co2 + and Cr6 + ions through the cooperative operation of SQE and IFE.

This chapter has elucidated the various mechanisms underpinning the deployment of green Carbon Quantum Dots (CQDs) as fluorescent probes for metal ion sensing, with each mechanism offering unique advantages and nuances in detecting different metal ions.

3.2 Electrochemiluminescence (ECL) Sensors for Analytical Applications

Electrochemiluminescence (ECL) sensors represent a rapidly evolving class of analytical devices that combine the principles of electrochemistry and luminescence to offer remarkable sensitivity, selectivity, and versatility in detecting a wide range of analytes. ECL sensors have found applications in several fields, comprising clinical diagnostics, environmental monitoring, and drug discovery. This note explores the recent advancements and key features of ECL sensors for analytical purposes.

One key advancement is the utilization of novel nanomaterials in ECL sensor design. The incorporation of nanomaterials, such as metal nanoparticles and carbon-based nanotubes, has improved the electrocatalytic properties of ECL sensors. This enhancement leads to increased sensitivity, reduced detection limits, and the ability to notice trace quantities of heavy metal ions in various sample matrices. These advancements are particularly valuable in environmental monitoring, where stringent regulations necessitate the detection of heavy metal pollutants at extremely low concentrations.

Miniaturization and portability have also made ECL sensors for heavy metal analysis more accessible and practical. Smaller, portable ECL platforms with integrated microfluidics and user-friendly interfaces have been developed. These miniaturized systems enable on-site testing and field applications, which are crucial for real-time monitoring of heavy metal contamination in water sources, soil, and industrial settings.

Furthermore, the integration of specific recognition elements, such as molecularly imprinted polymers (MIPs) and aptamers, into ECL sensor surfaces has improved selectivity. This selectivity is essential for distinguishing between different heavy metal ions and reducing false positive results.

The detection of a wide variety of heavy metals, such as cadmium (Cd), lead (Pb), arsenic (As), mercury (Hg), and chromium (Cr), is now possible with ECL sensors, which are used in environmental monitoring activities. Their high sensitivity and low detection limits enable the rapid identification of heavy metal pollutants in environmental samples, ensuring compliance with stringent regulations and protecting ecosystems.

ECL sensors have also found applications in food safety, where they are employed to detect heavy metal contaminants in food products, such as seafood and vegetables. The ability to perform rapid, on-site testing for heavy metal residues ensures the safety and quality of food consumed by the public.

The recent advances in ECL sensors for heavy metal analysis have expanded their capabilities, making them powerful tools for detecting and quantifying heavy metal pollutants. These sensors offer improved sensitivity, selectivity, and portability, and they find applications in environmental monitoring, food safety, and industrial quality control. The continued development of ECL sensors is expected to further enhance their performance and contribute to the efficient management of heavy metal contamination in various sectors [33].

3.3 Colorimetric Sensing

Colorimetric assays based on green-synthesized Carbon Quantum Dots (CQDs) have gained significant attention for their simple and interpretable method of quickly detecting heavy metal ions. The appeal of these assays lies in their simplicity, providing a visible output that capitalizes on the unique optical properties of CQDs, resulting in a discernible color change in response to specific heavy metal ions.

Recent advancements in the field have focused on developing green synthesis methods for CQDs, utilizing environmentally friendly and sustainable sources such as plant extracts or agricultural byproducts. Green-synthesized CQDs exhibit exceptional performance in colorimetric sensing of heavy metals.

A key advantage of green-synthesized CQDs is their biocompatibility and low toxicity, making them suitable for diverse applications in environmental monitoring, food safety, and medical diagnostics. The use of renewable and eco-friendly sources aligns with the growing emphasis on sustainable and green chemistry practices.

The mechanisms behind colorimetric detection with CQDs involve selective interactions between heavy metal ions and CQDs, resulting in changes to their optical properties. This interaction induces alterations in the CQDs' surface charge or energy band structure, leading to shifts in absorption or emission spectra and a visually detectable color change corresponding to the presence and concentration of target heavy metal ions.

Green-synthesized CQDs have been successfully applied for the detection of various heavy metal ions, including lead (Pb), mercury (Hg), cadmium (Cd), and copper (Cu), showcasing their potential in environmental monitoring, industrial quality management, and food safety assurance. Their suitability for on-site and rapid screening procedures further enhances their applicability, protecting the security and integrity of our environment, food chain, and industrial processes. Ongoing developments in this field are expected to contribute to the continued progress of environmentally friendly and efficient sensing technologies (Fig. 6).

4 Applications of Green-Synthesized CQDs for Heavy Metal Ion Detection

The utilization of green-synthesized Carbon Quantum Dots (CQDs) for heavy metal ion detection transcends various domains, benefiting from their eco-friendly nature, low toxicity, and distinctive optical properties. These properties make them versatile tools for tackling environmental, safety, and healthcare challenges. Below, we delve into the extensive applications of green-synthesized CQDs in the detection of heavy metal ions, with a focus on the nature of the functional groups and selectivity for specific heavy metal ions [35]

4.1 Environmental Monitoring

Green-synthesized CQDs often possess abundant surface functional groups such as carboxyl, hydroxyl, and amine. These functional groups enable strong chelation with heavy metal ions.

The functional groups on CQD surfaces allow for selective binding to specific heavy metal ions. For example, carboxyl-functionalized CQDs exhibit a high affinity for heavy metals like lead (Pb) and cadmium (Cd). Hydroxyl-functionalized CQDs can selectively detect mercury (Hg) ions due to the strong affinity between hydroxyl groups and Hg^{2+}.

4.2 Food Safety

Green-synthesized CQDs used in food safety applications are often surface-functionalized with groups such as carboxyl and amino groups. These functional groups facilitate the selective adsorption of heavy metal ions. CQDs' surface functionalization imparts selectivity, enabling the discrimination of heavy metal ions. For instance, amino-functionalized CQDs exhibit excellent selectivity for copper (Cu) ions, ensuring the safety of copper-contaminated food products.

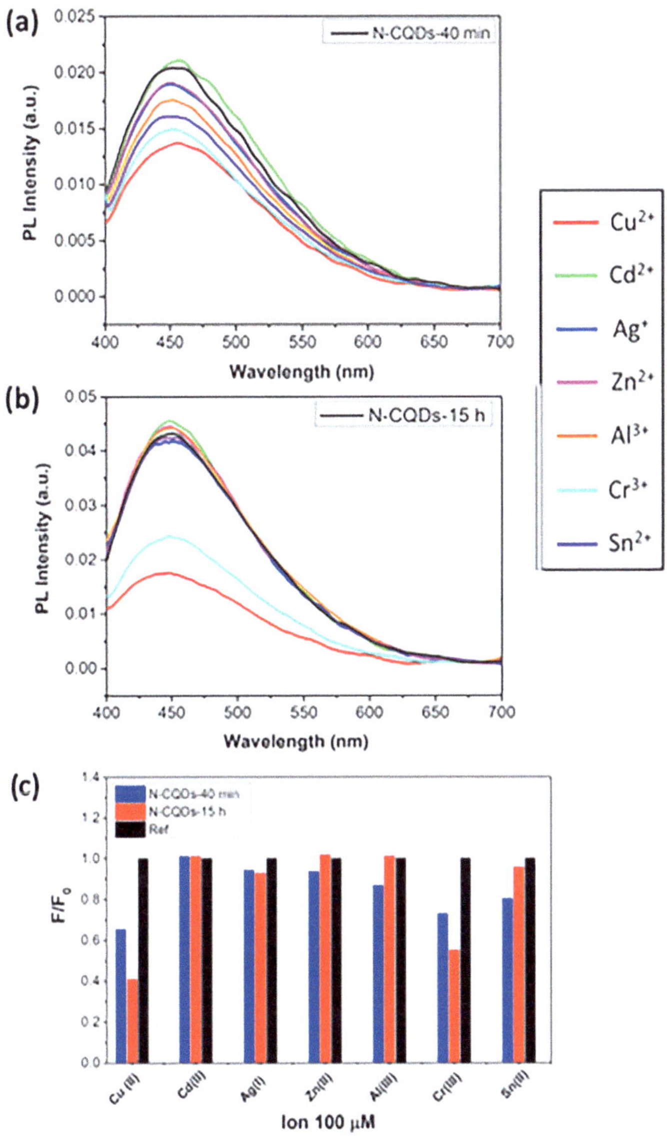

Fig. 6 The fluorescence spectra were obtained by illuminating the N-CQDs at a wavelength of 360 nm for a period of forty minutes (**a**) and 15 h (**b**), respectively, while the N-CQDs were in the presence of a total of one 100 micromolar (μM) metal ions. As shown in (**c**), the fluorescence quenching responses of N-CQDs-40 min (represented by blue bars) and N-CQDs-15 h (represented by red bars) to various metal ions at the same concentration (100 μM) are depicted. In the experiment that served as a control, N-CQDs were used without the presence of metal ions. (Adapted with permission from [34], Copyright 2023 MDPI)

4.3 Biomedical Applications

Biocompatible Carbon Quantum Dots (CQDs) with amino and hydroxyl surface functionalization are extensively applied in biomedicine, particularly for detecting heavy metal ions in biological samples. The functional groups on CQDs facilitate interactions with heavy metal ions, enhancing selectivity in biomedical applications. Amino-functionalized CQDs, in particular, exhibit notable selectivity in detecting heavy metals such as arsenic (As) and lead (Pb) in biological fluids. This selectivity ensures the accuracy and reliability of diagnostic outcomes. The specific interactions between functionalized CQDs and target heavy metal ions contribute to precise detection, making these particles promising for biomedical applications. The biocompatibility and tailored surface functionalization of CQDs offer avenues for improved selectivity in heavy metal ion detection, promising accurate and dependable diagnostic results in biomedical contexts.

4.4 Industrial Quality Control

CQDs employed in industrial quality control may have various functional groups, tailored to specific heavy metal ion targets in manufacturing processes.

Selectivity: The selectivity of CQDs in industrial settings is achieved through the customization of functional groups. For instance, CQDs with specific functional groups can selectively detect chromium (Cr) or nickel (Ni) ions in industrial solutions, ensuring product quality and regulatory compliance.

5 On-Site and Remote Monitoring:

CQDs designed for on-site and remote monitoring applications are often engineered to have surface functional groups that enhance interactions with heavy metal ions. The selectivity of CQDs in these applications is achieved by tailoring the functional groups to target specific heavy metal ions. This selectivity ensures accurate measurements in challenging environments.

The applications of green-synthesized CQDs for heavy metal ion detection span across various fields, from environmental monitoring and food safety to biomedical and industrial quality control. The unique nature of the surface functional groups on CQDs imparts selectivity, making them capable of discerning specific heavy metal ions. These eco-friendly and efficient sensing platforms are poised to have a significant impact on various aspects of our lives, ensuring a safer and more sustainable future.

6 Challenges and Future Directions

While the application of green-synthesized Carbon Quantum Dots (CQDs) for heavy metal ion detection has shown great promise, there are several challenges and exciting avenues for future research and development that should be considered:

6.1 Selectivity

Enhancing the selectivity of CQD-based sensors to discriminate between different heavy metal ions remains a challenge, especially in complex sample matrices. Achieving higher specificity is crucial for reliable and accurate detection in diverse applications. The developing innovative surface functionalization techniques and modifying the CQD structures to improve selectivity. It is also possible that the addition of molecular recognition components, such as molecularly imprinted polymers oraptamers, could improve the ability to differentiate between particular heavy metal ions that are present in complicated mixtures.

6.2 Sensitivity

It is vital for the practical application of CQD-based sensors in trace-level detection that efforts be made to improve their sensitivity as well as their detection limitations. By increasing sensitivity, it is possible to detect heavy metal ions at quantities that could potentially be hazardous to the health of humans or the environment themselves. It is essential to investigate a variety of approaches in order to increase sensitivity. These may include optimizing CQD synthesis methods to produce highly luminescent and responsive CQDs. Additionally, engineering CQDs with tailored functional groups and exploring advanced analytical techniques can help push the limits of detection to even lower concentrations of heavy metal ions.

6.3 Field Deployability

Developing portable and field-deployable CQD-based sensors is essential for on-site monitoring in remote or resource-limited areas. These sensors need to be user-friendly, robust, and capable of delivering rapid results in real-world conditions. The development of miniaturized and cost-effective CQD-based sensor platforms that can be used in the field is a priority. These sensors should be designed to withstand challenging environmental conditions and require minimal sample preparation. Integration with smartphones or other portable devices for data collection and transmission could further enhance their utility for on-site monitoring.

6.4 Sustainability

Ensuring the sustainability of green-synthesized CQDs, from both an environmental and economic perspective, is essential. Researchers must consider the scalability of green synthesis methods and the long-term environmental impact. Future research should aim to optimize and scale up the green synthesis of CQDs using eco-friendly precursors. Investigating the life cycle assessment of CQD production and exploring recycling or repurposing options for CQDs at the end of their lifecycle can contribute to sustainability.

6.5 Multimodal Sensing

Integrating multiple sensing modalities, such as fluorescence, electrochemistry, and colorimetry, into a single CQD-based sensor can be complex. The development of multimodal sensors that provide complementary information for heavy metal ion detection is an exciting but challenging direction. The designing of CQDs that can perform multiple types of analytical detection. By combining different sensing mechanisms, such as fluorescence quenching and electrochemical responses, the sensors can offer enhanced accuracy and reliability. In addition, the utilisation of sophisticated data analysis methods, such as machine learning and artificial intelligence, can be of assistance in the process of interpreting the outputs of multimodal sensors.

7 Conclusion

Carbon Quantum Dots (CQDs) have emerged as pivotal entities in the nanomaterials landscape, showcasing unique characteristics and diverse applications. With a carbon core adorned with oxygen-containing functional groups, CQDs stand out as game-changing agents. Overcoming the limitations of traditional semiconductor Quantum Dots (QDs) by avoiding heavy metals, CQDs offer low toxicity and robust chemical stability, particularly beneficial in heavy metal ion detection.

The versatility of CQDs extends across sectors, including bioimaging, sensors, drug and gene delivery, photocatalysis, solar energy conversion, photoelectrochemical water splitting, and light-emitting diodes (LEDs). Their seamless integration into various scientific domains underscores their adaptability and potential impact on advancing technologies. In sensor development, CQDs have achieved impressive sensitivity with picomolar and femtomolar Limits of Detection (LODs) while maintaining high selectivity. However, challenges persist, notably in enhancing fluorescence intensity and Quantum Yield (QY) to compete with semiconductor QDs. Persistent research is required to unlock the full potential of high QY CQDs, crucial for the development of highly sensitive sensors. Examining the impact of diverse functional groups on CQD optical characteristics and selectivity is essential for tailoring sensors. Surface coating and heteroatom doping offer viable avenues for fine-tuning optical properties, adding adaptability to CQD-based sensor technologies.

CQDs provide a cost-effective and promising alternative to semiconductor QDs in sensor applications. As our understanding of CQDs grows, the development of highly sensitive sensors capable of detecting a wide range of substances becomes more imminent. Harnessing the potential of CQDs promises the creation of more sustainable, effective, and flexible sensing systems.

References

1. Rahman, Zeeshanur, & Ved Pal Singh. (2009). The relative impact of toxic heavy metals (THMs)(arsenic (As), cadmium (Cd), chromium (Cr)(VI), mercury (Hg), and lead (Pb)) on the total environment: an overview. *Environmental Monitoring and Assessment, 191,* 1–21.
2. Raju, Chikkili Venkateswara, Chae Hwan Cho, Gokana Mohana Rani, Venkatesan Manju, Reddicherla Umapathi, Yun Suk Huh, & Jong Pil Park. (2023). Emerging insights into the use of carbon-based nanomaterials for the electrochemical detection of heavy metal ions. *Coordination Chemistry Reviews, 476*, 214920.
3. Pourmadadi, Mehrab, Erfan Rahmani, Maryam Rajabzadeh-Khosroshahi, Amirmasoud Samadi, Razieh Behzadmehr, Abbas Rahdar, & Luiz Fernando Romanholo Ferreira. (2023). Properties and application of carbon quantum dots (CQDs) in biosensors for disease detection: a comprehensive review. *Journal of Drug Delivery Science and Technology,* 104156
4. Yao, Hai, Tong Liu, Yong Qiang Jia, Ying Jie Du, Bo Wen Yao, Jian Hua Xu, & Jia Jun Fu. (2023). Water-insensitive self-healing materials: from network structure design to advanced soft electronics. *Advanced Functional Materials,* 2307455.
5. Tejashwini, D. M. H. V., Harini, H. P., Nagaswarupa, Ramachandra Naik, V. V. Deshmukh, & N. Basavaraju. (2023). An in-depth exploration of eco-friendly synthesis methods for metal oxide nanoparticles and their role in photocatalysis for industrial dye degradation. *Chemical Physics Impact,* 100355.
6. Nadar, Nandini Robin, Richelle M. Rego, Gara Dheeraj Kumar, H. Jeevan Rao, Ranjith Krishna Pai, & Mahaveer D. Kurkuri. (2023). Demystifying the influence of design parameters of nature-inspired materials for supercapacitors. *Journal of Energy Storage, 72*, 108670
7. Deshmukh, Lalita, & Kadam, S. L. (2023). Effect of microwave annealing on Tin Oxide nanomaterials. *Materials Today: Proceedings.*
8. Kumari, Khushboo, & Md Ahmaruzzaman. (2023). SnO2 quantum dots (QDs): synthesis and potential applications in energy storage and environmental remediation. *Materials Research Bulletin*, 112446.
9. Yi, Wenhui, Asif Khalid, Naila Arshad, M. Sohail Asghar, Muhammad Sultan Irshad, Xianbao Wang, Yueyang Yi, Jinhai Si, Xun Hou, & Hong Rong Li (2023) Recent progress and perspective of an evolving carbon family from 0D to 3D: Synthesis, biomedical applications, and potential challenges. *ACS Applied Bio Materials*

10. Nasrollahzadeh, M., Sajjadi, M., Iravani, S., & Varma, R. S. (2021). Carbon-based sustainable nanomaterials for water treatment: State-of-art and future perspectives. *Chemosphere, 263*, 128005.
11. Vardhan, Kilaru Harsha, Ponnusamy Senthil Kumar, & Rames C. Panda. (2019). A review on heavy metal pollution, toxicity and remedial measures: Current trends and future perspectives. *Journal of Molecular Liquids, 290*, 111197.
12. Martínez-Alcalá, Isabel, & Maria Pilar Bernal. (2020). Environmental impact of metals, metalloids, and their toxicity. *Metalloids in Plants: Advances and Future Prospects*, 451–488.
13. Makhdoumi, P., Karimi, H., & Khazaei, M. (2020). Review on metal-based nanoparticles: Role of reactive oxygen species in renal toxicity. *Chemical Research in Toxicology, 33*(10), 2503–2514.
14. Devi, P., Rajput, P., Thakur, A., Kim, K.-H., & Kumar, P. (2019). Recent advances in carbon quantum dot-based sensing of heavy metals in water. *TrAC Trends in Analytical Chemistry, 114*, 171–195.
15. Li, Pingjing, & Sam F. Y. Li. (2020). Recent advances in fluorescence probes based on carbon dots for sensing and speciation of heavy metals. *Nanophotonics, 10*(2), 877–908.
16. Ghosh, Saikat, Priyanka Dheer, Nilaya Kumar Panda, Soumya Biswas, Sourav Das, Pankaj Kumar Parhi, Sumira Malik, & Rahul Kumar. (2023). Emphasizes the role of nanotechnology in bioremediation of pollutants. In Industrial Wastewater Reuse: Applications, Prospects and Challenges, pp. 469–504. Singapore: Springer Nature Singapore.
17. Mansouri, F., Chouchene, K., Roche, N., & Ksibi, M. (2021). Removal of pharmaceuticals from water by adsorption and advanced oxidation processes: State of the art and trends. *Applied Sciences, 11*(14), 6659.
18. Yin, H., Truskewycz, A., & Cole, I. S. (2020). Quantum dot (QD)-based probes for multiplexed determination of heavy metal ions. *Microchimica Acta, 187*, 1–25.
19. Jia, Xicheng, Wasim Khan, Zhijie Wu, Jungkyu Choi, & Alex CK Yip. (2019). Modern synthesis strategies for hierarchical zeolites: Bottom-up versus top-down strategies. *Advanced Powder Technology, 30*(3), 467–484.
20. Berktas, Ilayda, Marjan Hezarkhani, Leila Haghighi Poudeh, & Burcu Saner Okan. (2020). Recent developments in the synthesis of graphene and graphene-like structures from waste sources by recycling and upcycling technologies: a review. *Graphene Technology, 5*, 59–73.
21. Xia, C., Zhu, S., Feng, T., Yang, M., & Yang, B. (2019). Evolution and synthesis of carbon dots: From carbon dots to carbonized polymer dots. *Advanced Science, 6*(23), 1901316.
22. Rasal, Akash S., Sudesh Yadav, Anchal Yadav, Anil A. Kashale, Subrahmanya Thagare Manjunatha, Ali Altaee, & Jia-Yaw Chang. (2021). Carbon quantum dots for energy applications: a review. *ACS Applied Nano Materials, 4*(7), 6515–6541.
23. Xue, B., Yang, Y., Tang, R., Sun, Y., Sun, S., Cao, X., Li, P., Zhang, Z., & Li, X. (2020). One-step hydrothermal synthesis of a flexible nanopaper-based Fe 3+ sensor using carbon quantum dot grafted cellulose nanofibrils. *Cellulose, 27*, 729–742.
24. Prekodravac, Jovana, Bojana Vasiljević, Zoran Marković, Dragana Jovanović, Duška Kleut, Zdenko Špitalský, Matej Mičušik, Martin Danko, Danica Bajuk–Bogdanović, & Biljana Todorović–Marković. (2019). Green and facile microwave assisted synthesis of (metal-free) N-doped carbon quantum dots for catalytic applications. *Ceramics International, 45*(14), 17006–17013.
25. Wang, C., Yang, M., Shi, H., Yao, Z., Liu, E., Hu., Xiaoyun, Guo, P., Xue, W., & Fan, J. (2022). Carbon quantum dots prepared by pyrolysis: Investigation of the luminescence mechanism and application as fluorescent probes. *Dyes and Pigments, 204*, 110431.
26. Yadav, Pradeep Kumar, Subhash Chandra, Vivek Kumar, Deepak Kumar, & Syed Hadi Hasan. (2023). Carbon quantum dots: Synthesis, structure, properties, and catalytic applications for organic synthesis. *Catalysts, 13*(2), 422.
27. Yang, Lei, Jiaxin Wen, Kunjian Li, Lu Liu, & Wei Wang. (2021). Carbon quantum dots: Comprehensively understanding of the internal quenching mechanism and application for catechol detection. *Sensors and Actuators B: Chemical 333*, 129557.

28. Nazri, Nur Afifah Ahmad, Nur Hidayah Azeman, Yunhan Luo, & Ahmad Ashrif A. Bakar. (2021). Carbon quantum dots for optical sensor applications: A review. *Optics & Laser Technology, 139*, 106928.
29. Molaei, M. J. (2020). Principles, mechanisms, and application of carbon quantum dots in sensors: A review. *Analytical Methods, 12*(10), 1266–1287.
30. Sekar, A., Yadav, R., & Basavaraj, N. (2021). Fluorescence quenching mechanism and the application of green carbon nanodots in the detection of heavy metal ions: A review. *New Journal of Chemistry, 45*(5), 2326–2360.
31. Yi., Li, Zhang, W., Jiang, X., Kou, Y., Lu., Jiajia, & Tan, L. (2019). Investigation of photo-induced electron transfer between amino-functionalized graphene quantum dots and selenium nanoparticle and its application for sensitive fluorescent detection of copper ions. *Talanta, 197*, 341–347.
32. Dos Santos, Marcelina Cardoso, W. Russ Algar, Igor L. Medintz, & Niko Hildebrandt. (2020). Quantum dots for Förster resonance energy transfer (FRET). *TrAC Trends in Analytical Chemistry, 125*, 115819.
33. Zhao, W., Ma, Y., Ye, J., & Jin, J. (2020). A closed bipolar electrochemiluminescence sensing platform based on quantum dots: A practical solution for biochemical analysis and detection. *Sensors and Actuators B: Chemical, 311*, 127930.
34. Limosani, Francesca, Elvira Maria Bauer, Daniele Cecchetti, Stefano Biagioni, Viviana Orlando, Roberto Pizzoferrato, Paolo Prosposito, & Marilena Carbone. (2021). Top-down N-doped carbon quantum dots for multiple purposes: Heavy metal detection and intracellular fluorescence. *Nanomaterials, 11*(9), 2249.
35. Manikandan, Velu, & Nae Yoon Lee. (2022). Green synthesis of carbon quantum dots and their environmental applications. *Environmental Research, 212*, 113283.

Degradation of Organic Pollutants Present in Water Using Green Synthesized Carbon Quantum Dots

Permender Singh, Neeru Rani, Krishan Kumar, Sandeep Kumar, Parmod Kumar, and Vinita Bhankar

Abstract Accessibility to fresh, hygienic potable water and a clean-living surrounding are basic needs that facilitate a healthy life. But, the widespread and persistent use of harmful chemicals, particularly cancer-causing organic pollutants (OPs) including dyes like methyl orange, rose Bengal, malachite green, methylene blue, and other OPs like tetracycline hydrochloride, 2,4 dichlorophenol, etc., at the domestic and commercial scale has resulted in a shortage of drinking water bodies and ecological issues. It causes a significant impact on aquatic life, ecologically sound and long-term world growth, and human civilization. To solve this particular issue, scientists have concentrated on more sophisticated ways than traditional wastewater management procedures for the elimination/decontamination of harmful OPs. The photocatalytic cleaning of wastewater containing OPs utilizing green synthesized carbon quantum dots (CQDs) produced from biomass is effective, secure, and affordable method for degrading potentially dangerous organic contaminants from wastewater. The eco-friendly CQDs-based photocatalysts for the remediation of contaminated wastewater have several advantages over conventional techniques, like low cost, non-toxicity, simple components requirement, rapid reactions, simple processes, and simple further processing stages. As a result, this chapter offers a summary of recent studies on photocatalytic wastewater treatment, together with an analysis of related problems and potential solutions.

P. Singh · N. Rani · K. Kumar (✉)
Department of Chemistry, Deenbandhu Chhotu Ram University of Science and Technology, Murthal, Sonepat, Haryana 131039, India
e-mail: krishankumar.chem@dcrustm.org

S. Kumar (✉)
Department of Chemistry, J. C. Bose University of Science and Technology, YMCA, Faridabad, Haryana 121006, India
e-mail: sandeepkumar@jcboseust.ac.in

P. Kumar
Department of Physics, J. C. Bose University of Science and Technology, YMCA, Faridabad, Haryana 121006, India

V. Bhankar
Department of Biochemistry, Kurukshetra University, Kurukshetra, Haryana 136119, India

V. Kumar et al. (eds.), *Green Carbon Quantum Dots*, Engineering Materials,
https://doi.org/10.1007/978-981-97-6203-3_8

1 Introduction

Growing industrialization has made the contamination of the environment a serious health risk to living creatures. Concerns about dangerous and deadly pollution in water resources have garnered attention on an international level [1]. Chemical waste disposal has contaminated groundwater aquifers, rivers, lakes, and oceans. "Numerous hazardous pollutants, including herbicides, insecticides, heavy metals, textile dyes, persistent organic pollutants (POPs), volatile organic compounds (VOCs), pharmaceuticals, personal care products (PPCPs), bacteria/fungi and surfactants, have been found in the aquatic system". A few instances of poorly treated water include the inadvertent discharge of solid or chemical waste from dyeing industries, hospitals, and universities that eventually contaminates water, as well as the mixing of industrial waste into water bodies that contains hazardous substances [2]. Such water poisons or kills plants and trees, preventing them from growing. When water having harmful contaminants is consumed, it can cause numerous ailments, including reproductive, neurological, and digestive issues. It may additionally result in severe illnesses like cancer. On the other hand, direct contact can cause skin conditions like psoriasis, allergies, and hair loss. Importantly, the main cause of serious health problems for plants and animals is the erratic discharge of industrial residue that contains organic dyes, pharmaceuticals, POPs, VOCs, and PPCPs into water bodies. High level of these pollutants in water bodies affects BOD and COD levels, inhibit the process of photosynthesis and the development of plants, have an adverse effect on animals and marine systems, provide recalcitrance, and cause biological accumulation. According to their chemical structures, the main OPs i.e., dyes can be categorized as basic, acidic, dispersion, azo, anthraquinone, anionic and cationic. "Rhodamine-B (RhB), Methyl Orange (MO), Methylene Blue (MB), Congo Red (CR), Crystal Violet (CV), azo dyes, and other textile dyes are among the dyes which raise the toxic content of water and can lead to serious health problems like vomiting, diarrhea, and itchiness in the eyes [3]".

Numerous attempts have been made for water decontamination by the use of a variety of conventional procedures, including sedimentation, reverse osmosis, coagulation, adsorption, filtration, chemical and biological processes, etc. Nevertheless, these technologies' efficiency is insufficient for treating wastewater that includes a variety of pollutants, including pesticides, pharmaceutical wastes, organic solvents, text-tile dyes and household chemicals [4]. When it comes to adsorption, activated carbons are typically employed to adsorb a variety of water contaminants; however, as the number of cycles grows, the adsorbent capacity of the carbons steadily decreases [5]. Moreover, a vacuum or steam source is needed for the regeneration of conventional adsorbents, and the process's effectiveness may be reduced by the large amount of side-products produced. Furthermore, ion-exchange generates a large quantity of hazardous wastes, and the treatment procedure is limited due to its high materials and energy consumption [6]. Due to numerous drawbacks of alternative wastewater remediation techniques, as well as the several benefits of photocatalysis including its greater catalytic performance, quick reaction times, lack of additional harmful side-product generation, economic feasibility, reuse ability, and the potential to use renewable solar energy to activate catalyst for dye elimination and nearly 100% dye decontamination, it is emerging as an appealing technique for improving the biodegradability of POPs and getting rid of microbial pathogens [7]. Solar or other energy-driven reactions known as "photocatalytic oxidation" employ a catalyst and are dependent on the production of strong reactive radical such as $O_2^{\bullet -}$, H_2O_2, and most importantly, the •OH radicals [8]. These radicals systematically break down all hazardous organic matter found in wastewater due to their strong oxidizing potential [8–10]. Numerous photocatalysts have shown great capability in the recent several years for the breakdown of harmful OPs especially dyes that are present in wastewater. However, those derived from biomass or biomass waste (or natural sources) by ecological synthesis have received a huge affection from the scientific world in the recent past. It has the benefit of not just stopping the harmful emissions produced by burning or dumping of biomass waste directly, but also allowing the waste to be employed for environmentally beneficial objectives. Biomass is a multifaceted, diverse, easily accessible, recyclable, and bio-organic substance that comes from a range of naturally occurring sources, such as organic household trash, perennial grass, and leftovers from animal husbandry, farming, poultry, and related organizations. It is mostly made up of proteins, cellulose, hemicellulose, ash, lignin, and other elements. It is a naturally occurring resource of organic carbon. For instance, cellulose accounts up 30–60% of plant biomass waste, with hemicellulose (20–40%) and lignin (15–25%) making up the remaining portions [4, 11–13].

Among various green synthesized photocatalysts, CQDs, mainly either doped or in form of composites, have remarkable potential for eradication of OPs such as hazardous textile dyes, pharmaceuticals and other toxic substances present in wastewater. A few of the remarkable properties of CQDs include the different surface functionalities like halogens, $-NH_2$, $-SO_3H$, –COOH, etc. their small size, excellent photo- and chemical stability, excellent biocompatibility, fluorescence emission, water solubility, microscopic dimensions, ease of fabrication, photoinduced electron transfer, adjustable and small emission spectra [14]. The environmentally friendly fabrication of CQDs is advantageous due to economically affordability, non-hazardous, easy and excessive availability of green precursors, easy procedures, and sustainable resources. "A few instances of the different types of green carbon precursors are fruits, peels of fruit and juices, animals and materials that originate from animals, including eggs from and chickens, vegetables, spices, leftover kitchen goods like frying oil or waste paper, leaves of plants and their derivatives, etc. [4, 15]". For example, utilizing onion waste as a raw material, Bandi et al. examined the fabrication of extremely luminous CQDs utilizing a simple greener approach. The resultant CQDs were useful for tracking, delivering drugs, sensitive and selective probing of Fe^{3+} ions, and cell imaging applications [16]. Sahu et al. produced CQD using a single-step hydrothermal process from orange juice. Owing to their excellent photostability as well as little toxicity, the resultant CQDs have been employed in solution state optoelectronics. The CQDs were also effective as cell imaging agents [17]. In addition, CQDs have remarkable role in sensing of radioactive ions [18].

As far as we are aware about literature, the present literature on green synthesized CQDs explains the usefulness of CQDs as photocatalyst for eradication of organic contaminants present in wastewater [2, 3, 14, 19, 20]. Therefore, our main focus in this chapter will be to provide the most recent advancement for utilizations of green synthesized CQDs as photocatalyst for treatment of OPs contaminated wastewater such as deep insight into the mechanistic routes of photocatalytic dye degradation, various factors influencing OPs degradation, associated problems and their probable solutions.

2 Functional Entities of CQDs and Their Properties as Catalyst

CQDs typically have a diameter of 1–10 nm and a quasi-spherical morphology. The components that constitute a CQD are all basic and easily accessible [21]. "There are both top-down and bottom-up methods for fabricating CQDs, including hydrothermal, microwave, laser ablation, solvothermal ultrasonication, electrochemical oxidation, and other synthetic processes". In the top-down process, various carbonaceous bulk materials, including as coal, graphite, and carbon black, may split or cleave. On the other flip, a bottom-up technique is thought to be more environmentally friendly, straightforward, and economical to generate CQDs in relatively larger quantities because it typically involves the carbonization or pyrolyzing of biomass and tiny organic molecules [22]. The essential characteristics as well as certain major drawbacks of CQDs are depicted in Fig. 1. Graphitic or turbostratic carbon (sp^2 hybrid), graphene, and GO sheets make up the majority of the amorphous nano-crystalline core of CQDs. These sheets are glued together by sp^3 hybridized carbon insertions that resemble diamonds. Oxidised surface groups consisting of carbonyl, –COOH, –C–O–C–, and –OH groups stabilize the core [21, 23, 24]. On the basis of synthesis technique, the oxygen proportion of oxidised CQDs varies from 5 to 50% (weight) [24]. Despite, there is still a degree of uncertainty regarding the exact oversight over the surface and inner structure. Because most CQDs include functional groups on their surface which include oxygen, they are hydrophilic and thus highly compatible with water-based chemistry [21, 23].

CQDs are fluorescent and having intense UV–visible absorption band. "CQDs exhibit a significant absorption peak in the UV spectrum and a tail that reaches into the visible range [21]. The π–π (C = C bond) or n–π (C = O bond) transitions, among other transitions, are responsible for the absorption peaks of CQDs". Additionally, it has been observed that CQDs can suppress the emission of photoluminescence while acting electron donors as well as acceptors simultaneously. Therefore, CQDs can be conceptualized as a "superposition state" that combines different amounts of organic and carbon-material-like "states" in a variable ratio [23]. However, the production procedure can result in noticeable changes in the structure and composition of carbon nanoparticles, which in turn affects their photophysical properties

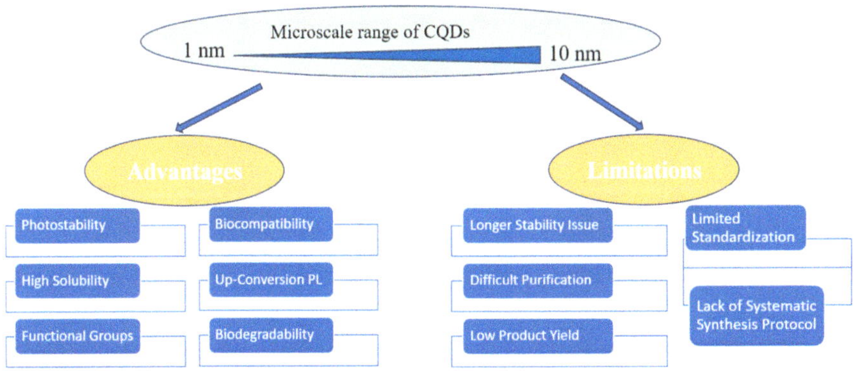

Fig. 1 Properties of green synthesized CQDs

[21]. Despite lacking inherent photocatalytic capability CQDs may be used to design exceptionally effective CQD-based photocatalytic materials because of their special electronic and optical characteristics, which include intense light harvesting, up-conversion PL, photo-induced e^- transfer capacity, adjustable bandgap, and inhibiting e^-/h^+ reunion. When considering light harvesting, CQDs offer significant advantages over known cutting-edge photosensitizers. CQDs have the ability to be employed for several remediation procedures, such as water remediation, where contaminants are broken down into inorganic compounds and CO_2 [4].

By influencing the fundamental photophysical characteristics of the CQDs, innovative synthetic approaches for producing heteroatom-doped CQDs, such as nitrogen, may enable additional developments [21]. Many CQDs have been found to have a noticeable photo current in visible light, whereas others are also known to have a minor photocurrent. The "quantum effect" of CQDs ought to be utilized with caution regardless of their microscopic size because their spectrum variations don't necessarily correspond to a real quantum confinement effect. However, to ascertain superior crystalline CQD structure and characteristics, an understanding from a "quantum dot" perspective is required. The continual improvement of CQDs continues to be complicated by the need for extremely precise synthetic control, chemical composition determination, and structure–property relationships [23]. Precise synthesis management continues to encounter difficulties in producing CQDs with consistent size, composition, and surface synthesis, which calls for a thorough grasp of synthetic chemistry, development mechanisms, and in situ assessment [25].

3 Photocatalytic Eradication of OPs

The research on CQDs and CQDs-based photocatalysts that have been conducted, wherein their photocatalytic efficiency is examined against the photocatalytic decontamination of several OPs such as organic dyes, POPs, VOCs, and pharmaceuticals,

PPCPs and bacteria/fungi is compiled and highlighted in the following section. The capability of CQDs solely to photo-decontaminate OPs is primarily related to their surface states and intrinsic abilities, including electron mediation, and there have been a few studies that concentrate on utilizing sole CQDs as photocatalysts for the breakdown of OPs.

3.1 Photocatalytic Elimination of Dyes

Dyes are chemical compounds which absorb light and add color to the visible world. They can resist water, light, oxidation, and bacteria without fading [26]. "As per an updated approximation, the yearly manufacturing of dyes is approximately 7×10^8 kg. The textile sector discharges up to 2×10^8 kg of these dyes to wastewater yearly through dyeing and finishing works due to inefficient dyeing techniques [27, 28]". The release of such dyes in water bodies pose a serious threat to both the public's health and the environment. The direct release of unprocessed wastewater carrying hazardous dyes into natural water resources has an adverse influence on the of marine lives also [29]. This wastewater affects marine life and fish species in a mutagenic or teratogenic manner because of the presence of harmful dyes [30, 31].

3.1.1 Photocatalytic Decomposition of Cationic Dyes

The important cationic dyes are RhB, MG, CV, and MB. Despite their many applications, these dyes pose a risk to both the natural world and living organisms. For instances, RhB is highly adaptable water-soluble dye and may be employed for an extensive range of applications, including textiles, photosensitizers, textile markers, water tracing, etc. [32]. Due to its mutagenic and cancer-causing properties, it has been shown to have deleterious impacts on living things. It may cause cancer when it enters the human body. MB is broadly employed in many industries, like the food, paint, dyeing, textile, and pharmaceutical sectors [33]. The research community finds it to be an intriguing chemical because of its significant therapeutic value in the "treatment of anaemia, malaria, and *Barrett's oesophagus* [34]". Nevertheless, exposure to it at higher concentrations higher can have serious harmful effects. MB dye can induce a number of illnesses, including respiratory difficulties, mental health problems, eyesight, and stomach troubles [35]. Therefore, decomposition of these dyes highly important. Nugraha et al. revealed the production of WO_3/amino-functionalized CQDs (WO_3/N-CQDs) from sugarcane bagasse composite. The resultant composite showed MB photocatalytic decontamination with a decontamination efficacy of 96.86% [36]. Employing palm powder as a precursor, Zhu et al. fabricated S and Cl-doped CDs from palm powder and the obtained CQDs showed 71.7% photocatalytic eradication of RhB dye [37]. Table 1 shows the recent studies for the photocatalytic elimination of cationic dyes employing green CQDs/CQDs-based photocatalyst.

Table 1 Photocatalytic decontamination of cationic dyes by green CQDs-based photocatalysts

Sr. no.	Raw-materials	Photocatalyst	Synthesis condition	Dye eradicated	Source of light	Morphology	Performance	Refs.
1	Pear juice	CQDs	Hydrothermal 180 °C, 36 h	MB	Vis	Spherical 3–6 nm	99.5% in 130 min	Das et al. [38]
2	Aqua mesophase pitch	N-CQDs	Hydrothermal, 120 °C, 24 h	RhB	Vis	Sphere 2.8 nm	97% in 4 h	Cheng et al. [39]
3	Rice bran	CDs	Hydrothermal, 200 °C, 4 h	MB	Vis	Spherical 2.96 nm	89.20% in 30 min	Jothi et al. [40]
4	D-fructose	ZnO/CQDs	Hydrothermal	RhB	Vis	124 ± 6 nm	94% in 105 min	Bozetine et al. [41]
5	*Bovine serum albumin*	PANI-NCDs	Ultra-sonication	CR, RhB, CV, MB	Vis	Non-uniform spherical 20–70 nm	Complete eradication of CR in 20 min, 60%, 20%,3% for MB, RhB, CV resp.	Maruthapandi et al. [42]
6	Peels of potato	Fe_3O_4/CDs		MG, RhB, S-1		Spherical 4.47 ± 0.87 nm	80% of MG in 2 h, 70% of RhB in 4 h	Shivalkar et al. [43]
7	*Polyalthia longifolia*	CDs	Ultra-sonication	MB, CR	Vis	–	CR and MB dyes	Zaib et al. [44]
8	Gum ghatti	CDs/ZnO	Microwave	MG	Vis	Nearly spherical 2.6 nm	94.8% in 60 min	Sekar and Yadav [45]

(continued)

Table 1 (continued)

Sr. no.	Raw-materials	Photocatalyst	Synthesis condition	Dye eradicated	Source of light	Morphology	Performance	Refs.
9	*Citrus limetta* (Au, Ag CQD/ TiO_2)	Au, Ag-CQD/TiO_2	–	MB	UV	Spherical >10 nm	Complete eradication in 20 min, 30 min for Au and Ag composite resp.	Thakur et al. [46]
10	Peels pf lemon	TiO_2-CDs	Hydrothermal (TiO_2 CQD)	MB	UV	Spherical 1–3 nm	–	Tyagi et al. [47]

3.1.2 Photocatalytic Decontamination of Anionic Dyes

Anionic dyes including malachite red (CR), MO, and congo red (CR) are commonly utilized. Anionic dyes are highly versatile but can pose a considerable environmental hazard, necessitating their degradation. For instance, MO is a carcinogenic, toxic and mutagenic azo dye [48]. It is a rather common waste product emitted by the paper, research, food, textile, printing, and pharmaceutical industries [49]. Similarly, CR is phytotoxic, raises COD, mutagenic and carcinogenic [50]. Table 2 presents a summary of the use of green CQDs for photocatalytic eradication of several hazardous anionic dyes.

3.2 Photocatalytic Decontamination of POPs and VOCs

In the past few years, the primary area of photocatalytic environmental remediation study has been on improving the spectrum response range by addressing the redox ability of photoexcited carriers. For instance, C-dots/$BiSbO_4$ composite was developed by Wang et al. and it demonstrated the increased deterioration of RhB and CIP (ciprofloxacin) in comparison to separate constituents in the presence of Xe lamp radiation and solar radiations [56]. Dadigala et al. developed g-C_3N_4/CQDs/Ag composite for effective photocatalytic eradication of MO and p-nitrophenol (PNP) in the presence of solar radiations [57]. Song et al. used surface deposition to coat C_3N_4 with B-doped C-Dots. Under the influence of visible irradiations, the obtained composite exhibited exceptional photocatalytic decontamination of RhB and tetracycline hydrochloride (TC) in aqueous media. Moreover, the photocatalytic efficiency of C-dots/C_3N_4 for RhB and TC elimination was 4.80 and 7.21 times higher in comparison to that of C_3N_4 and C-dots individually, sequentially [58]. A WO_3 nanoplate paired with C-dots was developed by Huang et al. [59] to photo-decontaminate HCHO and CH_3COCH_3under the effect of visible radiations. The CO_2 generation rate increased by 411 and 188 μmol g^{-1} h^{-1} with this composite, correspondingly [59]. Peng and colleagues demonstrated a solvothermal technique to fabricate Bi_2MoO_6/CQDs/Bi_2S_3 (BMO/CQDs/BIS). Under the impact of visible radiations, the degrading efficacy of TC by BMO/CQDs/BIS was determined to be 92.8%, which was significantly greater than every individual constituent [60]. Hak et al. fabricated CQDs from the leaves of water hyacinth and constructed CQDs/g-C_3N_4 binary composite [61]. The composites having with 20 weight percent (20CQDs/g-C_3N_4) and 40 weight percent (40CQDs/g-C_3N_4) demonstrated decontamination of 2,4-DCP with 1.7 folded elimination capability of pristine g-C_3N_4.

Table 2 Photocatalytic decontamination of anionic dyes by green CQDs-based photocatalyst

Sr. no.	Precursor	Photocatalyst	Synthesis conditions	Dye eradicated	Light source	Morphology	Efficiency	Refs.
1	Yerba mate	CD-decorated magnetite	Acid carbonization and hydrothermal	MO	Visible	4.9 ± 1.5 and 4.1 ± 1.2 nm	98% in 7 h	Monje et al. [51]
2	Pine fruit	N-CQDs	Hydrothermal	MB, MO, EY, AB, EBT	UV	<10 nm	Eradication of AB, EBT, EY, MB and MO	Shahba and Sabet [52]
3	Grass	N-CQDs	Hydrothermal, 180 °C, 120 min	MB, MO, AB, AR, EY	UV	<10 nm	AR and AB 100% in 0.5 h, MB and EY 100% in 90 min, MO and eriochrome 100% decontaminated	Sabet and Mahdavi [53]
4	*Aloe barbadensis*	CQDs	Microwave	Eosin yellow	Visible	Spherical <5 nm	Complete eradication in 100 min	Malavika et al. [54]
5	*Opuntia Ficus Indica*	N-CDs	Hydrothermal	MO				Rajapandi et al. [55]
6	Gourd flesh	F/N-CDs	Hydrothermal	CR, MO	Sunrays	4–6 nm	CR 98% eradicated in 20 min, MB 100% in 90 min	Wang et al. [98]

4 Factors Affecting the Photocatalytic Decomposition of OPs

Many distinct factors, including pH, the dosage of CQDs, the initial concentration of pollutants, light exposure period, dopant, water matrix etc., control the degree at which pollutants degrade. These operational variables that affect the photo-decontamination process have been investigated as part of several studies [62].

4.1 Photocatalyst Dosage

Usually, the rate of photocatalytic eradication of OPs initially increases followed by a decrement when the dosage of photocatalyst in the reaction mixture increases [4]. With an increase in photocatalyst quantity, the reaction rate initially increased quickly. It's because the overall surface area as well as number of surface-active sites of the reaction are now raised through the addition of photocatalyst, which accelerated its rate. Under specific light conditions, the free electrons do not significantly change as the catalyst concentration increases until it reaches a particular value. Reaction rates are slowed down when an excessive amount of photocatalysts is utilized because they accumulate and cover one another, preventing light absorption. Feng et al. degraded MB by adding 15, 25, and 35 mg photocatalyst CQDs/Cu_2O, correspondingly. They reported that using 25 mg has the best degrading impact [63]. The preceding assertions is correctly proved by the fact that excessive or insufficient amount does not promote the photocatalytic process.

4.2 Effect of pH

pH is another essential parameter influencing photocatalytic performance; an excess of either acid or alkali will alter the photocatalytic rate of OPs decomposition. Photocatalytic efficiency is primarily impacted by pH in three ways: (a) catalyst synthesis is impacted by pH, (b) the effect on the system of reactions. Overabundance of H^+ capture free carriers $O_2^{\bullet-}$ on the photocatalyst surface. Both the H^+ in the solution and the h^+ generated by photon will be consumed by too much OH^-. The decomposition of MB by N, P-CQDs/NW-TiO_2 at various pH values was examined by Bai et al. [64]. It was observed that at pH = 7, catalytic performance is at its maximum. On the other hand, when pH was acidic, the rate of MB breakdown drastically slowed, indicating that too much H^+ had eaten the reactive material $O_2^{\bullet-}$.However, not every study has an appropriate pH = 7 because it depends on how contaminants and photocatalysts interact. For instance, for the reduction of Cr(VI), Wu et al. noticed that the composite photocatalyst CQDs/$NaBiO_3$ had the greatest catalytic performance at pH = 3 [65].

According to Behnood et al., an alkaline pH of 11 was ideal for breakdown of MB utilizing N-CQDs/Ni-ZnO [66]. (3) The catalyst material's effect. The charge and shape of the catalyzed substance are mostly influenced by pH. Tetracycline antibiotics, for instance, will degrade at distinct pH values, resulting in tetracycline surfaces with varying charges. In basic solutions, the majority of tetracyclines are negatively charged, but in acidic solutions, they are positively charged. Consequently, the catalyst and the intended contaminant will have an identical charge in a solution containing strong base or strong acid, resulting in electrostatic repulsive force that is detrimental to the catalyst's ability to adsorb pollutants. The existing form of Cr(VI) changes with pH when it is reduced by composite photocatalyst, i.e., Cr(VI) takes the form of $Cr_2O_7^{2-}$ at pH 9 [19].

4.3 *Effect of Temperature*

Semiconductor photocatalytic reactions typically have low activation energies, making them less susceptible to systemic temperature changes. The primary reaction stages that are impacted by temperature are surface migrations, adsorption, and desorption; however, they are not the essential steps in figuring out the rate of photocatalytic process. Consequently, it is typically decided to be done at room temperature to reduce costs. However, the fabrication of the photocatalyst is greatly influenced by temperature [67]. In an effort to maximize photocatalytic efficiency, Li et al. examined the rate at which CIP (ciprofloxacin) degrades by photocatalyst CQDs@In_2S_3/SWNTs under various synthesis circumstances [68]. The temperature (120, 150, 180, 200 °C) and reaction time (2, 10, 16, and 24 h) were varied to control the rate of reaction. The findings demonstrated that when temperature and time increased, the CIP decomposition rate first rises and subsequently falls. For photocatalyst, the ideal reaction circumstances were found to be T = 180 °C and t = 10 h. Precursors cannot completely react to produce photocatalysts if the temperature is very low. In addition, the resulting photocatalysts decompose or change into other substances when the temperature is highly elevated.

4.4 *Impacts of O_2 and H_2O_2 Presence*

One of the key parameters influencing the speed of the photocatalytic process and a significant oxidant in the whole process is O_2. As an electron recipient, O_2 easily absorbs on the photocatalyst surface, preventing the reunion of electrons with holes. Furthermore, contaminants can be degraded by using the created $O_2^{\bullet-}$ as an active species. Therefore, the reaction is generally accelerated by adding O_2. Another substance that traps electrons is H_2O_2. H_2O_2 addition promotes the production of •OH and the breakdown of contaminants. Bozetine fabricated ZnO/CQDs/AgNPs composite catalysts using a hydrothermal process to degrade MB and to examine the

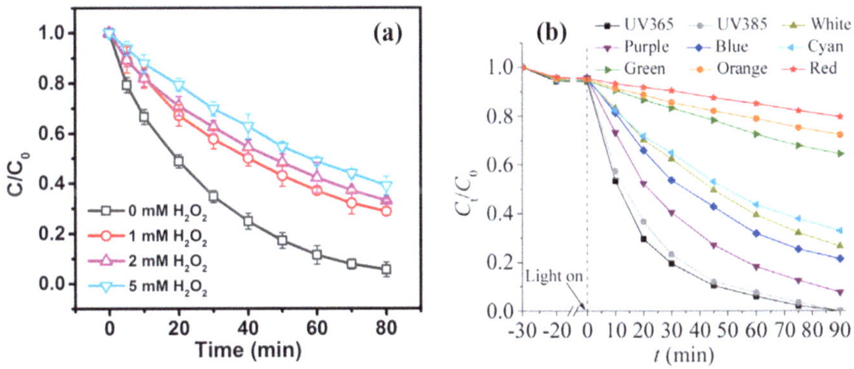

Fig. 2 **a** SMZ photocatalytic efficacy of CDCN-10 at various introduced H_2O_2 concentrations (0, 1, 2, and 5 mM)"Reproduced with permission, copyright © 2020 Elsevier Ltd, (license number—5,655,610,549,966). All rights reserved [70]", **b** Photocatalytic decontamination of AMX using CQDs/$K_2Ti_6O_{13}$ hybrid material under the influence of distinct wavelengths of light. "Reproduced with permission, copyright © 2019 Elsevier Ltd, (license number—5,655,620,393,161). All rights reserved [72]."

influence of introducing H_2O_2 on rate of MB photocatalytic breakdown [69]. The findings demonstrated that H_2O_2 can greatly increase ZnO/CQDs/AgNPs' catalytic efficiency and boost the rate of reaction. The addition of H_2O_2 shortened the MB's overall discoloration time by 20 min. This is due to the fact that H_2O_2 produces ·OH, an active radical for MB eradication, when it combines with photoinduced e^- by the photocatalyst. On the other hands, a few photocatalytic activities are likewise inhibited by H_2O_2. Di et al. investigated the role of various H_2O_2 concentrations for photocatalytic decontamination of sulfamethazine (SMZ) by CQD/g-C_3N_4 [70]. The findings showed that the degree of SMZ photo catalytic breakdown progressively lowered as the conc. of H_2O_2 increased (1, 2, and 5 mM) as shown in Fig. 2a. This is because the photocatalytic efficiency can be inhibited, as H_2O_2 has the ability to poison g-C_3N_4.

4.5 *Light Intensity and Its Wavelength*

Photocatalytic reactions are primarily fueled by source of light irradiations. The relationship between photocatalytic performance and its structure, luminous mode, and light source life is strong. The rate of the photocatalytic process is highly dependent on light intensity. The rate of the photocatalytic process usually gets greater under brighter light (greater intensity). Obviously, variations in the wavelengths of light energy also affect the photocatalytic activity. Guo et al. examined that the photodegradation of MB under ultraviolet and visible light using CQDs/UYTMs and observed that its degradation was more effective in 20 min exposure to UV light in comparison to 6 h exposure to visible light [71]. Chen and his colleagues employed a range of nine distinct light wavelengths as illumination sources to facilitate the photocatalytic breakdown of amoxicillinas displayed in Fig. 3b [72]. It was noticed that the photocatalytic eradication rate rises with decreasing wavelength. Additionally, in visible light (white light), this photocatalyst $CQDs/K_2Ti_6O_{13}$ demonstrated nice photocatalytic capability.

4.6 Initial Pollutants' Concentration

Most often, photocatalysis is used to remediate pollutants at concentrations ranging from mol/L and mmol/L. As the number of pollutants increases, the degree of photocatalytic deterioration generally decreases [19]. Following are some explanations: (1) when the dosage of catalysts is constant, the catalytic efficiency will be comparatively lower because contaminants have competition for a limited number of catalytic sites, (2) weak adsorption activity on the catalyst surface was caused by an elevated number of contaminants aggregating with one another, (3) the absorption of light is impacted by a decline in the solution's ability to transmit light as the level of contaminants rises. Along with investigating how catalyst input and CQD dosage affect reaction rate, four MB dye solutions with varying beginning concentrations were investigated by Hu et al. for their impact on the adsorption and photocatalyst's eradication efficiency for contaminants [73]. The adsorption effect is enhanced by a lower initial MB concentration. The decomposition rate increases with decreasing concentration of MB. The photocatalytic decomposition efficiency may naturally enhance initially and subsequently decline as the number of contaminants rises. The speed of every stage in the photocatalytic breakdown of contaminants may be influenced by these two factors.

The slowest step determines the overall reaction rate, in accordance with the concept of assessing chemical kinetics. According to Vinodgopal et al., the interaction between the pollutant and the catalyst or the pollutant's degrading response by the photocatalyst may control the reaction rate [74]. This occurs when the former's rate is slower than the latter's. Due to the fact that, under some circumstances, the concentrations of e^-, h^+, $^{\bullet}OH$, and $O_2^{\bullet-}$ do not change, the likelihood of colliding with these species rises with the concentration of pollutants. A concentration that is very high, on the other hand, surpasses the content of activated reaction species, and the photocatalytic eradication rate falls.

4.7 Impact of Water Matrix

The water network is also crucial factor to the photocatalysts' ability to breakdown harmful OPs [75]. Both soluble and suspended materials in wastewater have a substantial impact on the photocatalyst's dye eradication effectiveness in strongly charged water. However, the photocatalytic wastewater remediation process may be negatively, neutrally, or favorably affected by dissolved chemicals [76, 77]. If impurities exist in the aqueous solutions, scavengers can also impede their removal. Inorganic species such as bicarbonate, Cl^-, CO_3^{2-}, HCO_3^-, Br^- can function as scavengers of •OH in neutral water. Furthermore, the photocatalyst's efficacy in treating wastewater containing harmful OPs decreases with an increase in water matrix complexity [60, 61]. When inorganic salts such as chlorides of alkali, alkaline and transition metals are present in large enough amounts in wastewater, they can completely suppress photocatalytic activities and have a detrimental effect on the breakdown of OPs through photocatalysis [78–80].

4.8 *Impact of Hetero-Atom Doping*

To increase the capacity for sorption and photocatalytic performance of CQDs produced from biomass, hetero-atom-doping is one of the most commonly used techniques. Stated otherwise, several chemical functional moieties with O, N, or S atoms get attached to activated CQDs particles. Surface functionalization employing these hybrid atoms has a major impact on the surface morphology of CQDs. It boosts the surface area, which enhances the CQDs' capacity to absorb harmful dyes. Consequently, the photocatalytic decomposition efficacy of CQDs gets noticeably enhanced [81, 82]. Zhu and his colleagues reported the improved photocatalytic dye eradication performance of S/Cl co-doped CQDs derived from palm powder [37]. This is because groups containing S and Cl preferably cause more electronic transition channels within the CQDs' structure, giving them unique photocatalytic properties. Through high-temperature pyrolysis and oxidation in concentrated H_2SO_4, Hu et al. synthesized N and S co-doped CQDs from the leftover green tea leaves. The resultant CQDs showed a fluorescence QY of up to 14.8% that was 3 times more as compared to undoped CQDs [83].

5 Mechanism of Photocatalytic OPs Degradation

Harvesting of light, charge creation and detachment, and catalytic reaction are the three main phases of a photocatalytic process. The net effectiveness of the photocatalytic process is determined by the balance among the thermodynamics and the kinetics of the three main phases aforementioned. With the help of effective catalytic interactions, high charge carrier separating effectiveness, and superior light-harvesting capabilities, researchers have achieved noticeable advancement in the area of photocatalysts [84]. As light is the primary prerequisite for a subsequent photocatalytic reaction, all the conditions of the new research that uses CQDs made from green precursors as catalysts must be satisfied.

5.1 Light-Harvesting

Harvesting of light energy i.e., photons capturing is the initial step of the three main stages that assesses the system's ability to produce pairs of electrons and holes by capturing photons for a later catalytic reaction [85]. With direct sunlight as the energy source, the spectrum light is mostly composed of 43% visible, 52% IR, and only 5% UV light. When CQDs are attached with the dopants, the energy or electron band structure is altered, which causes the absorption of incident light to change, particularly in the visible and IR spectrum. In solar energy, visible and IR light absorption is most commonly used [86]. When semiconductors are doped with CQDs, the energy band structure is used to calculate the light-harvesting ability of the materials.

5.2 Charge Transfer Dynamics

Charge transfer kinetics is a critical factor for improving the effectiveness of the photo catalytic process. Effective separation of charges and transmission of photo-induced e^-s and h^+s at the CQDs (or CQDs and dopant) interface is necessary to achieve a greater effectiveness of the photocatalytic procedure [87]. The CQDs and their dopants ought to be separated by a modest charge movement distance. Higher photog-originated charge carriers recombination results in lower photocatalytic efficiency, which is reflected in the impossibility of commercial platform applications. A suitable photocatalyst should have two essential characteristics: a wide spectrum of photo absorption and elevated capability in separating charge carriers [88, 89].

5.3 Surface Catalytic Reaction

The catalytic reaction occurring on the surface of a system dictates both its level of selectivity and photocatalytic efficiency. In photocatalytic processes, the two characteristics that control the surface interaction are morphology and electronic structure. These characteristics change based on the surface structure of the dopant [90–92].

The different functions that CQDs play in effective photocatalysis can be divided into [93]:

(1) Acceptor and mediator of photoinduced electron from conduction band (CB).
(2) Improving the visible light output of a broad band gap photocatalyst by photosensitization techniques.
(3) Reducing agent used in the surface plasmon resonance (SPR) phenomena during the production of different metal NPs.

(4) Effectively using a wide range of the solar spectrum through the phenomena of up-conversion photoluminescence (UCPL), which releases light with a shorter wavelength than what is required to excite CQDs.

5.4 Probable Mechanism

When a suitable light source is present, meaning its energy is either equivalent to or higher than the band-gap energy of the photocatalyst, a minimum of two reactions occur simultaneously: one is an oxidation caused by holes, while the second is a reduction brought on by photoinduced electrons. The most common mechanistic route of photocatalytic organic pollutant eradication involves capturing of photons by the photocatalyst from light source which results into the creation of e^-/h^+ pair. These e^-/h^+ pairs react with H_2O and O_2 to generate intensively reactive radicals such as $OH^•$ and $O_2^{•-}$ radicals. Then these radicals interact with OPs adsorbed on the surface of CQDs (or its composite) and consequently degrade it as represented by Eqs. 1–11 and Fig. 3 [4, 7, 9, 10].

$$\text{Photoinduction of e} - /\text{h} + \text{ pair : Photocatalyst} + hv \rightarrow h_{VB}^{+} + e_{CB}^{-} \tag{1}$$

$$\text{Ionization of water : } H_2O \rightleftharpoons H^+ + OH^- \tag{2}$$

$$OH^- + h_{VB}^{+} \rightarrow HO^{\cdot} \tag{3}$$

$$O_2\text{ionosorption : } (O_2)_{ads} + e_{CB}^{-} \rightarrow O_2^{-} \tag{4}$$

$$\text{rotonationofsuperoxide : } O_2^{-} + H^+ \rightarrow HOO^{\cdot} \tag{5}$$

$$\text{ProductionofHydrogenperoxide : } 2HO_2 \rightarrow H_2O_2 + O_2 \tag{6}$$

$$\text{Formationofhydroxyl radical : } H_2O_2 + e_{CB}^{-} \rightarrow HO^{\cdot} + OH^- \tag{7}$$

$$OH^- + h_{VB}^{+} \rightarrow HO \tag{8}$$

Decontamination of OPs:

$$\text{pollutant} + h_{VB}^{+} \rightarrow \text{pollutant}^{+} \rightarrow \text{oxidationofpollutant} \tag{9}$$

$$\text{pollutant} + e_{CB}^{-} \rightarrow \text{pollutant}^{-} \rightarrow \text{reductionofpollutant} \tag{10}$$

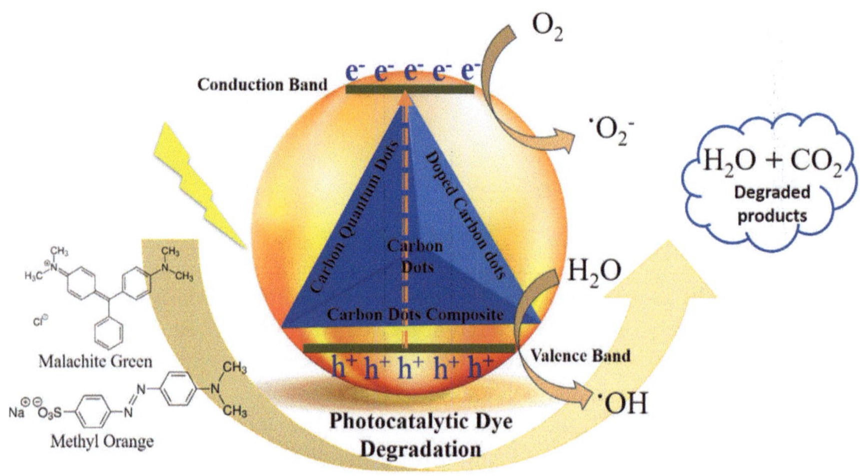

Fig. 3 Photocatalytic eradication of OPs present in wastewater. "Reproduced with permission, copyright 4 © 2023 Elsevier Ltd, (license number—5,654,860,649,093). All rights reserved []."

$$HO^{\cdot} + \text{pollutant} \rightarrow CO_2 + H_2O + \text{degradedproduct} \quad (11)$$

5.5 *Scavenging Test*

To examine the role of different radicals/ions in photocatalytic OPs decomposition, scavenger test can be performed. The photocatalytic OPs breakdown involves reactive radicals like h^+, e^-, $OH^{\bullet}$, and $O_2^{\bullet-}$ radicals. To examine the involvement and role of these radicals, scavengers are usually used in trapping tests. Several investigations utilizing CQDs (or CQDs/composite) as the photocatalyst and a very small quantity of scavengers were carried out to examine the involvement of these radicals in the photocatalytic elimination of different OPs [94]. "For figuring out sole participation of these active species in CQDs derived from green precursors, the most widely used scavengers are (a) para-benzo quinone (p-BZQ), ascorbic acid (AA)(to trap $O_2^{\bullet-}$ radicals), and (b) tertiary butyl alcohol (t-BA), iso-propyl alcohol (IPA), and C_2H_5OH (to trap OH• radicals).When exposed to light radiations, the highly inert intermediates are formed by aliphatic alcohols such as ethanol, t-BA, IPA, etc. which are used to trap $OH^{\bullet}$radicals [95]". It all begins with an interaction between the $OH^{\bullet}$ radical and the OPs. EDTA is the most frequently used trapping reagent for h^+ because of its capacity to oxidize and decompose into glyoxylic acid and ethylenediamine-N, N' diacetic acid. Because p-BZQ can readily reduce $BZQ^{\bullet}$ with continuous light exposure, it is often employed as a scavenger to capture $O_2^{\bullet-}$ radicals. The product of the reaction between $O_2^{\bullet-}$ and p-BZQ is hydroquinone [2].

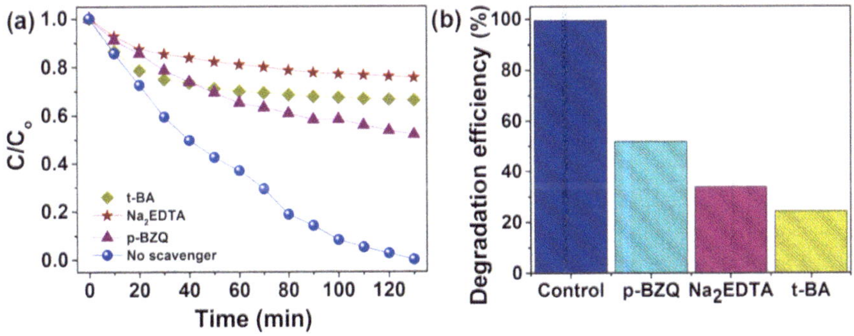

Fig. 4 **a** MB dye degradation with irradiation time in presence of distinct scavengers, **b** % reduction in MB degradation by scavengers. "Reproduced with CC BY license from Das et al. [38]"

Aggarwal et al. performed capturing investigation employing scavengers such as Na_2-EDTA, p-BZQ, and t-BA in order to trap h^+, $O_2^{\bullet -}$, and $OH^{\bullet}$, respectively [96]. Compared to the less noticeable impacts of other trapping agents, the Na_2-EDTA complex's capacity to trap h^+ causes a significant reduction in photocatalytic decontamination. Das et al. conducted a related work for decontaminating MB dye by CDs synthesized from pear juice employing the identical kind of scavengers (10 mM). It was observed that Na_2-EDTA and t-BA captured h^+ and $^{\bullet}OH$ radicals and reduced the percentage of MB dye degradation to 33% and 25%, respectively, were found to be the primary active species in photocatalytic dye decontamination. However, when p-BZQ was added as an $O_2^{\bullet -}$ trap, the percentage of degradation of MB dropped to 50% as displayed in Fig. 4 [38]. A detailed investigation is required on the radicals and scavengers. A specific mechanism may be demonstrated by conducting investigations such as charge carrier separation and EPR intensity signal. Furthermore, studies employing infrared Fourier transform spectroscopy with Scan diffuse-reflectance can be utilized to assess the strength of the H-bond associated with the holes.

6 Conclusion and Future Outlooks

The primary focus of this book chapter is to provide recent advancements for CQDs, doped-CQDs and their composites as photocatalysts derived from sustainable natural resources. These days, catalysts may be tailored to specific uses, and photocatalysis can be made to function better and have more activity. This method's environmental friendliness renders CQDs a viable candidate for photocatalytic deterioration. Nevertheless, this approach is not yet widely used in industries and has only been researched in research labs. The scientific world has assessed the use of cutting-edge methods that produce improved catalytic performances in the manufacturing of CQDs generated from green precursors. The 3 major processes for efficient photocatalytic processes were covered in this article: harvesting of light, charge transfer kinetics, and photocatalytic reactions occurring on the surface. Afterwards, the parameters affecting the photocatalytic process were investigated for CQDs and its composite produced from green precursors in order to determine the ideal pH, photocatalyst dosage, conc. of dye, light intensity, irradiation period, and temperature. Subsequently, we emphasized how radical scavengers contribute to the better understanding of mechanism providing a clear insight into the mechanism. A few research gaps still exist, despite the fact that our understanding of CQDs' role as photocatalysts in OPs degradation has improved. There are strategies available to address these issues. Some of them are compared with earlier studies that used other kinds of materials as photocatalysts. The description of every analysis aim, the facts of specific misunderstandings about each purpose, and potential remedies to the issues are provided below.

1. Selecting and tailing the appropriate substance to be used as a dopant for CQDs can improve the performance of surface catalytic reaction, charge transfer kinetics, and light-harvesting. Additionally, by conducting additional researches on these factors, the functioning of the catalytic reaction can be examined and interpreted.
2. Establishing the border with ideal pH, dopant/CQD conc., photocatalyst dosage and dye conc., intensity of light, irradiation period, and temperature is crucial when addressing the parameters impacting photocatalytic dye decontamination. Regretfully, while several studies have achieved photocatalytic activity by establishing limits, none have determined the ideal temperature for the catalytic system.
3. Moreover, a misperception exists regarding the relationship between dopants/CQDs and the photocatalyst phenomenon. While the impact of photocatalyst dosage is the conc. of the entire composite for carrying out the photocatalytic process, the impact of dopants/CQDs is the concentration in individuals. Selecting the ideal ratio of each in turn increases the photocatalytic activity's performance.
4. In order to determine which active species are in accountable for efficient catalysis, scavengers trap studies are carried out. During the process, some species may have little selectivity, and additional active radical species may be formed. As such, the use of radical scavengers needs to be carefully analyzed. It is important to understand how radical effect affects the proportion of dye molecule breakdown when scavengers are used. Findings that exclusively depend on this technique can lead to misconceptions when examining the response mechanism in detail.
5. Up until now, the basic photocatalytic decontamination mechanistic route has been illustrated for CQDs produced using green materials. Flash radiolysis should be employed to figure out the entrapped carriers of charge to have a deeper comprehension of the mechanism. In order to better clarify the mechanism, an assessment of studies conducted prior to and subsequent to the photocatalytic decomposition of OPs is strongly advised.

6. The most prevalent problem, aside from the goals outlined in this analysis, is using as-obtained CQDs photocatalysts to degrade OPs found in industrial effluents in labs scale only. As such, this study needs to be expanded to industrial level outside of labs.

Author's Contributions The first original-draft writing, methodology, resources, data collection, investigation, conceptualization, editing and formal analysis have been performed by Permender Singh. Editing and formal analysis and investigation have been done by Neeru Rani, Parmod Kumar and Vinita Bhankar and writing-review, formal analysis and supervision of this book chapter have been done by Sandeep Kumar and Krishan Kumar.

References

1. Raizada, P., Sudhaik, A., Singh, P., et al. (2020). Visible light assisted photodegradation of 2,4-dinitrophenol using Ag_2CO_3 loaded phosphorus and sulphur co-doped graphitic carbon nitride nanosheets in simulated wastewater. *Arabian Journal of Chemistry, 13*, 3196–3209. https://doi.org/10.1016/j.arabjc.2018.10.004
2. Basavaraj, N., Sekar, A., & Yadav, R. (2021). Review on green carbon dot-based materials for the photocatalytic degradation of dyes: Fundamentals and future perspective. *Materials Advanced, 2*, 7559–7582. https://doi.org/10.1039/D1MA00773D
3. Rani, U. A., Ng, L. Y., Ng, C. Y., & Mahmoudi, E. (2020). A review of carbon quantum dots and their applications in wastewater treatment. *Advances in Colloid and Interface Science, 278*, 102124. https://doi.org/10.1016/j.cis.2020.102124
4. Singh, P., Rani, N., Kumar, S., et al. (2023). Assessing the biomass-based carbon dots and their composites for photocatalytic treatment of wastewater. *Journal of Cleaner Production, 413*, 137474. https://doi.org/10.1016/j.jclepro.2023.137474
5. Priya, B., Shandilya, P., Raizada, P., et al. (2016). Photocatalytic mineralization and degradation kinetics of ampicillin and oxytetracycline antibiotics using graphene sand composite and chitosan supported BiOCl. *Journal of Molecular Catalysis A: Chemical, 423*, 400–413. https://doi.org/10.1016/j.molcata.2016.07.043
6. Singh, P., Sonu, R. P., et al. (2019). Enhanced photocatalytic activity and stability of AgBr/BiOBr/graphene heterojunction for phenol degradation under visible light. *Journal of Saudi Chemical Society, 23*, 586–599. https://doi.org/10.1016/j.jscs.2018.10.005
7. Singh, P., Mohan, B., Madaan, V., et al. (2022). Nanomaterials photocatalytic activities for waste water treatment: A review. *Environmental Science and Pollution Research, 29*(46), 69294–69326. https://doi.org/10.1007/s11356-022-22550-7
8. Singh, P., Kumar, S., & Kumar, K. (2023). Biogenic synthesis of Allium cepa derived magnetic carbon dots for enhanced photocatalytic degradation of methylene blue and rhodamine B dyes. *Biomass Conversion and Biorefinery*. https://doi.org/10.1007/s13399-023-05047-2
9. Madaan, V., Mohan, B., & Bhankar, V., et al. (2022). Metal-decorated CeO_2 nanomaterials for photocatalytic degradation of organic pollutants. *Inorganic Chemistry Communications, 146*.https://doi.org/10.1016/j.inoche.2022.110099
10. Singh, P., Kumar, S., & Kumar, K., et al. (2023). Photocatalytic technologies in the removal of pharmaceuticals from aquatic systems. *Pharmaceuticals in the Aquatic Environment*, 173–201.https://doi.org/10.1201/9781003436607-9
11. Arpita, K. P., Kataria, N., et al. (2023). Plastic waste-derived carbon dots: Insights of recycling valuable materials towards environmental sustainability. *Current Pollution Reports*. https://doi.org/10.1007/s40726-023-00268-5
12. Lu, W., Qin, X., & Asiri, A. M., et al. (2013). Green synthesis of carbon nanodots as an effective fluorescent probe for sensitive and selective detection of mercury(II) ions. *Journal of Nanoparticle Research, 15*https://doi.org/10.1007/s11051-012-1344-0

13. Singh, P., & Kumar, S., et al. (2023). Assessment of biomass-derived carbon dots as highly sensitive and selective template for sensing of hazardous ions. *Nanoscale*.https://doi.org/10.1039/d3nr01966g
14. Cruz-Cruz, A., Gallareta-Olivares, G., Rivas-Sanchez, A., et al. (2022). Recent advances in carbon dots based biocatalysts for degrading organic pollutants. *Current Pollution Reports, 8*, 384–394. https://doi.org/10.1007/s40726-022-00228-5
15. Rani, N., Singh, P., Kumar, S., et al. (2023). Plant-mediated synthesis of nanoparticles and their applications: A review. *Materials Research Bulletin, 163*, 112233. https://doi.org/10.1016/j.materresbull.2023.112233
16. Bandi, R., Gangapuram, B. R., Dadigala, R., et al. (2016). Facile and green synthesis of fluorescent carbon dots from onion waste and their potential applications as sensor and multicolour imaging agents. *RSC Advances, 6*, 28633–28639. https://doi.org/10.1039/c6ra01669c
17. Sahu, S., Behera, B., Maiti, T. K., & Mohapatra, S. (2012). Simple one-step synthesis of highly luminescent carbon dots from orange juice: Application as excellent bio-imaging agents. *Chemical Communications, 48*, 8835–8837. https://doi.org/10.1039/c2cc33796g
18. Rani, N., Singh, P., Kumar, S., et al. (2023). Recent advancement in nanomaterials for the detection and removal of uranium: A review. *Environmental Research, 234*, 116536. https://doi.org/10.1016/j.envres.2023.116536
19. Gao, W., Zhang, S., Wang, G., et al. (2022). A review on mechanism, applications and influencing factors of carbon quantum dots based photocatalysis. *Ceramics International, 48*, 35986–35999. https://doi.org/10.1016/j.ceramint.2022.10.116
20. Pirsaheb, M., Asadi, A., Sillanpää, M., & Farhadian, N. (2018). Application of carbon quantum dots to increase the activity of conventional photocatalysts: A systematic review. *Journal of Molecular Liquids, 271*, 857–871. https://doi.org/10.1016/j.molliq.2018.09.064
21. Hutton, G. A. M., Martindale, B. C. M., & Reisner, E. (2017). Carbon dots as photosensitisers for solar-driven catalysis. *Chemical Society Reviews, 46*, 6111–6123. https://doi.org/10.1039/c7cs00235a
22. Dhenadhayalan, N., Lin, K. C., & Saleh, T. A. (2020). Recent advances in functionalized carbon dots toward the design of efficient materials for sensing and catalysis applications. *Small (Weinheim an der Bergstrasse, Germany), 16*, 1–24. https://doi.org/10.1002/smll.201905767
23. Kang, Z., & Lee, S. T. (2019). Carbon dots: Advances in nanocarbon applications. *Nanoscale, 11*, 19214–19224. https://doi.org/10.1039/c9nr05647e
24. Lim, S. Y., Shen, W., & Gao, Z. (2015). Carbon quantum dots and their applications. *Chemical Society Reviews, 44*, 362–381. https://doi.org/10.1039/c4cs00269e
25. Liu, J., Li, R., & Yang, B. (2020). Carbon dots: A new type of carbon-based nanomaterial with wide applications. *ACS Central Science, 6*, 2179–2195. https://doi.org/10.1021/acscentsci.0c01306
26. Khan, I., Khan, I., Usman, M., et al. (2020). Nanoclay-mediated photocatalytic activity enhancement of copper oxide nanoparticles for enhanced methyl orange photodegradation. *Journal of Materials Science: Materials in Electronics, 31*, 8971–8985. https://doi.org/10.1007/s10854-020-03431-6
27. Hassan, M. F., Sabri, M. A., Fazal, H., et al. (2020). Recent trends in activated carbon fibers production from various precursors and applications—A comparative review. *Journal of Analytical and Applied Pyrolysis, 145*, 104715. https://doi.org/10.1016/j.jaap.2019.104715
28. Ogugbue, C. J., & Sawidis, T. (2011). Bioremediation and detoxification of synthetic wastewater containing triarylmethane dyes by aeromonas hydrophila isolated from industrial effluent. *Biotechnology Research International, 2011*, 1–11. https://doi.org/10.4061/2011/967925
29. Nasar, A., & Mashkoor, F. (2019). Application of polyaniline-based adsorbents for dye removal from water and wastewater—A review. *Environmental Science and Pollution Research, 26*, 5333–5356. https://doi.org/10.1007/s11356-018-3990-y
30. Deering, K., Spiegel, E., Quaisser, C., et al. (2020). Exposure assessment of toxic metals and organochlorine pesticides among employees of a natural history museum. *Environmental Research, 184*, 109271. https://doi.org/10.1016/j.envres.2020.109271

31. Von Lau, E., Gan, S., Ng, H. K., & Poh, P. E. (2014). Extraction agents for the removal of polycyclic aromatic hydrocarbons (PAHs) from soil in soil washing technologies. *Environmental Pollution, 184*, 640–649. https://doi.org/10.1016/j.envpol.2013.09.010
32. Al-buriahi, A. K., Al-gheethi, A. A., Senthil, P., et al. (2022). Chemosphere Elimination of rhodamine B from textile wastewater using nanoparticle photocatalysts: A review for sustainable approaches. *Chemosphere, 287*, 132162. https://doi.org/10.1016/j.chemosphere.2021.132162
33. Sarkar Phyllis, A. K., Tortora, G., & Johnson, I. (2022). Photodegradation. *Fairchild Books Dictionary of Textiles, 10*(5040/9781501365072), 12105.
34. Dao, H. M., Whang, C. H., Shankar, V. K., et al. (2020). Methylene blue as a far-red light-mediated photocleavable multifunctional ligand. *Chemical Communications, 56*, 1673–1676. https://doi.org/10.1039/c9cc08916k
35. Santoso, E., Ediati, R., Kusumawati, Y., et al. (2020). Review on recent advances of carbon based adsorbent for methylene blue removal from waste water. *Materials Today Chemistry, 16*, 100233. https://doi.org/10.1016/j.mtchem.2019.100233
36. Nugraha, M. W., Zainal Abidin, N. H., & Supandi Sambudi, N. S. (2021). Synthesis of tungsten oxide/ amino-functionalized sugarcane bagasse derived-carbon quantum dots (WO_3/N-CQDs) composites for methylene blue removal. *Chemosphere, 277*.https://doi.org/10.1016/j.chemosphere.2021.130300
37. Zhu, Z., Yang, P., & Li, X., et al. (2019). Green preparation of palm powder-derived carbon dots co-doped with sulfur/chlorine and their application in visible-light photocatalysis. *Spectrochimica Acta Part A Molecular and Biomolecular Spectroscopy*, 117659. https://doi.org/10.1016/j.saa.2019.117659
38. Das, G. S., Shim, J. P., Bhatnagar, A., et al. (2019). Biomass-derived carbon quantum dots for visible-light-induced photocatalysis and label-free detection of Fe(III) and ascorbic acid. *Science and Reports, 9*, 1–9. https://doi.org/10.1038/s41598-019-49266-y
39. Cheng, Y., Bai, M., Su, J., et al. (2019). Synthesis of fluorescent carbon quantum dots from aqua mesophase pitch and their photocatalytic degradation activity of organic dyes. *Journal of Materials Science and Technology, 35*, 1515–1522. https://doi.org/10.1016/j.jmst.2019.03.039
40. Jothi, V. K., Ganesan, K., Natarajan, A., & Rajaram, A. (2021). Green synthesis of self-passivated fluorescent carbon dots derived from rice bran for degradation of methylene blue and fluorescent ink applications. *Journal of Fluorescence, 31*, 427–436. https://doi.org/10.1007/s10895-020-02652-6
41. Bozetine, H., Wang, Q., Barras, A., et al. (2016). Green chemistry approach for the synthesis of ZnO-carbon dots nanocomposites with good photocatalytic properties under visible light. *Journal of Colloid and Interface Science, 465*, 286–294. https://doi.org/10.1016/j.jcis.2015.12.001
42. Maruthapandi, M., Saravanan, A., & Manohar, P., et al. (2021). Photocatalytic degradation of organic dyes and antimicrobial activities by polyaniline–nitrogen-doped carbon dot nanocomposite. *Nanomaterials, 11*https://doi.org/10.3390/nano11051128
43. Shivalkar, S., Gautam, P. K., Verma, A., et al. (2021). Autonomous magnetic microbots for environmental remediation developed by organic waste derived carbon dots. *Journal of Environmental Management, 297*, 113322. https://doi.org/10.1016/j.jenvman.2021.113322
44. Zaib, M., Arshad, A., Khalid, S., & Shahzadi, T. (2021). One pot ultrasonic plant mediated green synthesis of carbon dots and their application invisible light induced dye photocatalytic studies: A kinetic approach. *International Journal of Environmental Analytical Chemistry, 00*, 1–19. https://doi.org/10.1080/03067319.2021.1934463
45. Sekar, A., & Yadav, R. (2021). Optik Green fabrication of zinc oxide supported carbon dots for visible light-responsive photocatalytic decolourization of Malachite Green dye: Optimization and kinetic studies. *Optik (Stuttg), 242*, 167311. https://doi.org/10.1016/j.ijleo.2021.167311
46. Thakur, A., Kumar, P., Kaur, D., et al. (2020). TiO_2 nanofibres decorated with green-synthesized PAu/Ag@CQDs for the efficient photocatalytic degradation of organic dyes and pharmaceutical drugs. *RSC Advances, 10*, 8941–8948. https://doi.org/10.1039/c9ra10804a

47. Tyagi, A., Tripathi, K. M., Singh, N., et al. (2016). Green synthesis of carbon quantum dots from lemon peel waste: Applications in sensing and photocatalysis. *RSC Advances, 6*, 72423–72432. https://doi.org/10.1039/c6ra10488f
48. Ali, I., Burakova, I., Galunin, E., et al. (2019). High-speed and high-capacity removal of methyl orange and malachite green in water using newly developed mesoporous carbon: Kinetic and isotherm studies. *ACS Omega, 4*, 19293–19306. https://doi.org/10.1021/acsomega.9b02669
49. Dutta, S. K., Amin, M. K., & Ahmed, J., et al. (2022). Removal of toxic methyl orange by a cost-free and eco-friendly adsorbent: Mechanism, phytotoxicity, thermodynamics, and kinetics. *South African Journal of Chemical Engineering, 40*, 195–208. https://doi.org/10.1016/j.sajce.2022.03.006
50. Oladoye, P. O., Bamigboye, M. O., Ogunbiyi, O. D., & Akano, M. T. (2022). Toxicity and decontamination strategies of Congo red dye. *Groundwater for Sustainable Development, 19*, 100844. https://doi.org/10.1016/j.gsd.2022.100844
51. Monje, D. S., Mercado, D. F., & Peñuela, G. A., et al. (2022) Carbon dots decorated magnetite nanocomposite obtained using yerba mate useful for remediation of textile wastewater through a photo—fenton treatment: Ilex paraguariensis as a platform of environmental interest—Part 2. https://doi.org/10.1007/s11356-022-22405-1
52. Shahba, H., & Sabet, M. (2020). Two-step and green synthesis of highly fluorescent carbon quantum dots and carbon nanofibers from pine fruit. *Journal of Fluorescence, 30*, 927–938. https://doi.org/10.1007/s10895-020-02562-7
53. Sabet, M., & Mahdavi, K. (2019). Green synthesis of high photoluminescence nitrogen-doped carbon quantum dots from grass via a simple hydrothermal method for removing organic and inorganic water pollutions. *Applied Surface Science, 463*, 283–291. https://doi.org/10.1016/j.apsusc.2018.08.223
54. Malavika, J. P., Shobana, C., Ragupathi, M., et al. (2021). A sustainable green synthesis of functionalized biocompatible carbon quantum dots from Aloe barbadensis Miller and its multi-functional applications. *Environmental Research, 200*, 111414. https://doi.org/10.1016/j.envres.2021.111414
55. Rajapandi, S., Pandeeswaran, M., & Kousalya, G. N. (2022). Novel green synthesis of N-doped carbon dots from fruits of Opuntia ficus Indica as an effective catalyst for the photocatalytic degradation of methyl orange dye and antibacterial studies. *Inorganic Chemistry Communications, 146*, 110041. https://doi.org/10.1016/j.inoche.2022.110041
56. Wang, Z., Cheng, Q., Wang, X., et al. (2021). Carbon dots modified bismuth antimonate for broad spectrum photocatalytic degradation of organic pollutants: Boosted charge separation, DFT calculations and mechanism unveiling. *Chemical Engineering Journal, 418*, 129460. https://doi.org/10.1016/j.cej.2021.129460
57. Dadigala, R., Bandi, R. K., Gangapuram, B. R., & Guttena, V. (2017). Carbon dots and Ag nanoparticles decorated g-C_3N_4 nanosheets for enhanced organic pollutants degradation under sunlight irradiation. *Journal of Photochemistry and Photobiology, A: Chemistry, 342*, 42–52. https://doi.org/10.1016/j.jphotochem.2017.03.032
58. Song, B., Wang, Q., Wang, L., et al. (2020). Carbon nitride nanoplatelet photocatalysts heterostructured with B-doped carbon nanodots for enhanced photodegradation of organic pollutants. *Journal of Colloid and Interface Science, 559*, 124–133. https://doi.org/10.1016/j.jcis.2019.10.015
59. Huang, G., Liu, L., Chen, L., et al. (2022). Unique insights into photocatalytic VOCs oxidation over WO_3/carbon dots nanohybrids assisted by water activation and electron transfer at interfaces. *Journal of Hazardous Materials, 423*, 127134. https://doi.org/10.1016/j.jhazmat.2021.127134
60. Peng, P., Chen, Z., Li, X., et al. (2022). Biomass-derived carbon quantum dots modified Bi_2MoO_6/Bi_2S_3 heterojunction for efficient photocatalytic removal of organic pollutants and Cr (VI). *Separation and Purification Technology, 291*, 120901. https://doi.org/10.1016/j.seppur.2022.120901
61. Hak, C. H., Leong, K. H., Chin, Y. H., et al. (2020). Water hyacinth derived carbon quantum dots and g-C_3N_4 composites for sunlight driven photodegradation of 2,4-dichlorophenol. *SN Applied Science, 2*, 1–14. https://doi.org/10.1007/s42452-020-2840-y

62. Tripti, T., Singh, P., Rani, N., et al. (2023). Carbon dots as potential candidate for photocatalytic treatment of dye wastewater. *Environmental Science and Pollution Research.* https://doi.org/10.1007/s11356-023-31437-0
63. Feng, H., Zhang, Y., & Cui, F. (2022). Enhanced photocatalytic activity of Cu_2O for visible light-driven dye degradation by carbon quantum dots. *Environmental Science and Pollution Research, 29*, 8613–8622. https://doi.org/10.1007/s11356-021-16337-5
64. Bai, L., Liu, L., Pang, J., et al. (2022). N, P-codoped carbon quantum dots-decorated Ti_O2 nanowires as nanosized heterojunction photocatalyst with improved photocatalytic performance for methyl blue degradation. *Environmental Science and Pollution Research, 29*, 9932–9943. https://doi.org/10.1007/s11356-021-16295-y
65. Wu, Y., Chen, C., He, S., et al. (2021). In situ preparation of visible-light-driven carbon quantum dots/$NaBiO_3$ hybrid materials for the photoreduction of Cr(VI). *Journal of Environmental Sciences (China), 99*, 100–109. https://doi.org/10.1016/j.jes.2020.06.016
66. Behnood, R., & Sodeifian, G. (2020). Synthesis of N doped-CQDs/Ni doped-ZnO nanocomposites for visible light photodegradation of organic pollutants. *Journal of Environmental Chemical Engineering, 8*https://doi.org/10.1016/j.jece.2020.103821
67. Mahmood, A., Wang, X., Shi, G., et al. (2020). Revealing adsorption and the photodegradation mechanism of gas phase o-xylene on carbon quantum dots modified TiO_2 nanoparticles. *Journal of Hazardous Materials, 386*, 121962. https://doi.org/10.1016/j.jhazmat.2019.121962
68. Li, J., Ma, Y., Ye, Z., et al. (2017). Fast electron transfer and enhanced visible light photocatalytic activity using multi-dimensional components of carbon quantum dots@3D daisy-like In_2S_3/single-wall carbon nanotubes. *Applied Catalysis B: Environmental, 204*, 224–238. https://doi.org/10.1016/j.apcatb.2016.11.021
69. Bozetine, H., Meziane, S., & Aziri, S., et al. (2021). Facile and green synthesis of a ZnO/CQDs/AgNPs ternary heterostructure photocatalyst: Study of the methylene blue dye photodegradation. *Bulletin of Materials Science, 44*.https://doi.org/10.1007/s12034-021-02353-1
70. Di, G., Zhu, Z., Dai, Q., et al. (2020). Wavelength-dependent effects of carbon quantum dots on the photocatalytic activity of g-C_3N_4 enabled by LEDs. *Chemical Engineering Journal, 379*, 122296. https://doi.org/10.1016/j.cej.2019.122296
71. Guo, Z., Wang, Q., Shen, T., et al. (2019). Synthesis of 3D CQDs/urchin-like and yolk-shell TiO_2 hierarchical structure with enhanced photocatalytic properties. *Ceramics International, 45*, 5858–5865. https://doi.org/10.1016/j.ceramint.2018.12.052
72. Chen, Q., Chen, L., Qi, J., et al. (2019). Photocatalytic degradation of amoxicillin by carbon quantum dots modified $K_2Ti_6O_{13}$ nanotubes: Effect of light wavelength. *Chinese Chemical Letters, 30*, 1214–1218. https://doi.org/10.1016/j.cclet.2019.03.002
73. Hu, X., Han, W., Zhang, M., et al. (2022). Enhanced adsorption and visible-light photocatalysis on TiO_2 with in situ formed carbon quantum dots. *Environmental Science and Pollution Research, 29*, 56379–56392. https://doi.org/10.1007/s11356-022-19810-x
74. Vinodgopal, K. (1992). Photochemistry on surfaces—photodegradation of 1,3- diphenylisobenzofuran over metal-oxide particles. *Journal of Physical Chemistry* 5053–5059
75. Hosseini, F., Assadi, A. A., Nguyen-Tri, P., et al. (2022). Titanium-based photocatalytic coatings for bacterial disinfection: The shift from suspended powders to catalytic interfaces. *Surfaces and Interfaces, 32*, 102078. https://doi.org/10.1016/j.surfin.2022.102078
76. García-Fernández, I., Fernández-Calderero, I., Inmaculada Polo-López, M., & Fernández-Ibáñez, P. (2015). Disinfection of urban effluents using solar TiO_2 photocatalysis: A study of significance of dissolved oxygen, temperature, type of microorganism and water matrix. *Catalysis Today, 240*, 30–38. https://doi.org/10.1016/j.cattod.2014.03.026
77. García-Fernández, I., Miralles-Cuevas, S., Oller, I., et al. (2019). Inactivation of E. coli and E. faecalis by solar photo-Fenton with EDDS complex at neutral pH in municipal wastewater effluents. *Journal of Hazardous Materials, 372*, 85–93. https://doi.org/10.1016/j.jhazmat.2018.07.037
78. Dimitroula, H., Daskalaki, V. M., Frontistis, Z., et al. (2012). Solar photocatalysis for the abatement of emerging micro-contaminants in wastewater: Synthesis, characterization and

testing of various TiO_2 samples. *Applied Catalysis B: Environmental, 117–118*, 283–291. https://doi.org/10.1016/j.apcatb.2012.01.024

79. Real, F. J., Acero, J. L., Benitez, F. J., et al. (2010). Oxidation of hydrochlorothiazide by UV radiation, hydroxyl radicals and ozone: Kinetics and elimination from water systems. *Chemical Engineering Journal, 160*, 72–78. https://doi.org/10.1016/j.cej.2010.03.009
80. Rioja, N., Zorita, S., & Peñas, F. J. (2016). Effect of water matrix on photocatalytic degradation and general kinetic modeling. *Applied Catalysis B: Environmental, 180*, 330–335. https://doi.org/10.1016/j.apcatb.2015.06.038
81. Ma, G., Yang, Q., Sun, K., et al. (2015). Bioresource technology nitrogen-doped porous carbon derived from biomass waste for high-performance supercapacitor. *Bioresource Technology, 197*, 137–142. https://doi.org/10.1016/j.biortech.2015.07.100
82. Thanh Son, B., Viet Long, N., Thi Nhat Hang, N. (2021). The development of biomass-derived carbon-based photocatalysts for the visible-light-driven photodegradation of pollutants: A comprehensive review. https://doi.org/10.1039/d1ra05079f
83. Hu, Z., Jiao, X., & Xu, L. (2020). The N , S co-doped carbon dots with excellent luminescent properties from green tea leaf residue and its sensing of gefitinib *154*.https://doi.org/10.1016/j.microc.2019.104588
84. Gao, C., Low, J., Long, R., et al. (2020). Heterogeneous single-atom photocatalysts: Fundamentals and applications. *Chemical Reviews, 120*, 12175–12216. https://doi.org/10.1021/acs.chemrev.9b00840
85. Wang, Q., & Domen, K. (2020). Particulate photocatalysts for light-driven water splitting: Mechanisms, challenges, and design strategies. *Chemical Reviews, 120*, 919–985. https://doi.org/10.1021/acs.chemrev.9b00201
86. Chen, X., & Burda, C. (2008). The electronic origin of the visible-light absorption properties of C-, N- and S-doped TiO_2 nanomaterials. *Journal of the American Chemical Society, 130*, 5018–5019. https://doi.org/10.1021/ja711023z
87. Bai, S., Jiang, J., Zhang, Q., & Xiong, Y. (2015). Steering charge kinetics in photocatalysis: Intersection of materials syntheses, characterization techniques and theoretical simulations. *Chemical Society Reviews, 44*, 2893–2939. https://doi.org/10.1039/c5cs00064e
88. Habibi-Yangjeh, A., & Akhundi, A. (2016). Novel ternary g-C_3N_4/Fe_3O_4/Ag_2CrO_4 nanocomposites: Magnetically separable and visible-light-driven photocatalysts for degradation of water pollutants. *Journal of Molecular Catalysis A: Chemical, 415*, 122–130. https://doi.org/10.1016/j.molcata.2016.01.032
89. Yang, S., Gong, Y., Zhang, J., et al. (2013). Exfoliated graphitic carbon nitride nanosheets as efficient catalysts for hydrogen evolution under visible light. *Advanced Materials, 25*, 2452–2456. https://doi.org/10.1002/adma.201204453
90. Bai, S., Wang, L., Li, Z., & Xiong, Y. (2017). Facet-engineered surface and interface design of photocatalytic materials. *Advanced Science, 4*.https://doi.org/10.1002/advs.201600216
91. Kubacka, A., Fernández-García, M., & Colón, G. (2012). Advanced nanoarchitectures for solar photocatalytic applications. *Chemical Reviews, 112*, 1555–1614. https://doi.org/10.1021/cr100454n
92. White, J. L., Baruch, M. F., Pander, J. E., et al. (2015). Light-driven heterogeneous reduction of carbon dioxide: Photocatalysts and photoelectrodes. *Chemical Reviews, 115*, 12888–12935. https://doi.org/10.1021/acs.chemrev.5b00370
93. Sharma, S., Dutta, V., Singh, P., et al. (2019). Carbon quantum dot supported semiconductor photocatalysts for efficient degradation of organic pollutants in water: A review. *Journal of Cleaner Production, 228*, 755–769. https://doi.org/10.1016/j.jclepro.2019.04.292
94. Antunes, C. S. A., Bietti, M., Salamone, M., & Scione, N. (2004). Early stages in the TiO_2-photocatalyzed degradation of simple phenolic and non-phenolic lignin model compounds. *Journal of Photochemistry and Photobiology, A: Chemistry, 163*, 453–462. https://doi.org/10.1016/j.jphotochem.2004.01.018
95. Terzian, R., Serpone, N., Draper, R. B., et al. (1991). Pulse radiolytic studies of the reaction of pentahalophenols with OH radicals: Formation of pentahalophenoxyl, dihydroxypentahalocyclohexadienyl, and semiquinone radicals. *Langmuir, 7*, 3081–3089. https://doi.org/10.1021/la00060a030

96. Aggarwal, R., Saini, D., Singh, B., et al. (2020). Bitter apple peel derived photoactive carbon dots for the sunlight induced photocatalytic degradation of crystal violet dye. *Solar Energy, 197*, 326–331. https://doi.org/10.1016/j.solener.2020.01.010
97. Ismida, Y., & Amin, M. Effect of contact time adsorption of rhodamine B , methyl orange and methylene blue colours on langsat shell with batch methods effect of contact time adsorption of rhodamine B , methyl orange and methylene blue colours on langsat shell with batch methods. https://doi.org/10.1088/1742-6596/1788/1/012008
98. Wang, C., Xu, J., Li, H., & Zhao, W. (2021). Alkali-assisted synthesis of gourd flesh derived tricolor F/N co-doped carbon dots for efficient degradation of organic dyes. *Optical Materials (Amst), 114*https://doi.org/10.1016/j.optmat.2021.110950

Green Carbon Quantum Dots—Environmental Applications

Sensing Activity of Green Carbon Quantum Dots and Environmental Monitoring

P. Venugopalan and N. Vidya

Abstract Carbon quantum dots (CQDs) have recently gained popularity in environmental research due to their intriguing chemical and physical properties, as well as their environmental friendliness. Greener synthesised CQDs, CQDs synthesised from bio precursors using green procedures, have a better acceptability due to their more eco-friendly nature, as opposed to damaging inorganic quantum dots and CQDs made from chemical precursors. The photoluminescence (PL) or fluorescence properties of CQDs provide obvious sensing potential, and they account for the vast majority of sensing applications in CQD-based systems. CQDs have a wide range of sensing applications, as indicated by the well-known fluorescence processes in systems such as energy transfer, fluorescence quenching, and the chemical sensitivity of fluorescent spectra. The detection of pollutants in environmental systems may be regarded as the most important phase in environmental analysis, the introduction of green CQDs to this field of environmental monitoring gives value addition to this nano-carbon candidate. This chapter describes CQDs' sensing activity toward various contaminants, including organic pollutants and heavy metal ions. It also examines the potential sensing process that underpins the usage of green CQDs in sensing applications for detecting harmful contaminants. Apart from the fluorescent sensing, chapter also dealing with the colorimetric and electrochemical sensing by CQDs, which were also came in to picture with a greater acceptance.

Keywords Carbon quantum dots (CQDs) · Sensors · Fluorescence · Environmental monitoring · Organic pollutants · Heavy metal ions

P. Venugopalan (✉) · N. Vidya
Department of Chemistry, Sree Neelakanta Government Sanskrit College (University of Calicut), Pattambi, Kerala 679306, India
e-mail: venugpamrita@gmail.com

V. Kumar et al. (eds.), *Green Carbon Quantum Dots*, Engineering Materials,
https://doi.org/10.1007/978-981-97-6203-3_9

1 Introduction

Carbon quantum dots (CQDs), a fluorescent star from the carbon nanomaterial family acquire the significant importance in almost every field owing to its versatile potentiality and environmental friendliness. The commendable biocompatibility and water solubility of this tiny luminary make it more users friendly. The biocompatibility of the CQDs can further be increased by the introduction of green carbon quantum dots [53, 104].

Green carbon quantum dots are CQDs with biological origin that is CQDs derived from natural resources. There has recently been an increasing interest in one-pot CQDs synthesis, mostly to decrease waste and by-products [60]. Many natural carbon sources, including food, plant biomass, biowaste, and animal products, have been investigated as potential raw materials for the production of CQDs [37, 61]. In general, plant parts such as flowers, fruits, seeds, and stems that contain multiple acidic, basic, and neutral bioactive compounds are intriguing as powerful and sustainable biosources for the production of CQDs in aqueous media. Eco-friendly raw materials, in contrast to synthetic precursors, are rich in carbon and nitrogen supplies in the form of proteins and carbohydrates and also act as self-passivating agents to produce surface-functionalized CQDs [104]. Green sources provide various advantages, including simple availability, low cost, relatively clean reactions, and non-toxicity. Furthermore, since green sources are made up of various organic molecules and may be thought of as organic reservoirs, there is no need for an external dopant in their case [46]. Furthermore, various phytochemical constituents present in a certain precursor not only influence reaction kinetics, but also the surface functional groups on the CQDs, and hence their reactivity. Besides that, the quantum yield of CQDs is affected by the kind of phytoconstituents, particle size, solvent, and dopants used [53, 104]. These types of CQDs production not only make it economical and eco-friendly but also give value addition to the precursor. Thus, further investigation into natural resources can also be conducted. Green resource-based CQDs can better handle almost all of the CQDs application area since they have essentially equivalent characteristics and greater biocompatibility to their synthetic versions [61, 88].

As mentioned earlier, there are several reports in the literature discussing about the synthesis and applications of CQDs from natural products. In general, vegetables, fruits, leaves and rhizomes etc. are widely used as a carbon source to prepare CQDs (Fig. 1). Almost of them are successfully employed in various significant applications. Some of the most recent and relevant works and corresponding applications are enlisted in the Table 1.

The aforementioned table makes it abundantly clear that the sensing potential of green CQDs is the most intriguing and well-studied application. This sensing capacity is mostly used for environmental monitoring [79]. To protect the ecosystem from various pollutants and ensure better living conditions for all living organisms, adequate environmental surveillance is essential.

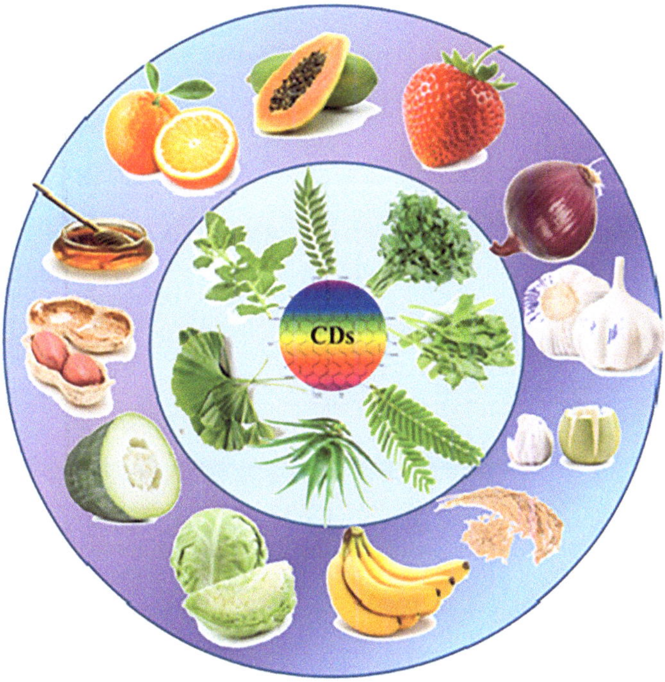

Fig. 1 Different natural products used for the synthesis of CQDs

2 Green CQDs for Pollutant Sensing

With the advantage of enhanced biocompatibility and a greener nature, green precursor-based CQDs find promising applications akin to those produced from chemical precursors. They are extensively employed in the field of sensing because of their remarkable optical characteristics, particularly their excellent photoluminescence. The mechanism of sensing relies on the modifications made to the probe by the addition of particular analytes. These modifications can affect the structure and electronic distribution of CQDs, resulting in notable observations about their properties. Owing to the eco-friendly nature of the green precursor based CQDs, it can be directly opted for the sensing application in various field without any risk. The lesser or null toxicity of these CQDs make it as an excellent candidate for the environmental monitoring applications. The hydrophilic functionalities present in the surface of CQDs made it water soluble so it can easily applicable in water medium based sensing. The sensing ability of CQDs is often utilized in environmental pollutant screening applications. The sensing of heavy metal ions and other organic pollutants from environmental sources by green CQDs are the most renounced application from this area. As per the available literature till date, there are three main methods for detecting these pollutants using green CQDs: fluorescence-based sensing, colorimetric detection and

Table 1 CQDs from natural resources and applications

Source	Method	Size (nm)	Quantum yield (%)	Application	References
Coriander leaves	Hydrothermal	1.5–2.98	6.48	Detecting Fe^{3+}	Sachdev and Gopinath [75]
Papaya	Hydrothermal	3.4	18.98	Detecting Fe^{3+}	Wang et al. [101]
Pomelo peel	Hydrothermal	2–4	6.9	Detecting Hg^{2+}	Lu et al. [58]
Soy milk	Hydrothermal	25	2.6	Electrocatalyst	Zhu et al. [118]
Banana juice	Hydrothermal	3	8.95	Not reported	De and Karak [16]
Peach gum	Hydrothermal	2–5	28.46	Detecting Au^{3+}	Liao et al. [50]
Linseed	Hydrothermal	4–8	14.2	Biosensor and bioimaging	Song et al. [81]
Black tea	Hydrothermal	4.6	x	Detecting Fe^{3+}	Song et al. [80]
Prawn shell	Hydrothermal	6	54	Drug delivery	Gedda et al. [21]
Bamboo leaves	Hydrothermal	2–6	7.1	Detecting Cu^{2+}	Liu et al. [55]
Honey	Hydrothermal	2	19.8	Detecting Fe^{3+}	Yang et al. [108]
Willow bark	Hydrothermal	1–4	6	Biosensing and photoctalyst	Qin et al. [70]
Lemon peel	Hydrothermal	1–3	14	Detecting Cr(VI)	Tyagi et al. [89]
Cornflour	Hydrothermal	2–6	7.7	Bioimaging and detecting Cu^{2+}	Wei et al. [105]
Eggshell membrane	Microwave	5	14	Biosensor	Wang et al. [102]
Goose feather	Microwave-hydrothermal	21.5	17.1	Detecting Fe^{3+}	Liu et al. [54]
Coconut water	Microwave-hydrothermal	1–6	54	Detecting thiamine and Cu^{2+}	Purbia and Paria [68]
Jackfruit seeds	Microwave	3–7	17.91	Bioimaging and detecting Au^{3+}	Raji et al. [73]
Cow manure	Chemical oxidation	4	65	Bioimaging	D'Angelis do et al. [14]

(continued)

Table 1 (continued)

Source	Method	Size (nm)	Quantum yield (%)	Application	References
Urine	Pyrolysis	20.6	14	Detecting $Hg^{2+}$and Cu^{2+}	Essner et al. [18]
Konjac flour	Pyrolysis	3.37	22	Bioimaging and detecting Fe^{3+} and lysine	Teng et al. [87]

x: Not reported

electrochemical sensing. Nonetheless, the majority of research is focused on fluorescence sensing. Research in carbon quantum dots electrochemical and colorimetric sensing is still in the early stages of development.

2.1 *Fluorescence Based Sensing*

As mentioned before, fluorescence is the charming feature of CQDs effectively used for the sensing applications, according to the differences in fluorescence (reduction or enhancement of fluorescence intensity) by the presence of foreign species. It has been broadly used as fluorescent nanoprobe for the detection of several analytes in biological as well as environmental samples. Substantial fluorescence, fast responses, high selectivity and sensitivity, low cost, users friendliness, higher biocompatibility, water solubility and abundant surface functional groups are some of the reasons by which CQDs can act as efficient fluorescent sensors. Generally, different types of analytes can be detected through this fluorescent sensing with good level of selectivity and sensitivity, including anions, metal ions, organic contaminants, etc. [5, 13, 15, 104].

The mechanism of this reduction or enhancement of fluorescence, by the addition of analyte depends up on the structure and optical properties of CQDs [8]. These CQD sensing applications work on the principle that when analytes and CQDs interact, fluorescence is either suppressed or enhanced. However, the majority of CQDs sensing research is based on fluorescence quenching. In general, there are five major types of sensing mechanisms (quenching mechanisms), which are static quenching mechanisms, dynamic quenching mechanisms, inner filter effect (IFE), Förster resonance energy transfer (FRET), and photo induced electron transfer (PET) [38, 120].

(1) *Static quenching mechanism:* It arises when quenchers and CQDs combine to create a ground state complex that is not luminescent [113].
(2) *Dynamic quenching mechanism*: When the quencher and the CQDs collide and either energy or electrons are transferred, the excited state of the CQDs returns to the ground state, which explains this kind of quenching [53, 63, 64].

(3) *Inner filter effect (IFE)*: According to observations, it occurs when the emission or excitation spectra of the CQDs and the quencher's absorption spectrum coincide. By using a shield action, this mechanism works [66].
(4) *Förster resonance energy transfer (FRET)*: When the emission spectra of the CQDs and the quencher's absorption spectrum coincide, an electrodynamic process takes place between the CQDs in the excitation state and the quenchers in the ground state [120].
(5) *Photo induced electron transfer (PET)*: It is explained by the fact that quencher and CQDs take part in an electron transfer reaction that produces cation and anion radicals. This process leads to fluorescence quenching because a complex formed between the electron donor and acceptor can return to the ground state without producing photons [103].

The literature includes an array of intriguing studies based on CQDs' use in optical sensing, and according to publishing data, it is the most explored application of CQDs.

Based on their fluorescence behaviour, CQDs obtained from natural resources have a wide range of applications in the sensing field. The majority of these sensors used the fluorescence emission quenching or enhancing method. The sensing sector includes several analyte molecules like metals and organic pollutants etc. [26, 35, 59]. Sensing of these substances has gained much attention since they are very compatible in aquatic environments and directly and indirectly leads to several health issues to all beings. Some of the applications from the field of sensing of metal and organic pollutants are discussed in the subsequent sections.

2.1.1 Sensing of Heavy Metal Ions

Industrialization and contemporaneous agricultural practices play a major role in the contamination of the planet, and the input of heavy metal ions is inevitable. Generally Hg(II), Pb(II), Cd(II) and Cr(VI) are considered as the most hazardous metal ions under use [5]. The extensive usage of these metals adversely affects the environmental system, accumulation even at lower concentration leads to several serious health hazards in all beings. Thus, it is crucial to screen the contaminants, particularly in water bodies, since nearly all of the industrial effluents are directly released into water resources without any kind of pre-treatment, and the compatibility of these pollutants with water intensifies the adverse effects. And hence, for the identification of these metal ions, an efficient and quick-response analytical technique is crucial [94].

The typical methods for the detection of metal ions, including colorimetric analysis [33], photo-electrochemical method [78], inductively coupled plasma mass spectrometry (ICP-MS) [83], headspace gas chromatography [116], electrocatalytic detection [44], and surface-enhanced Raman scattering [117]. Some of them lack sensitivity, but the majority are pricey and complex [110]. Recently, the acceptance of fluorescence spectroscopy over the other detection methods increased with higher rate, since it involves comparatively simple operational procedures, cost-effectiveness, fast responses and high sensitivity [85].

Organic molecules, quantum dots, and metal nanoparticles are the most commonly used fluorescent sensor probes for metal ion sensing [51, 67, 100]. Although the lesser water solubility, minor biocompatibility, and complicated synthetic procedures of these probes make them difficult to use, they lead to the search for alternatives that can resolve all the defects. Since this novel candidates can handle most of the aforementioned challenges, biocompatible fluorescent carbon quantum dots can be mounted for this purpose with ease [94].

There are numerous reports in the literature on toxic metal ion sensing with green CQDs, some of which are discussed below.

Hg(II)

Several types of effective fluorescent CQDs made from natural precursors are used for mercury sensing; the continuous reports of mercury contamination in water and other environmental systems make it one of the most important dangerous metal ions. Most of the CQDs system for the sensing of mercuric ion focussed on the surface state alteration of the probe by the addition of the ions, consequently the corresponding surface state energy levels get modulated resulting change in the fluorescence intensity. Hg(II) has been successfully detected using CQDs made from flour using microwave-assisted synthesis, as reported in 2013 by Qin et al. with a limit of detection (LOD) of 0.5 nM [71]. Similarly, in 2014 Wang et al. reported using CQDs from cucumber juice to detect Hg(II) by a decrease in fluorescence upon interaction with an ion, with a limit of detection of 180 nM [96]. In the very next year, rapid and sensitive of Hg(II) with CQDs was also reported by Yu et al. by using finger citron bergamot fruit as the carbon precursor via hydrothermal heating with a detection limit of 5.5 nM. The luminescence turn-off behaviour of the CQDs with Hg(II) is investigated by the team and finally attributed to the dynamic quenching mechanism [109]. Likewise, within the same year, Li et al. used Chinese yams for the production of highly fluorescent CQDs and were effectively utilized as sensing probe towards Hg(II) in aqueous medium with LOD 1.26 nM [49]. By 2016, a nano sensor was created by Gu et al. to detect Hg(II) from readily available lotus root, enriched with various amino acids through microwave heating. Due to the presence of N-containing groups from the precursor and their function as surface passivation reagents, these CQDs exhibited a relatively greater quantum yield. The CQDs showed excellent sensitiveness as well as selectivity with Hg(II) with detection limit of 18.7 nM and it was attributed to relatively higher and faster chelating kinetics of the Hg(II) with surface functionalities of CQDs and it leads to static fluorescence quenching [22]. Zhao et al. have designed nanohybrids in 2017 based on CQDs derived from corn bract exhibiting bright fluorescence characteristics due to the presence of chlorophyll groups, which are composed of porphyrin ring systems, used as the sensor for Hg(II) detection by fluorescence quenching method with a detection limit of 9 nM [114]. In the same year, casein, a milk protein based CQDs were developed and effectively utilized as a fluorescent probe for the sensing of Hg(II) with LOD of 6.5 nM by Xu et al. [107]. In the same way, in 2018 fluorimetric sensing of

Hg(II) was also carried out by using CQDs developed from pineapple peel as carbon source with LOD of 4.5 nm [91]. Blue fluorescent CQDs were reported in the same year with excellent fluorescence quantum yield, good stability and water solubility. It was synthesised from tamarind leaves comprised with several proteins; carbohydrates and vitamin-C; the research group have adopted hydrothermal treatment for the synthesis. The prepared CQDs displays efficient binding capacity towards Hg(II) by the sulphur atoms in the surface of CQDs resulting the fluorescence turn-off sensing with detection limit of 6 nM [4]. Likewise, the next year there is an another work utilizing ethanolic extract of bamboo leaves, for preparation of CQDs by Liu et al. CQDs nano hybrid was used as dual ratiometric sensor for Hg(II) and Pb(II) ions. Flavonoids and chlorophyll contents in the extract successfully converted into multi-emissive nanohybrids and its specific binding of Hg(II) and Pb(II) to the porphyrin and flavonoid moieties leads to fluorescence quenching with LODs of 0.22 nM (Hg(II)) and 0.14 nM (Pb(II)) [56]. There are several other reports from the same area and all are tabulated in Table 1.

Pb(II)

Lead (Pb) is a metal ion that is toxic and well-known for causing severe problems in humans [86, 88]. Lead exists in three different oxidation states; Pb(0), Pb(II), and Pb(IV), and is mostly found in the environment as Pb(II) [1, 88]. Trace level concentration of Pb (II) in blood is highly hazardous and even cause death. Mental disability, anemia, memory loss, and migraine are the other serious health issues due to this toxic element. The highest permitted level of lead intake in drinking water, according to the Environmental Protection Agency (EPA), should be less than 72 nM (15.0 ppb) [12, 24, 27].

The lead detection by green precursor derived CQDs are very common in literature, the same principle of sensing applies here: the addition of Pb(II) causes the probe's fluorescence intensity to vary. The chelation of this ion with the surface functionalities of CQDs resulting the changes in the luminescence characteristics of the system. The most commonly known herb source tulsi leaves derived CQDs was synthesised by Kumar et al. in 2017 by means of hydrothermal method. The starting material enriched with different functional groups including carvacrol, several acids (rosmarinic, oleanolic and ursolic), alcohols, aldehydes and ketones, and were attached in to the surface of CQDs. Through fluorescence quenching, the green fluorescent system as acquired was utilized for detecting lead (Pb(II)) ions with LOD of 0.59 nM and it was successfully implemented in water as well as live cells [45]. Bandi and co-workers demonstrated the production of bright fluorescent N-doped CQDs by utilizing lantana berries with ethylenediamine as carbon and nitrogen precursors respectively by the year 2018. It was used as a promising sensing probe towards Pb(II) through fluorescence turn-off mechanism with LOD of 9.64 nM. Moreover, this CQDs with polar surface functionalities was employed for the sensing of Pb(II) in the real samples including urine, human serum and water [3]. The major works detailing about the Pb (II) sensing via green precursor based CQDs have been tabulated in Table 1.

Cd(II)

As mentioned before, like other heavy metals Cd(II) can adversely affect the health system especially the CNS and other organs and leads to several diseases, like high blood pressure, neuro disorders and kidney related issues [36].

Sensitive and specific detection of Cd(II) ions was effectively carried out by fluorescent CQDs, from curry leaves via hydrothermal treatment was reported in 2020 with LOD of 0.29 nM. The original fluorescence of CQDs was reduced by the Cd(II) through dynamic fluorescence quenching by ligand-to-metal charge transfer (LMCT) mechanism [65]. In the same year, Chauhan and co-workers reported a coconut waste-derived fluorescent carbon quantum dots prepared through thermal calcinations and successfully demonstrated it as a fluorescent turn-on sensing probe for Cd(II) with detection limit of 0.18 nM. The developed system was effectively checked over different water sources with satisfactory parameters [11]. Later on, Sariga et al. fabricated an effective fluorescence turn-off probe from CQDs for Cd(II) detection in water samples by utilizing Ashoka tree leaves as starting material through a green and facile method. The suggested method exhibits a LOD of 2.4 nM and it was efficiently implemented in real samples of water as well as industrial effluents [77]. There few more reports from the same area and are tabulated in Table 1.

Cr(VI)

Chromium(VI) is the another very important toxic metal ion considered as one of the primary aquatic pollutant and is carcinogenic in nature. In compliance with World Health Organization (WHO) regulations, the maximum amount of Cr(VI) in potable water need to be below 50 μg/L [86].

For the purpose of sensing Cr(VI) in aquatic environments, various green precursor-produced CQDs are frequently used; some of the reports are covered as follows. Lemon peel derived CQDs were reported by Tyagi et al. in 2016. The prepared system with an average size range of 1–3 nm and high photostability, tested for the sensing of Cr (VI) in drinking water and reported a LOD of 73 nM. The possible fluorescence quenching by Cr(VI) was accredited to the non-radiative recombination of electron–hole pairs because of low redox potentials and low-lying d-d transition states, including functional groups on the surface of CQD [89].

Similarly, tulsi leaves-derived CQDs were reported in 2018 by Bhatt et al. and successfully used as fluorescent probe against Cr(VI) with LOD of 86.54 nM and also employed in real samples of water with good level of recovery [7]. In 2019, Feng et al. developed a fluorescent sensor from natural kelp based CQDs and it was employed in environmental water samples for the determination of Cr(VI) with sufficient recoveries [19]. In the above two cases the fluorescence quenching mechanism was accredited to combination of IFE and static quenching, which is confirmed through different methods [7, 19, 88]. The main works describing Cr(VI) sensing via natural products derived CQDs have been tabulated in Table 2. The sensing mechanism is found to be different for different systems and is merely depending up on the optical and structural properties of the systems.

Table 2 Green precursor based CQDs for toxic metal ion sensing

Precursor	Analyte species	LOD (nM)	References
Flour	Hg(II)	0.5	Qin et al. [71]
Cucumber juice	Hg(II)	180	Wang et al. [96]
Finger citron bergamot fruit	Hg(II)	5.5	Yu et al. [109]
Chinese yam	Hg(II)	1.26	Li et al. [49]
Lotus root	Hg(II)	18.7	Gu et al. [22]
Corn bract	Hg(II)	9	Zhao et al. [114]
Casein	Hg(II)	6.5	Xu et al. [107]
Pineapple peel	Hg(II)	4.5	Vandarkuzhali et al. [91]
Tamarind leaves	Hg(II)	6	Bano et al. [4]
Strawberry juice	Hg(II)	3	Huang et al. [31]
Pomelo peel	Hg(II)	0.23	Lu et al. [58]
Honey	Hg(II)	1.02	Srinivasan et al. [84]
Coconut milk	Hg(II)	16.5	Roshni and Ottoor [74]
Muskmelon	Hg(II)	330	Desai et al. [17]
Tulsi leaves	Pb(II)	0.59	Kumar et al. [45]
Lantana berries and ethylenediamine	Pb(II)	9.64	Bandi et al. [3]
Ginkgo leaves	Pb(II)	0.055	Xu et al. [106]
Table sugar	Pb(II)	67	Ansi and Renuka [2]
Potato-dextrose agar	Pb(II)	0.11	Gupta et al. [24]
Curry leaves	Cd(II)	0.29	Pandey et al. [65]
Coconut waste	Cd(II)	0.18	Chauhan et al. [11]
Ashoka tree leaves	Cd(II)	2.4	Kolaprath et al. [77]
Tea residue	Cd(II)	x	Huang et al. [32]
Lemon peel	Cr(VI)	73	Tyagi et al. [89]
Tulsi leaves	Cr(VI)	86.54	Bhatt et al. [7]
Kelp	Cr(VI)	520	Feng et al. [19]
Shallot	Cr(VI)	3500	Sakaew et al. [76]
Flax straw	Cr(VI)	190	Hu et al. [30]
Shrimp shell	Cr(VI)	100	Tai et al. [85]
Groundnuts	Cr(VI)	1923	Roshni et al. [90]
Mangoginger	Cr(VI)	84	Venugopalan and Vidya [94]

x: Not reported

NB: to ensure consistency in the appropriate units with regard to other reports, some values were revised

The sensing of various toxic metal ions with several green precursor derived CQDs systems was discussed overall; however, the precise mechanism of this sensing remains unclear, despite the fact that in nearly all cases, the complex formation between the ionic moiety with the surface functionalities of CQDs leads to a change in the native fluorescence intensity of the CQDs probes. Apart from this static quenching, dynamic quenching also reported for these ions under discussion by the sensing with green CQDs, which involves the collision of the quenching species with CQDs system in the excited state leading to fluorescence reduction. Another important quenching process is due to inner filter effect, it is photo-physical process rather than a chemical phenomenon. In which the absorbance wavelength of the quenching species is gets overlapped with either excitation or emission wavelength of the CQDs probe system leading to the quenching of fluorescence. In some cases, the combination of these different mechanisms was also reported.

2.1.2 Sensing of Organic Pollutants

Certain non-biodegradable organic pollutants, similar to inorganic hazardous metal pollutants, can persist in the environment and make their way up the food chain, posing a risk to human health. Specifically, organic pollutants like nitrophenols widely used in various fields like in the manufacturing of pharmaceuticals, agro-chemicals, explosives and dyes [40]. As they are carcinogenic, its extensive usages and discharge to water resources without any pre-treatments leading to serious health issues. Similarly, the modern agricultural practices made use of neurotoxic pesticides to control the pests and there by minimize the crop loss. Superfluous usages of these toxic pesticides leads to water pollution since it can be easily reach to the water resources. As in the case of pesticides, several antibiotics are now used in the field of agriculture and livestocks to increase crop productivity and as growth promoters. The over usage of these antibiotics make it as primary water pollutant and it will reach in to the body of non-targeted beings, resulting severe health issues. The sensing of these hazardous organic water pollutants are of great importance.

Green precursor developed CQDs are successfully entered to the field of organic pollutants sensing by utilizing its luminescence ability. There are several reports in the literature from this area; some of them are discussed below.

Nitrophenols

Nitrophenols, a class of compounds with appended nitro and phenolic groups, are well-known for its explosive nature. In practical analytical applications, it is critical to detect nitrophenol compounds. There are several reports in the literature that discuss nitrophenol sensing with green CQDs, few of them are discussed below.

Walter's dogwood leaves derived CQDs was reported by Wang et al. and were effectively implemented as fluorescent CQDs for the identification of 4-nitrophenol (4-NP) with LOD of 17.5 nM [97]. Similarly, Chatzimarkou et al. described an efficient and sustainable method for producing CQDs from apple seeds. The resulted N-doped CQDs were shown to be a sensitive as well as effective fluorescent probe, its fluorescence were quenched by the addition of 4-nitrophenol via the FRET quenching processes [10]. The proposed approach outperforms previously reported approaches due to its good recovery percentages and lower limit of detection (13 nM).

In addition to 4-NP, other nitrophenols like 2-nitrophenol, 3-nitrophenol, 2,4-dinitrophenol (DNP) and 2,4,6-trinitrophenol (TNP) were also simultaneously detected by green precursor derived CQDs through fluorescence based sensing method. Such a work was reported in 2018 by Soni and co-workers, in which Palm shell powder and triflic acid based N, S co-doped CQDs was synthesised. The as obtained CQDs was effectively employed as a sensor for 4-NP, DNP, TNP with LOD in the nanomolar scale and the values are found to be 79 nM (4-NP), 165 nM (DNP) and 82 nM (TNP) [82].

There are some more related works in the literature and the important reports with details are listed in Table 3.

Pesticides

Some investigators have recently used natural product-based CQDs to create pesticide probes. Organophosphorus pesticides are considered to be a significant agrochemical capable of increasing crop yield. There are numbers of studies in the literature deal with pesticide sensing especially, organophosphorous pesticides using bioprecursor derived CQDs; some of them are mentioned as follows.

Bera and Mohapatra developed fluorescent CQDs by using chitosan as precursor, that were utilized for ultrasensitive glyphosate detection via effective photoelectron transfer between CQDTe and CQDs with a limit of detection of 0.002 nM. The CQDTe-CQD combined probe is appropriate for practical uses because of its ability to recognize glyphosate with specificity and selectivity, even when additional organophosphorus herbicides are present [6].

On the other hand, Hou et al. employed CQDs made from Japanese pagoda leaves to create a turn-on fluorescence sensor for glyphosate detection. Electron transfer was discovered between Fe^{3+} and the as-prepared CQDs. As a result, Fe^{3+} demonstrated a distinct dynamic-quenching behaviour towards CQDs. The addition of glyphosate, however, brought back the fluorescence of the quenched CQDs/Fe^{3+} system because it impeded the electron transfer pathway. The glyphosate molecule's functional groups and the significant complexation of Fe^{3+} were the cause of it. The LOD value of the CQDs/Fe^{3+} fluorescent probe as prepared is found to be 51.75 nM [29].

Apart from the discussed, several additional related works with different pesticides are available in the literature, and Table 3 lists the significant reports with specific information.

Antibiotics

Antibiotics are substances that prevent the growth of other cells and are commonly used to treat bacterial infections. One antibiotic that has been used extensively is tetracycline. Excessive doses of tetracycline in the human body can cause a variety of illnesses, including hepatotoxicity and nephrotoxicity [20, 53]. Fluorescent based

Table 3 Green precursor based CQDs for organic pollutant sensing

Precursor for CQDs	Analyte species	LOD (nM)	References
Apple seeds	4-NP	13	Chatzimarkou et al. [10]
Palm shell powder and triflic acid	TNP, DNP, 4-NP,	82(TNP), 165(DNP) 79 (4-NP)	Soni and Pamidimukkala [82]
Celery leaves and glutathione	4-NP, 3-NP, 2-NP	26(4-NP), 43(3-NP) 39 (2-NP)	Qu et al. [72]
Sweet chestnut rose	2-NP	15.2	Zhang et al. [112]
Sweet flag	4-NP	207	Venugopalan and Vidya [92]
Shaddock peel and HCl	TNP	37.1	Wang et al. [98]
Walter's dogwood leaves	4-NP	17.5	Wang et al. [97]
Chitosan	Glyphosate	0.002	Bera and Mohapatra [6]
Pork rib bones	Dimethoate	64	Liu et al. [52]
Bilimbi fruit	Quinalphos	510	Venugopalan and Vidya [95]
Flavonoid from ginkgo leaves (a) NSB-CQD-FLA (b) NS-CQD-FLA (c) N-CQD-FLA	Dinoseb Dithianon Fenitrothion	0.66 0.28 0.36	Li et al. [48]
Feather, H_2O_2 and NH_3	Dichlorvos	3.8	Hou et al. [28]
Jatropha fruits	Chlorpyrifos	7.701	Chandra et al. [9]
Japanese pagoda leaves	Glyphosate	51.75	Hou et al. [29]
Dried beet powder	Amoxicillin	475	Wang et al. [99]
Plum	Doxorubicin	120	Zhu et al. [119]
Rose flower and P_2O_5	Tetracycline	3.3	Feng et al. [20]
Crab shell	Tetracycline	0.011	Guo et al. [23]
Rice residue and lysine	Tetracycline Chlortetracycline Terramycin	236.7 279.1 373.9	Qi et al. [69]
Tobacco	Tetracycline, ChlotetracyclineOxytetracycline	5.18 14 6.06	Miao et al. [62]
Hawthorn powder	Chlortetracycline	152.44	Zhang et al. [111]
Wild lemon leaves	Tetracycline	420	Venugopalan and Vidya [93]

4-NP: 4-nitrophenol; 3-NP: 3-nitrophenol; 2-NP: 2-nitrophenol; DNP: 2,4-dinitrophenol; TNP: 2,4,6-trinitrophenol

NB: to ensure consistency in the appropriate units with regard to other reports, some values were revised sensing of antibiotics by biogenic CQDs is well documented in the literature. About a few of them are discussed below.

Feng et al. describe the fabrication of CQDs from rose flowers and P_2O_5. Based on the tetracycline-induced quenching in the fluorescence of CQDs, it was used as a tetracycline sensing probe. The interaction between tetracycline and the surface groups of CQDs was thought to account for this process [20].

Similar to this, Miao et al. achieved the quantitative detection by differentiating three tetracyclines using tobacco-derived CQDs as a sensor. Three tetracyclines reacted with CQDs in distinct ways. Tetracycline specifically quenched the fluorescence without causing a fluorescence shift because of the IFE quenching process. Meanwhile, CQDs responded to chlortetracycline, and as a result of the expanded energy band gap, their fluorescence displayed a blue shift. Additionally, the band gap narrowing brought about by the introduction of oxytetracycline resulted in a red-shift emission. The quantitative analytical method was constructed using the data from this investigation, and the LODs for oxytetracycline, chlotetracycline, and tetracycline were, 6.06 nM, 14 nM, and 5.18 nM respectively [62]. In a comparable manner, Guo et al. synthesised CQDs from crab shells in 2020. The CQDs had good fluorescence due to functional groups made of nitrogen that were generated from the precursor material. The researchers effectively detected tetracycline traces using the produced CQDs, which had high stability and a relatively small LOD value (0.011 nM). Tetracycline may be quantitatively detected in most acidic solutions using the developed CQDs, which also have significant potential for analyzing real sewage [23]. In addition to the ones mentioned, there are a number of biogenic CQDs in the literature that are utilised as fluorescence sensors for various types of antibiotics; Table 3 summarises the noteworthy reports with details.

Similar to heavy metal pollutants, different systems have distinct ways of exploiting green CQDs to detect organic pollutants in environmental sources. In addition to the inner filter effect and static and dynamic quenching mechanisms, the FRET mechanism of quenching has also been reported for green CQDs for the sensing of organic environmental pollutants [47]. This entails the overlap of the absorption spectra of the quencher with the emission spectrum of the CQD-based sensing probe under study. Energy transferring based fluorescence quenching was found to be more common in fluorescence sensing of organic pollutant with CQDs.

2.2 *Colorimetric Sensing*

The colorimetric sensor is one of the most straightforward, affordable, and useful analysis tools available. Transferring the detecting signal into a colour change is the fundamental idea underlying this sensing. As a result, CQDs have drawn plenty of attention for application in colorimetric sensors that detect environmental contaminants. Green synthesized CQDs are, nevertheless, little investigated for this kind of environmental pollutant sensing. In 2018 Zheng and co-workers prepared a green

CQDs system and it was employed for the colorimetric sensing of an organic pollutant. They used Lycii fructus derived CQDs for the preparation of CQDs–AgNPs system, which is successfully implemented for the colorimetric sensing of phoxim pesticide. It is worth noting that the existence of many functional groups on CQDs' surface could contribute to the stability of AgNPs. Because the scattered CQDs–AgNPs aggregated when phoxim was added, the solution colour is changed from yellow to red. The colorimetric sensor has shown to be an efficient tool for detecting phoxim in fruit and other environmental samples [115].

The CQDs-based colorimetric method is less used because of its susceptibility to external disturbances, while offering a few benefits over the fluorescence approach for detecting pollution. Since colorimetric and fluorescence dual-mode assays provide a highly sensitive fluorescence test and enable target observation with the naked eye, they are a good choice for guaranteeing the sensitivity and simplicity of environmental contaminant detection. Target pollution can be conveniently, sensitively, and selectively detected with the help of these colorimetric and fluorescent dual-mode sensors. [57].

2.3 Electrochemical Sensing

The fluorescence behaviour of green CQDs can be utilized to identify organic compounds and metal ions in environmental systems, as was discussed in the sections above. Due to their good conductivity, excellent electrical characteristics, low production cost, and ease of surface modification, CQDs have lately become a very desirable probe for electrochemical sensing applications [26]. To get desirable CQDs for electrochemical sensing applications, the stability and electrical conductivity of CQDs can be enhanced by modifying synthetic conditions the precursors. CQDs have large active areas and a large number of hydrophilic functional groups, both are critical for electrochemical based sensing [25]. Because of their desirable properties, CQDs are frequently used in sensing applications as modifiers for the generation of electrode materials that can increase the rate of electron transference [26, 39]. Regarding CQDs, a smaller particle size results in a higher surface-to-area ratio, thereby expanding the electrode contact area. CQDs are a great choice for electrochemical sensing because of this result. Nevertheless, there have only been a few reported investigations to date [34].

CQDs from mango ginger were reported by Korah and team and it was successfully implemented for the sensing of mercuric ions through electrochemical sensing. The as-produced NCD modified glassy carbon electrode detected mercuric ions electrochemically with a detection limit of 0.2 nM. Additionally, NCDs were utilized to analyze actual samples of tap and river water for mercuric ions [43]. The same group successfully modified a glassy carbon electrode using CQDs made from the leaves of mountain knotgrass, and the modified electrode demonstrated good selectivity for Cu(II) detection. They investigated at the sensing and found that selective sensing at the electrode surface results from the electrochemical reaction between CQDs and

Cu(II) ions. They finally came to the conclusion that the sensing is based on the electron transport mechanism because the surface of CQDs is enhanced with many special functions that have the ability to specifically react with Cu(II) ions [42]. There few reports on the electrochemical sensing of pollutants with CQDs. This area is in its pioneer stage and need to be expanded to explore the potential electrochemical applications of green CQDs.

3 Challenges and Future Perspectives

The current overview of green synthesised carbon quantum dots (CQDs) shows a noticeable acceleration of their growth in recent years, as seen by their breakthroughs in optical detection of important environmental contaminants, especially organic pollutants and heavy metal ions. However, a few problems still need to be fixed before they can be used as practically viable everyday life-sensing probes and maybe scaled up.

Despite extensive efforts, there are still numerous obstacles in the way of producing multifunctional carbon quantum dots from green precursors. First off, there is still much to be understood about the photoluminescence mechanism of CQDs. Expanding the use of CQDs from renewable green precursors requires a thorough understanding of their photoluminescence mechanism.

Second, achieving large-scale, high-quality CQDs fabrication from green precursors is still a major challenge. The functionalities on the CQDs are intimately related to their solvent solubility and possible applications. Regretfully, precisely controlling the surface functionalization of CQDs is a frustrating problem that arises during the synthetic process. Green CQDs made from biomass typically have their surfaces functionalized with carboxyl, hydroxyl, and amino groups. This contributes to their excellent water solubility and rapid precipitation in other solvents, but it may also shorten the fluorescence.

For certain applications and higher sensitivity, a narrow bandwidth of the fluorescence signal and emission from the whole visible spectrum are required. In addition, better synthesis techniques are required to yield incredibly stable and productive green CQDs. On comparing with chemically produced quantum dots, the quantum yield and fluorescence intensity of green CQDs stay modest; this issue pulls these CQDs backward from several applications. It is necessary to evaluate a range of less researched sustainable precursors, including recycled waste, biomaterials, and residuals, in order to generate naturally doped CQDs with high quantum yield.

Another important challenges faced by the CQDs from green precursors is the lesser clarity about the formation mechanism. It is still obscure in nature, proper understanding about the formation mechanism leads to the better designing of the CQDs. Apart from the necessary mechanistic comprehension of the green CQDs formation, precursor based selectivity of the CQDs for particular metal ions are essential. However, for improved applicability, a simple and simultaneous surface alteration might boost optical signal. Detailed research is needed to create ratiometric and reusable sensing probes with strong fluorescence emission.

For quantitative study of metal ions or other chemicals, the fluorescence-quenching mechanism underpins the use of CQDs as fluorescence probes. At the same time, the fluorescence intensity of CQDs can also be quenched by a variety of substances besides the analytes of interest. Thus, there is a great need for CQDs with superior selectivity for the target analytes. More specifically the green CQDs exert potential selectivity for detecting different metal ions. It is necessary to use suitable, simple functionalization processes that preserve all of the major features of the nanomaterials in order to increase the selectivity and sensitivity of CQDs to heavy metal ions.

In addition, developing finger-printed response patterns for optical sensing can be aided by the use of genetically modified biomass that is sustainable and has certain metabolic pathways. The direct way to increase the sensitivity of optical sensors is through signal amplification, just like with selectivity. Passivating the CQDs using various polymer-based functionalization procedures is a popular and simple method of increasing the signal strength. Thus, it is possible to strike a requirement-based compromise between the sensitivity and selectivity processes [88].

Additionally, the combination of metal/semiconductor nanoparticles and CQDs may yield the desired optical properties; in this case, the nanoparticles can be used as transporters to load several active substances. These nanoparticles may enhance signals and function as an extremely sensitive optical sensor for the detection of analytes [41].

There is a tendency toward miniaturization and detection facilitation through the use of compact, portable devices that offer quick, precise responses and have the potential to be significant in point-of-care technology. For detection procedures like fluorescent assays and colorimetric testing to work properly, hardware and software can be integrated with wearable, mobile, and handheld devices like, lateral flow test strips and microfluidic strips, etc.

4 Conclusions

In summary the sensing activity of green CQDs were discussed in this chapter and it is found that the green precursor derived CQDs are highly effective towards the sensing of several environmental pollutants ranging from traces of heavy metal ions to various organic pollutants including toxic pesticides, antibiotics and so on. The domains of pollution sensing will pay close attention to the development of sensitive optical detection systems utilizing CQDs due to their affordability, biocompatibility, and ease of use. Biomass-based materials are becoming more and more well-liked as an abundant, economical, and environmentally friendly substitute for chemicals in the manufacturing of a variety of products with added value. The efficacy of manufacturing processes is determined by the types of biomass used as raw materials, which vary in carbon content, composition, and functional groups. Additionally, the efficiency of carbon quantum dot production can be increased by the presence of diversity in the chemical composition of biomass. Furthermore, the use of green precursors containing certain metal ions can increase the quantum efficiency of CQDs in addition to heteroatom-rich biomass.

The selective and sensitive detection of hazardous pollutant had been successfully carried out by several interesting green CQDs and almost all of them implemented in real environmental samples indicate its efficiency. The environmental monitoring is highly demandable during this time because almost all the purity of natural resources are under questioning due to this contamination with several non-biodegradable materials presents in environmental sources. Heavy metal ions, pesticides, antibiotics, pharmaceuticals are some from the list. The accumulation of such pollutants makes the sources unsafe and proper monitoring of these contaminants especially in water, soil and air are highly demandable. The evaluations of such pollutants by means of more eco-friendly probes are getting more appraisals. Green precursor based CQDs can easily handle the role of environmental monitoring against major contaminants. The encouraging intrinsic qualities of CQDs can be used to support this functionality. While the precise mechanism of how pollutants interact with CQDs for sensing remains unclear, one potential explanation is that certain interactions between pollutant species and the surface functionalities of CQDs alter the characteristics of the CQDs, which may lead to changes in the intensity of fluorescence and other properties that can be directly used for detection.

Major research on this environmental sensing of the aforementioned pollutants by CQDs carried out by fluorescent sensing process. The fluorescence quenching and enhancement is the main area of interest and the presence of surface functionalities plays vital role in this type of sensing capability. Being natural-based, the surface of green precursor developed CQDs are enriched with variety of functional groups and the interaction of analyte species with the functionalities can induce fluorescence alteration. Apart from this, the matching optical characteristics of the analyte with the optical features of CQDs also made it used as a sensor for the same listed pollutants. Electrochemical and colorimetric sensing studies are on the developing pattern advanced research in these areas is required for the more exploration of green CQDs. However, combination of all of the mentioned sensing methods of pollutants can effectively enhance the value of the excellent nanoprobe for the environmental monitoring functions. All things considered, green precursor produced CQDs have been viewed as promising nanomaterials for the past few decades and are essential for a sustainable future due to their environmental significance.

References

1. Abadin, H., Ashizawa, A., Stevens, Y. -W., Llados, F., Diamond, G., Sage, G., Citra, M., Quinones, A., Bosch, S. J., & Swarts, S. G. (2007). Toxicological Profile for Lead, Agency for Toxic Substances and Disease Registry (ATSDR) Toxicological Profiles. Agency for Toxic Substances and Disease Registry (US), Atlanta (GA).

2. Ansi, V. A., & Renuka, N. K. (2018). Table sugar derived Carbon dot—A naked eye sensor for toxic Pb2+ ions. *Sensors and Actuators, B: Chemical, 264*, 67–75. https://doi.org/10.1016/j.snb.2018.02.167
3. Bandi, R., Dadigala, R., Gangapuram, B. R., & Guttena, V. (2018). Green synthesis of highly fluorescent nitrogen—Doped carbon dots from Lantana camara berries for effective detection of lead(II) and bioimaging. *Journal of Photochemistry and Photobiology B: Biology, 178*, 330–338. https://doi.org/10.1016/j.jphotobiol.2017.11.010
4. Bano, D., Kumar, V., Singh, V. K., & Hasan, S. H. (2018). Green synthesis of fluorescent carbon quantum dots for the detection of mercury(II) and glutathione. *New Journal of Chemistry, 42*, 5814–5821. https://doi.org/10.1039/C8NJ00432C
5. Batool, M., Junaid, H. M., Tabassum, S., Kanwal, F., Abid, K., Fatima, Z., Shah, A. T. (2020). Metal ion detection by carbon dots—A review. *Critical Reviews in Analytical Chemistry*, 1–12. https://doi.org/10.1080/10408347.2020.1824117
6. Bera, M. K., & Mohapatra, S. (2020). Ultrasensitive detection of glyphosate through effective photoelectron transfer between CdTe and chitosan derived carbon dot. *Colloids and Surfaces Physicochemical and Engineering Aspects, 596*, 124710. https://doi.org/10.1016/j.colsurfa.2020.124710
7. Bhatt, S., Bhatt, M., Kumar, A., Vyas, G., Gajaria, T., & Paul, P. (2018). Green route for synthesis of multifunctional fluorescent carbon dots from Tulsi leaves and its application as Cr (VI) sensors, bio-imaging and patterning agents. *Colloids and Surfaces. B, Biointerfaces, 167*, 126. https://doi.org/10.1016/j.colsurfb.2018.04.008
8. Chahal, S., Macairan, J.-R., Yousefi, N., Tufenkji, N., & Naccache, R. (2021). Green synthesis of carbon dots and their applications. *RSC Advances, 11*, 25354–25363. https://doi.org/10.1039/D1RA04718C
9. Chandra, S., Bano, D., Sahoo, K., Kumar, D., Kumar, V., Kumar Yadav, P., & Hadi Hasan, S. (2022). Synthesis of fluorescent carbon quantum dots from Jatropha fruits and their application in fluorometric sensor for the detection of chlorpyrifos. *Microchemical Journal, 172*, 106953. https://doi.org/10.1016/j.microc.2021.106953
10. Chatzimarkou, A., Chatzimitakos, T. G., Kasouni, A., Sygellou, L., Avgeropoulos, A., & Stalikas, C. D. (2018). Selective FRET-based sensing of 4-nitrophenol and cell imaging capitalizing on the fluorescent properties of carbon nanodots from apple seeds. *Sensors and Actuators, B: Chemical, 258*, 1152–1160. https://doi.org/10.1016/j.snb.2017.11.182
11. Chauhan, P., Dogra, S., Chaudhary, S., & Kumar, R. (2020). Usage of coconut coir for sustainable production of high-valued carbon dots with discriminatory sensing aptitude toward metal ions. *Materials Today Chemistry, 16*, 100247. https://doi.org/10.1016/j.mtchem.2020.100247
12. Chen, Y.-Y., Chang, H.-T., Shiang, Y.-C., Hung, Y.-L., Chiang, C.-K., & Huang, C.-C. (2009). Colorimetric assay for lead ions based on the leaching of gold nanoparticles. *Analytical Chemistry, 81*, 9433–9439. https://doi.org/10.1021/ac9018268
13. Cui, L., Ren, X., Sun, M., Liu, H., & Xia, L. (2021). Carbon dots: Synthesis, properties and applications. *Nanomaterials, 11*, 3419. https://doi.org/10.3390/nano11123419
14. D'Angelis do E. S., Barbosa, C., Corrêa, J. R., Medeiros, G. A., Barreto, G., Magalhães, K. G., de Oliveira, A. L., Spencer, J., Rodrigues, M. O., Neto, B. A. D. (2015). Carbon dots (C-dots) from cow manure with impressive subcellular selectivity tuned by simple chemical modification. *Chemistry—European Journal, 21*, 5055–5060. https://doi.org/10.1002/chem.201406330
15. Das, G. S., Shim, J. P., Bhatnagar, A., Tripathi, K. M., & Kim, T. (2019). Biomass-derived carbon quantum dots for visible-light-induced photocatalysis and label-free detection of Fe(III) and ascorbic acid. *Science and Reports, 9*, 15084. https://doi.org/10.1038/s41598-019-49266-y
16. De, B., & Karak, N. (2013). A green and facile approach for the synthesis of water soluble fluorescent carbon dots from banana juice. *RSC Advances, 3*, 8286. https://doi.org/10.1039/c3ra00088e
17. Desai, M. L., Jha, S., Basu, H., Singhal, R. K., Park, T.-J., & Kailasa, S. K. (2019). Acid oxidation of muskmelon fruit for the fabrication of carbon dots with specific emission colors

for recognition of Hg2+ ions and cell imaging. *ACS Omega, 4*, 19332–19340. https://doi.org/10.1021/acsomega.9b02730
18. Essner, J. B., Laber, C. H., Ravula, S., Polo-Parada, L., & Baker, G. A. (2015). Pee-dots: Biocompatible fluorescent carbon dots derived from the upcycling of urine. *Green Chemistry, 18*, 243–250. https://doi.org/10.1039/C5GC02032H
19. Feng, S., Gao, Z., Liu, H., Huang, J., Li, X., & Yang, Y. (2019). Feasibility of detection valence speciation of Cr(III) and Cr(VI) in environmental samples by spectrofluorimetric method with fluorescent carbon quantum dots. *Spectrochimica Acta. A. Molecular and Biomolecular Spectroscopy, 212*, 286–292. https://doi.org/10.1016/j.saa.2018.12.055
20. Feng, Y., Zhong, D., Miao, H., & Yang, X. (2015). Carbon dots derived from rose flowers for tetracycline sensing. *Talanta, 140*, 128–133. https://doi.org/10.1016/j.talanta.2015.03.038
21. Gedda, G., Lee, C.-Y., Lin, Y.-C., & Wu, H. (2016). Green synthesis of carbon dots from prawn shells for highly selective and sensitive detection of copper ions. *Sensors and Actuators, B: Chemical, 224*, 396–403. https://doi.org/10.1016/j.snb.2015.09.065
22. Gu, D., Shang, S., Yu, Q., & Shen, J. (2016). Green synthesis of nitrogen-doped carbon dots from lotus root for Hg(II) ions detection and cell imaging. *Applied Surface Science, 390*, 38–42. https://doi.org/10.1016/j.apsusc.2016.08.012
23. Guo, F., Zhu, Z., Zheng, Z., Jin, Y., Di, X., Xu, Z., & Guan, H. (2020). Facile synthesis of highly efficient fluorescent carbon dots for tetracycline detection. *Environmental Science and Pollution Research, 27*, 4520–4527. https://doi.org/10.1007/s11356-019-06779-3
24. Gupta, A., Verma, N. C., Khan, S., Tiwari, S., Chaudhary, A., & Nandi, C. K. (2016). Paper strip based and live cell ultrasensitive lead sensor using carbon dots synthesized from biological media. *Sensors and Actuators, B: Chemical, 232*, 107–114. https://doi.org/10.1016/j.snb.2016.03.110
25. Hassanvand, Z., Jalali, F., Nazari, M., Parnianchi, F., & Santoro, C. (2021). Carbon nanodots in electrochemical sensors and biosensors: A review. *ChemElectroChem, 8*, 15–35. https://doi.org/10.1002/celc.202001229
26. Hatimuria, M., Phukan, P., Bag, S., Ghosh, J., Gavvala, K., Pabbathi, A., & Das, J. (2023). Green carbon dots: Applications in development of electrochemical sensors, assessment of toxicity as well as anticancer properties. *Catalysts, 13*, 537. https://doi.org/10.3390/catal13030537
27. Hou, C., Xiong, Y., Fu, N., Jacquot, C. C., Squier, T. C., & Cao, H. (2011). Turn-on ratiometric fluorescent sensor for Pb2+ detection. *Tetrahedron Letters, 52*, 2692–2696. https://doi.org/10.1016/j.tetlet.2011.03.075
28. Hou, J., Dong, G., Tian, Z., Lu, J., Wang, Q., Ai, S., & Wang, M. (2016). A sensitive fluorescent sensor for selective determination of dichlorvos based on the recovered fluorescence of carbon dots-Cu(II) system. *Food Chemistry, 202*, 81–87. https://doi.org/10.1016/j.foodchem.2015.11.134
29. Hou, J., Wang, X., Lan, S., Zhang, C., Hou, C., He, Q., & Huo, D. (2020). A turn-on fluorescent sensor based on carbon dots from Sophora japonica leaves for the detection of glyphosate. *Analytical Methods, 12*, 4130–4138. https://doi.org/10.1039/D0AY01241F
30. Hu, G., Ge, L., Li, Y., Mukhtar, M., Shen, B., Yang, D., & Li, J. (2020). Carbon dots derived from flax straw for highly sensitive and selective detections of cobalt, chromium, and ascorbic acid. *Journal of Colloid and Interface Science, 579*, 96–108. https://doi.org/10.1016/j.jcis.2020.06.034
31. Huang, H., Lv, J.-J., Zhou, D.-L., Bao, N., Xu, Y., Wang, A.-J., & Feng, J.-J. (2013). One-pot green synthesis of nitrogen-doped carbon nanoparticles as fluorescent probes for mercury ions. *RSC Advances, 3*, 21691–21696. https://doi.org/10.1039/C3RA43452D
32. Huang, Z.-Y., Wu, W.-Z., Li, Z.-X., Wu, Y., Wu, C.-B., Gao, J., Guo, J., Chen, Y., Hu, Y., & Huang, C. (2022). Solvothermal production of tea residue derived carbon dots by the pretreatment of choline chloride/urea and its application for cadmium detection. *Industrial Crops and Products, 184*, 115085. https://doi.org/10.1016/j.indcrop.2022.115085
33. Jian-feng, G., Chang-jun, H., Mei, Y., Dan-qun, H., Jun-jie, L., Huan-bao, F., Hui-bo, L., & Ping, Y. (2016). Colorimetric sensing of chromium(VI) ions in aqueous solution based on

the leaching of protein-stabled gold nanoparticles. *Analytical Methods, 8*, 5526–5532. https://doi.org/10.1039/C6AY01200K

34. Jiang, Z., Guan, L., Xu, X., Wang, E., & Wang, C. (2022). Applications of carbon dots in electrochemical energy storage. *ACS Applied Electronic Materials, 4*, 5144–5164. https://doi.org/10.1021/acsaelm.2c01152
35. Jing, H. H., Bardakci, F., Akgöl, S., Kusat, K., Adnan, M., Alam, M. J., Gupta, R., Sahreen, S., Chen, Y., Gopinath, S. C. B., & Sasidharan, S. (2023). Green carbon dots: Synthesis, characterization, properties and biomedical applications. *Journal of Functional Biomaterials, 14*, 27. https://doi.org/10.3390/jfb14010027
36. Johri, N., Jacquillet, G., & Unwin, R. (2010). Heavy metal poisoning: The effects of cadmium on the kidney. *BioMetals, 23*, 783–792. https://doi.org/10.1007/s10534-010-9328-y
37. Kang, C., Huang, Y., Yang, H., Yan, X. F., & Chen, Z. P. (2020). A review of carbon dots produced from biomass wastes. *Nanomaterials, 10*, 2316. https://doi.org/10.3390/nano10112316
38. Kanwal, A., Bibi, N., Hyder, S., Muhammad, A., Ren, H., Liu, J., & Lei, Z. (2022). Recent advances in green carbon dots (2015–2022): Synthesis, metal ion sensing, and biological applications. *Beilstein Journal of Nanotechnology, 13*, 1068–1107. https://doi.org/10.3762/bjnano.13.93
39. Karimian, N., Fakhri, H., Amidi, S., Hajian, A., Arduini, F., & Bagheri, H. (2019). A novel sensing layer based on metal–organic framework UiO-66 modified with TiO_2–graphene oxide: Application to rapid, sensitive and simultaneous determination of paraoxon and chlorpyrifos. *New Journal of Chemistry, 43*, 2600–2609. https://doi.org/10.1039/C8NJ06208K
40. Kaur, I., Batra, V., Kumar Reddy Bogireddy, N., Torres Landa, S. D., & Agarwal, V. (2023). Detection of organic pollutants, food additives and antibiotics using sustainable carbon dots. *Food Chemistry, 406*, 135029. https://doi.org/10.1016/j.foodchem.2022.135029
41. Khanal, B. P., Pandey, A., Li, L., Lin, Q., Bae, W. K., Luo, H., Klimov, V. I., & Pietryga, J. M. (2012). Generalized synthesis of hybrid metal-semiconductor nanostructures tunable from the visible to the infrared. *ACS Nano, 6*, 3832–3840. https://doi.org/10.1021/nn204932m
42. Korah, B. K., Murali, A., Chacko, A. R., Thara, C. R., Mathew, J., George, B., & Mathew, B. (2022a). Bio-inspired novel carbon dots as fluorescence and electrochemical-based sensors and fluorescent ink. Biomass Conversion and Biorefinery https://doi.org/10.1007/s13399-022-03294-3
43. Korah, B. K., Punnoose, M. S., Thara, C. R., Abraham, T., Ambady, K. G., & Mathew, B. (2022). Curcuma amada derived nitrogen-doped carbon dots as a dual sensor for tetracycline and mercury ions. *Diamond and Related Materials, 125*, 108980. https://doi.org/10.1016/j.diamond.2022.108980
44. Korshoj, L. E., Zaitouna, A. J., & Lai, R. Y. (2015). Methylene Blue-mediated electrocatalytic detection of hexavalent chromium. *Analytical Chemistry, 87*, 2560–2564. https://doi.org/10.1021/acs.analchem.5b00197
45. Kumar, A., Chowdhuri, A. R., Laha, D., Mahto, T. K., Karmakar, P., & Sahu, S. K. (2017). Green synthesis of carbon dots from Ocimum sanctum for effective fluorescent sensing of Pb2+ ions and live cell imaging. *Sensors and Actuators, B: Chemical, 242*, 679–686. https://doi.org/10.1016/j.snb.2016.11.109
46. Kurian, M., & Paul, A. (2021). Recent trends in the use of green sources for carbon dot synthesis—A short review. *Carbon Trends, 3*, 100032. https://doi.org/10.1016/j.cartre.2021.100032
47. Li, M., Chen, T., Gooding, J. J., & Liu, J. (2019). Review of carbon and graphene quantum dots for sensing. *ACS Sensors, 4*, 1732–1748. https://doi.org/10.1021/acssensors.9b00514
48. Li, W.-K., Feng, J.-T., & Ma, Z.-Q. (2020). Nitrogen, sulfur, boron and flavonoid moiety co-incorporated carbon dots for sensitive fluorescence detection of pesticides. *Carbon, 161*, 685–693. https://doi.org/10.1016/j.carbon.2020.01.098
49. Li, Z., Ni, Y., & Kokot, S. (2015). A new fluorescent nitrogen-doped carbon dot system modified by the fluorophore-labeled ssDNA for the analysis of 6-mercaptopurine and Hg (II). *Biosensors & Bioelectronics, 74*, 91–97. https://doi.org/10.1016/j.bios.2015.06.014

50. Liao, J., Cheng, Z., & Zhou, L. (2016). Nitrogen-doping enhanced fluorescent carbon dots: Green synthesis and their applications for bioimaging and label-free detection of Au3+ ions. *ACS Sustainable Chemistry & Engineering, 4*, 3053–3061. https://doi.org/10.1021/acssuschemeng.6b00018
51. Lin, Y.-S., Chiu, T.-C., & Hu, C.-C. (2019). Fluorescence-tunable copper nanoclusters and their application in hexavalent chromium sensing. *RSC Advances, 9*, 9228–9234. https://doi.org/10.1039/C9RA00916G
52. Liu, H., Ding, J., Chen, L., & Ding, L. (2020). A novel fluorescence assay based on self-doping biomass carbon dots for rapid detection of dimethoate. *Journal of Photochemistry and Photobiology Chemistry, 400*, 112724. https://doi.org/10.1016/j.jphotochem.2020.112724
53. Liu, H., Ding, J., Zhang, K., & Ding, L. (2019). Construction of biomass carbon dots based fluorescence sensors and their applications in chemical and biological analysis. *TrAC, Trends in Analytical Chemistry, 118*, 315–337. https://doi.org/10.1016/j.trac.2019.05.051
54. Liu, R., Zhang, J., Gao, M., Li, Z., Chen, J., Wu, D., & Liu, P. (2014). A facile microwave-hydrothermal approach towards highly photoluminescent carbon dots from goose feathers. *RSC Advances, 5*, 4428–4433. https://doi.org/10.1039/C4RA12077A
55. Liu, Y., Zhao, Y., & Zhang, Y. (2014). One-step green synthesized fluorescent carbon nanodots from bamboo leaves for copper(II) ion detection. *Sensors and Actuators, B: Chemical, 196*, 647–652. https://doi.org/10.1016/j.snb.2014.02.053
56. Liu, Z., Jin, W., Wang, F., Li, T., Nie, J., Xiao, W., Zhang, Q., & Zhang, Y. (2019). Ratiometric fluorescent sensing of Pb2+ and Hg2+ with two types of carbon dot nanohybrids synthesized from the same biomass. *Sensors and Actuators, B: Chemical, 296*, 126698. https://doi.org/10.1016/j.snb.2019.126698
57. Long, C., Jiang, Z., Shangguan, J., Qing, T., Zhang, P., & Feng, B. (2021). Applications of carbon dots in environmental pollution control: A review. *Chemical Engineering Journal, 406*, 126848. https://doi.org/10.1016/j.cej.2020.126848
58. Lu, W., Qin, X., Liu, S., Chang, G., Zhang, Y., Luo, Y., Asiri, A. M., Al-Youbi, A. O., & Sun, X. (2012). Economical, Green synthesis of fluorescent carbon nanoparticles and their use as probes for sensitive and selective detection of mercury(II) ions. *Analytical Chemistry, 84*, 5351–5357. https://doi.org/10.1021/ac3007939
59. Manikandan, V., & Lee, N. Y. (2022). Green synthesis of carbon quantum dots and their environmental applications. *Environmental Research, 212*, 113283. https://doi.org/10.1016/j.envres.2022.113283
60. Mansi, M., Bhikhu, M., & Gaurav, S. (2023). Chapter 24—Synthesis and applications of carbon dots from waste biomass. In S. K. Kailasa, & C. M. Hussain (Eds.), *Carbon dots in analytical chemistry* (pp. 319–328). Elsevier. https://doi.org/10.1016/B978-0-323-98350-1.00008-6
61. Meng, W., Bai, X., Wang, B., Liu, Z., Lu, S., & Yang, B. (2019). Biomass-derived carbon dots and their applications. *Energy & Environmental Materials, 2*, 172–192. https://doi.org/10.1002/eem2.12038
62. Miao, H., Wang, Y., & Yang, X. (2018). Carbon dots derived from tobacco for visually distinguishing and detecting three kinds of tetracyclines. *Nanoscale, 10*, 8139–8145. https://doi.org/10.1039/C8NR02405G
63. Molaei, M. J. (2020). Principles, mechanisms, and application of carbon quantum dots in sensors: A review. *Analytical Methods, 12*, 1266–1287. https://doi.org/10.1039/C9AY02696G
64. Molahalli, V., Sharma, A., Bijapur, K., Soman, G., Chattham, N., & Hegde, G. (2024). Low-cost bio-waste carbon nanocomposites for sustainable electrochemical devices: A systematic review. *Materials Today Communications, 38*, 108034. https://doi.org/10.1016/j.mtcomm.2024.108034
65. Pandey, S. C., Kumar, A., & Sahu, S. K. (2020). Single step green synthesis of carbon dots from Murraya koenigii leaves; A unique turn-off fluorescent contrivance for selective sensing of Cd (II) ion. *Journal of Photochemistry and Photobiology Chemistry, 400*, 112620. https://doi.org/10.1016/j.jphotochem.2020.112620

66. Pereira Lopes Moreira, R., ul Islam, S. (2023). Carbon-dots based sensors for detection of pollutants from soil. In *Green carbon materials for environmental analysis: Emerging research and future opportunities, ACS symposium series* (pp. 139–162). American Chemical Society. https://doi.org/10.1021/bk-2023-1441.ch006
67. Phan, L. M. T., Baek, S. H., Nguyen, T. P., Park, K. Y., Ha, S., Rafique, R., Kailasa, S. K., & Park, T. J. (2018). Synthesis of fluorescent silicon quantum dots for ultra-rapid and selective sensing of Cr(VI) ion and biomonitoring of cancer cells. *Materials Science and Engineering C, 93*, 429–436. https://doi.org/10.1016/j.msec.2018.08.024
68. Purbia, R., & Paria, S. (2016). A simple turn on fluorescent sensor for the selective detection of thiamine using coconut water derived luminescent carbon dots. *Biosensors & Bioelectronics, 79*, 467–475. https://doi.org/10.1016/j.bios.2015.12.087
69. Qi, H., Teng, M., Liu, M., Liu, S., Li, J., Yu, H., Teng, C., Huang, Z., Liu, H., Shao, Q., Umar, A., Ding, T., Gao, Q., & Guo, Z. (2019). Biomass-derived nitrogen-doped carbon quantum dots: Highly selective fluorescent probe for detecting Fe3+ ions and tetracyclines. *Journal of Colloid and Interface Science, 539*, 332–341. https://doi.org/10.1016/j.jcis.2018.12.047
70. Qin, X., Lu, W., Asiri, A. M., Al-Youbi, A. O., & Sun, X. (2013). Green, low-cost synthesis of photoluminescent carbon dots by hydrothermal treatment of willow bark and their application as an effective photocatalyst for fabricating Au nanoparticles–reduced graphene oxide nanocomposites for glucose detection. *Catalysis Science & Technology, 3*, 1027–1035. https://doi.org/10.1039/C2CY20635H
71. Qin, X., Lu, W., Asiri, A. M., Al-Youbi, A. O., & Sun, X. (2013). Microwave-assisted rapid green synthesis of photoluminescent carbon nanodots from flour and their applications for sensitive and selective detection of mercury(II) ions. *Sensors and Actuators, B: Chemical, 184*, 156–162. https://doi.org/10.1016/j.snb.2013.04.079
72. Qu, Y., Yu, L., Zhu, B., Chai, F., & Su, Z. (2020). Green synthesis of carbon dots by celery leaves for use as fluorescent paper sensors for the detection of nitrophenols. *New Journal of Chemistry, 44*, 1500–1507. https://doi.org/10.1039/C9NJ05285B
73. Raji, K., Ramanan, V., & Ramamurthy, P. (2019). Facile and green synthesis of highly fluorescent nitrogen-doped carbon dots from jackfruit seeds and its applications towards the fluorimetric detection of Au3+ ions in aqueous medium and in in vitro multicolor cell imaging. *New Journal of Chemistry, 43*, 11710–11719. https://doi.org/10.1039/C9NJ02590A
74. Roshni, V., & Ottoor, D. (2015). Synthesis of carbon nanoparticles using one step green approach and their application as mercuric ion sensor. *Journal of Luminescence, 161*, 117–122. https://doi.org/10.1016/j.jlumin.2014.12.048
75. Sachdev, A., & Gopinath, P. (2015). Green synthesis of multifunctional carbon dots from coriander leaves and their potential application as antioxidants, sensors and bioimaging agents. *The Analyst, 140*, 4260–4269. https://doi.org/10.1039/C5AN00454C
76. Sakaew, C., Sricharoen, P., Limchoowong, N., Nuengmatcha, P., Kukusamude, C., Kongsri, S., & Chanthai, S. (2020). Green and facile synthesis of water-soluble carbon dots from ethanolic shallot extract for chromium ion sensing in milk, fruit juices, and wastewater samples. *RSC Advances, 10*, 20638–20645. https://doi.org/10.1039/D0RA03101A
77. Kolaprath, M. K. A., Benny, L., & Varghese, A. (2023). A facile, green synthesis of carbon quantum dots from Polyalthia longifolia and its application for the selective detection of cadmium. *Dyes Pigments, 210*, 111048. https://doi.org/10.1016/j.dyepig.2022.111048
78. Siavash Moakhar, R., Goh, G. K. L., Dolati, A., & Ghorbani, M. (2017). Sunlight-driven photoelectrochemical sensor for direct determination of hexavalent chromium based on Au decorated rutile TiO2 nanorods. *Applied Catalysis B Environment, 201*, 411–418. https://doi.org/10.1016/j.apcatb.2016.08.026
79. Singh, P., Kumar, S., Kumar, P., Kataria, N., Bhankar, V., Kumar, K., Kumar, R., Hsieh, C. -T., Khoo, & K. S. (2023). Assessment of biomass-derived carbon dots as highly sensitive and selective templates for the sensing of hazardous ions. *Nanoscale, 15*, 16241–16267. https://doi.org/10.1039/D3NR01966G
80. Song, P., Zhang, L., Long, H., Meng, M., Liu, T., Yin, Y., & Xi, R. (2017). A multianalyte fluorescent carbon dots sensing system constructed based on specific recognition of Fe(III) ions. *RSC Advances, 7*, 28637–28646. https://doi.org/10.1039/C7RA04122E

81. Song, Y., Yan, X., Li, Z., Qu, L., Zhu, C., Ye, R., Li, S., Du, D., & Lin, Y. (2018). Highly photoluminescent carbon dots derived from linseed and their applications in cellular imaging and sensing. *Journal of Materials Chemistry B, 6*, 3181–3187. https://doi.org/10.1039/C8TB00116B
82. Soni, H., & Pamidimukkala, P. S. (2018). Green synthesis of N, S co-doped carbon quantum dots from triflic acid treated palm shell waste and their application in nitrophenol sensing. *Materials Research Bulletin, 108*, 250–254. https://doi.org/10.1016/j.materresbull.2018.08.033
83. Spanu, D., Monticelli, D., Binda, G., Dossi, C., Rampazzi, L., & Recchia, S. (2021). One-minute highly selective Cr(VI) determination at ultra-trace levels: An ICP-MS method based on the on-line trapping of Cr(III). *Journal of Hazardous Materials, 412*, 125280. https://doi.org/10.1016/j.jhazmat.2021.125280
84. Srinivasan, K., Subramanian, K., Murugan, K., & Dinakaran, K. (2016). Sensitive fluorescence detection of mercury(II) in aqueous solution by the fluorescence quenching effect of MoS_2 with DNA functionalized carbon dots. *The Analyst, 141*, 6344–6352. https://doi.org/10.1039/C6AN00879H
85. Tai, D., Liu, C., & Liu, J. (2019). Facile synthesis of fluorescent carbon dots from shrimp shells and using the carbon dots to detect chromium(VI). *Spectroscopy Letters, 52*, 194–199. https://doi.org/10.1080/00387010.2019.1607879
86. Tchounwou, P. B., Yedjou, C. G., Patlolla, A. K., Sutton, D. J. (2012). Heavy metal toxicity and the environment. In A. Luch (Ed.), *Molecular, clinical and environmental toxicology: Volume 3: Environmental toxicology, Experientia Supplementum* (pp. 133–164). Springer, Basel. https://doi.org/10.1007/978-3-7643-8340-4_6
87. Teng, X., Ma, C., Ge, C., Yan, M., Yang, J., Zhang, Y., Morais, P. C., & Bi, H. (2014). Green synthesis of nitrogen-doped carbon dots from konjac flour with "off–on" fluorescence by Fe3+ and L-lysine for bioimaging. *Journal of Materials Chemistry B, 2*, 4631–4639. https://doi.org/10.1039/C4TB00368C
88. Torres Landa, S. D., Reddy Bogireddy, N. K., Kaur, I., Batra, V., & Agarwal, V. (2022a). Heavy metal ion detection using green precursor derived carbon dots. *iScience, 25*, 103816. https://doi.org/10.1016/j.isci.2022.103816
89. Tyagi, A., Tripathi, K. M., Singh, N., Choudhary, S., & Gupta, R. K. (2016). Green synthesis of carbon quantum dots from lemon peel waste: Applications in sensing and photocatalysis. *RSC Advances, 6*, 72423–72432. https://doi.org/10.1039/C6RA10488F
90. Roshni, V., Misra, S., Santra, M. K., & Ottoor, D. (2019). One pot green synthesis of C-dots from groundnuts and its application as Cr(VI) sensor and in vitro bioimaging agent. *Journal of Photochemistry and Photobiology Chemistry, 373*, 28–36. https://doi.org/10.1016/j.jphotochem.2018.12.028
91. Vandarkuzhali, S. A. A., Natarajan, S., Jeyabalan, S., Sivaraman, G., Singaravadivel, S., Muthusubramanian, S., & Viswanathan, B. (2018). Pineapple peel-derived carbon dots: Applications as sensor, molecular keypad lock, and memory device. *ACS Omega, 3*, 12584–12592. https://doi.org/10.1021/acsomega.8b01146
92. Venugopalan, P., & Vidya, N. (2023). Microwave assisted green synthesis of carbon dots from sweet flag (Acorus calamus) for fluorescent sensing of 4-nitrophenol. *Journal of Photochemistry and Photobiology Chemistry, 439*, 114625. https://doi.org/10.1016/j.jphotochem.2023.114625
93. Venugopalan, P., & Vidya, N. (2023). Microwave-assisted green synthesis of carbon dots derived from wild lemon (Citrus pennivesiculata) leaves as a fluorescent probe for tetracycline sensing in water. *Spectrochimica Acta. A. Molecular and Biomolecular Spectroscopy, 286*, 122024. https://doi.org/10.1016/j.saa.2022.122024
94. Venugopalan, P., & Vidya, N. (2022). Green synthesis of mango ginger (Curcuma amada) derived fluorescent carbon dots—A potent label-free probe for hexavalent chromium sensing in water. *Spectroscopy Letters, 55*, 373–388. https://doi.org/10.1080/00387010.2022.2082483
95. Venugopalan, P., & Vidya, N. (2022b). Bilimbi (Averrhoa bilimbi) fruit derived carbon dots for dual sensing of Cu(II) and quinalphos. *International Journal of Environmental Analytical Chemistry, 0*, 1–14. https://doi.org/10.1080/03067319.2022.2149331

96. Wang, C., Sun, D., Zhuo, K., Zhang, H., & Wang, J. (2014). Simple and green synthesis of nitrogen-, sulfur-, and phosphorus-co-doped carbon dots with tunable luminescence properties and sensing application. *RSC Advances, 4*, 54060–54065. https://doi.org/10.1039/C4RA10885J
97. Wang, C., Xu, J., Zhang, R., & Zhao, W. (2022). Facile and low-energy-consumption synthesis of dual-functional carbon dots from Cornus walteri leaves for detection of p-nitrophenol and photocatalytic degradation of dyes. *Colloids and Surfaces Physicochemical Engineering Aspects, 640*, 128351. https://doi.org/10.1016/j.colsurfa.2022.128351
98. Wang, H., Zhang, L., Guo, X., Dong, W., Wang, R., Shuang, S., Gong, X., & Dong, C. (2019). Comparative study of Cl, N-Cdots and N-Cdots and application for trinitrophenol and ClO− sensor and cell-imaging. *Analytica Chimica Acta, 1091*, 76–87. https://doi.org/10.1016/j.aca.2019.09.019
99. Wang, K., Ji, Q., Xu, J., Li, H., Zhang, D., Liu, X., Wu, Y., & Fan, H. (2018). Highly sensitive and selective detection of amoxicillin using carbon quantum dots derived from beet. *Journal of Fluorescence, 28*, 759–765. https://doi.org/10.1007/s10895-018-2237-0
100. Wang, L., Qian, B., Chen, H., Liu, Y., & Liang, A. (2008). A novel terbium composite nanoparticles: Preparation and selective fluorescence determination of chromium(VI). *Journal of Luminescence, 128*, 1952–1956. https://doi.org/10.1016/j.jlumin.2008.06.005
101. Wang, N., Wang, Y., Guo, T., Yang, T., Chen, M., & Wang, J. (2016). Green preparation of carbon dots with papaya as carbon source for effective fluorescent sensing of Iron (III) and Escherichia coli. *Biosensors & Bioelectronics, 85*, 68–75. https://doi.org/10.1016/j.bios.2016.04.089
102. Wang, Q., Liu, X., Zhang, L., & Lv, Y. (2012). Microwave-assisted synthesis of carbon nanodots through an eggshell membrane and their fluorescent application. *The Analyst, 137*, 5392–5397. https://doi.org/10.1039/C2AN36059D
103. Wang, Y., Hu, X., Li, W., Huang, X., Li, Z., Zhang, W., Zhang, X., Zou, X., & Shi, J. (2020). Preparation of boron nitrogen co-doped carbon quantum dots for rapid detection of Cr(VI). *Spectrochimica Acta. A. Molecular and Biomolecular Spectroscopy, 243*, 118807. https://doi.org/10.1016/j.saa.2020.118807
104. Wareing, T. C., Gentile, P., & Phan, A. N. (2021). Biomass-based carbon dots: Current development and future perspectives. *ACS Nano, 15*, 15471–15501. https://doi.org/10.1021/acsnano.1c03886
105. Wei, J., Zhang, X., Sheng, Y., Shen, J., Huang, P., Guo, S., Pan, J., & Feng, B. (2014). Dual functional carbon dots derived from cornflour via a simple one-pot hydrothermal route. *Materials Letters, 123*, 107–111. https://doi.org/10.1016/j.matlet.2014.02.090
106. Xu, J., Jie, X., Xie, F., Yang, H., Wei, W., & Xia, Z. (2018). Flavonoid moiety-incorporated carbon dots for ultrasensitive and highly selective fluorescence detection and removal of Pb2+. *Nano Research, 11*, 3648–3657. https://doi.org/10.1007/s12274-017-1931-6
107. Xu, S., Liu, Y., Yang, H., Zhao, K., Li, J., & Deng, A. (2017). Fluorescent nitrogen and sulfur co-doped carbon dots from casein and their applications for sensitive detection of Hg2+ and biothiols and cellular imaging. *Analytica Chimica Acta, 964*, 150–160. https://doi.org/10.1016/j.aca.2017.01.037
108. Yang, X., Zhuo, Y., Zhu, S., Luo, Y., Feng, Y., & Dou, Y. (2014). Novel and green synthesis of high-fluorescent carbon dots originated from honey for sensing and imaging. *Biosensors & Bioelectronics, 60*, 292–298. https://doi.org/10.1016/j.bios.2014.04.046
109. Yu, J., Song, N., Zhang, Y.-K., Zhong, S.-X., Wang, A.-J., & Chen, J. (2015). Green preparation of carbon dots by Jinhua bergamot for sensitive and selective fluorescent detection of Hg2+ and Fe3+. *Sensors and Actuators, B: Chemical, 214*, 29–35. https://doi.org/10.1016/j.snb.2015.03.006
110. Zhang, H., Huang, Y., Hu, Z., Tong, C., Zhang, Z., & Hu, S. (2017). Carbon dots codoped with nitrogen and sulfur are viable fluorescent probes for chromium(VI). *Microchimica Acta, 184*, 1547–1553. https://doi.org/10.1007/s00604-017-2132-4

111. Zhang, H., Zhou, Q., Han, X., Li, M., Yuan, J., Wei, R., Zhang, X., Wu, M., & Zhao, W. (2021). Nitrogen-doped carbon dots derived from hawthorn for the rapid determination of chlortetracycline in pork samples. *Spectrochimica Acta. A. Molecular and Biomolecular Spectroscopy, 255*, 119736. https://doi.org/10.1016/j.saa.2021.119736
112. Zhang, Q., Liang, J., Zhao, L., Wang, Y., Zheng, Y., Wu, Y., & Jiang, L. (2020). Synthesis of novel fluorescent carbon quantum dots from Rosa roxburghii for rapid and highly selective detection of o-nitrophenol and cellular imaging. *Frontiers in Chemistry, 8*.
113. Zhang, W., Zhong, H., Zhao, P., Shen, A., Li, H., & Liu, X. (2022). Carbon quantum dot fluorescent probes for food safety detection: Progress, opportunities and challenges. *Food Control, 133*, 108591. https://doi.org/10.1016/j.foodcont.2021.108591
114. Zhao, J., Huang, M., Zhang, L., Zou, M., Chen, D., Huang, Y., & Zhao, S. (2017). Unique approach to develop carbon dot-based nanohybrid near-infrared ratiometric fluorescent sensor for the detection of mercury ions. *Analytical Chemistry, 89*, 8044–8049. https://doi.org/10.1021/acs.analchem.7b01443
115. Zheng, M., Wang, C., Wang, Y., Wei, W., Ma, S., Sun, X., & He, J. (2018). Green synthesis of carbon dots functionalized silver nanoparticles for the colorimetric detection of phoxim. *Talanta, 185*, 309–315. https://doi.org/10.1016/j.talanta.2018.03.066
116. Zhong, G.-Z., Hu, Z.-B., Tu, X.-J., Chai, X.-S., & Chen, G. (2020). Determination of hexavalent chromium in solid waste hypochlorite treated leachates by headspace gas chromatography. *Microchemical Journal, 153*, 104494. https://doi.org/10.1016/j.microc.2019.104494
117. Zhou, W., Yin, B.-C., & Ye, B.-C. (2017). Highly sensitive surface-enhanced Raman scattering detection of hexavalent chromium based on hollow sea urchin-like TiO2@Ag nanoparticle substrate. *Biosensors & Bioelectronics, 87*, 187–194. https://doi.org/10.1016/j.bios.2016.08.036
118. Zhu, C., Zhai, J., & Dong, S. (2012). Bifunctional fluorescent carbon nanodots: Green synthesis via soy milk and application as metal-free electrocatalysts for oxygen reduction. *Chemical Communications, 48*, 9367–9369. https://doi.org/10.1039/C2CC33844K
119. Zhu, J., Chu, H., Shen, J., Wang, C., & Wei, Y. (2021). Green preparation of carbon dots from plum as a ratiometric fluorescent probe for detection of doxorubicin. *Optical Materials, 114*, 110941. https://doi.org/10.1016/j.optmat.2021.110941
120. Zu, F., Yan, F., Bai, Z., Xu, J., Wang, Y., Huang, Y., & Zhou, X. (2017). The quenching of the fluorescence of carbon dots: A review on mechanisms and applications. *Microchimica Acta, 184*, 1899–1914. https://doi.org/10.1007/s00604-017-2318-9

Application of Carbon Quantum Dots in the Food Industry

Linlin Zhao and Min Zhang

Abstract Carbon quantum dots (CQDs) have recently attracted extensive attention due to their excellent physical and chemical properties, including abundant surface functional groups, ultra-small size, tunable high-fluorescence, good biocompatibility, excellent antibacterial activity, and nontoxic/low-toxicity etc. These advantages provide opportunities for the application of CQDs in the food industry. The CQDs have been used to develop new technologies that can improve shelf life or freshness of food samples, design methods or tools for rapid food analysis, and fabricate functional and environmentally friendly food packaging. This chapter systematically reviews the applications and detecting principles of CQDs used as fluorescent probes in food safety and quality detection, and reviews the applications of CQDs used as value-added multifunctional fillers in protective and active intelligent packaging films. In addition, it also summarizes the applications and effects of CQDs used as nanoscale food additives and coating agents in food preservation field and discusses the security of CQDs.

Keywords Carbon quantum dots · Food preservation · Active intelligent packaging · Security · Antioxidant

---

L. Zhao
College of Tourism and Culinary Science, Yangzhou University, Yangzhou 225127, Jiangsu, China

Key Laboratory of Chinese Cuisine Intangible Cultural Heritage Technology Inheritance, Ministry of Culture and Tourism, Yangzhou 225127, China

M. Zhang (✉)
State Key Laboratory of Food Science and Resources, Jiangnan University, Wuxi 214122, Jiangsu, China
e-mail: min@jiangnan.edu.cn

Jiangsu Province Key Laboratory of Advanced Food Manufacturing Equipment and Technology, Jiangnan University, Wuxi 214122, Jiangsu, China

V. Kumar et al. (eds.), *Green Carbon Quantum Dots*, Engineering Materials,
https://doi.org/10.1007/978-981-97-6203-3_10

1 Introduction

Ensuring food quality and safety is one of the biggest challenges faced by the global community. Every country spends a significant portion of its economy producing food for its people. However, a significant portion of food is wasted due to spoilage before it reaches consumers, resulting in huge economic losses. Besides, the chemical contamination of foodstuffs, for example, heavy metal ions, pesticides, veterinary drugs and banned additives, has caused serious health problems. Therefore, it is necessary to develop new processing, preservation, and detection technologies in order to detect harmful substances in food, inhibit the growth of spoilage bacteria, extend food shelf life, and ensure food quality and safety.

As for the application of carbon quantum dots (CQDs) in food processing and preservation, due to their fluorescence quenching effect, many studies have been devoted to the use of CQDs for detecting foreign compounds and pathogenic bacteria in foods. Food spoilage during storage and distribution is mainly due to the growth of microorganisms or biochemical actions such as the oxidation of food ingredients. Several studies have confirmed that CQDs can be used as an antimicrobial and antioxidant agent to improve freshness or shelf life and maintain the quality of foods. As one of the food packaging technologies to secure food safety and extend shelf life, active packaging attempts to prevent food spoilage and reduce food loss after processing, gaining popularity recently. Notably, as a new nanomaterial with plenty of oxygen containing functional groups, CQDs have the potential to product active packaging materials with new functions, such as UV-blocking properties, antioxidant properties, and antimicrobial properties.

This chapter systematically reviews the applications of CQDs in the food industry, including used as fluorescent probes to detect food safety and quality, used as value-added multifunctional fillers to develop protective and active intelligent packaging, and used as food additives and coating agents to extend the shelf life of food samples. Figure 1 shows the various applications of CQDs in the food industry. In addition, the security and migration behavior of CQDs are briefly discussed.

2 Application of Carbon Quantum Dots in Food Quality and Safety Detection

Several biosensors based on CQDs have been developed for applications related to food quality and safety. The fundamental principle underlying the fluorescence sensor involves the interaction between the target component and the recognition substance. This interaction leads to alterations in the fluorescence properties of CQDs, and these alterations can be quantitatively correlated with the concentration or structure of the target analyte. The principles may be categorized into four distinct categories, (1) direct fluorescence enhancement (turn on); (2) direct fluorescence quenching (turn

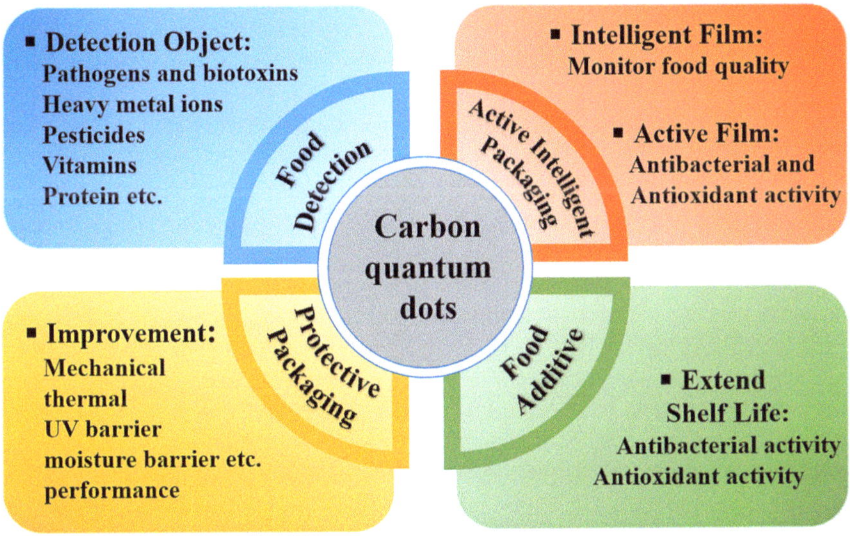

Fig. 1 Various applications of carbon quantum dots in the food industry

off); (3) fluorescence recovery (off–on); 4) fluorescence quenching again (off–on–off) [45]. The fluorescence quenching processes typically include inner filter effect (IFE), fluorescence resonance energy transfer (FRET), static quenching and dynamic quenching [82].

2.1 Detection of Chemical Contaminant in Food

The presence of chemical contaminants in food products, such as heavy metal ions, pesticides, veterinary drugs, and prohibited additions, has resulted in significant health concerns. The implementation of an effective testing approach is necessary to ensure the safety of food. In contrast to conventional techniques such as complexometric titrations, chemiluminescence, and atomic spectroscopy, emerging approaches using the fluorescence of CQDs provide advantages in terms of simplified procedure, cost-effectiveness, and enhanced sensitivity.

2.1.1 Detection of Heavy Metal Ions

Some metal ions, such as Al^{3+}, Hg^{2+}, Cu^{2+}, Fe^{3+}, etc. are not biodegradable, they accumulate in the human body through the food chain and cause various health problems. Therefore, metal ions' detection is an essential part of food safety monitoring, especially toxic heavy metal ions' detection. The fluorescence analysis method based on CQDs has been widely used for the detection of metal ions because of its advantages of simple operation, low cost, sensitivity and accuracy.

A fluorescence quenching nanosensor with CQDs to measure Hg^{2+} in tap water was described by Zhou et al. [81]. The interaction between Hg^{2+} and –COOH/–OH on the surfaces of CQDs causes fluorescence quenching. They obtained good linearity for detecting Hg^{2+} over the concentration from 0 to 3 μM with a detection limit of 4.2 nM. According to Huang et al. [25], CQDs synthesized using sugarcane molasses shown potential as a sensing probe for the detection of Fe^{3+}. The decline in fluorescence intensity of CQDs exhibited a strong linear correlation with varying concentrations of Fe^{3+} throughout the range of 0 to 100 μM. Several metal ions, such as Cu^{2+}, Cr^{6+}, Ag^{+}, Al^{3+}, Pb^{2+} and Co^{2+}, have been accurately measured using fluorescent probes that rely on CQDs. The enhanced selectivity of CQDs towards metal ions may be attributed to a potentially greater affinity or more rapid coordination process between the surface functional groups of the CQDs and the metal ions.

2.1.2 Detection of Pesticide Residues

Pesticides, particularly organophosphorus pesticides, have been widely used in the agricultural sector for several years, resulting in high levels of pesticide residues in food products, particularly vegetables and fruits. The detection of pesticides with high sensitivity often depends on nanoparticles, metal ions, or enzymes. Several research projects have examined the viability of CQDs for detecting pesticides in food and agricultural samples.

Li et al. [32, 33] prepared a nano-sensor for pesticides based on CQDs with an "off–on" mechanism, which is often seen in such sensors. The enzymatic action of acetylcholinesterase (AChE) facilitates the conversion of acetylcholine into choline. Subsequently, choline undergoes additional oxidation into betaine by the action of choline oxidase, resulting in the generation of the fluorescence quencher H_2O_2. Since a number of pesticides inhibit the activity of AChE, the fluorescence recovered due to the absence of H_2O_2, enables sensitive detection of pesticides. Similar methods can be used to identify paragon and chlorpyrifos. In addition, a number of pesticides have been detected with "off–on " or "off–on-off" strategies, including glyphosate in spiked cereal [52], chlortoluron in irrigation water [50], and paraoxon in apple juice [19].

2.1.3 Detection of Veterinary Drug Residues

Veterinary drug residues refer to the residues of veterinary drug prototypes or their metabolites contained in the edible parts of animal products. Long-term consumption of animal derived food (such as meat, milk, eggs) with excessive veterinary drug residues may cause accumulation and chronic poisoning. Therefore, the development of simple, rapid and reliable detection method based on CQDs is of great significance for the detection of veterinary drug residues in food.

Tetracycline (TC) and its derivatives are frequently-used antibiotics. The sensitive quantification of TC was accomplished by using the fluorescence quenching effect of TC towards CQDs, with a detection limit of 7.5 nM [67]. This approach is also applicable for the identification of doxycycline, chlortetracycline, and oxytetracycline. Fluorescence quenching arises from the interplay between CQDs and antibiotics, whereby the binding of antibiotics to CQDs leads to the energy limitation of CQDs. In addition to TC and its derivatives, sulfasalazine [74] in drinking water, norfloxacin [24] and cephalexin [2] in raw milk have also been identified by the fluorescence enhancing or quenching of CQDs.

2.1.4 Detection of Food Additives

Melamine, as a chemical raw material, is commonly used in the production of plastics, adhesives and food packaging materials. Because of its low cost and high nitrogen content, melamine may be illegally mixed into milk, infant formula, and pet food. Over the past years, several nanosensors based on CQDs have been developed for sensitive detection of melamine. Daia et al. [9] developed a FRET system using gold nanoparticles and CQDs for the detection of melamine, with a detection limit of 36 nM. The suggested technique was then effectively applied to milk samples with satisfactory results.

The detection of nitrite is crucial in the food industry. Excessive intake of nitrite can cause serious health hazards or even death. Zhang et al. [71] determined the amount of nitrite in water using prawn-shell derived N-doped CQDs. The result showed that there was a strong linear correlation between the concentration of NO^- (0–1.0 Mm) and fluorescence intensity.

Food products incorporate artificial colorants in order to ensure the color consistency and enhance their visual appeal. Certain food colorants have shown carcinogenic or genotoxic properties, leading to their prohibition in food applications. In the investigation conducted by Xu et al. [61], the artificial colorant tartrazine was found to be a quencher molecule that formed a ground-state complex with aloe CQDs. The concentration of tartrazine in the candy sample was determined to be 4.8 μM using the CQDs sensing method. Su et al. [48] produced fluorescent CQDs using dried lemon peel for the detection of carmine in beverages. The calibration curve exhibited a linear range spanning from 0.20 to 30.00 mg/L under ideal conditions. Additionally, the limit of detection for carmine using CQDs was 0.16 mg/L. In addition to carmine, CQDs could be used for the detection of tannic acid [1] and sunset yellow [70].

2.2 *Detection of Nutritional and Functional Components in Food*

Nutrients in food can provide enough energy and nutrients for the human body, such as protein, lipid, carbohydrates, vitamins and minerals, etc., which are essential for normal life activities. Therefore, it is crucial for determining the nutritional content of food. In the field of food analysis, CQDs has been used as a sensing probe to detect and analyze the functional components of food, such as proteins, vitamins, phenolic compounds, etc.

2.2.1 Detection of Vitamins

Vitamins, being a crucial class of necessary nutrients, have a substantial impact on the regulation of bodily functions. The use of excessive or inadequate amounts of vitamins may lead to significant harm. Various vitamins are quantified based on CQDs-sensor, which perform well in real samples.

It was discovered that Vitamin B_9 has the ability to quench the fluorescence of CQDs due to the interaction of the functional groups between Vitamin B9 and CQDs [8]. Additionally, using CQDs to build a FRET system allowed for the determination of vitamin B_2 [29] and vitamin B_{12} [56]. Moreover, instead of connecting with CQDs directly, Vitamin B1 and Vitamin C were proved to recover fluorescence by removing or reducing metal ions that mediated CQDs, resulting in selective *Escherichia coli* detection. Liu et al. [33] developed a sensing platform using CQDs and MnO_2 for the purpose of assessing Vitamin C levels. This platform was then used to effectively detect Vitamin C in various fresh fruits, vegetables, and commercially available fruit juices. Purbia and Paria [37] developed a green and blue fluorescence CQDs for the detection of vitamin B_1, with detection limit of 280 $nmol \cdot L^{-1}$. The fluorescence of CQDs can be quenched by Cu^{2+} and restored after the addition of thiamine due to the formation of a soluble cuprum-thiamine complex, thus achieving rapid detection of the target component.

2.2.2 Detection of Proteins and Amino Acids

The content of ovalbumin (OVA) is used as a reference to evaluate protein quality. Fu et al. [17] synthesized a new type of N, O, and P co-doped CQDs (NOP-CQDs) by one-step hydrothermal synthesis method and applied it to the quantitative detection of OVA in egg products. In the fluorescence resonance energy transfer system composed of NOP-CQDs, graphene oxide and OVA antibodies, the "on–off" sensing probe realizes the selective recognition and capture of OVA based on the specific antigen–antibody interaction, and the detection limit is 153 $\mu g \cdot L^{-1}$.

Glutathione (GSH) is added to foods as a functional element. Wang et al. [58] prepared CQDs from eggshell membrane and established CQDS-Cu^{2+} system for quantitative analysis of GSH. The fluorescence of CQDs was quenched by Cu^{2+}, and the fluorescence intensity was partially recovered after the addition of GSH, which was attributed to the formation of Cu^{2+}-S bond and the reduction of the amount of Cu^{2+} on the surface of CQDs.

Cysteine is an amino acid that contains sulfur and has significant importance as a crucial substrate in the process of protein synthesis. Amjadi et al.[3] devised an innovative sensor for assessing cysteine using silver nanoparticles (AgNPs) and CQDs. The researchers discovered that the fluorescence of CQDs was suppressed due to the presence of AgNPs owing to the formation of FRET system. Nevertheless, the introduction of cysteine contributed to the recovery of fluorescence based on the competitive adsorption of cysteine onto AgNPs, thereby achieving the sensitive and selective detection of cysteine.

2.2.3 Detection of Carbohydrates

In addition, fluorescent probes based on CQDs can be used to detect carbohydrates. Shan et al. [44] found that B-doped CQDs could be used to detect glucose in the presence of glucose oxidase (GOx). The catalytic action of GOx is associated with the production of H_2O_2, and the charge transfer between CQDs and H_2O_2 is the main reason for the fluorescence quenching of CQDs. Similarly, Wang et al. [53] reported that mixing GOx, Fe^{2+}, and CQDs into a glucose solution also causes CQDs fluorescence quenching due to the oxidation of CQDs by hydroxyl radicals produced by the reaction of Fe^{2+} and H_2O_2. A galactose biosensing system based on boronic acid functionalized CQDs as the fluorescent probe was reported by Yang et al. [66]. The interaction between the boronic acid on the surface of CQDs and the cis-conformational diol units in galactose was used to assess galactose using the fluorescence quenching effect.

2.2.4 Detection of Other Functional Components

Some small molecule substances in plant-derived foods have special antibacterial, anti-inflammatory or antioxidant properties, thus giving the food special medicinal properties. These small molecules are defined as "functional components". The qualitative or quantitative analysis of functional components in functional foods is the main method to evaluate their quality. Fluorescent CQDs provides an effective, convenient and accurate method for the analysis and detection of these functional components.

Yang et al. [65] synthesized water-soluble CQDs for fluorescence detection of chlorogenic acid in honeysuckle. When chlorogenic acid concentration increased in the range of 0.15–60 $mol{\cdot}L^{-1}$, the fluorescence of CQDs could be effectively quenched, and the detection limit was 45 $nmol{\cdot}L^{-1}$. Compared with the results of chlorogenic acid detection by HPLC method, the CQDs-based detection method offers improved sensitivity and a much higher detection rate. Ahmed et al. [1] used

CQDs to directly quantify the presence of tannic acid (TA) in both red and white wine samples. The limit of detection for TA was determined to be 0.018 mg/L, as TA exhibited the capability to quench fluorescence signals. Several phytochemicals, including kaempferol [34], catechol [68], caffeic acid [49], and curcumin [5], were detected with high sensitivity based on dynamic quenching, static quenching, and FRET.

2.3 Detection of Pathogens and Biotoxins

The majority of bacteria have a low surface isoelectric point, usually with a rich negative charge. The CQDs can be absorbed onto the surface of bacteria, resulting in a substantial enhancement in their fluorescence intensity. In addition, the fluorescence quenching effect caused by the aggregation of CQDs on the surface of bacteria can also be used for the quantitative detection of bacteria. The schematic illustration of CQDs based sensors for pathogens detection as shown in Fig. 2. Therefore, bacterial fluorescence sensors based on CQDs have attracted the attention of many researchers.

Wang et al. [59] reported a typical composite CQDs sensor based on aptamer modification, which could identify a specific membrane protein on the surface of *Salmonella typhimurium* and quantitatively detect the concentration of *Salmonella typhimurium* according to the fluorescence intensity of the solution. Using a similar principle, CQDs modified by amicacin [6], colistin [7], and mannose [30], were successfully used for quantitative detection of *Escherichia coli*. On the other hand, CQDs can be used to detect biotoxins in food products. Biotoxins are toxic chemicals secreted by the organism, metabolized or semi-biosynthetic, which cannot be self-replicating. Food poisoning caused by biotoxins has attracted great attention in many

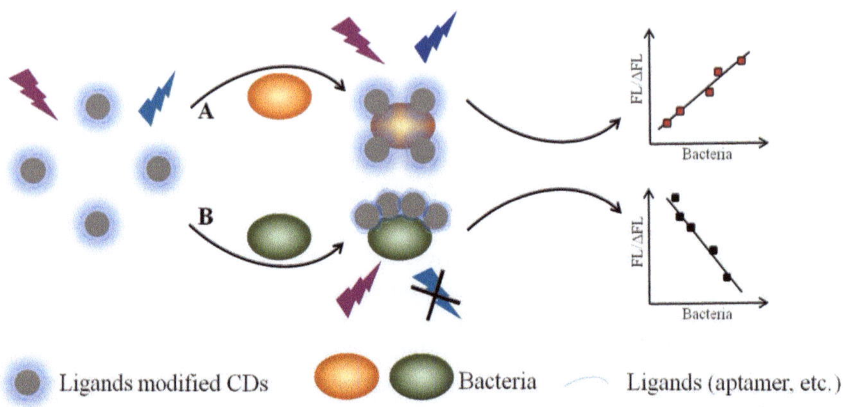

Fig. 2 Schematic illustration of CQDs based sensors for pathogens detection. A and B refer to fluorescence increasing and fluorescence quenching, respectively. [45]

countries. Based on the fluorescence on–off-on principle, Wang et al. [52] construct a system comprising of CQDs and gold nanoparticles with the assistance of aptamer. This system was designed for the targeted detection of aflatoxin B1, resulting in a limit of detection of 5 pg/mL. This proposed strategy has proved to be successful in real samples and has been expanded to the detection of other objects by changing the corresponding aptamer. These researches demonstrated that the use of CQDs is a viable approach for detecting harmful microorganisms and their secreted toxins. In general, it is recommended to include aptamers or antibodies into the sensing system to improve its specificity.

3 Application of Carbon Quantum Dots in Food Packaging

By preventing adverse factors or conditions such as spoilage microorganisms, chemical pollutants, oxygen, moisture, light, and external forces, food packaging can delay the quality deterioration of food during storage and transportation and reduce food safety risks. To achieve these functions, packaging materials need to provide physical protection and create appropriate physicochemical conditions, thereby maintaining food quality and safety to obtain a satisfactory shelf life [39]. Petroleum-based plastics have excellent strength, transparency, stability, permeability and flexibility, are widely used in the field of food packaging [72]. However, the non-degradation of petroleum-based plastics and the gradual depletion of petroleum have promoted the rapid development of biodegradable materials, making them gradually become a substitute for synthetic plastic packaging materials. According to the different production methods, biopolymers can be divided into natural biopolymers and synthetic biopolymers. Compared with petroleum-derived packaging materials, packaging materials derived from natural biopolymers such as polysaccharides and proteins have the potential to reduce environmental pollution. However, the absence of suitable UV protection, mechanical strength, antioxidant, and antibacterial characteristics in biopolymer materials restricts the use of these materials in food packaging.

Recently, CQDs have been employed to prepare food packaging as bioactive additive and polymer structure enhancer, serving a variety of functions in food packaging film. The presence of various functional groups and miscellaneous materials on the surface of CQDs contributes to their ability to improve the UV shielding, mechanical strength, antioxidant, and antibacterial properties of food packaging. Scholars have conducted numerous researches in the field of CQDs-based food packaging, wherein the efficacy of its use in food preservation has been duly confirmed.

3.1 Develop Protective Food Packaging

3.1.1 Mechanical, Moisture Barrier, and Thermal Performance Improvement

The fundamental characteristics of food packaging materials include mechanical performance, thermal stability, and water barrier performance of polymer films. The improvement of moisture barrier qualities contributes to the extension of the shelf life of packed goods. The improvement in mechanical performance facilitates the use of thinner film packing, hence resulting in a reduction in material consumption and expense. Thermal stability enhancements are implemented with the objective of mitigating flammability and enhancing thermal protection. These enhancing benefits of CQDs have been seen in some packaging materials, such as soy protein isolate (SPI), polyvinyl alcohol (PVA), cellulose nanofiber (CNF), chitosan, and carboxymethyl cellulose (CMC), among others.

The protein macromolecule known as SPI has many notable attributes, including its satisfactory film-forming capacity, good biocompatibility, natural degradability, and easy availability. Due to these inherent attributes, SPI is a suitable substance for the production of edible films and packaging materials. Nevertheless, the economic viability of SPI films is hindered by their extreme susceptibility to water. To enhance the mechanical properties and reduce water absorption in SPI-based materials, researchers have incorporated CQDs into the SPI matrix. This addition aims to optimize the performance of SPI films, encompassing improvements in water resistance, thermal stability, mechanical strength, and physicochemical characteristics. The presence of significant quantities of –COOH, $–NH_2$, and –OH functional groups on the surface of CQDs enables facile hydrogen bonding interactions with –COOH, –OH, and $–NH_2$ groups in SPI. Consequently, the formation of a network structure via physical interactions results in a coarser and more robust cross section of the composite film. This, in turn, enhances the overall performance of the SPI film. Li et al. [32, 33] prepared a composite film with SPI and CQDs. The tensile strength and modulus of composite film with 5 g CQDs were increased by 82.97 and 79.74% compared with pure SPI film, respectively. In addition, the water vapor transmittance of composite film was reduced by 48.36% compared with SPI film.

PVA, a kind of polymer that is capable of biodegradation, has been extensively used in several fields such as food packaging, film production, and clinical applications. However, the barrier characteristics of PVA such as thermal resistance and water resistance are insufficient, hence restricting its potential applications. The incorporation of CQDs into PVA matrix has the potential to improve the mechanical characteristics and water resistance of PVA films. The presence of many surface functional groups and appropriate dimensions provide CQDs a very effective nucleating agent for PVA crystallization. This characteristic facilitates the formation of a more uniform and compact cross-linking network inside PVA hydrogel, thus leading to improved mechanical characteristics in PVA film. The formation model of CQDs-PVA gel as shown in Fig. 3. Furthermore, the reduction in porosity will concurrently

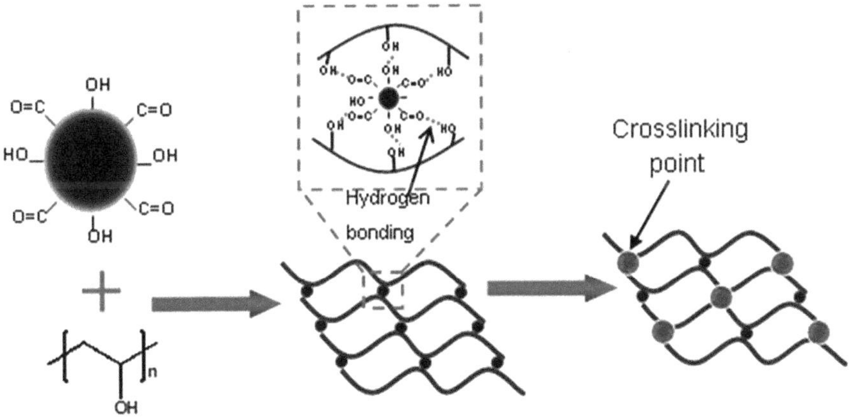

Fig. 3 Formation model of CQDs-PVA gel [22]

diminish the film's capacity for water absorption, hence enhancing its water barrier characteristics. According to Xu et al. [62], CQDs enhanced the mechanical characteristics of PVA/CNF/CQDs films. The breaking strength and tensile modulus considerably increased as CQD concentration increased. Additionally, the surface contact angle and water absorption of the produced films steadily reduced as the concentration of CQDs (>1 mL) increased.

Chitosan is a linear polysaccharide derived from chitin and can be obtained from the shells of crustaceans. Chitosan has good biocompatibility, film forming and antibacterial activity, and has been utilized to create food preservation film. However, due to the weak mechanical properties, it is difficult to prepare films with satisfactory strength and toughness. Konwar et al. [27] presented a composite film made of chitosan and CQDs. Using the interaction between the positive and negative charges on CQDs and chitosan, a strong and stable composite hydrogel film was effectively created. Addition of 0.5% CQDs enhanced the tensile strength of composite film to 18.6 MPa, whereas chitosan film was only 5.1 MPa. With 1.5% CQDs, the composite film's surface contact angle rose from 64.95° to 88.75°.

CMC is a water-soluble cellulose derivative with anionic carboxyl group, which is widely used in the preparation of biodegradable packaging film. CMC has excellent film forming ability and good gas barrier properties, and moderate tensile strength. You et al. [69] developed a transparent sunlight conversion film based on CQDs and CMC. The tensile strength of composite film was 55% higher than that of CMC film. This was mainly attributed to the good miscibility between CMC and CQDs and the increase of hydrogen bonds in CMC/CQDs film.

3.1.2 UV Barrier Performance Improvement

Polymers are widely used in food packaging due to their good recyclability, process ability and economic attractiveness. It is a pity that no polymer can completely block the gas, and most polymers permit UV light through. However, several food items not only need protection against gas exposure during storage but also require shielding from UV radiation. For example, many dairy products that contain riboflavin (also known as vitamin B_2) experience a significant loss of essential nutrients upon exposure to UV radiation. In order to mitigate the UV absorption capacity of polymers, it is suggested to include various UV blocking agents into food packaging materials. CQDs are a kind of materials with high UV absorption capabilities, which can convert light energy by absorbing certain characteristic wavelengths. Figure 4 shows a possible UV barrier mechanism for CQDs based film. Recently, the investigation of UV barrier characteristics of food packaging materials based on CQDs has received considerable attention.

Using bacterial nanocellulose (BNC), Salimi et al. [43] created an improved nanopaper. The UV barrier properties of CQDs-nanopaper were more than 99% below the wavelength of 410 nm. Hess et al. [21] fabricated a composite film composed of PVA and CQDs, which shown a high efficacy in blocking over 90% of UV radiation. Similarly, Xu et al. [62] demonstrated the ability of CQDs to augment the UV barrier characteristics of PVA films, and observed an inverse relationship between the UV transmittance and the concentration of CQDs present in the films.

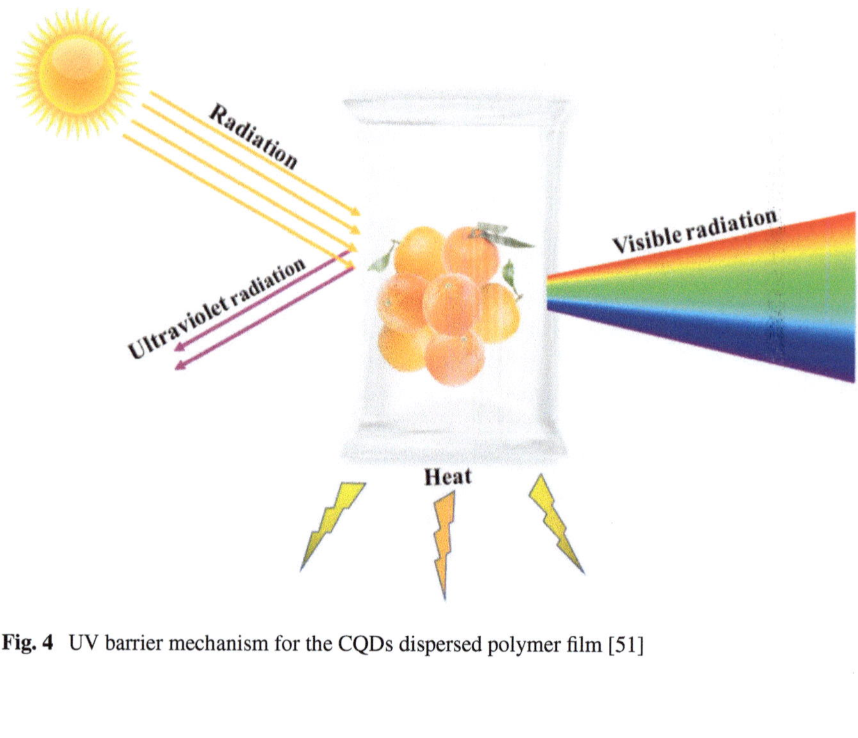

Fig. 4 UV barrier mechanism for the CQDs dispersed polymer film [51]

The UV blocking properties of CQDs composite films have been demonstrated to have the ability to extend the shelf life of fruit samples. Patil et al. [36] fabricated a composite film with tea residue CQDs (WTR-CQDs) and PVA, with a blocking efficiency ranging from 20 to 60% for UV-A (315–400 nm), and complete blocking (100%) for both UV-B (280–315 nm) and UV-C (230–280 nm). Grapes wrapped with PVA/WTR- CQDs film were irradiated under UV lamp for 30 h, and the browning and shrinkage of grapes were lower than those wrapped with PVA film, which verified the applicability of the composite film as a UV blocking material. The 0.5% CD/PVA film prepared by Zhao et al. [77] can be used as an excellent UV blocking material for jujube storage. The film can reduce the ultraviolet radiation by 81.22%, thus reducing the oxidative damage of jujube under ultraviolet irradiation and prolonging the storage period. The aforementioned studies show that CQDs-doped polymers may be employed in protective food packaging, particularly in UV blocking.

3.2 Develop Active Intelligent Food Packaging

Furthermore, with its fundamental roles of containment, protection, and provision of food information, modern food packaging exhibits other attributes categorized as "active" or "intelligent". The term "active" features of packaging refer to the enhancement of sensory attributes, preservation quality, and/or prolongation of the shelf life of food products. The "intelligent" aspect of packaging refer to its ability to monitor the condition of the product, packaging, or packaging environment without altering the qualities of the product [41]. Although numerous nanoparticles have been proposed for active intelligent packaging, CQDs undoubtedly provide great potential for the development of innovative packaging materials with amazing performance.

3.2.1 Develop Antibacterial Activity Packaging

With the increase in the research of CQDs, various functionalized CQDs have been developed and are increasingly used in antibacterial aspects. The antibacterial activities of CQDs are attributed to complex mechanisms, such as the production of reactive oxygen species (ROS), leaking of cytoplasm, disruption of cell structure, and fragmentation or condensation of genomic DNA. Besides, antibacterial characteristics of CQDs are associated with their size and shape, surface charge, and functional groups.

Antibacterial packaging refers to a kind of active packaging that interacts with the product or the space inside the packaging in order to diminish, inhibit or delay the proliferation of germs present on the food's surface, thereby achieving the objective of prolonging the shelf life of foods [41]. This objective might be accomplished by the introduction of an active agent into the packaging material or by applying a coating onto or inside the packaging material. The CQDs active films have been demonstrated to have the enhanced antibacterial and anti-biological contamination activities against

Pseudomonas aeruginosa, Staphylococcus aureus, Escherichia coli, and *Bacillus cereus, Listeria monocytogenes*, et al. [43, 47].

At present, CQDs based antibacterial films have been used to extend the shelf life of food samples. Riahi et al. [40] prepared a CQDs-CMC composite film with antibacterial properties, and coated CQDs-CMC film-forming solution on the surface of lemons. It was found that the lemons with a protective coating maintained their sensory quality and exhibited no signs of mold growth after a storage period of 21 days. Ezati et al. [12] prepared a CQDs/chitosan/gelatin composite film, which effectively inhibited the growth of mold on the surface of avocado, had a better antibacterial effect than the coating of chitosan/gelatin film, and extended the shelf life of avocado to more than 14 days. The antibacterial efficacy exhibited a direct correlation with the concentration of CQDs.

3.2.2 Develop Antioxidant Activity Packaging

It is known that ROS play a vital role in food spoilage, polymers and chemical products degradation, and biological structures destruction. Therefore, the use of antioxidants or free radical scavengers is imperative in several domains such as health, food, packaging, cosmetics, corrosion prevention and other related fields. In recent times, CQDs have been used as antioxidants for ROS elimination.

Numerous researches have shown the mechanisms related to the antioxidant activity of CQDs. The antioxidant activity of CQDs is influenced by many aspects, including electron transport, unpaired electrons caused by vacancies and defects, sp^2 hybrid carbon domain, hydrogen donor behavior, doping element type and surface functional group type [75]. The antioxidant activity of CQDs was also related to the antioxidant capacity of their precursors. Substances with antioxidant properties were shown to have a higher likelihood of synthesizing CQDs with the ability to scavenge free radicals. Son et al. [46] synthesized a new kind of CQDs with tannic acid as precursor, and its antioxidant and anti-aging properties were superior to quercetin, L-ascorbic acid and tannic acid, which were used as positive controls in the experiment.

Although many studies have fabricated CQDs based antioxidant films, the application of this active film in food preservation field is still in its infancy. Das et al. [10] packaged rapeseed oil with rapeseed protein-CQDs composite film bags to investigate the effect of this composite film on prolonging the shelf life of rapeseed oil. Upon completion of the 28-day storage period, the oil samples contained inside the rapeseed protein-CQDs bags exhibited a notable delay in oxidized rancidity as compared to the oil samples stored in regular protein bags. The levels of free fatty acid, thiobarbituric acid, and peroxide value shown a reduction of 1.4, 2, and 1.2 times, respectively. The findings demonstrated that the composite packaging including CQDs can improve the oxidative stability of lipid-based food products during the storage period.

Active food packaging materials based CQDs have been applied to the preservation of fruits and vegetables, meat and poultry products, aquatic products, prepared dishes, bread and other samples.

3.2.3 Develop Intelligent Packaging

Smart packaging is a new development in the field of packaging. Packaging materials contain smart materials that can indicate product freshness, temperature, safety, etc. This packaging material can let consumers know the quality of the food without opening the package. During the transportation and storage of products, changes in environmental indicators in the packaging, such as pH, humidity, O_2, CO_2, microbial number and other physical and chemical indicators, can be monitored using intelligent packaging indicators.

Kilic et al. [26] prepared anthocyanins doped fish gelatin (FG) film by using CQDs and ultraviolet irradiation as crosslinking agent, and monitored its color changes. It was found that the color response of the film had a good correlation with microbial growth and total volatile basic nitrogen (TVB-N) increase in chicken breast samples. This led to the development of an image-processing smartphone App for quantitative analysis of food spoilage (Fig. 5). Fu et al. [16] developed a pH-sensitive gelatin/chitosan/CQDs composite film and applied it as fish freshness indicator. Fish deterioration result in an increase in TVB-N value, which would lead to an increase in pH value. With the increase of storage time, the pH value of the fish increased and the brightness of the composite film decreased significantly. Koshy et al. [28] also prepared a highly sensitive starch-based biopolymer film with pH indicator function, which has the potential to monitor the freshness of packaged pork. By combining the CQDs and anthocyanins, the film exhibits color changes at different pH values. As the storage time increases, the visual color changes from purple to green.

Rahman and Chowdhury [38] prepared a guar gum-sodium alginate mixed glucose-glycerol CQDs composite film, and proved that the composite film could be used to detect relative humidity. The fluorescence quenching was observed when the composite film was wrapped on bread and placed at high humidity. Therefore, the prepared composite film can monitor the freshness of packaged food using only an ultraviolet light, without opening the package. Hu et al. [23] used acid-responsive CQDs to establish a flexible and visually readable sensing system for milk freshness detection. The fluorescence intensity of CQDs was highly correlated with milk freshness. As the freshness of milk decreases, the fluorescence intensity of CQDs exhibits a corresponding drop. Based on this principle, a fluorescence colorimetric card was developed. The sensor has the advantages of high sensitivity and intuitiveness, and can detect the spoiled milk without any pretreatment. Xu et al. [63] prepared an oxygen-sensitive nanocellulose films using ruthenium-doped CQDs, and the respiration of agricultural products could be monitored by identifying the fluorescence intensity of the films under ultraviolet light.

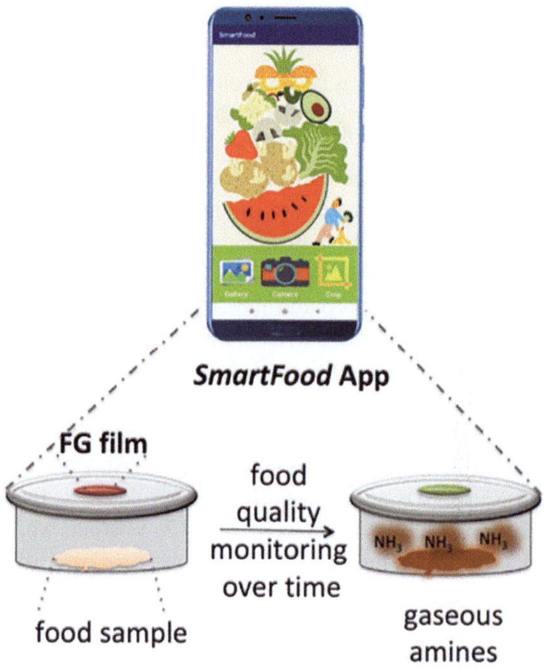

Fig. 5 The colorimetric analysis of food spoilage using image-processing smartphone App

4 Application of Carbon Quantum Dots as Coating Agent and Additive

4.1 Application as Coating Agent

Coatings were increasingly applied in food preservation areas owing to their biodegradable, non-toxic and edible characteristics. The use of edible coatings on food surfaces has the potential to regulate enzymatic browning, minimize water loss and gas exchange, and prevent microbiological spoilage by creating a semi-permeable barrier.

In many studies, the original intention of researchers is to prepare food packaging films with antioxidant or antibacterial activities. However, in actual food applications, these active film solutions often covered the sample in the form of a coating rather than a real food packaging film. Zhang et al. [73, 74] applied the CQDs/PVA mixed solution and the PVA solution onto the bananas, respectively. After stored for 3 days at room temperature, bananas that were treated with CQDs/PVA solution exhibited fewer black regions on their peels compared to bananas that were coated with PVA solution or without coating. The results demonstrated the functional antioxidant capacity of CQDs in fruit storage. The coating also provides a physical barrier, which inhibits respiration and evaporation of the peel, thereby delaying fruit ripening. Xiao et al. [60] coated a CQDs/chitosan mixture solution on the surface of fresh mango for storage experiment. The application of CQDs/chitosan coating shown a significant antimicrobial activity against both fungus and bacteria present on the surface of mango. It effectively reduced the water loss rate and decay rate, inhibited respiration of mango during storage.

In addition to the preservation of fresh fruits and vegetables, some studies have used the functionality of CQDs to prepare edible coatings for the preservation of fresh-cut fruits and vegetables. The research conducted by Fan et al. [15] used a coating film composed of CQDs and chitosan to prolong the shelf life of fresh-cut cucumber inside a modified environment packaging system. The application of a 4.5% CQDs/chitosan coating demonstrated positive effects in terms of mitigating weight loss, preserving ascorbic acid and total soluble solids levels, suppressing peroxidase activity, and retaining flavor. In another study by Fan et al. [13, 14], freshly cut cucumbers were treated with CQDs/chitosan coating combined with ultrasound (US).The results showed that the total number of colonies in the control group was 6.72 log CFU g^{-1} after 15 days of storage, and that in the combined treatment was 3.45 log CFU g^{-1}. The combined treatment improved the microbiological, physical and chemical quality of fresh-cut cucumbers, and its preservation effect was better than that of coating or US treatment alone.

4.2 Application as Food Additive

CQDs has the capacity to function as antioxidants and antibacterial agents in food products owing to their remarkable antioxidant and antibacterial characteristics, alongside their low-toxic or non-toxic nature.

Zhao et al. [78] prepared a CQDs that could inhibit the growth of *Escherichia coli*, *Staphylococcus aureus* and *Bacillus subtilis*. The CQDs in combination with chitosan could extend the shelf life of soy milk samples. After storage at room temperature for 4 days, the soy milk including 0.16% chitosan and 8% CQDs exhibited a significant decrease in total bacterial population compared to regular soy milk. Furthermore, the CQDs was employed in combination with radio frequency (RF) as a means to decrease the microbial count in the packed dishes [76]. Compared with high pressure steam sterilization, CQDs + RF pasteurization not only guaranteed the microbiological safety, but also kept the physical and sensory attributes of the dishes.

In order to investigate the antioxidant effects of CQDs, Zhao et al. [80] introduced CQDs to frying oil and continually cooked meatballs in this oil. After 35 batches of meatballs continually fried, the oxidative rancidity of oil was significantly inhibited by CQDs, as shown by the changes in acid value, peroxide value, total polar component, etc. It is noteworthy that the combination of CQDs and carnosic acid (0.05%) exhibited a comparable anti-oxidative rancidity effect to that of 0.02% TBHQ. Additionally, the oxidative rancidity of the fried meatballs during storage was delayed and their shelf life was extended by the incorporation of CQDs into the frying oil [79].

5 The Security Control of CQDs in the Application

Currently, numerous innovative nanofillers are now being studied for application in active food packaging. New packaging materials employing nanofillers must not only be effective at preserving food but also stable during storage and inert in terms of food sensory performance in order to be approved by consumers and the food industry. The attention on nanofillers is mostly owing to their potential toxicity, since high surface area to volume ratio provides high reactivity, hence the use of nanofillers in food packaging film needs more careful examination of their safety [35]. Therefore, in view of the application of CQDs in the field of food preservation, appropriate security control methods must be adopted to ensure food safety.

5.1 Toxicity Evaluation of CQDs

Although CQDs have promising potential for many applications, it is important to consider the potential biological toxicity associated with nanoparticles in general. Therefore, it is crucial to assess the toxicity of CQDs towards organisms. The present investigation demonstrates that the use of CQDs solutions at low concentrations does not exhibit substantial toxicity towards human cells, and in some instances, may even stimulate cell proliferation.

Stanković et al. [47] used a hydrothermal approach to manufacture CQDs. Subsequently, cytotoxic studies were conducted to evaluate the impact of CQDs on mouse embryonic fibroblast cell lines, revealing no observed harmful effects. Dehvari et al. [11] examined the cytotoxicity of MNNs:CQDs on HeLa and B16F1 cell lines, and the results showed that the viability of cells remained over 90% after exposure to several doses (ranging from 0 to 1000 μg/ml) of MNNs:CQDs for 24 h. The lutein-incorporated L-CQDs produced by Yang et al. [64] exhibited extremely low cytotoxicity. Following a 24-h period of NCI-H1299 cell exposure to low concentrations of L-CQDs (below 100 μg/mL), the cell viability was seen to be about 100%. Despite the presence of a high quantity of L-CQDs (1.5 mg/mL), the cell survival rate remained over 80%. The effects of N-doped CQDs on the L-929 and MCF-7 cell lines were investigated by Arul et al. [4]. N-CQDs had less impact on the L-929 cells than they did on the MCF-7 cells. After being exposed to 50 μL/mL of N-CQDs for 24 h, the survival rate of L-929 cells was > 90%, whereas the survival rate of MCF-7 cells was < 81%.

In brief, it can be concluded that at suitable concentrations, CQDs do not exhibit cytotoxicity. However, when the concentration of CQDs beyond a particular threshold, there is a notable increase in cytotoxicity. Therefore, to guarantee the safety of its use in the food industry, the concentration of CQDs in practical applications should be strictly controlled.

Various cytotoxicity studies have confirmed that CQDs show no toxicity or low toxicity and good biocompatibility. In some instances, CQDs have the potential to safeguard cells against oxidative stress-induced damage. CQDs derived from baked lamb by Wang et al. [54] exhibited low cytotoxicity towards HepG2 cells. The viability of these cells surpassed 90% even after being exposed to 2 mg/mL CQDs for 4 h. Furthermore, the CQDs had a protective effect on HepG2 cells by mitigating oxidative damage generated by H_2O_2. Wang et al. [55] found that the cytotoxicity of CQDs to J774A.1 cells was minimal within the concentration range of 0 to 0.46 mg/ml. Furthermore, this CQDs have shown efficacy as an intracellular ROS scavenger. It has the ability to mitigate inflammation generated by lipopolysaccharide in macrophages and safeguard cells from oxidative stress-related harm.

5.2 *Green Synthesis of CQDs*

Considering the demand for industrialization, the synthesis process of CQDs should be safe, green, and environmentally friendly to ensure the safety of products. In generally, the surface of CQDs synthesized by hydrothermal method does not require additional passivation, which can maximize safety and reduce toxicity [20].

In terms of the synthesis method of CQDs, its controlled synthesis is still in the early stage of development, and is usually divided into "top-down" and "bottom-up" routes. The "top-down" route usually adopts harsh experimental conditions, cumbersome operation steps and expensive equipment, which greatly limits its practical application [18]. The "bottom-up" route include hydrothermal method, pyrolysis method, solvothermal method and microwave assisted method. Hydrothermal method is the most common method to synthesize CQDs. The raw material and distilled water (used as a common solvent and reaction medium) are added to the polytetrafluoron-lined stainless steel autoclave, and reacted at a certain temperature (80~300 °C) and high pressure for a period of time to obtain the target CQDs.

In addition, in terms of synthetic precursors, a great deal of green precursors, which have benefits including wide material resources, renewable nature, no chemical pollution and environmental friendliness can be used as natural carbon sources for the synthesis of CQDs. Compared with some chemical raw materials, the CQDs prepared with green precursors have higher safety. Various green carbon precursors have been studied and applied, including sucrose, milk, lemon juice, fruit peel, tea residue, shrimp shell, pig skin, etc. [13].

5.3 *Controlled Release of CQDs*

For food packaging, specific materials can be designed to ensure that CQDs are released at a predetermined rate, to maintain a constant concentration for a certain period of time, while minimizing side effects. However, so far, only a few studies have investigated the release of CQDs from packaging materials, and no in-depth exploration has been conducted on how to achieve controlled release of CQDs.

At present, the release characteristics of CQDs are tested by simulated solvent stripping test. For instance, the release rate of CQDs in the Gelatin/Carrageenan composite film was depended on the specific kind of food simulant and the concentration of CQDs. The release rates of CQDs in water and 10% ethanol solution were obviously higher compared to those in 50% and 95% ethanol solutions. When the concentration of CQDs grew from 1 to 5%, the release rate in all simulated environments also increased correspondingly [42]. The release of active component from the composite film to food simulant is related to many factors, including the solubility of polymer matrix, the type of food simulation solution, the concentration of incorporated component and the affinity between solvent and incorporated component, etc. Researchers can develop novel film materials that can effectively control the release of CQDs from the above perspectives.

6 Conclusion

CQDs, a novel carbon-based nanomaterial, has attracted increasing attention from researchers and emerged as a research hotspot attributed to its excellent fluorescence properties, good biocompatibility, low toxicity, etc. These advantages provide opportunities for the application of CQDs in the food industry. Currently, CQDs have been used to develop new technologies to improve shelf life or freshness of food samples, to design methods or tools for rapid food analysis, and to manufacture functional and environmentally friendly food packaging.

At present, the development and application of CQDs in the food industry are still in the preliminary research stage, and the following problems need to be solved: (1) The fluorescence mechanism of CQDs is still controversial. Because of the abundance and complexity of precursor, the synthesized CQDs exhibit different fluorescence characteristics. It is crucial to identify the fluorescence mechanism of CQDs for selecting effective synthesis routes and application scenarios. (2) In food packaging field, the preservation mechanism of CQD based active film has not been extensively studied. To give a theoretical basis for food preservation, for instance, omics research may be used to examine the physiological mechanism of the active film that delays the aging of fruits and vegetables and the spoiling of meat. In addition, the combination of active film and intelligent film shows the great potential of packaging, and the integration with information, image recognition and other disciplines requires further investigation. (3) Researches on migration and degradation of CQDs is still in its infancy. Additional studies are required to investigate the potential migratory patterns of CQDs from sensing or packaging films into various food matrices, develop methods for regulating the release of CQDs. The impact of CQDs on gastrointestinal tract, and the toxicological assessment of CQDs should be conducted utilizing in vivo and in vitro models. It is believed that in the foreseeable future, with the diligent endeavors of researchers, the aforementioned challenges may be effectively addressed, thereby facilitating the widespread use of CQDs in the food industry.

References

1. Ahmed, G. H. G., Laino, R. B., Calzon, J. A. G., & Garcia, M. E. D. (2015). Fluorescent carbon nanodots for sensitive and selective detection of tannic acid in wines. *Talanta, 132*, 252–257. https://doi.org/10.1016/j.talanta.2014.09.028
2. Akhgari, F., Samadi, N., Farhadi, K., & Akhgari, M. (2017). A green one-pot synthesis of nitrogen and sulfur co-doped carbon quantum dots for sensitive and selective detection of cephalexin. *Canadian Journal of Chemistry, 95*(6), 641–648. https://doi.org/10.1139/cjc-2016-0531
3. Amjadi, M., Abolghasemi-Fakhri, Z., & Hallaj, T. (2015). Carbon dots-silver nanoparticles fluorescence resonance energy transfer system as a novel turn-on fluorescent probe for selective determination of cysteine. *Journal of Photochemistry and Photobiology a-Chemistry, 309*, 8–14. https://doi.org/10.1016/j.jphotochem.2015.04.016
4. Arul, V., Edison, T. N. J. I., Lee, Y. R., & Sethuraman, M. G. (2017). Biological and catalytic applications of green synthesized fluorescent N-doped carbon dots using Hylocereus undatus. *Journal of Photochemistry Photobiology B: Biology, 168*, 142–148.
5. Baig, M. M. F., & Chen, Y.-C. (2017). Bright carbon dots as fluorescence sensing agents for bacteria and curcumin. *Journal of Colloid and Interface Science, 501*, 341–349. https://doi.org/10.1016/j.jcis.2017.04.045
6. Chandra, S., Chowdhuri, A. R., Mahto, T. K., Samui, A., & Sahu, S. K. (2016). One-step synthesis of amikacin modified fluorescent carbon dots for the detection of Gram-negative bacteria like *Escherichia coli. RSC Advances, 6*(76), 72471–72478. https://doi.org/10.1039/c6ra15778e
7. Chandra, S., Mahto, T. K., Chowdhuri, A. R., Das, B., & Sahu, S. K. (2017). One step synthesis of functionalized carbon dots for the ultrasensitive detection of *Escherichia coli* and iron (III). *Sensors and Actuators B-Chemical, 245*, 835–844. https://doi.org/10.1016/j.snb.2017.02.017
8. Chen, Z., Wang, J., Miao, H., Wang, L., Wu, S., & Yang, X. (2016). Fluorescent carbon dots derived from lactose for assaying folic acid. *Science China-Chemistry, 59*(4), 487–492. https://doi.org/10.1007/s11426-015-5536-1
9. Daia, H., Shi, Y., Wang, Y., Sun, Y., Hu, J., Ni, P., & Li, Z. (2014). A carbon dot based biosensor for melamine detection by fluorescence resonance energy transfer. *Sensors and Actuators B-Chemical, 202*, 201–208. https://doi.org/10.1016/j.snb.2014.05.058
10. Das, B., Dadhich, P., Pal, P., Srivas, P. K., Bankoti, K., & Dhara, S. (2014). Carbon nanodots from date molasses: New nanolights for the *in vitro* scavenging of reactive oxygen species. *Journal of Materials Chemistry B, 2*(39), 6839–6847. https://doi.org/10.1039/c4tb01020e
11. Dehvari, K., Chiu, S.-H., Lin, J.-S., Girma, W. M., Ling, Y.-C., & Chang, J.-Y. (2020). Heteroatom doped carbon dots with nanoenzyme like properties as theranostic platforms for free radical scavenging, imaging, and chemotherapy. *Acta Biomaterialia, 114*, 343–357.
12. Ezati, P., Rhim, J.-W., Molaei, R., & Rezaei, Z. (2022). Carbon quantum dots-based antifungal coating film for active packaging application of avocado. *Food Packaging and Shelf Life, 33*. https://doi.org/10.1016/j.fpsl.2022.100878
13. Fan, H., Zhang, M., Bhandari, B., & Yang, C.-H. (2020). Food waste as a carbon source in carbon quantum dots technology and their applications in food safety detection. *Trends in Food Science & Technology, 95*, 86–96. https://doi.org/10.1016/j.tifs.2019.11.008
14. Fan, K., Zhang, M., & Chen, H. (2020). Effect of ultrasound treatment combined with carbon dots coating on the microbial and physicochemical quality of fresh-cut cucumber. *Food and Bioprocess Technology, 13*(4), 648–660.
15. Fan, K., Zhang, M., Fan, D., & Jiang, F. (2019). Effect of carbon dots with chitosan coating on microorganisms and storage quality of modified-atmosphere-packaged fresh-cut cucumber. *Journal of the Science of Food Agriculture, 99*(13), 6032–6041.
16. Fu, B., Liu, Q., Liu, M., Chen, X., Lin, H., Zheng, Z., & Yang, D.-P. (2022). Carbon dots enhanced gelatin/chitosan bio-nanocomposite packaging film for perishable foods. *Chinese Chemical Letters, 33*(10), 4577–4582. https://doi.org/10.1016/j.cclet.2022.03.048

17. Fu, X., Sheng, L., Yu, Y., Ma, M., Cai, Z., & Huang, X. (2018). Rapid and universal detection of ovalbumin based on N, O, P-co-doped carbon dots-fluorescence resonance energy transfer technology. *Sensors and Actuators B: Chemical, 269*, 278–287.
18. Gao, J., Zhu, M., Huang, H., Liu, Y., & Kang, Z. (2017). Advances, challenges and promises of carbon dots. *Inorganic Chemistry Frontiers, 4*(12), 1963–1986. https://doi.org/10.1039/c7qi00614d
19. Gong, N. C., Li, Y. L., Jiang, X., Zheng, X. F., Wang, Y. Y., & Huan, S. Y. (2016). Fluorescence resonance energy transfer-based biosensor composed of nitrogen-doped carbon dots and gold nanoparticles for the highly sensitive detection of organophosphorus pesticides. *Analytical Sciences, 32*(9), 951–956. https://doi.org/10.2116/analsci.32.951
20. He, J., Chen, J., & Qiu, H. (2023). Synthesis of traditional Chinese medicines-derived carbon dots for bioimaging and therapeutics. *Progress in Chemistry, 35*(5), 655–682.
21. Hess, S. C., Permatasari, F. A., Fukazawa, H., Schneider, E. M., Balgis, R., Ogi, T., & Stark, W. (2017). Direct synthesis of carbon quantum dots in aqueous polymer solution: One-pot reaction and preparation of transparent UV-blocking films. *Journal of Materials Chemistry A, 5*(10), 5187–5194.
22. Hu, M., Gu, X., Hu, Y., Deng, Y., & Wang, C. (2016). PVA/carbon dot nanocomposite hydrogels for simple introduction of Ag nanoparticles with enhanced antibacterial activity. *Macromolecular Materials and Engineering, 301*(11), 1352–1362. https://doi.org/10.1002/mame.201600248
23. Hu, X., Zhang, X., Li, Y., Shi, J., Huang, X., Li, Z., Zou, X. (2022). Easy-to-use visual sensing system for milk freshness, sensitized with acidity-responsive N-doped carbon quantum dots. *Foods, 11*(13). https://doi.org/10.3390/foods11131855
24. Hua, J., Jiao, Y., Wang, M., & Yang, Y. (2018). Determination of norfloxacin or ciprofloxacin by carbon dots fluorescence enhancement using magnetic nanoparticles as adsorbent. *Microchimica Acta, 185*(2). https://doi.org/10.1007/s00604-018-2685-x
25. Huang, G., Chen, X., Wang, C., Zheng, H., Huang, Z., Chen, D., & Xie, H. (2017). Photoluminescent carbon dots derived from sugarcane molasses: Synthesis, properties, and applications. *RSC Advances, 7*(75), 47840–47847.
26. Kilic, B., Dogan, V., Kilic, V., & Kahyaoglu, L. N. (2022). Colorimetric food spoilage monitoring with carbon dot and UV light reinforced fish gelatin films using a smartphone application. *International Journal of Biological Macromolecules, 209*, 1562–1572. https://doi.org/10.1016/j.ijbiomac.2022.04.119
27. Konwar, A., Gogoi, N., Majumdar, G., & Chowdhury, D. (2015). Green chitosan–carbon dots nanocomposite hydrogel film with superior properties. *Carbohydrate Polymers, 115*, 238–245. https://doi.org/10.1016/j.carbpol.2014.08.021
28. Koshy, R. R., Koshy, J. T., Mary, S. K., Sadanandan, S., Jisha, S., & Pothan, L. A. (2021). Preparation of pH sensitive film based on starch/carbon nano dots incorporating anthocyanin for monitoring spoilage of pork. *Food Control, 126*. https://doi.org/10.1016/j.foodcont.2021.108039
29. Kundu, A., Nandi, S., Das, P., & Nandi, A. K. (2016). Facile and green approach to prepare fluorescent carbon dots: Emergent nanomaterial for cell imaging and detection of vitamin B_2 *Journal of Colloid and Interface Science, 468*, 276–283. https://doi.org/10.1016/j.jcis.2016.01.070
30. Lai, I.P.-J., Harroun, S. G., Chen, S.-Y., Unnikrishnan, B., Li, Y.-J., & Huang, C.-C. (2016). Solid-state synthesis of self-functional carbon quantum dots for detection of bacteria and tumor cells. *Sensors and Actuators B-Chemical, 228*, 465–470. https://doi.org/10.1016/j.snb.2016.01.062
31. Li, H., Sun, C., Vijayaraghavan, R., Zhou, F., Zhang, X., & MacFarlane, D. R. (2016). Long lifetime photoluminescence in N, S co-doped carbon quantum dots from an ionic liquid and their applications in ultrasensitive detection of pesticides. *Carbon, 104*, 33–39. https://doi.org/10.1016/j.carbon.2016.03.040
32. Li, Y., Chen, H., Dong, Y., Li, K., Li, L., & Li, J. (2016). Carbon nanoparticles/soy protein isolate bio-films with excellent mechanical and water barrier properties. *Industrial Crops and Products, 82*, 133–140.

33. Liu, J., Chen, Y., Wang, W., Feng, J., Liang, M., Ma, S., & Chen, X. (2016). Switch-On" fluorescent sensing of ascorbic acid in food samples based on carbon quantum dots-MnO_2 probe. *Journal of Agricultural and Food Chemistry, 64*(1), 371–380. https://doi.org/10.1021/acs.jafc.5b05726
34. Liu, L., Feng, F., Paau, M. C., Hu, Q., Liu, Y., Chen, Z., & Choi, M. M. F. (2015). Sensitive determination of kaempferol using carbon dots as a fluorescence probe. *Talanta, 144*, 390–397. https://doi.org/10.1016/j.talanta.2015.07.004
35. Omerović, N., Djisalov, M., Živojević, K., Mladenović, M., Vunduk, J., Milenković, I., & Vidić, J. (2021). Antimicrobial nanoparticles and biodegradable polymer composites for active food packaging applications. *Comprehensive Reviews in Food Science and Food Safety, 20*(3), 2428–2454.
36. Patil, A. S., Waghmare, R. D., Pawar, S. P., Salunkhe, S. T., Kolekar, G. B., Sohn, D., & Gore, A. H. (2020). Photophysical insights of highly transparent, flexible and re-emissive PVA @ WTR-CDs composite thin films: A next generation food packaging material for UV blocking applications. *Journal of Photochemistry and Photobiology a-Chemistry, 400*. https://doi.org/10.1016/j.jphotochem.2020.112647
37. Purbia, R., & Paria, S. (2016). A simple turn on fluorescent sensor for the selective detection of thiamine using coconut water derived luminescent carbon dots. *Biosensors and Bioelectronics, 79*, 467–475.
38. Rahman, S., & Chowdhury, D. (2022). Guar gum-sodium alginate nanocomposite film as a smart fluorescence-based humidity sensor: A smart packaging material. *International Journal of Biological Macromolecules, 216*, 571–582. https://doi.org/10.1016/j.ijbiomac.2022.07.008
39. Rhim, J.-W., Park, H.-M., & Ha, C.-S. (2013). Bio-nanocomposites for food packaging applications. *Progress in Polymer Science, 38*(10–11), 1629–1652.
40. Riahi, Z., Rhim, J.-W., Bagheri, R., Pircheraghi, G., & Lotfali, E. (2022). Carboxymethyl cellulose-based functional film integrated with chitosan-based carbon quantum dots for active food packaging applications. *Progress in Organic Coatings, 166*. https://doi.org/10.1016/j.porgcoat.2022.106794
41. Rossi, M., Passeri, D., Sinibaldi, A., Angjellari, M., Tamburri, E., Sorbo, A., & Dini, L. (2017). Nanotechnology for food packaging and food quality assessment. *Advances in Food and Nutrition Research, 82*, 149–204. https://doi.org/10.1016/bs.afnr.2017.01.002
42. Roy, S., Ezati, P., & Rhim, J. W. (2021). Gelatin/carrageenan-based functional films with carbon dots from Enoki mushroom for active food packaging applications. *ACS Applied Polymer Materials, 3*(12), 6437–6445. https://doi.org/10.1021/acsapm.1c01175
43. Salimi, F., Moradi, M., Tajik, H., & Molaei, R. (2021). Optimization and characterization of eco-friendly antimicrobial nanocellulose sheet prepared using carbon dots of white mulberry (Morus alba L.). *Journal of the Science of Food and Agriculture, 101*(8), 3439–3447. https://doi.org/10.1002/jsfa.10974
44. Shan, X., Chai, L., Ma, J., Qian, Z., Chen, J., & Feng, H. (2014). B-doped carbon quantum dots as a sensitive fluorescence probe for hydrogen peroxide and glucose detection. *The Analyst, 139*(10), 2322–2325. https://doi.org/10.1039/c3an02222f
45. Shi, X., Wei, W., Fu, Z., Gao, W., Zhang, C., Zhao, Q., & Lu, X. (2019). Review on carbon dots in food safety applications. *Talanta, 194*, 809–821. https://doi.org/10.1016/j.talanta.2018.11.005
46. Son, M. H., Park, S. W., & Jung, Y. K. (2021). Antioxidant and anti-aging carbon quantum dots using tannic acid. *Nanotechnology, 32*(41). https://doi.org/10.1088/1361-6528/ac027b
47. Stanković, N. K., Bodik, M., Šiffalovič, P., Kotlar, M., & Mičušik, M. (2018). Antibacterial and antibiofouling properties of light triggered fluorescent hydrophobic carbon quantum dots Langmuir–Blodgett thin films. *ACS Sustainable Chemistry Engineering, 6*(3).
48. Su, A., Wang, D., Shu, X., Zhong, Q., Chen, Y., Liu, J., & Wang, Y. (2018). Synthesis of fluorescent carbon quantum dots from dried lemon peel for determination of carmine in drinks. *Chemical Research in Chinese Universities, 34*(2), 164–168. https://doi.org/10.1007/s40242-018-7286-z

49. Sun, Q., Long, Y., Li, H., Pan, S., Yang, J., Liu, S., & Hu, X. (2018). Fluorescent carbon dots as cost-effective and facile probes for caffeic acid sensing via a fluorescence quenching process. *Journal of Fluorescence, 28*(2), 523–531. https://doi.org/10.1007/s10895-018-2213-8
50. Tao, H., Liao, X., Sun, C., Xie, X., Zhong, F., Yi, Z., & Huang, Y. (2015). A carbon dots-CdTe quantum dots fluorescence resonance energy transfer system for the analysis of ultra-trace chlortoluron in water. *Spectrochimica Acta Part a-Molecular and Biomolecular Spectroscopy, 136*, 1328–1334. https://doi.org/10.1016/j.saa.2014.10.020
51. Uthirakumar, P., Devendiran, M., Kim, T. H., & Lee, I.-H. (2018). A convenient method for isolating carbon quantum dots in high yield as an alternative to the dialysis process and the fabrication of a full-band UV blocking polymer film. *New Journal of Chemistry, 42*(22), 18312–18317. https://doi.org/10.1039/c8nj04615h
52. Wang, B., Chen, Y., Wu, Y., Weng, B., Liu, Y., Lu, Z., & Yu, C. (2016). Aptamer induced assembly of fluorescent nitrogen-doped carbon dots on gold nanoparticles for sensitive detection of AFB_1. *Biosensors & Bioelectronics, 78*, 23–30. https://doi.org/10.1016/j.bios.2015.11.015
53. Wang, H., Xie, Y., Liu, S., Cong, S., Song, Y., Xu, X., & Tan, M. (2017). Presence of fluorescent carbon nanoparticles in baked lamb: their properties and potential application for sensors. *Journal of Agricultural and Food Chemistry, 65*(34), 7553–7559. https://doi.org/10.1021/acs.jafc.7b02913
54. Wang, H., Xie, Y., Na, X., Bi, J., Liu, S., Zhang, L., & Tan, M. (2019). Fluorescent carbon dots in baked lamb: Formation, cytotoxicity and scavenging capability to free radicals. *Food Chemistry, 286*, 405–412.
55. Wang, H. B., Zhang, M. L., Ma, Y. R., Wang, B., Huang, H., Liu, Y., & Kang, Z. H. (2020). Carbon dots derived from citric acid and glutathione as a highly efficient intracellular reactive oxygen species scavenger for alleviating the lipopolysaccharide-induced inflammation in macrophages. *ACS Applied Materials & Interfaces, 12*(37), 41088–41095. https://doi.org/10.1021/acsami.0c11735
56. Wang, J., Wei, J., Su, S., & Qiu, J. (2015). Novel fluorescence resonance energy transfer optical sensors for vitamin B_{12} detection using thermally reduced carbon dots. *New Journal of Chemistry, 39*(1), 501–507. https://doi.org/10.1039/c4nj00538d
57. Wang, L., Bi, Y., Hou, J., Li, H., Xu, Y., Wang, B., & Ding, L. (2016). Facile, green and clean one-step synthesis of carbon dots from wool: Application as a sensor for glyphosate detection based on the inner filter effect. *Talanta, 160*, 268–275. https://doi.org/10.1016/j.talanta.2016.07.020
58. Wang, Q., Liu, X., Zhang, L., & Lv, Y. (2012). Microwave-assisted synthesis of carbon nanodots through an eggshell membrane and their fluorescent application. *The Analyst, 137*(22), 5392–5397. https://doi.org/10.1039/c2an36059d
59. Wang, R., Xu, Y., Zhang, T., & Jiang, Y. (2015). Rapid and sensitive detection of *Salmonella typhimurium* using aptamer-conjugated carbon dots as fluorescence probe. *Analytical Methods, 7*(5), 1701–1706. https://doi.org/10.1039/c4ay02880e
60. Xiao, D., Pu, H., Tian, H., Yang, D., Yang, Y., & Li, H. (2019). Application of carbon dots-chitosan coating in preservation of mango. *Food and Fermentation Industries, 45*(22), 130–135.
61. Xu, H., Yang, X., Li, G., Zhao, C., & Liao, X. (2015). Green synthesis of fluorescent carbon dots for selective detection of tartrazine in food samples. *Journal of Agricultural and Food Chemistry, 63*(30), 6707–6714.
62. Xu, L., Zhang, Y., Pan, H., Xu, N., Mei, C., Mao, H., & Xu, C. (2019). Preparation and performance of radiata-pine-derived polyvinyl alcohol/carbon quantum dots fluorescent films. *Materials, 13*(1), 67.
63. Xu, Y., Yang, D., Huo, S., Ren, J., Gao, N., Chen, Z., & Qu, X. (2021). Carbon dots and ruthenium doped oxygen sensitive nanofibrous membranes for monitoring the respiration of agricultural products. *Polymer Testing, 93*, 106957. https://doi.org/10.1016/j.polymertesting.2020.106957
64. Yang, D., Li, L., Cao, L., Chang, Z., Mei, Q., Yan, R., & Dong, W.-F. (2020). Green synthesis of lutein-based carbon dots applied for free-radical scavenging within cells. *Materials, 13*(18), 4146.

65. Yang, H., Yang, L., Yuan, Y., Pan, S., Yang, J., Yan, J., & Hu, X. (2018). A portable synthesis of water-soluble carbon dots for highly sensitive and selective detection of chlorogenic acid based on inner filter effect. *Spectrochimica Acta Part A: Molecular and Biomolecular Spectroscopy, 189*, 139–146.
66. Yang, J., He, X., Chen, L., & Zhang, Y. (2016). The selective detection of galactose based on boronic acid functionalized fluorescent carbon dots. *Analytical Methods, 8*(47), 8345–8351. https://doi.org/10.1039/c6ay02530g
67. Yang, X., Luo, Y., Zhu, S., Feng, Y., Zhuo, Y., & Dou, Y. (2014). One-pot synthesis of high fluorescent carbon nanoparticles and their applications as probes for detection of tetracyclines. *Biosensors & Bioelectronics, 56*, 6–11. https://doi.org/10.1016/j.bios.2013.12.064
68. Ye, Q., Yan, F., Kong, D., Zhang, J., Zhou, X., Xu, J., & Chen, L. (2017). Constructing a fluorescent probe for specific detection of catechol based on 4-carboxyphenylboronic acid-functionalized carbon dots. *Sensors and Actuators B-Chemical, 250*, 712–720. https://doi.org/10.1016/j.snb.2017.03.081
69. You, Y., Zhang, H., Liu, Y., & Lei, B. (2016). Transparent sunlight conversion film based on carboxymethyl cellulose and carbon dots. *Carbohydrate Polymers, 151*, 245–250. https://doi.org/10.1016/j.carbpol.2016.05.063
70. Yuan, Y., Zhao, X., Qiao, M., Zhu, J., Liu, S., Yang, J., & Hu, X. (2016). Determination of sunset yellow in soft drinks based on fluorescence quenching of carbon dots. *Spectrochimica Acta Part a-Molecular and Biomolecular Spectroscopy, 167*, 106–110. https://doi.org/10.1016/j.saa.2016.05.038
71. Zhang, H., Kang, S., Wang, G., Zhang, Y., & Zhaou, H. (2016). Fluorescence determination of nitrite in water using prawn-shell derived nitrogen-doped carbon nanodots as fluorophores. *ACS Sensors, 1*(7), 875–881. https://doi.org/10.1021/acssensors.6b00269
72. Zhang, J., Xu, W.-R., Zhang, Y.-C., Han, X.-D., Chen, C., & Chen, A. (2020). In situ generated silica reinforced polyvinyl alcohol/liquefied chitin biodegradable films for food packaging. *Carbohydrate Polymers, 238*, 116182.
73. Zhang, X., Wang, H., Ma, C., Niu, N., Chen, Z., Liu, S., & Li, S. (2018). Seeking value from biomass materials: Preparation of coffee bean shell-derived fluorescent carbon dots via molecular aggregation for antioxidation and bioimaging applications. *Materials Chemistry Frontiers, 2*(7), 1269–1275. https://doi.org/10.1039/c8qm00030a
74. Zhang, Z., Chen, J., Duan, Y., Liu, W., Li, D., Yan, Z., & Yang, K. (2018). Highly luminescent nitrogen-doped carbon dots for simultaneous determination of chlortetracycline and sulfasalazine. *Luminescence, 33*(2), 318–325. https://doi.org/10.1002/bio.3416
75. Zhao, L., Wang, Y., & Li, Y. (2019). Antioxidant activity of graphene quantum dots prepared in different electrolyte environments. *Nanomaterials, 9*(12). https://doi.org/10.3390/nano9121708
76. Zhao, L., Zhang, M., Bhandari, B., & Bai, B. (2020). Microbial and quality improvement of boiled gansi dish using carbon dots combined with radio frequency treatment. *International Journal of Food Microbiology, 334*. https://doi.org/10.1016/j.ijfoodmicro.2020.108835
77. Zhao, L., Zhang, M., Mujumdar, A. S., Adhikari, B., & Wang, H. (2022). Preparation of a novel carbon dot/polyvinyl alcohol composite film and its application in food preservation. *ACS Applied Materials and Interfaces, 14*(33), 37528–37539.
78. Zhao, L., Zhang, M., Wang, H., & Devahastin, S. (2020). Effect of carbon dots in combination with aqueous chitosan solution on shelf life and stability of soy milk. *International Journal of Food Microbiology, 326*, 108650. https://doi.org/10.1016/j.ijfoodmicro.2020.108650
79. Zhao, L., Zhang, M., Wang, H., & Devahastin, S. (2022). Effect of addition of carbon dots to the frying oils on oxidative stabilities and quality changes of fried meatballs during refrigerated storage. *Meat Science, 185*, 108715.
80. Zhao, L., Zhang, M., Wang, H., & Devahastin, S. J. F. C. (2021). Effects of carbon dots in combination with rosemary-inspired carnosic acid on oxidative stability of deep frying oils. *125*, 107968.
81. Zhou, L., Lin, Y., Huang, Z., Ren, J., & Qu, X. (2012). Carbon nanodots as fluorescence probes for rapid, sensitive, and label-free detection of Hg^{2+} and biothiols in complex matrices. *Chemical Communications, 48*(8), 1147–1149. https://doi.org/10.1039/c2cc16791c

82. Zu, F., Yan, F., Bai, Z., Xu, J., Wang, Y., Huang, Y., & Zhou, X. (2017). The quenching of the fluorescence of carbon dots: A review on mechanisms and applications. *Microchimica Acta, 184*(7), 1899–1914. https://doi.org/10.1007/s00604-017-2318-9

Green Carbon Quantum Dots for Efficient Sensing of Heavy Metal Ions

Pradeep Kumar Yadav and Vellaichamy Ganesan

Abstract Owing to their unique fluorescence properties, carbon quantum dots (CQDs), the new zero-dimensional carbon nanomaterials have intrigued many research interests. CQDs have attracted massive attention worldwide because of their stunning properties like high solubility, high fluorescence intensity, biocompatibility, low toxicity, and easy surface functionalization. As a result, CQDs can cover various possible applications in many fields including catalysis, medical, light-emitting diodes, energy-related fields, and sensing. The selective and sensitive detection of heavy metal ions using CQDs involves exploiting the unique properties of these nanomaterials, particularly their fluorescence behavior. The use of green carbon quantum dots (CQDs) for detecting heavy metal ions is an area of growing interest due to the eco-friendly nature of the synthesis process and the unique properties of CQDs. Green synthesis typically involves using natural extracts, such as plant extracts or waste materials to produce CQDs. In this chapter, we will summarize and discuss the synthesis methods, surface passivation and functionalization, stability, mechanism of fluorescence sensing, and applications of CQDs for selective and sensitive detection of heavy metal ions.

Keywords Carbon quantum dots · Fluorescence · Green synthesis · Sensing · Heavy metal ions

1 Introduction

Heavy metal ions are metallic elements with high atomic weight and density. Heavy metals include lead (Pb), cadmium (Cd), mercury (Hg), iron (Fe), arsenic (As), chromium (Cr), etc. They can be toxic to living organisms in towering concentrations and result in environmental and health risks. These heavy metals can enter the environment through various anthropogenic activities such as industrial discharges,

P. K. Yadav · V. Ganesan (✉)
Department of Chemistry, Institute of Science, Banaras Hindu University, Varanasi, UP 221005, India
e-mail: velgan@bhu.ac.in

V. Kumar et al. (eds.), *Green Carbon Quantum Dots*, Engineering Materials,
https://doi.org/10.1007/978-981-97-6203-3_11

mining, and improper waste disposal. They can accumulate in soil, water, and air, leading to contamination and posing risks to ecosystems and human health. These heavy metals are usually necessary in small amounts for typical growth, development, and physiological functions; but their presence in excessive concentrationscan be toxic to living organisms [1, 2]. Though, heavy metals are identified for their affinity to make complexes with other molecules, ions, or particularly with ligands, these messages may provideinformation about the carcinogenic and toxicological effects of heavy metals similar to those influencing the central nervous system (Hg^{2+}, Pb^{2+}, As^{3+}); liver or kidneys (Cd^{2+}, Cu^{2+}, Pb^{2+}, Hg^{2+}) or teeth, skin or bones, (Ni^{2+}, Cu^{2+}, Cd^{2+}, Cr^{3+}) [3–6].

Poisoning of lead can cause neurological and behavioral exertion, particularly in children who have developmental delays, learning disabilities, and impaired cognitive function. On the other hand, kidney damage, high blood pressure, and reproductive issues are the major problemsassociated with adults [7–11]. Mercury affects the nervous system and causes symptoms such as memory problems, mood disorders, and tremors. Minamata disease is a neurological disarray caused by mercury poisoning, particularly methylmercury exposure [12–18]. Cadmium exposure can damage the bones, lungs, and kidneys. Lung cancer and prostate cancer also occur by long-term exposure to Cd. Cadmium poisoning causes Itai-Itai disease [19–21]. Arsenic exposure causes lung, skin, and bladder canceras well as neurological problems, skin lesions, and cardiovascular diseases [22–25]. Cr(VI) is carcinogenic in nature and when inhaled, causes lung cancer. The respiratory system and skin can also be harmed by Chromium. The toxicity of Cr(III) is less and at low levels, itis used as an essential nutrient [26, 27]. Nickel exposure causes respiratory issues and skin allergies. Lung and nasal cancer can be caused by prolonged exposure [28, 29].

Heavy metals can persevere in ecological systems for extensive periods, and their company can have enduring consequences on ecosystems and organisms enlightening very high levels of pollution. For the reason that many organizations have recognized safe limits or maximum contaminant levels for a variety of substances in drinking water. World Health Organization (WHO) and the U.S. Environmental Protection Agency (EPA) have suggested guidelines and standards for certain heavy metals in consumption water (Table 1) [30–34].

Table 1 Strategy and standard for heavy metals in consumption water suggested by the WHO and EPA

Metals	WHO (mg/L)	EPA (mg/L)
Cd	0.003	0.005
Hg	0.001	0.002
Pb	0.010	0.015
Zn	3	5
Cu	2	1.3
As	0.01	0.010
Ni	0.07	0.04

Thus, monitoring and detecting the presence of heavy metals in environmental samples is crucial for the protection of the environment as well as public health. Technologies like sensors, including those based on CQDs, are being developed for efficient and rapid detection of heavy metal ions in diverse settings.

CQDs were first reported by Xu's Group in 2004, accidentally all over the purification of single-walled carbon nanotubes [35]. CQDs are a group of newly emerged charming fluorescent nanomaterials which are typically quasi-spherical and core contain mainly sp^2 and sp^3 hybridized carbon atoms, whereas the external surface consists of a variety of functional groups, mostly hydroxyl, carboxylic, carbonyl, amine, etc. The functional groups are accountable for outstanding water solubility, excellent quantum yield, chemical stability, surface passivation (doping), and functionalization with different organic, inorganic, polymeric, and biomolecules [12, 17, 33]. CQDs have high water solubility, low cyto-toxicity, ease of synthesis, high photo-response, high photo-stability, facile surface functionalization, efficient catalysis properties, and tunable excitation-emission. That's why CQDs are extensively used in medical diagnosis, sensing, photovoltaic devices, catalysis, photocatalysis, drug delivery, optronic devices, bio-imaging, laser, solar cells, and LEDs [36, 37].

2 Sensing Activity of CQDs for Detecting Heavy Metal Ions

To sense heavy metals, several techniques such as inductively coupled plasma mass spectrometry (ICP-MS), atomic absorption spectroscopy (AAS), electrochemical sensors, colorimetric assays, surface plasmon resonance (SPR), flame atomic absorption spectroscopy (FAAS), colorimetric assays, surface plasmon resonance (SPR) and nanoparticles such as silver nanoparticles (AgNPs),gold nanoparticles (AuNPs) and CQDs have been utilized [38–44]. Among them, CQDs are efficiently used for sensing heavy metal ions due to their unique optical and surface properties. The sensing activity involves the interaction between the CQDs and specific metal ions, resulting in detectable changes in the fluorescence properties of the CQDs. The sensing of heavy metals by CQDs involves leveraging the unique properties of these nanomaterials to detect heavy metals selectively and sensitively [45–47]. The sensing process involves following steps.

2.1 Synthesis of Carbon Quantum Dots (CQDs)

CQDs are synthesized from natural as well as chemical precursors. The chemical precursor comprises tartaric acid, ascorbic acid, citric acid, glucose, sucrose, glycerol, glycol, etc. The natural precursor comprises seeds, plant leaves, plant latex, orange juice, jatropha fruits, pomelo peel, cocoon silk, potato, soy milk, milk, soybean, banana, etc. [48–51]. There are two methods utilized for the preparation of CQDs, top-down and bottom-up synthesis.

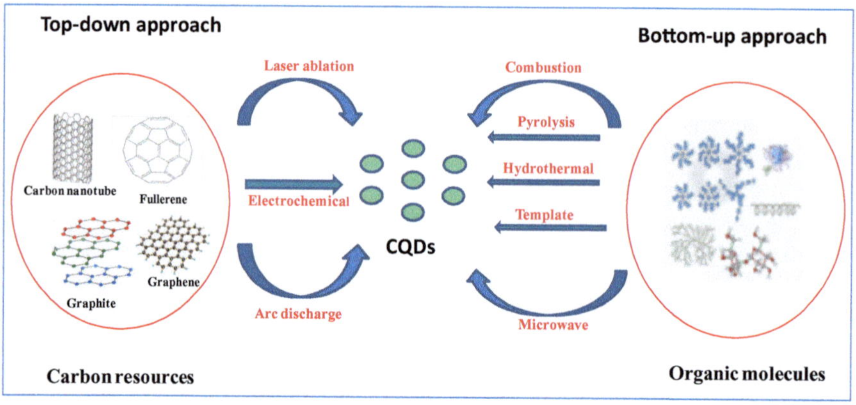

Scheme 1 Diagrammatic illustration for the synthesis of CQDs

In the top-down synthesis, the larger carbon sources like carbon soot, fullerene, activated carbon, graphene, carbon nanotubes, graphite, etc. are crashed into smaller ingredients by facilitating various techniques like arch discharge, laser ablation, and electrochemical methods [36, 45].

In bottom-up synthesis, the minor carbon sources like atoms and molecules unite to outline CQDs by a range of methods like combustion, pyrolysis, microwave irradiation, and hydrothermal. Among them, hydrothermal synthesis has concerned much interest worldwide due to low cost, ease of operation, one-step, non-toxic, and eco-friendly. CQDs synthesized by hydrothermal methods have various advantageous properties such as water-soluble, mono-dispersed, homogeneous, salt tolerance, controllable particle size, photo stable, and show high quantum yield (QY) without surface passivation [17, 33] (Scheme 1).

2.2 *Surface Passivation and Functionalization*

Surface functionalization refers to the modification of surface of a material or substrate to impart specific properties or functionalities. The emission efficiency of CQDs, synthesized by various methods, is relatively low compared to conventional semiconductor QDs. Therefore, to improve the emission efficiency of CQDs, surface passivation (doping) and functionalization are carried out. The doping component could be either metals or non-metals. Mostly, non-metals (heteroatom) are used as doping components because metals may cause toxicity in CQDs and thus may not be suitable for biological applications. The heteroatom doping (single or co-doped) in CQDs can enhance the optical property and their interaction with heavy metal ions and improve selectivity [36, 45].

2.2.1 Single Heteroatom Doping

Nitrogen (N)-doped CQDs

Nitrogen-doping in CQDs not only enhances the fluorescence efficiency, but also increases the applications of CQDs in bioimaging and sensing. Arul et al. have synthesized nitrogen-doped CQDs through the hydrothermal method using the aqueous extract of *Phyllanthus emblica* fruit as a carbon source and aq NH_3 as a nitrogen source [52]. Nitrogen-doped CQDs were also prepared by several groups and they stated that N doped CQDs have more QY than un-doped CQDs [53–56].

Sulfur (S)-Doped CQDs

The doping of sulfur also increases the fluorescence properties of CQDs. S-doping enlarges the catalytic activity as well as sensing applications of CQDs. The S-containing precursors such as thiourea, hydrogen sulfide, sulfuric acid, etc., are used in the doping of Sulfur [57]. Chandra et al. have reported that the S-doped CQDs show fluorescence QY approximately 11.8% higher compared to the un-doped CQDs [58]. Hence, to increase the electronic properties and other large-scale applications of CQDs, doping is an admirable and straightforward approach.

Phosphorus (P)-Doped CQDs

Phosphorus-doped CQDs have been effectively synthesized by numerous research groups. Han et al. have reported that the doping of phosphorus changes the electronics and fluorescence properties of CQDs [59]. Feng and coworkers have prepared P-doped CQDs using hydroquinone as the C source and phosphorous bromide as the P source [60].

Boron (B)-Doped CQDs

The catalytic activities and the electronic properties of CQDs have been increased by doping of boron element [61]. Feng and co-workers have synthesized B-doped CQDs from hydroquinone as the C source and BBr_3 as the B source. These CQDs yielded a QY of 14.8% and displayed an intense blue fluorescence [62]. In addition, using microwave methods, Bourlinos et al. have prepared B-doped CQDs with QYs of 10–15% [63].

2.2.2 Multiple Heteroatom Doping

N and P Co-Doped CQDs

Liao et al. have reported N, and P co-doped CQDs using hydrothermal method for the sensing of Co^{2+} a limit of detection (LOD) of 0.053 μM [64]. Gong et al. synthesized fluorescent N, and P co-doped CQDs, which have higher QYs than un-doped CQDs [65]. Heteroatom-doped CQDs display significant QY compared to un-doped CQDs mainly because of the combined effect [66].

N and S-doped CQDs

Dong et al. prepared fluorescent CQDs by doping N and S from L-cytokine and citric acid by hydrothermal method and obtained a very high QY (73%) [67]. N and S passivated CQDs were further reported by Zhao and colleagues by using ammonium thiocyanate as the carbon, sulfur and nitrogen source with a high QY of 74.15%. These results revealed that N, and S-doped CQDs not only improve the QYs but also broaden their applications like the fields of nanomedicine, catalysis, bioimaging, and biosensing [68, 69].

N and B Doped CQDs

The doping of B and N atoms from chemical and natural sources has significantly influenced the optical properties of CQDs and enhanced their fields of applications [70]. Kanwal and co-workers synthesized fluorescent CQDs through the hydrothermal method by doping N-(4-hydroxyphenyl) glycine as the N source and boric acid as the B source with a QY of 11.44% [71].

2.3 *Determination of Stability of CQDs*

The stability of CQDs refers to their ability to maintain their structural and optical properties over time. The stability of CQDs is crucial for their practical applications in various fields, including optoelectronics, photovoltaics, bioimaging, and sensing [72]. Several factors can affect the stability of CQDs, and understanding and controlling these factors are essential for ensuring the long-term performance of CQD-based devices. To determine the stability of CQDs, researchers often conduct long-term stability studies under relevant conditions. This involves monitoring the changes in size, shape, optical properties, and dispersibility over time. Various analytical techniques, including spectroscopy, microscopy, and dynamic light scattering, can be employed to assess the stability of CQDs. Additionally, stability can be evaluated in the context of specific applications, such as in solar cells or light-emitting devices, to ensure that the performance is maintained over the desired operational lifetime. The stability of CQDs can also be performed in highly ionic strength condition [73]. Bano et al. have synthesized fluorescent CQDs and checked the stability in different pH and highly concentrated NaCl solutions. They reported that CQDs were stable in a wide pH range and there was nearly no outcome on the fluorescence emission intensity on changing NaCl salt concentration from 20 to 100 mM. The photostability of CQDs was also checked after exposure to UV light on CQDs for 24 h, demonstrating that CQDs are highly photostable [74].

2.4 Selection of Heavy Metal Ion Target

When selecting a heavy metal ion target for sensing applications, various factors must be considered to ensure the effectiveness and relevance of the sensing system. Common targets include Pb, Hg, Cd, As, Fe, and Cr as these metals are often associated with environmental pollution and can pose serious health risks.

2.4.1 Sensing of Fe(III)

Fe is regarded as the largely significant transition metal and is the fourth highest abundant element in the earth's crust. It is a crucial component of hemoglobin which assists blood circulation in the body of the individual. The scarcity of Fe results in a lot of health troubles for individuals such as frequent fatigue, loss of appetite, anemia, decreased immunity, and reduced work routine while the excess concentration of iron in the blood causes various biological disorders like tissue damage of organs, and deterioration of lipids, proteins, and nucleic acids. Hence, it is necessary to determine the concentrations of iron in biological fluids. Ferric ion (Fe^{3+}) is roughly a ubiquitous metal ion in biological and environmental systems. To detect Fe^{3+}, Zhu et al. synthesized a CQDs-based nanoprobe with LOD of 0.55 ppm [75]. In addition to this, N-doped CQDs were prepared by Chen and co-workers for the detection of Fe^{3+} with an LOD of 2.5 nM. Subsequently, highly fluorescent CQDs were synthesized by doping of N and P to detect Fe^{3+} in cancer cells [76]. Fluorescent green-blue CQDs were synthesized from *Artocarpus lakoocha* seeds using one-pot hydrothermal method to detect Fe^{3+}. The calculated LOD was 0.62 μM. Further, the Fe^{3+} was detected in real samples such as human blood serum and river water [33]. The N-doped fluorescent CQDs were also prepared from rose-heart radish via a one-step hydrothermal method for 3 h at 180 °C. Synthesized N-doped CQDs exhibited high fluorescent QYs of 13.6%. In addition, Liu and co-workers investigated that the N-doped CQDs powerfully sensed Fe^{3+}resulting in the quenching of fluorescence [77].

Khan and co-workers have prepared N-doped fluorescent CDs from red lentils using the hydrothermal method. The QY obtained was 13.2%. The N-doped CQDs showed selective and sensitive detection of Fe^{3+} ions. The quenching mechanism reveals that the electrons are transferred from excited states of CQDs to half-filled 3d orbitals of Fe^{3+} ions, resulting in fluorescence quenching [78].

2.4.2 Sensing of Hg(II)

Hg is renowned as a hazardous metal and is one of the most toxic pollutants. Hg badly influences the whole ecosystem, particularly the human body and other flora and fauna. The major route of Hg poisoning in human beings is the utilization of Hg-contaminated food especially fish. In Japan (1956), mercury-contaminated fish caused worldwide known Minamata disease. Even low concentrations of Hg cause liver damage expiratory dyspnoea, headaches, gastrointestinal problems, neurological disorders, microtubule destruction, hyperspasmia, muscle destruction, kidney failure, renal failure, and so on [79–81]. Therefore, selective and sensitive detection of Hg is necessary. Singh et al. have prepared N-doped fluorescent CQDs for the selective and sensitive detection of Hg^{2+} by one-pot hydrothermal method. The LOD was found to be 0.085 μM. The as-synthesized N-CQDs exhibited excitation-dependent emission spectra with fluorescent QY of 41%. Further, N-CQDs + Hg^{2+} solution was used for the turn-on sensing of glutathione. This on–off-on sensing system was used for the implication of the logic gate. In addition to this, N-CQDs were also utilized in cell imaging [17]. Xie et al. have prepared N-CQDs using ethylenediamine as the N source and barley as the C source for the sensing of Hg^{2+}. The fluorescence quenching mechanisms were elucidated static quenching mechanism because of the electron transfer process [82].

2.4.3 Sensing of Cr(VI)

Chromium (VI) is extremely toxic and carcinogenic to a human being even in very small concentrations [83]. Presently, CQDs have been widely used for the sensing of Cr(VI). Zheng et al. have prepared CQDs by doping-nitrogen for the selective and sensitive detection of Cr(VI) which involves the inner filter effect (IFE) mechanism for fluorescence quenching [84]. Omer group has also synthesized nitrogen and phosphorous co-doped CQDs that selectivity detects Cr(II) ions through a quenching mechanism. The LOD was calculated to be 0.1 μM [85]. Our research group has also synthesized fluorescent CQDs by doping N and P via a one-step hydrothermal method for the detection of Cr(VI) selectively and sensitively. The as-prepared N, P-CQDs exhibited excellent fluorescence properties with a high QY of 73% [86].

2.4.4 Sensing of Co(II)

Co is the active center of a group of co-enzymes called cobalamin (Vitamin B_{12}). It plays an important role in the implementation of red blood cells (RBC) as well as the maintenance of nerve cells. Hence the selective and sensitive detection of Co is imperative [87]. In addition to this, Bano and co-workers have also synthesized N-doped CQDs from glycine and polyethyleneimine by hydrothermal method for the detection of Co^{2+} ions. The LOD was found to be 0.12 mM [88].

2.4.5 Sensing of Pb(II)

Rawat et al. have synthesized fluorescent CQDs from watermelon juice to detect Pb^{2+} in cancer cells and contaminated water. They synthesized watermelon-ethanolamine (WMEA)-CQDs and watermelon-ethylenediamine (WMED)-CQDs with fluorescence QYs of 8 and 7%, respectively. WMED-CQDs were utilized for the sensing of Pb^{2+} with an extraordinarily low LOD of 190 pM. Further sensing was also performed in a cancerous HeLa cell [89]. Sheikhzadeh and co-workers have synthesized a novel colorimetric sensor from curcumin extract for the sensing of Pb^{2+} in rice [90]. Babu et al. have prepared fluorescent CQDs from delonix regia leaves by hydrothermal method for the selective and sensitive detection of Pb^{2+} in polluted water samples [91].

2.4.6 Sensing of Cd

Chauhan and co-workers have synthesized CQDs from coconut coir via the hydrothermal method. The as synthesized CQDs exhibited a high QY of 48.0%. The prepared CQDs showed a turn-on sensor for Cd^{2+} and a turn-off sensor for Cu^{2+}. The prepared CQDs exhibited LOD of 0.28 nm and 0.18 for Cu^{2+} and Cd^{2+}, respectively [92].

2.4.7 Sensing of As(III)

Few efforts have been tried to use green-prepared CQDs for the detection of As(III). Pooja et al. have synthesized fluorescent CQDs by rotten prickly pear cactus and tomatoes for the sensing of As(III) in pond, tap, industrial, and river watersamples [93]. Ramezani et al. have synthesized CQDs-using quince fruit (Cydonia oblonga) powder as the carbon source for the sensing of As(III) [94].

2.5 *Fluorescence Quenching Enhancement*

When the emission intensity of fluorescent molecules decreases, then this phenomenon is called quenching. Quenching occurs when fluorescent molecules interact with quencher molecules i.e., the analytes. The most commonly used quencher molecules are iodide ions, metal ions, molecular oxygen, amino acids and acrylamide. The quenching procedure occurs depending upon the interaction of fluorescent molecules (CQDs) with quencher molecules including dynamic quenching, static quenching, and IFE [95].

2.5.1 Inner Filter Effect (IFE)

When a fine and accurate spectral overlapping occurs among the absorption spectra of fluorescent molecules and quencher molecules, then the quencher molecules absorb the emission light of fluorescent molecules, which results in the quenching of fluorescence. This process is called the IFE. IFE mechanism is generally shown by toxic metal ions because they have a wide absorption spectrum. Our research group has synthesized a simplistic, green, and eco-friendly hydrothermal method for the preparation of GB-CQDs from *Artocarpous lakoocha* seeds. The QY obtained was 38.5%. The GB-CQDs were highly photostable, based on which the material was effectively applied as a sensitive nanoprobe for the sensing of Fe^{3+}. Fe^{3+}detectionlinearity from 2 to 6 μM and LOD was calculated to be 0.62 μM. The fluorescence quenching mechanisms were because of the communication among the fluorescent molecule (GB-CQDs) and quencher molecule (Fe^{3+} ion) which were due to the inner filter effect (IFE) [33].

2.5.2 Static Quenching

In static quenching, the emission intensity decreases when ground-state interaction occurs between fluorescent molecules and quencher molecules. Ground state interaction is confirmed by fluorescent life time measurement experiment. In this experiment, the life-time of fluorescent molecules (CQDs) is recorded in the presence and absence of quencher molecules. If no obvious shift in the average life-time, then ground-state interaction occurs between the CQDs and quencher molecules. This reveals the static quenching. Yadav and co-workers have synthesized G-CQDs by using Ficus benghalensis latex as the C source and polyethyleneimine as the N source. The CQDs displayed green fluorescence with a 41.2% QY and exhibited excitation-dependent emission. The CQDs were utilized for the sensing of tyrosine (Tyr) with a LODof 0.13 mM. The fluorescence quenching mechanisms were due to the static quenching [96]. Static quenching can also be confirmed by optimizing temperature. The fluorescence intensity of CQDs gradually decreases with increasing temperature, confirming the static quenching [97].

2.5.3 Dynamic Quenching

In dynamic quenching, the emission intensity decreases when excited state interaction occurs between fluorescent and quencher molecules. During lifetime measurement experiments, if there is an obvious shift in the average life-time, it indicates the excited state interaction between the CQDs and quencher molecules, resulting in the dynamic quenching. This is also known as collisional quenching. Singh and co-workers have synthesized N-CQDs via a one-step hydrothermal method. The fluorescent QY was 41%concerningto quinine sulphate standard. The outstanding optical properties of N-CQDs made it a fluorescent probe for the selective and sensitive detection of toxic heavy metal ion, Hg^{2+}. The LOD was calculated to be 0.08 μM [17].

2.6 *Optimization of Experimental Conditions*

2.6.1 pH

The pH affects the sensing activity of fluorescent molecules *i.e.,* the CQDs. The different functional groups on the surface of CQDs at pH 7 underwent deprotonation to enhance binding among fluorescent CQDs and heavy metal ions. On the contrary, the metal ions co-ordinated more powerfully with hydroxyl groups than the surface of CQDs at higher pH (>–8) [93]. The fluorescence of CQDs is generally stable in a wide pH range (4–11) [98]. CQDs often exhibit pH-dependent fluorescence properties. So optimize the pH to maximize the sensitivity of the fluorescence signal to changes in heavy metal concentration.

2.6.2 Temperature

The optimization of temperature is a critical factor in the sensing of heavy metals using CQDs. The temperature can affect the reaction kinetics, binding affinity, and overall performance of the sensor. Some sensing mechanisms, such as fluorescence quenching, may be temperature-sensitive. Temperature can influence the optical properties like fluorescence intensity, emission spectra, and quantum yield of CQDs. Excessive temperatures might lead to degradation or changes in the CQD structure, affecting their sensing properties. Xu et al. synthesized fluorescent CQDs from wolfberry stem, and optimized the temperature. Issa et al. have synthesized fluorescent CQDs from carboxymethyl cellulose in the occurrence of linear polyethyleneimine and achieved a high QY of up to 44% by optimizing the temperature [97].

2.7 *Selectivity and Sensitivity*

Selectivity and sensitivity are crucial parameters in the sensing field, and they play pivotal roles in determining the effectiveness and reliability of a sensor. Selectivity refers to the ability of a sensor to exclusively respond to the target analyte of interest while ignoring interference from other substances present in the sample. High selectivity is essential to avoid false positives or inaccurate readings caused by substances with similar properties to the target analyte. It ensures that the sensor responds specifically to the intended substance.

Sensitivity is a measure of how well a sensor can detect and respond to changes in the concentration of the target analyte. High sensitivity is crucial for detecting low concentrations of the target analyte, especially in applications where trace amounts are significant. It allows for the detection of subtle changes in analyte concentration. The sensitivity of the CQD-based sensor can be calculated by evaluating the slope of the calibration curve. Using the standard deviation of the blank and slope of the calibration curve, the LOD is calculated, which is the lowest concentration of heavy metal ions that can be reliably detected by the sensor. Dhandapani et al. prepared CQDs by hydrothermal method and utilized them for the sensing of Hg^{2+} selectively and sensitively with an LOD of ~6.2 nM. The QY obtained was ~77.2%.The CQDs selectively sensed Hg^{2+} among Cd^{2+}, Zn^{2+}, Mn^{2+},Ni^{2+}, Fe^{3+}, Pb^{2+}, and Hg^{2+} [99]. Sariga et al. have synthesized red fluorescent CQDs via the hydrothermal method and applied them for selective and sensitive detection of Cd^{2+} with a low LOD of 2.4 nM. The interference experiments were performed for several metal ions (K^+, Na^+, Cu^{2+}, Mg^{2+}, Zn^{2+}, Ca^{2+}, Fe^{2+}, Fe^{3+}, Ni^{2+}, Pb^{2+}, Ag^{2+}, Co^{2+}, Mn^{2+}, and Hg^{2+}) [100]. In medical diagnostics, a sensor must be selective to detect specific biomarkers and sensitive enough to identify early stages of diseases.

2.8 Real Sample Analysis

Real sample analysis using CQDs involves applying these nanomaterials as sensors to detect and quantify specific analytes in complex, real-world samples. This could include environmental samples (water, soil), biological samples (blood, urine), or others based on the target analyte. Oliveira et al. have synthesized a fluorescent nanoprobe with high selectivity for multiway sensing of Co^{2+} and also utilized it in blood plasma samples with good recovery [101]. Kundu et al. have prepared CQDs from orange pomaca for rapid detection of Cr^{6+}. To evaluate the practical feasibility, real sample analyses were conducted with satisfactory a recovery rate [102]. Dong and co-workers have prepared CQDs by functionalization of polyamine (BPEI-CQDs) for the selective and sensitive detection of Cu^{2+}. The fluorescence quenching mechanism of CQDs revealed the IFE. The LOD was 6 nM. Further, they have also investigated for the sensing of Cu^{2+} in a Min river water sample [103]. Wang et al. have prepared CQDs@Cu-IIP as a fluorescence probe for the detection of Cu^{2+} by combining fluorescence analysis technology and sensing. The real sample analyses were further performed for the tap water and river water with satisfactory recovery rates [104]. Further, CQD-based hydrogel (HG) nanocomposite material (CQDsHG) was prepared by Wu et al. for the sensing of Fe^{3+} and achieved an LOD of 0.24 μM. In addition, the sensing was also established in lake water and tap water [105]. Sun et al. have synthesized CQDs from Ophiopogon japonicus for sensing Fe^{3+}in actual water samples with high recovery [106]. Lai and co-workers have synthesized nitrogen-doped CQDs by hydrothermal method for the detection of Fe(III) and imaging towards cancerous cells. The sensing of Fe(III) was also performed in tap water with satisfactory results [107]. Babu et al. prepared CQDs from Delonix regia that were also applied in the construction of Pb^{2+} sensors. The developed sensors for Pb^{2+} detection showed an LOD of 3.3 nM within the linear response range of 10–180 μM. The fluorescence quenching of CQDs was due to fluorescence resonance energy transfer (FRET) among the donor fluorophore (CQDs) and non-fluorescent acceptor (Pb^{2+}).Moreover, the sensing was also performed in river, lake, and tap water samples with 99–101% recovery rate [91]. Kumar et al. have prepared fluorescent CQDs from *Ocimum sanctum* for the detection of Pb^{2+} with a LOD of 0.59 nM. Further, the investigation was performed in real pond water [108]. Tabaraki et al. have synthesized nitrogen-doped CQDs for the sensing of Pb^{+2}in mineral and tap water samples [109]. Yan and co-workers have prepared nitrogen and boron co-doped CQDs for Cd sensing and utilized them for real sample sensing in human serum [110]. Wang et al. have synthesized S and N co-doped CQDs from waste foam and used for the detection of Cr^{3+} ions in mineral/tap water [111]. Oliveira et al. used fluorescent CQDs to determine Co^{2+} ions in blood samples [112]. Dastidar and co-workers prepared CQDs from the extract of onion for the selective sensing of Zn^{2+} in blood plasma [113]. These experiments revealed the practical applicability of CQDs in biological samples analyses.

3 Conclusion

Heavy metal (Pb, Cd, Hg, Cr, As, etc.) exposure can result in an extensive range of health problems, including cardiovascular diseases, neurological disorders, cancer, kidney damage, and developmental issues, especially in children. Heavy metal contamination poses significant environmental threats, affecting soils, water bodies, and air causing ecological disruptions, harm to aquatic life, and reduced agricultural productivity. Therefore, sensing of heavy metals is crucial. This chapter summarized the synthesis of fluorescent CQDs by top-down and bottom-up approaches. In the bottom-up approach, the hydrothermal method is superior because it is single-step, ecofriendly, non-toxic, and has an ease of operation. Further, we have also explained the synthesis of CQDs by surface passivation/functionalization. The stability of CQDs was also examined in highly ionic salt concentrations, in different pH, and in UV light exposure. The fluorescence quenching mechanism is also explained in detail. This chapter also explains the optimization of different experimental conditions such as pH and temperature. Finally, we have discussed real sample analysis using CQDs as sensors to detect and quantify specific analytes in complex, real-world samples. In conclusion, this systematic approach ensures a detailed investigation of the CQD-based sensing platform for heavy metal ions, addressing key aspects such as sensitivity, selectivity, and real-world applicability.

Acknowledgements P.K.Y. acknowledges IoE, BHU for the Raja Jwala Prasad Post Doctoral Fellowship. V.G. acknowledges financial support from SERB ((CRG/2022/003370), New Delhi and IoE, BHU (incentive grant for faculty, scheme number: 6031).

References

1. Abd Elnabi, M. K., Elkaliny, N. E., Elyazied, M. M., Azab, S. H., Elkhalifa, S. A., Elmasry, S., & Mahmoud, Y. A. G. (2023). Toxicity of heavy metals and recent advances in their removal: A review. *Toxics, 11*(7), 580.
2. Li, M., Shi, Q., Song, N., Xiao, Y., Wang, L., Chen, Z., & James, T. D. (2023). Current trends in the detection and removal of heavy metal ions using functional materials. *Chemical Society Reviews.*
3. Aragay, G., Pons, J., & Merkoçi, A. (2011). Recent trends in macro-, micro-, and nanomaterial-based tools and strategies for heavy-metal detection. *Chemical Reviews, 111*(5), 3433–3458.
4. Vallee, B. L., & Ulmer, D. D. (1972). Biochemical effects of mercury, cadmium, and lead. *Annual Review of Biochemistry, 41*(1), 91–128.
5. Hamilton, J. W., Kaltreider, R. C., Bajenova, O. V., Ihnat, M. A., McCaffrey, J., Turpie, B. W., Rowell, E. E., Oh, J., Nemeth, M. J., Pesce, C. A., Lariviere, J. P. (1998). Molecular basis for effects of carcinogenic heavy metals on inducible gene expression. *Environ Health Perspect*, *106*, 1005-1015
6. Partanen, T., Heikkilä, P., Hernberg, S., Kauppinen, T., Moneta, G., & Ojajärvi, A. (1991). Renal cell cancer and occupational exposure to chemical agents. *Scandinavian Journal of Work, Environment & Health*, 231–239.
7. Acosta, I., Rodríguez, A., Cárdenas, J. F., Martínez, V. M., & Contreras, D. (2023). Lead Removal from Aqueous Solutions Using Different Biosorbents. In *lead toxicity: Challenges and solution* (pp. 227–245). Springer Nature Switzerland.
8. Srinivasan, S., Ranganathan, V., McConnell, E. M., Murari, B. M., & DeRosa, M. C. (2023). Aptamer-based colorimetric and lateral flow assay approaches for the detection of toxic metal ions, thallium (i) and lead (ii). *RSC Advances, 13*(29), 20040–20049.
9. Aziz, K. H. H., Mustafa, F. S., Omer, K. M., Hama, S., Hamarawf, R. F., & Rahman, K. O. (2023). Heavy metal pollution in the aquatic environment: Efficient and low-cost removal approaches to eliminate their toxicity: A review. *RSC Advances, 13*(26), 17595–17610.
10. Ahmed, A. S., Mohamed, M. B. I., Bedair, M. A., El-Zomrawy, A. A., & Bakr, M. F. (2023). A new Schiff base-fabricated pencil lead electrode for the efficient detection of copper, lead, and cadmium ions in aqueous media. *RSC Advances, 13*(23), 15651–15666.
11. Jaffari, Z. H., Abbas, A., Umer, M., Kim, E. S., & Cho, K. H. (2023). Crystal graph convolution neural networks for fast and accurate prediction of adsorption ability of Nb 2 CT x towards Pb (ii) and Cd (ii) ions. *Journal of Materials Chemistry A, 11*(16), 9009–9018.
12. Chaghaghazardi, M., Kashanian, S., Nazari, M., Omidfar, K., Joseph, Y., & Rahimi, P. (2023). Nitrogen and sulfur co-doped carbon quantum dots fluorescence quenching assay for detection of mercury (II). *Spectrochimica Acta Part A: Molecular and Biomolecular Spectroscopy, 293*, 122448.
13. Habila, M. A., ALOthman, Z. A., Hakami, H. M., & Alanazi, A. G. (2023). Influence of synthesis-heating conditions on the porosity and performance of a carbon nanotube/sds-alumina nanocomposite for effective wastewater treatment: fabrication, characterization, and adsorption modeling. *Industrial & Engineering Chemistry Research.*
14. Chen, Z., Zhang, Z., Qi, J., You, J., Ma, J., & Chen, L. (2023). Colorimetric detection of heavy metal ions with various chromogenic materials: Strategies and applications. *Journal of Hazardous Materials, 441*, 129889.

15. Zhang, C., Liang, M., Shao, C., Li, Z., Cao, X., Wang, Y., & Lu, S. (2023). Visual detection and sensing of mercury ions and glutathione using fluorescent copper nanoclusters. *ACS Applied Bio Materials, 6*(3), 1283–1293.
16. Huang, Y., Li, Y., Li, Y., Zhong, K., & Tang, L. (2023). An "AIE+ ESIPT" mechanism-based benzothiazole-derived fluorescent probe for the detection of Hg^{2+} and its applications. *New Journal of Chemistry, 47*(14), 6916–6923.
17. Singh, V. K., Singh, V., Yadav, P. K., Chandra, S., Bano, D., Koch, B., & Hasan, S. H. (2019). Nitrogen doped fluorescent carbon quantum dots for on-off-on detection of Hg^{2+} and glutathione in aqueous medium: Live cell imaging and IMPLICATION logic gate operation. *Journal of Photochemistry and Photobiology A: Chemistry, 384*, 112042.
18. do Nascimento, F. H., & Masini, J. C. (2023). Porous-polymer monolith decorated with dithizone for greener visual and spectrophotometric sensing of Hg (II). *Sensors and Actuators B: Chemical, 381*, 133385.
19. Liu, W., Cui, H. L., Zhou, J., Su, Z. T., Zhang, Y. Z., Chen, X. L., & Yue, E. L. (2023). Synthesis of a Cd-MOF fluorescence sensor and its detection of Fe3+, Fluazinam, TNP, and sulfasalazine enteric-coated tablets in aqueous solution. *ACS Omega, 8*(27), 24635–24643.
20. Rahimzadeh, M. R., Rahimzadeh, M. R., Kazemi, S., & Moghadamnia, A. A. (2017). Cadmium toxicity and treatment: An update. *Caspian Journal of Internal Medicine, 8*(3), 135.
21. Garg, N., Deep, A., & Sharma, A. L. (2023). *Sensitive detection of cadmium using amine-functionalized Fe-MOF by anodic stripping voltammetry*. Industrial & Engineering Chemistry Research.
22. Takemura, K., Motomura, T., Iwasaki, W., & Morita, N. (2023). Fabrication of highly active nanoneedle gold electrode using rose petals for electrochemical detection of arsenic (III). *ACS Applied Nano Materials.*
23. Jose, A., Jana, A., Gupte, T., Nair, A. S., Unni, K., Nagar, A., & Pradeep, T. (2023). Vertically aligned nanoplates of atomically precise Co6S8 cluster for practical arsenic sensing. *ACS Materials Letters, 5*(3), 893–899.
24. Sudarman, F., Shiddiq, M., Armynah, B., & Tahir, D. (2023). Silver nanoparticles (AgNPs) synthesis methods as heavy-metal sensors: A review. *International Journal of Environmental Science and Technology*, 1–18.
25. Rauf, M., Shah, S. K., Algahtani, A., Tirth, V., Alghtani, A. H., Al-Mughanam, T., & Amin, M. A. (2023). Application of ZnO-NRs@ Ni-foam substrate for electrochemical fingerprint of arsenic detection in water. *RSC Advances, 13*(21), 14530–14538.
26. Menon, S., Usha, S. P., Manoharan, H., Kishore, P. V. N., & Sai, V. V. R. (2023). Metal–organic framework-based fiber optic sensor for chromium (VI) detection. *ACS Sensors, 8*(2), 684–693.
27. Bezuneh, T. T., Fereja, T. H., Li, H., & Jin, Y. (2023). Solid-phase pyrolysis synthesis of highly fluorescent nitrogen/sulfur codoped graphene quantum dots for selective and sensitive diversity detection of Cr(VI). *Langmuir, 39*(4), 1538–1547.
28. Sadia, M., Khan, J., Khan, R., Shah, S. W. A., Zada, A., Zahoor, M., & Ali, E. A. (2023). Synthesis and computational study of an optical fluorescent sensor for selective detection of Ni^{2+} ions. *ACS Omega, 8*(30), 27500–27509.
29. Zare, H., Ghalkhani, M., Akhavan, O., Taghavinia, N., & Marandi, M. (2017). Highly sensitive selective sensing of nickel ions using repeatable fluorescence quenching-emerging of the CdTe quantum dots. *Materials Research Bulletin, 95*, 532–538.
30. Jiang, C., Zhao, Q., Zheng, L., Chen, X., Li, C., & Ren, M. (2021). Distribution, source and health risk assessment based on the Monte Carlo method of heavy metals in shallow groundwater in an area affected by mining activities, China. *Ecotoxicology and Environmental Safety, 224*, 112679.
31. Bafe Dilebo, W., Desta Anchiso, M., & Eskezia Ayalew, M. (2023). Assessment of selected heavy metals concentration level of drinking water in Gazer Town and Selected Kebele, South Ari District, Southern Ethiopia. *International Journal of Analytical Chemistry,* 2023.

32. Ramos, E., Bux, R. K., Medina, D. I., Barrios-Piña, H., & Mahlknecht, J. (2023). Spatial and multivariate statistical analyses of human health risk associated with the consumption of heavy metals in groundwater of monterrey metropolitan area. *Mexico Water, 15*(6), 1243.
33. Yadav, P. K., Singh, V. K., Kumar, C., Chandra, S., Jit, S., Singh, S. K., & Hasan, S. H. (2019). A facile synthesis of green-blue carbon dots from Artocarpus lakoocha seeds and their application for the detection of iron (III) in biological fluids and cellular imaging. *Chemistry Select, 4*(42), 12252–12259.
34. Kandić, I., Kragović, M., Petrović, J., Janaćković, P., Gavrilović, M., Momčilović, M., & Stojmenović, M. (2023). Heavy metals content in selected medicinal plants produced and consumed in Serbia and their daily intake in herbal infusions. *Toxics, 11*(2), 198.
35. Xu, X., Ray, R., Gu, Y., Ploehn, H. J., Gearheart, L., Raker, K., & Scrivens, W. A. (2004). Electrophoretic analysis and purification of fluorescent single-walled carbon nanotube fragments. *Journal of the American Chemical Society, 126*(40), 12736–12737.
36. Yadav, P. K., Chandra, S., Kumar, V., Kumar, D., & Hasan, S. H. (2023). Carbon quantum dots: Synthesis, structure, properties, and catalytic applications for organic synthesis. *Catalysts, 13*(2), 422.
37. Deng, Y., Long, Y., Song, A., Wang, H., Xiang, S., Qiu, Y., & Weng, Q. (2023). Boron dopants in red-emitting B and N Co-doped carbon quantum dots enable targeted imaging of lysosomes. *ACS Applied Materials & Interfaces, 15*(13), 17045–17053.
38. Surucu, O. (2022). Electrochemical removal and simultaneous sensing of mercury with inductively coupled plasma-mass spectrometry from drinking water. *Materials Today Chemistry, 23*, 100639.
39. Gogoi, N., Barooah, M., Majumdar, G., & Chowdhury, D. (2015). Carbon dots rooted agarose hydrogel hybrid platform for optical detection and separation of heavy metal ions. *ACS Applied Materials & Interfaces, 7*(5), 3058–3067.
40. Liu, C., Deng, Z., Deng, Y., Yuan, K., Hu, M., Yuan, L., & Zhang, Y. (2023). Simple, universal colorimetric strategy for amplified detection of nanomolar toxic heavy metal ions based on multi-component complexation and Tyndall effect. *Sensors and Actuators B: Chemical, 380*, 133333.
41. Jiang, G., Miao, Y., Wang, J., Shao, H., Chen, H., Tao, P., & Zhou, X. (2023). Specific detection of mercury ions based on surface plasmon resonance sensor modified with 1, 6-hexanedithiol. *Sensors and Actuators A: Physical, 356*, 114343.
42. Huang, R., Lv, J., Chen, J., Zhu, Y., Zhu, J., Wågberg, T., & Hu, G. (2023). Three-dimensional porous high boron-nitrogen-doped carbon for the ultrasensitive electrochemical detection of trace heavy metals in food samples. *Journal of Hazardous Materials, 442*, 130020.
43. Geleta, G. S. (2023). A colorimetric aptasensor based on two dimensional (2D) nanomaterial and gold nanoparticles for detection of toxic heavy metal ions: a review. *Food Chemistry Advances*, 100184.
44. Yoo, D., Park, Y., Cheon, B., & Park, M. H. (2019). Carbon dots as an effective fluorescent sensing platform for metal ion detection. *Nanoscale Research Letters, 14*, 1–13.
45. Tian, J., An, M., Zhao, X., Wang, Y., & Hasan, M. (2023). Advances in fluorescent sensing carbon dots: An account of food analysis. *ACS Omega, 8*(10), 9031–9039.
46. Yang, Z., Xu, T., Li, H., She, M., Chen, J., Wang, Z., & Li, J. (2023). Zero-dimensional carbon nanomaterials for fluorescent sensing and imaging. *Chemical Reviews.*
47. Xu, Y., Lan, J., Wang, B., Bo, C., Ou, J., & Gong, B. (2023). Simple fabrication of carbon quantum dots and activated carbon from waste wolfberry stems for detection and adsorption of copper ion. *RSC Advances, 13*(31), 21199–21210.
48. Torrinha, Á., Oliveira, T. M., ul Islam, S., & Morais, S. (2023). Green carbon (Nano) materials-based sensors for analysis of hazardous metal ions. In *Green carbon materials for environmental analysis: emerging research and future opportunities* (pp. 91–138). American Chemical Society.
49. Pourmadadi, M., Rahmani, E., Rajabzadeh-Khosroshahi, M., Samadi, A., Behzadmehr, R., Rahdar, A., & Ferreira, L. F. R. (2023). Properties and application of carbon quantum dots (CQDs) in biosensors for disease detection: a comprehensive review. *Journal of Drug Delivery Science and Technology*, 104156.

50. Bilge, S., Karadurmus, L., Sınağ, A., & Ozkan, S. A. (2021). Green synthesis and characterization of carbon-based materials for sensitive detection of heavy metal ions. *TrAC Trends in Analytical Chemistry, 145*, 116473.
51. Rasal, A. S., Yadav, S., Yadav, A., Kashale, A. A., Manjunatha, S. T., Altaee, A., & Chang, J. Y. (2021). Carbon quantum dots for energy applications: A review. *ACS Applied Nano Materials, 4*(7), 6515–6541.
52. Arul, V., & Sethuraman, M. G. (2019). Hydrothermally green synthesized nitrogen-doped carbon dots from Phyllanthus emblica and their catalytic ability in the detoxification of textile effluents. *ACS Omega, 4*(2), 3449–3457.
53. Li, L., Yu, B., & You, T. (2015). Nitrogen and sulfur co-doped carbon dots for highly selective and sensitive detection of Hg(II) ions. *Biosensors and Bioelectronics, 74*, 263–269.
54. Zhu, H., Wang, X., Li, Y., Wang, Z., Yang, F., & Yang, X. (2009). Microwave synthesis of fluorescent carbon nanoparticles with electrochemiluminescence properties. *Chemical Communications, 34*, 5118–5120.
55. Wu, Z. L., Zhang, P., Gao, M. X., Liu, C. F., Wang, W., Leng, F., & Huang, C. Z. (2013). One-pot hydrothermal synthesis of highly luminescent nitrogen-doped amphoteric carbon dots for bioimaging from Bombyx mori silk–natural proteins. *Journal of Materials Chemistry B, 1*(22), 2868–2873.
56. Bhakare, M. A., Bondarde, M. P., Lokhande, K. D., Dhumal, P. S., & Some, S. (2023). Quick transformation of polymeric waste into high valuable N-self doped carbon quantum dot for detection of heavy metals from wastewater. *Chemical Engineering Science, 281*, 119150.
57. Dalui, A., Pradhan, B., Thupakula, U., Khan, A. H., Kumar, G. S., Ghosh, T., & Acharya, S. (2015). Insight into the mechanism revealing the peroxidase mimetic catalytic activity of quaternary CuZnFeS nanocrystals: Colorimetric biosensing of hydrogen peroxide and glucose. *Nanoscale, 7*(19), 9062–9074.
58. Chandra, S., Patra, P., Pathan, S. H., Roy, S., Mitra, S., Layek, A., & Goswami, A. (2013). Luminescent S-doped carbon dots: An emergent architecture for multimodal applications. *Journal of Materials Chemistry B, 1*(18), 2375–2382.
59. Han, Y., Tang, D., Yang, Y., Li, C., Kong, W., Huang, H., & Kang, Z. (2015). Non-metal single/dual doped carbon quantum dots: A general flame synthetic method and electro catalytic properties. *Nanoscale, 7*(14), 5955–5962.
60. Zhou, J., Shan, X., Ma, J., Gu, Y., Qian, Z., Chen, J., & Feng, H. (2014). Facile synthesis of P-doped carbon quantum dots with highly efficient photoluminescence. *RSC Advances, 4*(11), 5465–5468.
61. Ozkasapoglu, S., Caglayan, M. G., Akkurt, F., Ensarioğlu, H. K., Vatansever, H. S., & Celikkan, H. (2023). Boron-doped carbon nanodots as a theranostic agent for colon cancer stem cells. *ACS Omega, 8*(33), 30285–30293.
62. Shan, X., Chai, L., Ma, J., Qian, Z., Chen, J., & Feng, H. (2014). B-doped carbon quantum dots as a sensitive fluorescence probe for hydrogen peroxide and glucose detection. *The Analyst, 139*(10), 2322–2325.
63. Bourlinos, A. B., Stassinopoulos, A., Anglos, D., Zboril, R., Georgakilas, V., & Giannelis, E. P. (2008). Photoluminescent carbogenic dots. *Chemistry of Materials, 20*(14), 4539–4541.
64. Liao, S., Zhu, F., Zhao, X., Yang, H., & Chen, X. (2018). A reusable P, N-doped carbon quantum dot fluorescent sensor for cobalt ion. *Sensors and Actuators B: Chemical, 260*, 156–164.
65. Gong, Y., Yu, B., Yang, W., & Zhang, X. (2016). Phosphorus and nitrogen co-doped carbon dots as a fluorescent probe for real-time measurement of reactive oxygen and nitrogen species inside macrophages. *Biosensors and Bioelectronics, 79*, 822–828.
66. Dong, G., Lv, Q., Hao, L., Zhang, W., Zhang, Z., Chai, D. F., & Li, J. (2023). Integration of N, P-doped carbon quantum dots with hydrogel as a solid-phase fluorescent probe for adsorption and detection of Fe^{3+}. *Nanotechnology, 34*(46), 465702.
67. Dong, Y., Pang, H., Yang, H. B., Guo, C., Shao, J., Chi, Y., & Yu, T. (2013). Carbon-based dots co-doped with nitrogen and sulfur for high quantum yield and excitation-independent emission. *Angewandte Chemie, 125*(30), 7954–7958.

68. Xue, M., Zhang, L., Zhan, Z., Zou, M., Huang, Y., & Zhao, S. (2016). Sulfur and nitrogen binary doped carbon dots derived from ammonium thiocyanate for selective probing doxycycline in living cells and multicolor cell imaging. *Talanta, 150*, 324–330.
69. Ambade, R. B., Ali, M., Lee, K. H., Jeong, W., Jeong, S. H., & Han, T. H. (2023). Nitrogen and sulfur co-doped carbon quantum dot-engineered TiO_2 graphene on carbon fabric for photocatalysis applications. *ACS Applied Nano Materials, 6*(17), 15782–15794.
70. Jana, D., Sun, C. L., Chen, L. C., & Chen, K. H. (2013). Effect of chemical doping of boron and nitrogen on the electronic, optical, and electrochemical properties of carbon nanotubes. *Progress in Materials Science, 58*(5), 565–635.
71. Jahan, S., Mansoor, F., Naz, S., Lei, J., & Kanwal, S. (2013). Oxidative synthesis of highly fluorescent boron/nitrogen co-doped carbon nanodots enabling detection of photosensitizer and carcinogenic dye. *Analytical Chemistry, 85*(21), 10232–10239.
72. Chen, W., Zhou, J., Zhang, F., Li, X., & Guo, J. (2023). Stability study of low-cost carbon quantum dots nanofluids with saline water and their application investigation for the performance improvement of solar still. *Diamond and Related Materials, 138*, 110194.
73. Atchudan, R., Edison, T. N. J. I., Shanmugam, M., Perumal, S., Somanathan, T., & Lee, Y. R. (2021). Sustainable synthesis of carbon quantum dots from banana peel waste using hydrothermal process for in vivo bioimaging. *Physica E: Low-dimensional Systems and Nanostructures, 126*, 114417.
74. Bano, D., Kumar, V., Singh, V. K., Chandra, S., Singh, D. K., Yadav, P. K., & Hasan, S. H. (2019). A facile and simple strategy for the synthesis of label free carbon quantum dots from the latex of Euphorbia milii and its peroxidase-mimic activity for the naked eye detection of glutathione in a human blood serum. *ACS Sustainable Chemistry & Engineering, 7*(2), 1923–1932.
75. Yang, P., Zhao, J., Wang, J., Cao, B., Li, L., & Zhu, Z. (2014). Light-induced synthesis of photoluminescent carbon nanoparticles for Fe^{3+} sensing and photocatalytic hydrogen evolution. *Journal of Materials Chemistry A, 3*(1), 136–138.
76. Chandra, S., Laha, D., Pramanik, A., Ray Chowdhuri, A., Karmakar, P., & Sahu, S. K. (2016). Synthesis of highly fluorescent nitrogen and phosphorus doped carbon dots for the detection of Fe3+ ions in cancer cells. *Luminescence, 31*(1), 81–87.
77. Liu, W., Diao, H., Chang, H., Wang, H., Li, T., & Wei, W. (2017). Green synthesis of carbon dots from rose-heart radish and application for Fe^{3+} detection and cell imaging. *Sensors and Actuators, B: Chemical, 241*, 190–198.
78. Khan, Z. M., Rahman, R. S., Islam, S., & Zulfequar, M. (2019). Hydrothermal treatment of red lentils for the synthesis of fluorescent carbon quantum dots and its application for sensing Fe^{3+}. *Optical Materials, 91*, 386–395.
79. Gajalakshmi, S., Iswarya, V., Ashwini, R., Divya, G., Mythili, S., & Sathiavelu, A. (2012). Evaluation of heavy metals in medicinal plants growing in Vellore District. *European Journal of Experimental Biology, 2*(5), 1457–1461.
80. Gong, Y., Liu, Y., Xiong, Z., & Zhao, D. (2014). Immobilization of mercury by carboxymethyl cellulose stabilized iron sulfide nanoparticles: Reaction mechanisms and effects of stabilizer and water chemistry. *Environmental Science & Technology, 48*(7), 3986–3994.
81. Jaison, A. M. C., Vasudevan, D., Ponmudi, K., George, A., & Varghese, A. (2023). One pot hydrothermal synthesis and application of bright-yellow-emissive carbon quantum dots in Hg2+ detection. *Journal of Fluorescence*, 1–14.
82. Xie, Y., Cheng, D., Liu, X., & Han, A. (2019). Green hydrothermal synthesis of N-doped carbon dots from biomass highland barley for the detection of Hg^{2+}. *Sensors, 19*(14), 3169.
83. Gupta, R., Ganesan, V., Sonkar, P. K., Yadav, D. K., & Yadav, M. (2023). Phenosafranine encapsulated mesoporous silica as efficient electrocatalyst for Cr(VI) reduction and its subsequent sensitive determination. *Microchemical Journal, 193*, 109193.
84. Zheng, M., Xie, Z., Qu, D., Li, D., Du, P., Jing, X., & Sun, Z. (2013). On–off–on fluorescent carbon dot nanosensor for recognition of chromium (VI) and ascorbic acid based on the inner filter effect. *ACS Applied Materials & Interfaces, 5*(24), 13242–13247.

85. Omer, K. M., Tofiq, D. I., & Ghafoor, D. D. (2019). Highly photoluminescent label free probe for Chromium (II) ions using carbon quantum dots co-doped with nitrogen and phosphorous. *Journal of Luminescence, 206*, 540–546.
86. Singh, V. K., Singh, V., Yadav, P. K., Chandra, S., Bano, D., Kumar, V., & Hasan, S. H. (2018). Bright-blue-emission nitrogen and phosphorus-doped carbon quantum dots as a promising nanoprobe for detection of Cr (VI) and ascorbic acid in pure aqueous solution and in living cells. *New Journal of Chemistry, 42*(15), 12990–12997.
87. Zhang, Y., Zhu, C., Zhang, Y., Jing, N., & Wang, Y. (2021). Hydrothermal synthesis of polyethyleneimine modified carbon quantum dots for sensitively detection of cobalt ions. *Journal of Nanoscience and Nanotechnology, 21*(4), 2099–2108.
88. Bano, D., Kumar, V., Chandra, S., Singh, V. K., Mohan, S., Singh, D. K., & Hasan, S. H. (2019). Synthesis of highly fluorescent nitrogen-rich carbon quantum dots and their application for the turn-off detection of cobalt (II). *Optical Materials, 92*, 311–318.
89. Rawat, K. S., Singh, V., Sharma, C. P., Vyas, A., Pandey, P., Singh, J., & Goel, A. (2023). Picomolar detection of lead ions (Pb^{2+}) by functionally modified fluorescent carbon quantum dots from watermelon juice and their imaging in cancer cells. *Journal of Imaging, 9*(1), 19.
90. Sheikhzadeh, E., Naji-Tabasi, S., Verdian, A., & Kolahi-Ahari, S. (2022). Equipment-free and visual detection of Pb^{2+} ion based on curcumin-modified bacterial cellulose nanofiber. *Journal of the Iranian Chemical Society, 19*, 283–290.
91. Babu, A. M., Bijoy, G., Keerthana, P., & Varghese, A. (2022). Fluorescent detection of Pb^{2+} pollutant in water samples with the help of Delonix regia leaf-derived CQDs. *Synthetic Metals, 291*, 117211.
92. Chauhan, P., Dogra, S., Chaudhary, S., & Kumar, R. (2020). Usage of coconut coir for sustainable production of high-valued carbon dots with discriminatory sensing aptitude toward metal ions. *Mater Today Chem, 16*, 100247.
93. Devi, P., Rajput, P., Thakur, A., Kim, K. H., & Kumar, P. (2019). Recent advances in carbon quantum dot-based sensing of heavy metals in water. *TrAC Trends in Analytical Chemistry, 114*, 171–195.
94. Ramezani, Z., Qorbanpour, M., & Rahbar, N. (2018). Green synthesis of carbon quantum dots using quince fruit (Cydonia oblonga) powder as carbon precursor: Application in cell imaging and As^{3+} determination. *Colloids and Surfaces A: Physicochemical and Engineering Aspects, 549*, 58–66.
95. Yang, Z., Xu, T., Li, H., She, M., Chen, J., Wang, Z., & Li, J. (2023). Zero-dimensional carbon nanomaterials for fluorescent sensing and imaging. *Chemical Reviews, 123*(18), 11047–11136.
96. Yadav, P. K., Upadhyay, R. K., Kumar, D., Bano, D., Chandra, S., Jit, S., & Hasan, S. H. (2021). Synthesis of green fluorescent carbon quantum dots from the latex of Ficus benghalensis for the detection of tyrosine and fabrication of Schottky barrier diode. *New Journal of Chemistry, 45*(28), 12549–12556.
97. Issa, M. A., Abidin, Z. Z., Sobri, S., Rashid, S. A., Mahdi, M. A., & Ibrahim, N. A. (2020). Fluorescent recognition of Fe^{3+} in acidic environment by enhanced-quantum yield N-doped carbon dots: Optimization of variables using central composite design. *Scientific Reports, 10*(1), 11710.
98. Shen, Y., Rong, M., Qu, X., Zhao, B., Zou, J., Liu, Z., & Niu, L. (2022). Graphene oxide-assisted synthesis of N, S Co-doped carbon quantum dots for fluorescence detection of multiple heavy metal ions. *Talanta, 241*, 123224.
99. Dhandapani, E., Maadeswaran, P., Raj, R. M., Raj, V., Kandiah, K., & Duraisamy, N. (2023). A potential forecast of carbon quantum dots (CQDs) as an ultrasensitive and selective fluorescence probe for Hg(II) ions sensing. *Materials Science and Engineering: B, 287*, 116098.
100. Kolaprath, M. K. A., Benny, L., & Varghese, A. (2023). A facile, green synthesis of carbon quantum dots from Polyalthia longifolia and its application for the selective detection of cadmium. *Dyes and Pigments, 210*, 111048.

101. Oliveira, J. J. P., Carneiro, S. V., Cruz, A. A. C., Fechine, L. M. U. D., Michea, S., Antunes, R. A., & Fechine, P. B. A. (2023). Determination of Co^{2+} ions in blood samples: A multi-way sensing based on NH_2-rich carbon quantum dots. *Dyes and Pigments, 215*, 111253.
102. Kundu, A., Maity, B., & Basu, S. (2023). Orange pomace-derived fluorescent carbon quantum dots: Detection of dual analytes in the nanomolar range. *ACS Omega*
103. Dong, Y. Q., Wang, R. X., Li, G., Chen, C. Q., Chi, Y. W., & Chen, G. N. (2012). Polyamine-functionalized carbon quantum dots as fluorescent probes for selective and sensitive detection of copper ions. *Analytical Chemistry, 84*, 6220–6224.
104. Wang, Z. M., Zhou, C., Wu, S. W., & Sun, C. Y. (2021). Ion-imprinted polymer modified with carbon quantum dots as a highly sensitive copper (II) ion probe. *Polymers, 13*, 1376.
105. Wu, S. W., Zhou, C., Ma, C. M., Yin, Y. Z., & Sun, C. Y. (2022). Carbon quantum dots-based fluorescent hydrogel hybrid platform for sensitive detection of iron ions. *Journal of Chemistry, 2022*, 3737646.
106. Sun, Y. J., Tang, J. S., Xiang, L., Hu, X., Wei, J., & Song, X. J. (2022). Hydrothermal synthesis of high-performance nitrogen-doped carbon quantum dots from Ophiopogon japonicus and their application in sensing Fe (III) with a broad quantitative range. *Digest Journal of Nanomaterials & Biostructures (DJNB), 17*(4).
107. Lai, C. M., Lin, S. M., Xiong, L., Wu, Y. X., Liu, C., & Jin, Y. Q. (2023). High quantum yield nitrogen-doped carbon quantum dots: Green synthesis and application as "on-off" fluorescent sensors for specific Fe^{3+} ions detection and cell imaging. *Diamond and Related Materials, 133*, 109702.
108. Kumar, A., Chowdhuri, A. R., Laha, D., Mahto, T. K., Karmakar, P., & Sahu, S. K. (2017). Green synthesis of carbon dots from Ocimum sanctum for effective fluorescent sensing of Pb^{2+} ions and live cell imaging. *Sensors and Actuators, B: Chemical, 242*, 679–686.
109. Tabaraki, R., & Abdi, O. (2020). Fluorescent sensing of Pb2+ by microwave-assisted synthesized N-doped carbon dots: Application of response surface methodology and Doehlert design. *Journal of the Iranian Chemical Society, 17*, 839–846.
110. Yan, Z. H., Yao, W., Mai, K., Huang, J. Q., Wan, Y. T., Huang, L., Cai, B., & Liu, Y. (2022). A highly selective and sensitive "on–off" fluorescent probe for detecting cadmium ions and L-cysteine based on nitrogen and boron co-doped carbon quantum dots. *RSC Advances, 12*, 8202–8210.
111. Wang, C. L., Xu, J., Li, H. Z., & Zhao, W. L. (2020). Tunable multicolour S/N co-doped carbon quantum dots synthesized from waste foam and application to detection of Cr^{3+} ions. *Luminescence, 35*, 1373–1383.
112. Oliveira, J. J. P., Carneiro, S. V., Cruz, A. A. C., Fechine, L. M. U. D., Michea, S., Antunes, R. A., Neto, M. L. A., Moura, T. A., Cesar, C. L., & Carvalho, H. F. (2023). Determination of Co^{2+} ions in blood samples: A multi-way sensing based on NH_2-rich carbon quantum dots. *Dyes and Pigments, 215*, 11253.
113. Dastidar, D. G., Mukherjee, P., Ghosh, D., & Banerjee, D. (2021). Carbon quantum dots prepared from onion extract as fluorescence turn-on probes for selective estimation of Zn2+ in blood plasma. *Colloids and Surfaces A: Physicochemical and Engineering Aspects, 611*, 125781.

Sensing Activity of Green Synthesized Carbon Quantum Dots for Detecting Heavy Metal Ions

Prashant Dubey

Abstract The presence of hazardous contaminants, particularly, highly toxic heavy metal ions (Cr^{6+}, Hg^{2+}, Pb^{2+}, Cd^{2+}, Sn^{2+}, Cu^{2+}, and As^{3+}), is a big issue to access safe drinking water. Evidences say that the uptake of these metal ions above permissible limit may cause severe metabolic disorder, organ failure, and cancer growth in living organisms. Therefore, sensing of heavy metal ions in water is a vital task. Green carbon dots (GCDs) have shown immense potential to detect these metal ions in water as well as in biological systems. The biomass-based GCDs are advantageous over chemical counterparts due to the abundant and cheap availability of non-toxic starting precursor along with the simple, energy-saving, and rapid synthetic procedure. This chapter deals with a comprehensive overview of biomass-based GCDs in terms of synthesis, structural/optical properties, and sensing attributes towards heavy metal ions. Involvement of various mechanisms for the fluorescence quenching-based detection of metal ions along with the possibility of fluorescence enhancement with some metal ions is also discussed. Hence, we strongly believe that GCDs have an admissible opportunity for the development of efficient, selective, and sensitive sensor platform for toxic metal ions to identify the quality of water and tracking in biological medium.

1 Introduction

Needless to say that excessive industrialization and urbanization of modern society have severely impacted the quality of water and ecological imbalance [1]. Various hazardous contaminants, particularly, toxic heavy metal ions including Cr^{6+}, Hg^{2+}, Pb^{2+}, Cd^{2+}, and As^{3+} can adversely affect the living organism and human health [2]. Moreover, essential transition metal ions such as Fe^{3+}, Cu^{2+}, Co^{2+}, and Zn^{2+} are indispensible for living beings but may also be dangerous after a certain level [3]. The World Health Organization (WHO) has set the permissible limit for many of the heavy

P. Dubey (✉)
Centre of Material Sciences, Institute of Interdisciplinary Studies (IIDS), University of Allahabad, Uttar Pradesh, Prayagraj 211002, India
e-mail: pdubey@allduniv.ac.in; pdubey.au@gmail.com

V. Kumar et al. (eds.), *Green Carbon Quantum Dots*, Engineering Materials,
https://doi.org/10.1007/978-981-97-6203-3_12

metals (Table 1) [4]. Therefore, detecting a trace level of heavy metal ions in aqueous medium and biological system is paramount to assure the quality of water. There are several optical sensing platforms for the detection of heavy metal ions, which are based on various nanomaterials such as semiconductor quantum dots [5], organic dyes [6], aptamer-based biosensors [7], metal nanoparticles [8], and carbon quantum dots (CQDs) [9]. Among them, CQDs are highly explored fluorescent nanoprobes for the sensitive, selective, and quick detection of different metal ions in aqueous solution.

The quasi-spherical CQDs, with size typically below 10 nm, are not only limited to metal ions sensing [10], but can also be applied for many other fields such as bioimaging [11], electrocatalysis [12], photocatalysis [13], drug delivery [14], energy storage [15], etc. The existence of CQDs is noticed by Xu et al. in 2004 during electrophoretic purification of single-walled carbon nanotubes (SWCNTs) [16]. Later on, excellent physiochemical features of CQDs including tunable photoluminescence (PL) [17], high quantum yield (QY) [18] aqueous/organic solvent dispersibility [19], chemical inertness [20], photo-stability [21], and low cytotoxicity [22] have driven their wide research interest in scientific community. A schematic illustration of the various properties and diverse applications of CQDs is presented in Fig. 1.

Broadly, crystalline/amorphous carbon core terminated by oxygen/nitrogen-containing functional groups (–COOH, –OH, $–NH_2$, etc.), constitute a typical CQDs nanomaterial [23, 24]. Starting precursor and experimental condition strongly decide the surface functionality of CQDs, which can be further modified via covalent/non-covalent linkage [25]. The availability of abundant and renewable natural resources has made a possibility of green and sustainable synthesis of CQDs using cost effective and energy efficient synthetic protocols.

The present chapter is focused on the heavy metal ions sensing application of biomass/biomass-based waste derived CQDs, which we referred to as green carbon dots (GCDs). After highlighting the green/general synthesis of CQDs, the article proceeds with describing the synthesis of GCDs from various synthetic routes, particularly, bottom-up methods and their structural/optical features. Before comprehensively evaluating the heavy metal ions sensor performance of these GCDs, the involvements of various sensing mechanisms are discussed.

Table 1 Permissible limits of different heavy metals according to the WHO guideline [4]

Heavy metals	Permissible limit (μg/L)	Permissible limit (μM)	Permissible limit (ppm)
Copper (Cu)	2000	~31.47	02
Chromium (Cr)	50	~0.96	0.05
Mercury (Hg)	06	~0.03	0.006
Lead (Pb)	10	~0.05	0.01
Cadmium (Cd)	03	~0.03	0.003
Arsenic (As)	10	~0.13	0.01
Manganese (Mn)	80	~1.46	0.08

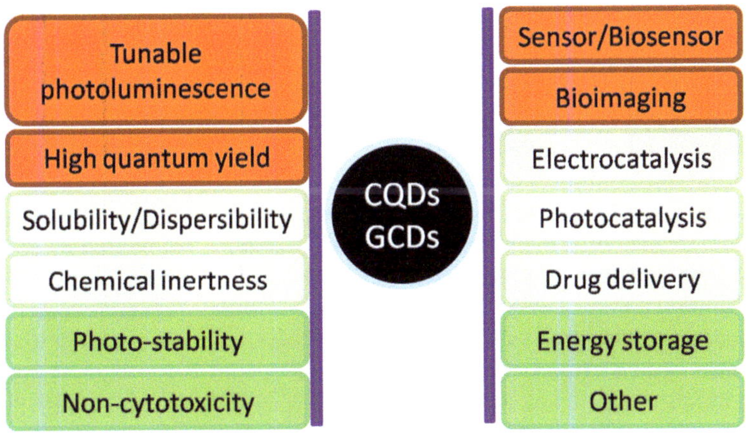

Fig. 1 Important properties and different applications of CQDs

2 Green Synthesis of CQDs (GCDs)

Development of sustainable and green synthesis is considered as a vital approach for any functional nanomaterials. In 1998, *Anastas* and *Warner* proposed twelve principles of green chemistry, which are as follows: minimization of waste generation, atom economy to maximize the involvement of all precursors, less hazardous chemical synthesis, design for safer chemical/product, use of safer solvents/auxiliary substances, minimum energy requirement, use of renewable feedstock, avoid of unnecessary derivatization, high efficient/selective catalysis, degradation possibility of final product, real-time monitoring to prevent the formation of hazardous side-products, and avoid accidental possibility (Fig. 2) [26]. These principles can provide a guideline to achieve GCDs.

Although, GCDs can be obtained from various non-toxic chemical precursors such as citric acid (CA) [27], ascorbic acid (AA) [28], sucrose [29], chitosan [30], and glucose [31], the synthesis of GCDs from biomass/biomass-based waste have surplus opportunity due to the extensive and cheap availability of resource material. Any ecological matter, which is living or lived-in some time before, can be classified as a "biomass", while discarded substances/things are called "biomass-based waste"

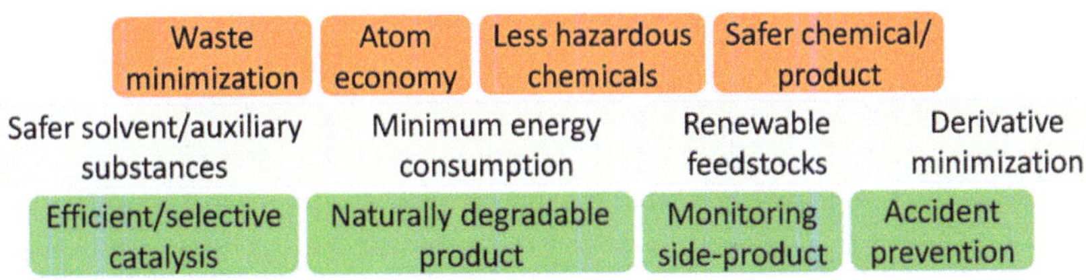

Fig. 2 Principles of green chemistry/synthesis [26]

[32, 33]. Biomass-based natural precursors are mainly acquired from plant/animal sources, whereas various discarded components of plant/animal/human, agriculture, and daily-life activity serve as biomass-based waste precursors for the fabrication of GCDs. The leaf/fruit/peel/seed/stem/flower/root of plants, egg/milk/silk/bones of animal, manure, hair/urine of human, various spices, and many more natural precursors have been successfully applied for the synthesis of GCDs (Tables 2, 3, 4, 5, 6, 7, 8, 9,10, 11, 12, 13, 14, 15, 16 and 17). These precursors are primarily composed of carbon, oxygen (O), and hydrogen; however depending on the starting source, it may also be enriched with heteroatoms like nitrogen (N), sulfur (S), phosphorous (P), calcium (Ca), etc. For example, *Bombyx mori* silk fiber contains N element as high as 18.65% [34]. Presence of Ca in animal bones can induce its self-doping in the synthesized GCDs [35]. Grass is probably the first biomass which is thermally treated to obtain photoluminescent polymer nanodots (PPNDs). These PPNDs have also shown a satisfactory sensing performance for Cu^{2+}, which opened a new era of biogenic GCDs [36]. Therefore, biomass-based materials possess a rich gallery of cheap, renewable, environmentally friendly, and non-toxic resource to produce GCDs and their doped-counterparts. Moreover, it can also provide a waste management strategy towards the production of value-added nanomaterial.

3 Synthetic Methods for CQDs/GCDs

3.1 Top-Down Methods

The synthesis of CQDs/GCDs via top-down methods involves breaking of bulk carbon precursors into nano-dimensional carbon particles. Graphite [215], activated carbon [216], vehicle exhaust soot [217], candle soot [218], etc. have been used as carbon precursors in the synthesis of CQDs. Moreover, arc-discharge [16], laser ablation [219, 220], electrochemical [215, 221], ball-milling (BM) [216], and chemical oxidation [217, 218] are typical top-down routes for the purpose of CQDs/GCDs (Fig. 3).

In fact the first accidental discovery of CQDs is based on chemical oxidation of arc-discharge soot obtained in the production of SWCNTs. The suspension of acid-treated soot is subjected to gel-electrophoresis to extract CQDs with blue-green/yellow/orange fluorescence under 365 nm UV light [16]. Subsequently, graphitic carbon target is irradiated with intense laser in the presence of water vapour and argon at 900 °C/75 kPa to prepare carbon soot via laser ablation. The resulting soot is refluxed with 2.6 M HNO_3 followed by surface passivation (diamine-terminated polyethylene glycol, PEG_{1500N}) to yield highly fluorescent CQDs [219]. In electrochemical method, high purity graphite anode is used for the preparation of CQDs [215, 221]. Zhou et al. [222] designed an electrochemical cell containing multiwalled carbon nanotubes-grown carbon paper as a working electrode, $Ag/AgClO_4$ as a reference electrode, and Pt-wire as a counter electrode. After applying potential across

Table 2 Biomass-derived GCDs: synthesis, size, QY, and sensing of Fe^{3+} ion

Biomass precursor	Synthesis condition	Size distribution/ average size (nm)[a]	QY (%)	Linear range (μM)	LOD (μM)	Refs.
Oriental plane leaves	P (350 °C, 2 h, N_2)	3.3–4.3/3.7	16.4	0–100	–	[37]
Potatoes	HT (180 °C, 2 h)	4–8/–	15	1–5, 5–50	0.025	[38][f]
Honey	HT (30% H_2O_2, 100 °C, 2 h)	–/2	19.8	0.005–100	0.0017	[39][f]
Yellow banana peels	P (200 °C, 2 h, air); MW (40 min)	4.5–8.5/5.9	–	2–16	0.211	[40]
Goose feathers	MW-HT (180 °C, 40 min)	15.5–26.5/21.5	17.1	2–7	0.196	[41][f]
Coriander leaves	HT (240 °C, 4 h)	1.5–2.98/2.39	6.48	0–6	0.4	[42]
Garlic	HT (180 °C, 10 h)	2–6/3.6	13	0.002–3	0.00022	[43][f, g]
Sheep wool	P (300 °C, 2 h, air); MW (700 W, 6 min)	2.7–9.3/6.05	22.5	0.1–10	0.01	[44][f]
Black tea	HT (200 °C, 5 h)	3.5–5.6/4.6	–	0.25–60	0.25	[45][f, g]
Otaheite gooseberry (*Phyllanthus acidus*) juice	HT (NH_3, 240 °C, 8 h)	3.5–5.5/4.5	14	2–25	0.9	[46][g]
Lycii fructus	HT (NH_3, 200 °C, 5 h)	2–5/3.3[b]	17.2	0–30	0.021	[47][f, g]
Cabbage juice	HT (200 °C, 5 h); HT (PEPA, 200 °C, 5 h); grafted over CaA	0.6–3.3/1.8	53.3	0–48	0.0014	[48][f]
Syringa obtata Lindl	HT (200 °C, 4 h)	1.5–4.5/2.76	12.4	0.5–80	0.11	[49][f, g]
Magnolia liliiflora flower	HT (240 °C, 12 h)	–/4	11	0–25	1.2	[50][g]
Borassus flabellifer flower	P (300 °C, 2 h, air)	3–8/–	13.97	0–0.03	0.01	[51][f]
Watermelon juice	HT (C_2H_5OH, 180 °C, 3 h)	3–7/–	10.6	0–300	0.16	[52]
Durian shell waste	HT (210 °C, 12 h)	1–6/3.2	6.2	0–20	0.128	[53][f]
Fish scales (*Crucian carp*)	HT (200 °C, 20 h)	5–10/–	6.9	1–78	0.54	[54][f]

(continued)

Table 2 (continued)

Biomass precursor	Synthesis condition	Size distribution/ average size (nm)[a]	QY (%)	Linear range (μM)	LOD (μM)	Refs.
Tomato (*Solanum lycopersicum*)	CO (H_2SO_4, 100 °C, 1 h)	5–10/–	12.7	0.1–1.25	0.016	[55][f]
	CO (H_3PO_4, 80 °C, 25 min)	"	4.21	"	0.072	
	CO (H_3PO_4, 80 °C, 20 min)	"	2.76	"	0.065	
Willow catkin	Comb. (H_2SO_4, urea, > 260 °C, air)	3–13/7.3	13.3	40–700	0.03	[56][g]
Miscanthus grass extract	Carbo. (EDA, 180 °C, 4 h)	–/4.6	11.65	0.02–200	0.02	[57]
Red lentils	HT (200 °C, 4 h)	4–10/6	13.2	2–20	0.1	[58]
Schisandra chinensis	EE; HT (200 °C, 8 h)	1–5/2.31	–	0–1000	0.125	[59]
Coffee beans	HT (180 °C, 12 h)	3.1–6/4.6	9.8	0–100[d] 0–100[e]	0.0154[d] 0.0163[e]	[60]
Lychee dust	ST (C_2H_5OH/H_2O, 180 °C, 5 h)	1.4–4.3/3.13	23.5	0.1–1.6	0.0236	[61][f, g]
Malus floribunda	HT (NH_3, 200 °C, 12 h)	3–5/3.5	18	10–50	2.5	[62][g]
Pineapple (*Ananas comosus*)	CO (H_2SO_4, 100 °C, 1 h)	1.25–3/2.08	18	0.05–100	0.03	[63][f]
Olive solid waste	P (600 °C, 1 h, air); CO (H_2O_2, 100 °C, 1.5 h); DA functionalization	1.3–4.3/3.2	–	0–5	0.32	[64][f]
Posidonia oceanica leaf litter	HT (180 °C, 6 h)	1–5/–	3.8	0–70	3.2	[65][f]
Green pepper seeds	HT (180 °C, 5 h)	2–6/3.2	8.7	1–500	0.1	[66][f, g]
Coconut water	Carbo. (70 °C, 5 h)	1–8/5	14.34	0–700	0.3	[67]

(continued)

Table 2 (continued)

Biomass precursor	Synthesis condition	Size distribution/ average size (nm)[a]	QY (%)	Linear range (μM)	LOD (μM)	Refs.
Mexican mint (*Plectranthus amboinicus*) leaves	MW-Reflux (1200 W, 100 °C, 4 min)	0.5–5.25/2.43	17	0–15	0.53	[68]
Alkali lignin	Heating (ABS, 90 °C, 1 h); HT (200 °C, 16 h)	2–9/4.86	23.68	0–300	0.77	[69]
Cranberry	HT (130 °C, 12 h)	4–6.5/5	14.5	0.0016–0.0228	0.00075	[70][f]
Black mulberry (*Morus nigra*)	HT (200 °C, 24 h)	3–6/4.5	24	5–30	0.47	[71][g]
Wolfberry	HT (180 °C, 8 h)	0.7–2.6/1.55	22	6–100[d] 3–100[e]	3[d] 3[e]	[72][f]
Bee pollen (water extract)	HT (180 °C, 24 h)	1–5/2.64	2.15[c]	50–100	27	[73]
Bee pollen (ethanol extract)	"	0.25–4.5/1.59	4.8[c]	20–100	15	
Kiwifruit juice	U (C_2H_5OH, 320 W, 25 °C, 3 h)	4.5–8.5/6.7	–	50–700	0.11	[74]
Chebulic myrobalan (*Terminalia chebula*)	HT (EDA, 180 °C, 6 h)	1–7/3.56	19.9	0.08–0.8	0.0045	[75]
Ice apple (*Borassus flabellifer*) endosperm	HT (180 °C, 12 h)	2–7/3	19.4	10–100	2.10	[76][f]
Red pitaya	HT (180 °C, 10 h)	4.7–7.5/5.8	13	0.002–0.04	0.0012	[77]
Mustard seeds	HT (180 °C, 10 h)	0.5–3.5/2	17.92[c]	0–75	0.51	[78][f, g]
Green beans	HT (180 °C, 12 h)	8–22/13	7.2	10–70	0.0036	[79][f, g]
Crescentia cujete pulp	HT (200 °C, 10 h)	1–10/4.36	1.57	0–25	0.257	[80][f]

(continued)

Table 2 (continued)

Biomass precursor	Synthesis condition	Size distribution/ average size (nm)[a]	QY (%)	Linear range (μM)	LOD (μM)	Refs.
Tagetes patula flower	HT (220 °C, 5 h)	3–8/5.15	29.88	0–4	0.32	[81][f]
Night jasmine (*Nyctanthes arbortristis*)	HT (175 °C, 24 h) HT (200 °C, 24 h)	1–6/3.58 0.5–4/2	8 13	3–100 ″	0.06 0.70	[82][f]
Lemon peels extract	HT (NH_4OH, 150 °C, 6 h)	–/7	35	0.02–0.1	1.7×10^{-5}	[83][f]
Blue-green algae	P (250 °C, 2 h, N_2); BM (180 rpm, 2 h); CO (2 mL H_2O_2, 0.2 g NaOH)	1–2.3/1.53	–	0–60	–	[84][f]
Human hair	MW (800 W, 8 min)	2–8/4.7	16	0–50	0.13	[85]
Prunus mume carbonisatus (traditional Chinese medicine)	CO (0.5 M NaOH/20wt% H_2O_2, 100 °C, 0.5 h)	1–4/2.5	9.1	0.5–60	0.03	[86]

Note EE: ethanol extraction, DA: dopamine, ABS: *o*-aminobenzenesulfonic acid
[a] Measured from TEM
[b] Measured from DLS
[c] Absolute QY
[d] Down-conversion fluorescence-based analyses
[e] Up-conversion fluorescence-based analyses
[f] Applicability for real sample analyses
[g] Applicability for analysis in biological system

Table 3 Biomass-derived GCDs: synthesis, size, QY, and sensing of Fe^{2+} ion

Biomass precursor	Synthesis condition	Size distribution/ average size (nm)[a]	QY (%)	Linear range (μM)	LOD (μM)	Refs.
Mango peel	P (300 °C, 2 h, air); U (15 min); CO (H_2SO_4, 25 °C, 6 h)	2–6/3	8.5	4–16	1.2	[87][e]
Mango leaves (*Mangifera indica*)	P (300 °C, 3 h, air); CO (H_2O_2, 12 h)	2–10/–	18.2	100–1000[b]	0.62[d]	[88][e]
Red Korean ginseng root	MW (700W, 7 × 2 min)	1–6/–	8	0.02–25	0.27	[89]
Bleached softwood kraft pulp	HT ($(NH_4)_2HPO_4$, 180 °C, 5 h)	–/ ~5	–	1356–4972 0–362[c]	3.070 1.278[c]	[90]
Wrightia tinctoria leaves	HT (180 °C, 8 h)	1–7/3.61	–	1–14	0.517	[91][e]
Lemon juice	HT (EDA, 200 °C, 3 h)	2.3–3.5/2.9	58.66	2–625	0.063	[92][e]

Note [a] Measured from TEM
[b] μL of 10 ppm Fe^{2+} solution
[c]Linear range/LOD of Fe^{3+}
[d] In ppm
[e] Applicability for real sample analyses

the electrodes, the colourless electrolyte solution (in acetonitrile, ACTN) gradually changed to yellow followed by dark brown colour, which showed blue fluorescence under UV light. The solid (after evaporation of ACTN) is dissolved in water and purified by dialysis to achieve CQDs.

Ge et al. [216] attempted a BM approach (rotation speed: 300 rpm, ball-to-powder ratio: 50:1, time: 40 h) on activated carbon/potassium carbonate to obtain CQDs in a size range of 1.6–6 nm. The PL characteristic of CQDs is found to be independent on excitation-wavelength (λ_{em}: ~430 nm corresponding to λ_{ex}: 260–380 nm) and showed 6% QY.

Chemical oxidation (CO) based breakdown of biomass or carbonaceous materials (derived from biomass) into GCDs are noticed in many reports (Tables 2, 3, 4, 8 and 10). The pyrolyzed mass of Olive solid waste (600 °C, 1 h) is refluxed with 0.45 wt% H_2O_2 for 90 min. and purified by centrifugation followed by filtration/dialysis to prepare N-GCDs (3.9 at% N, average size: 2.8 nm) [64]. Wang et al. [118] utilized *Ganoderma lucidum* bran (GB) for hydrothermal (HT) treatment, which resulted in soluble blue emitting BCDs and solid residue. The solid residue is treated with alkaline H_2O_2 to obtain green emitting GCDs (Fig. 4).

Table 4 Biomass-derived GCDs: synthesis, size, QY, and sensing of Cu^{2+} ion

Biomass precursor	Synthesis condition	Size distribution/ average size (nm)[a]	QY (%)	Linear range (μM)	LOD (μM)	Refs.
Pipe tobacco	HT (180 °C, 8 h)	1–4.6/–	3.2	0–0.07	0.01	[93]
Corn flour	HT (180 °C, 5 h)	2–6/3.5	7.7	0–1, 1–30	0.001	[94]
Prawn shells	Reflux (50% NaOH, 80 °C, 2 h); HT (200 °C, 8 h)	2–8/4	9	0.01–1.1	0.005	[95][e]
Pakchoi juice	HT (150 °C, 12 h)	1–3/1.8	37.5	0–0.1	0.00998	[96]
Oolong tea	HT (220 °C, 3 h)	1.7–5/3	4.9	0.01–75	0.002	[97][f]
Peanut shells	P (400 °C, 4 h, air)	1.8–4.2/3.3	10.58	0–5	4.8	[98][e]
Pear juice	HT (150 °C, 2 h)	–/10	–	0.1–50[c]	0.1[c]	[99][e]
Lily bulbs	MW (800 W, 6 min)	1.34–5.23/ 3.15	17.6	0.05–2	0.0128	[100][e]
Shikakai (*Acacia concinna* seeds)	MW (800 W, 2 min)	1.8–3.6/2.5	10.2	0.01–10	0.0043	[101][e]
Tuberose (*Polianthes tuberosa L.*) petals	P (300 °C, 8 h, air)	2.5–7/4.03	3	0–70	0.2	[102]
Coccinia indica	HT (L-cysteine, 180 °C, 7 h) HT (glycine, 180 °C, 7 h)	6.5–11/8.9 6.5–10.5/8.4	30 14.2	0–30 0–3.5[d]	0.045 6.2[d]	[103][e]
Finger millet ragi (*Eleusine coracana*)	P (300 °C, 3 h, air)	3–8/6	–	0–100	0.01	[104][e]
Grapes (*Vitis vinifera*) juice	TD (200 °C, 6 h)	5–12/8	32.1	0.07–60	0.02	[105][e]
Bran	ST (DMF, TA, 150 °C, 8 h)	2.5–8/4.85	46	0–500	0.0507	[106][e]
Smilax China	HT (200 °C, 5 h)	0.5–3/2.1	22.37	0.5–10, 75–225, 250–350	0.028	[107]
Carrot juice	HT (AAm, 220 °C, 12 h)	2–3[b]	5.8	60–200	0.187	[108]

(continued)

Table 4 (continued)

Biomass precursor	Synthesis condition	Size distribution/ average size (nm)[a]	QY (%)	Linear range (μM)	LOD (μM)	Refs.
Human fingernails	HT (200 °C, 3 h)	1.96–4.15/ 3.1	–	0–1	0.001	[109]
Empty fruit bunch biochar	MW-HT (IPA, 100 °C, 5 min)	2.5–7.5/4.5	–	0–400	0.42	[110]
Beef extract powder	HT (180 °C, 5 h)	1–6/3.11	61	0–20	0.086	[111][e, f]
Grape seeds	HT (TU, 200 °C, 8 h)	4–16/8.9	27.5	150–500[c]	0.048[c]	[112][e, f]
Palm kernel shell	MW (DEG, 400 W, 1.5 min)	2–8/5	43.07	100–500	50	[113][e]
Avocado seeds	P (600 °C, 2 h, air)	1.9–4.5/3.2	2–3	0.006–0.009, 1.5–4	0.0024	[114][e]
Wheat straw	P (250 °C, 2 h, air)	1.2–2.4/1.73	–	0–200	3.23	[115][e, f]
Milk powder	HT (Methionine, 220 °C, 3 h)	–/3.4	32.7	0.01–55	0.003	[116][e]
Plumeria alba	HT (20% NH_3, 200 °C, 6 h)	3–10/6.19	18.7	0.1–100	0.08	[117][e, f]
Ganoderma lucidum bran	HT (160 °C, 4 h) CO (H_2O_2/ NaOH, 25 °C, 24 h)	1.57–2.83/ 2.18 2.03–3.85/ 2.75	4.6 2.6	1–10 //	0.74 1.08	[118]
Wolfberry stems	HT (200 °C, 24 h)	1.25–3.5/ 2.25	59	0.01–0.08	0.0028	[119]

Note AAm: acrylamide, DEG: dethylene glycol
[a] Measured from TEM
[b] Height from AFM
[c] In μg/mL
[d] Linear range/LOD for Fe^{3+}
[e] Applicability for real sample analyses
[f] Applicability for analysis in biological system

3.2 *Bottom-Up Methods*

Bottom-up syntheses of CQDs/GCDs are based on dehydration/condensation of small organic molecules followed by polymerization, carbonization, and surface passivation [223]. There are a vast variety of carbon-containing organic precursors from laboratory chemicals [223] to biomass (Tables 2, 3, 4, 5, 6, 7, 8, 9,10, 11, 12,

Table 5 Biomass-derived GCDs: synthesis, size, QY, and sensing of Co^{2+} ion

Biomass precursor	Synthesis condition	Size distribution/ average size (nm)[a]	QY (%)	Linear range (μM)	LOD (μM)	Refs.
Pig skin	HT (250 °C, 2 h)	3.5–7/5.58	24.1	1–100	0.68	[120][c]
Kelp (seaweed plant)	MW-HT (EDA, 200 °C, 1.5 h)	3–4.4/3.7	23.5	1–200	0.39	[121][d]
Artemisia annua	HT (EDA, 180 °C, 2 h)	2–6/3.65	13.6	2.5–25	0.231	[122][c, d]
Flax straw	P (250 °C, 2 h, air); HT (EDA, 160 °C, 10 h)	1.1–4/2.2	20.7	0–500	0.38	[123][d]
Frozen tofu	HT (EDA + H_3PO_4, 210 °C, 4 h)	2–4/2.9	9.5	0–0.5	0.058	[124][c, d]
Alkali lignin	ST (C_2H_5OH, DETA, 180 °C, 12 h)	3–7.5/4.5	15.7	0–500 0–250[b]	0.45 0.27[b]	[125][d]
Sida cordifolia root	HT (TETA, 200 °C, 10 h)	3–10/6.3	17.8	0–4.7	0.11	[126][d]

Note DETA: diethylenetriamine
[a] Measured from TEM
[b] Linear range/LOD for Fe^{3+}
[c] Applicability for analysis in biological system
[d] Applicability for real sample analyses

Table 6 Biomass-derived GCDs: synthesis, size, QY, and sensing of Zn^{2+} ion

Biomass precursor	Synthesis condition	Size distribution/ average size (nm)[a]	QY (%)	Linear range (μM)	LOD (μM)	Refs.
Potato starch	U (HCl, 400 W, 6 h); Reflux (90 °C, 6 h)	1.75–5.5/ < 5	10	0–20	0.001	[127]
Cambuci (*Campomanesia phaea*) juice	HT (NH_4OH, 190 °C, 6 h)	2.25–5/3.77	21.26	0–125	5.4	[128]
Onion (*Allium cepa*) extract	HT (EDA, 180 °C, 2.5 h)	–/1.15	6.2	0–100	6.4	[129][b]

Note [a] Measured from TEM
[b] Applicability for real sample analyses

Table 7 Biomass-derived GCDs: synthesis, size, QY, and sensing of Cr^{6+}/Cr^{3+} ions

Biomass precursor	Synthesis condition	Size distribution/average size (nm)[a]	QY (%)	Linear range (μM)	LOD (μM)	Refs.
Lemon peel	HT (200 °C, 12 h)	1–3/–	14	2.5–50	0.073	[130]
Pomelo juice	HT (APS, 200 °C, 12 h)	5–8.7/6.7	18.7	1–40	0.52	[131][r]
Tulsi leaves	HT (200 °C, 4 h)	–/5	3.06	1.6–50	1.6	[132][r]
Papaya (*Carica papaya*) pulp	P (200 °C, 15 min); EDTA functionalization	–/~7	23.7	10–100[g, h]	0.708[g, h]	[133][r]
Groundnuts (peanuts)	HT (EDA, 250 °C, 6 h)	2–8/2.5	17.6	2–9	1.9	[134]
Denatured milk	HT (160 °C, 2 h)	1–5/2	15.34	5–100	14	[135][r]
Waste tea extract	HT (EDA, $Cu(Ac)_2.H_2O$, 150 °C, 6 h)	–/0.85	3.26	0.5–50[i, j] 0.8–30[i, k]	2.044[j] 1.418[k]	[136][r]
Jeera (*Cuminum cyminum*)	HT (250 °C, 6 h); Cystamine functionalization	1–18/6–9	5.33	1–10	1.57	[137][r]
Shallot extract	HT (180 °C, 4 h)	3.65–8.15/4.14	32.34	20–100[h]	3.5[h]	[138][r]
Flax straw	P (250 °C, 2 h, air); HT (EDA, 160 °C, 10 h)	1.1–4/2.2	20.7	0.5–80	0.19	[123][r]
Rice fried *Codonopsis pilosula*	U (CH_3OH, 25 °C, 4 h)	7.5–15.5/11.54	12.8	0.03–50	0.015	[139][r]
Panax notoginseng	HT (CA, EDA, 180 °C, 12 h)	5.58–7.75/6.66	18.41	0–6	0.000185	[140][r]
Sophora flavescens ait	HT (180 °C, 3 h)	1.5–3.5/2.4	35.2	0.06–40	0.02	[141][r]
Jute caddies	U (H_2SO_4, 30 min); BZC functionalization	4.5–8/6.5	14.5	1–140, 160–250	0.03	[142][r]

(continued)

Table 7 (continued)

Biomass precursor	Synthesis condition	Size distribution/average size (nm)[a]	QY (%)	Linear range (μM)	LOD (μM)	Refs.
Dried rose petals	HT (EDA, L-cysteine, 180 °C, 6 h)	–/6.9	28	0.000025–0.0001	0.000081 0.000121[l]	[143][r]
Ripe banana peel	″	–/9.3	27	0.000025–0.00025[l]		
Seaweed (*Sargassum carpophyllum*)	P (220 °C, 2 h, air)	0.77–4.62/2.19	4.37	0–120	1.04	[144][r]
Alkali lignin	Acidolysis (APBA, 90 °C, 1 h); HT (filtrate, 200 °C, 12 h)	1.5–6/3.82	0.095[c]/7.5[d]/4.5[e]	0–100[c, d, e]	0.054[c]/0.049[d]/0.077[e]	[145]
Wool keratin	HT (200 °C, 10 h)	2–6/–	8[f]	2.5–50 0.25–125[l]	0.0142 0.113[l]	[146][r]
Banana stem	HT (H_3PO_4, 180 °C, 6 h)	2–20/9.2	15.1	10–30	2.4	[147][r]
Annona reticulate leaves	HT (200 °C, 5 h)	4–9/6	12	0–40[m]	2[m]	[148][r]
Avocado seeds	P (600 °C, 2 h, air)	1.9–4.5/3.2	2–3	0.0005–0.0009, 30–200	0.00007	[114][r]
Banana pseudo-stem derived cellulose	HT (SO_3H-IL, PEG400, 200 °C, 6 h)	1–4.6/3.3	29.27	0.25–10, 15–30	0.017	[149][r]
Residual canola oil	Heating (100 °C, 25 min); Carbo. (H_2SO_4, 5 min)	4–8.8/9.7[b]	2.9	0.01–0.08	0.02	[150][r]

(continued)

Table 7 (continued)

Biomass precursor	Synthesis condition	Size distribution/average size (nm)[a]	QY (%)	Linear range (μM)	LOD (μM)	Refs.
Ligusticum wallichii + *Corydalis*	ST (C_2H_5OH, OPD, 180 °C, 9 h)	1.2–2.8/1.95	3.58	5–350[n]/250–600[o]/300–850[p]	0.28[n]/0.87[o]/0.90[p]	[151][r]
Allium macrostemon bunge polysaccharides	MW (L-cysteine, H_3PO_4, 630 W, 160 s)	1.25–3.5/2.34	16.4	50–500	5.25	[152][r]
Cinnamon	HT (ZA, 180 °C, 16 h)	2–7/5	7.47	0–20[i] 0–20[i, q]	3.97[i] 2.05[i, q]	[153][r]

Note APS: ammonium persulfate, EDTA: ethylenediaminetetraacetic acid
[a] Measured from TEM
[b] Measured from DLS
[c] Measured at 300 nm excitation
[d] Measured at 330 nm excitation
[e] Measured at 490 nm excitation
[f] Absolute QY
[g] In ppb
[h] Total Cr i.e. Cr^{6+} and Cr^{3+}
[i] In μg/mL
[j] Linear range/LOD of CrO_4^{2-}
[k] Linear range/LOD of $Cr_2O_7^{2-}$
[l] Linear range/LOD of Fe^{3+}
[m] Linear range/LOD of Cr^{3+}
[n] Measured at 326 nm excitation
[o] Measured at 380 nm excitation
[p] Measured at 578 nm excitation
[q] Linear range/LOD of Mn^{7+}
[r] Applicability for real sample analyses

Table 8 Biomass-derived GCDs: synthesis, size, QY, and sensing of Hg^{2+} ion

Biomass precursor	Synthesis condition	Size distribution/ average size (nm)[a]	QY (%)	Linear range (μM)	LOD (μM)	Refs.
Pomelo peel	HT (200 °C, 3 h)	2–4/–	6.9	0.0005–0.01, 0.5–40	0.00023	[154][h]
Sweet potato	HT (180 °C, 3 h)	1–3/–	2.8	0.001–0.05, 1–40	0.001	[155][h]
Strawberry juice	HT (180 °C, 12 h)	–/5.2	6.3	0.01–0.1, 1–50	0.003	[156][h]
Camphor	Comb.; CO (H_2SO_4/H_2O_2, 15 h)	4–8/–	21.16	0.5–5[d]	0.5	[157]
Jinhua bergamot	HT (200 °C, 5 h)	–/10	50.78	0.01–0.1, 1–100 0.025–1, 5–100[e]	0.0055 0.075[e]	[158]
Urine (unmodified diet)	P (200 °C, 12 h, air)	10–40/20.6	5.3	0–50/0–30[f]	2.7/3.4[f]	[159]
Urine (asparagus rich diet)	//	10–70/38.8	2.7	//	2.7/2.9[f]	
Urine (supplemented with vitamin C)	//	2–28/11.4	4.3	//	1.8/1.7[f]	
Pigeon egg white	P (300 °C, 3 h, air)	2.2–4.2/3.3	17.48	0.05–0.3	0.0346	[160][i]
Pigeon egg yolk	//	2.2–4.2/3.2	16.34	0–0.16	0.0349	
Pigeon feather	//	2.8–4.3/3.8	24.87	0–0.12/ 0–1.6[e]	0.0103/ 0.0609[e]	
Orange juice	HT (EDA, 200 °C, 11 h)	0.5–3/1.86	31.7	4–32	–	[161][h]
Mushroom (*Pleurotus spp.*)	HT (170 °C, 3 h); DHLA functionalization	2–6/4	25	0–0.1	1.74×10^{-5}	[162]
Bagasse waste	HT (urea, 180 °C, 12 h)	3–6/5	11.8	0.005–0.8	0.002	[163][h]
Table sugar	P (150 °C, air); U (NH_3, 12 h)	3–4/–	–	1×10^{-4}–7.5×10^{-4}	9.8×10^{-5}	[164][h]

(continued)

Table 8 (continued)

Biomass precursor	Synthesis condition	Size distribution/ average size (nm)[a]	QY (%)	Linear range (μM)	LOD (μM)	Refs.
Spider silk	HT (NaOH, 180 °C, 10 h)	1–6/3.65	21.5[c]	0–0.08	0.0053	[165][h, i]
Lemon juice	TD (100 °C, 1.5 h)	~8[b]	2.4	0–1820	36	[166]
Fish scale (China grass carp)	MW-HT (200 °C, 2 h)	1.4–3.4/2.6	19.92	0–30	0.014	[167][h]
Coccinia indica	HT (180 °C, 7 h)	2–9/5.2	5.6	0–0.025	0.0033	[103][h]
Ginkgo leaves	HT (220 °C, 10 h)	3.8–5.6/4.9	11	0.5–20	0.0124	[168][h]
Gardenia fruit	HT (180 °C, 5 h)	1.3–3.3/2.1	10.7[c]	0.2–20	0.32	[169]
Beef extract powder	HT (180 °C, 5 h)	1–6/3.11	61	0–20	0.079	[111][h, i]
Bitter tea oil residue	P (700 °C, 6 h, air); HT (urea, 160 °C, 12 h)	50–80/–	7.2	2–100[g] 2–100[e, g]	– –	[170]
Sugarcane bagasse	HT (CA, 200 °C, 12 h)	2–8/–	14.12	0–300	0.1	[171]
Patera (*Typha angustata Bory*)	HT (TU, 200 °C, 10 h)	2–7/5	83	0.01–60	0.0031	[172][h]
Chicken feather	HT (180 °C, 24 h)	3–9.5/5	77.2	0–500	0.0062	[173]
Kaempferol	HT (NaOH, 180 °C, 10 h)	2.5–3.8/3.2	–	5–50	1.6	[174][h]
Carboxymethylated nanocellulose	HT (urea, 190 °C, 6 h)	0.2–1.8/ 1.12	2.8	0–100	8.29	[175]

Note [a] Measured from TEM
[b] Height from AFM
[c] Absolute QY
[d] Dynamic range
[e] Linear range/LOD of Fe^{3+}
[f] Linear range/LOD of Cu^{2+}
[g] In ppm
[h] Applicability for real sample analyses
[i] Applicability for analysis in biological system

13, 14, 15, 16 and 17) and non-biodegradable waste such as plastics [224] that can serve as a starting material for the outcome of CQDs/GCDs. HT, solvothermal (ST), microwave (MW), microwave-assisted hydrothermal (MW-HT), ultrasonication (U), pyrolysis (P), and combustion (Comb.)/carbonization (Carbo.)/thermal decomposition (TD) are common bottom-up techniques, which are applied in the preparation of GCDs (Fig. 3, Tables 2, 3, 4, 5, 6, 7, 8, 9,10, 11, 12, 13, 14, 15, 16 and 17). It is

Table 9 Biomass-derived GCDs: synthesis, size, QY, and sensing of Pb^{2+} ion

Biomass precursor	Synthesis condition	Size distribution/ average size (nm)[a]	QY (%)	Linear range (μM)	LOD (μM)	Refs.
Sago industrial waste	P (400 °C, 1 h, air)	6–17/–	–	0–47.62 0–47.62[b]	7.49 7.78[b]	[176]
Chocolate	HT (NaOH, 200 °C, 8 h)	4.5–8.5/6.41	–	0.033–1.67	0.0127	[177][c]
Tulsi leaves (*Ocimum sanctum*)	HT (180 °C, 4 h)	1–4/3	9.3	0.01–1	0.00059	[178][c, d]
Lantana camara berries	HT (EDA, 180 °C, 3 h)	3–8/5	33.15	0–0.2	0.00964	[179][c, d]
Coccinia indica	HT (EDA, 180 °C, 7 h)	4.5–10/7.03	12.3	0–0.2	0.27	[103][c]
Spinach extract powder	HT (EDA, 120 °C, 8 h)	1–5/2.8	22.16	0.18–510	0.055	[180]
Cyano-bacteria	MW (NaOH, 700 W, 10 min)	2.5–6.4/4.10	14.03	0–50, 100–300	0.85	[181]

Note [a] Measured from TEM
[b] Linear range/LOD of Cu^{2+}
[c] Applicability for real sample analyses
[d] Applicability for analysis in biological system

worthwhile to mention that bottom-up approaches are advantageous over top-down methods in terms of simple/economical operation and possibility for the large scale synthesis of GCDs. Moreover, bottom-up route is greener and facile than traditional top-down methods.

3.2.1 HT Synthesis of GCDs

HT processing (water as a solvent) is the most common choice for synthesizing GCDs because it requires significantly low temperature (usually below 200 °C) along with simple experimental set-up (Tables 2, 3, 4, 5, 6, 7, 8, 9,10, 11, 12, 13, 14, 15, 16 and 17). He et al. [48] extracted Cabbage juice for the HT treatment at 200 °C (5 h). The resultant mixture is again treated in the presence of polyethylene polyamine (PEPA) to get brownish-yellow solution, which is centrifuged and

Table 10 Biomass-derived GCDs: synthesis, size, QY, and sensing of Cd^{2+} ion

Biomass precursor	Synthesis condition	Size distribution/ average size (nm)[a]	QY (%)	Linear range (μM)	LOD (μM)	Refs.
Camphor	Comb.; CO (H_2SO_4/ H_2O_2, 15 h)	4–8/–	21.16	0.5–6[b]	0.5	[157]
Scallions	MW (800 W, 4 min)	1.5–5.5/3.23	18.6	0.1–3, 5–30	0.015	[182][e]
Coconut coir	P (600 °C, 3 h, N_2)	< 10/–	48	–	0.00018	[183][e]
Rice husks	HT (EDA, 190 °C, 12 h)	1–7/1.94	–	50–300[c]	–	[184]
	HT (AA, 190 °C, 12 h)	1–10/3.89	//	//	//	
Tea residue	Stirring (DES, 100 °C, 2 h); ST (200 °C, 8 h)	< 10/–	16.99	1–20[d]	2.14[d]	[185][e]
Polyalthia longifolia leaves	HT (150 °C, 6 h)	1.5–6.5/3.33	22	0.0073–12	0.0024	[186][e]

Note DES: deep eutectic solvent
[a] Measured from TEM
[b] Dynamic range
[c] In ppm
[d] In μg/mL
[e] Applicability for real sample analyses

Table 11 Biomass-derived GCDs: synthesis, size, QY, and sensing of Sn^{2+} ion

Biomass precursor	Synthesis condition	Size distribution/ average size (nm)[a]	QY (%)	Linear range (μM)	LOD (μM)	Refs.
Rice husk	Carbo. (H_2SO_4, 120 °C, 30 min)	–	–	0–6130	18.7	[187]
Lignin (enzymatic hydrolyzed)	HT (Melamine + $La(NO_3)_3.6H_2O$, 180 °C, 5 h)	1.6–2.8/2.2	–	2–10 2–10[b]	1.1 0.99[b]	[188][c]

Note [a] Measured from TEM
[b] Linear range/LOD for Fe^{3+}
[c] Applicability for analysis in biological system

Table 12 Biomass-derived GCDs: synthesis, size, QY, and sensing of As^{3+} ion

Biomass precursor	Synthesis condition	Size distribution/ average size (nm)[a]	QY (%)	Linear range (μM)	LOD (μM)	Refs.
Quince fruit (*Cydonia oblonga*)	MW-HT (220 °C, 30 min)	< 10/4.85	8.55	0.1–2[d]	0.02[d]	[189][g]
Prickly pear cactus juice	Reflux (GSH, 100 °C, 1 h); HT (180 °C, 12 h)	3–8/5.6	12.7	0.002–0.012	0.0023	[190][g]
Banana leaves extract	HT (160 °C, 8 h)	2–5/3.5	~42.2	10–30 0.0001–0.0005[e]	1.53 0.000067[e]	[191][g]
Cynodon dactylon	MW (800 W, 5 min)	4.5–15.5/8.6	1[b], 8[c]	0.022–0.044 0.16–0.25[f]	0.019 0.10[f]	[192]

Note GSH: glutathione
[a] Measured from TEM
[b] In water
[c] In ethanol
[d] In μg/mL
[e] Linear range/LOD of Hg^{+}
[f] Linear range/LOD of Fe^{3+}
[g] Applicability for real sample analyses

dilated to yield N-GCDs. The N-GCDs is incorporated in calcium alginate (CaA) to obtain cross-linked CaA-N-GCDs film (size distribution: 0.6–3.3 nm, QY: 53.3%) for sensing application. Chattopadhyay et al. [111] dissolved Beef extract power in H_2O for HT treatment. Based on maximum production/QY of synthesized N-GCDs (63/61%, 3.7% N content), 180 °C is found to be optimum temperature for carbonization. Goswami et al. [147] demonstrated a phosphoric acid (H_3PO_4)-assisted HT synthesis of N,P-GCDs (N/P content: 5.78/8.08%) from Banana stems. The H_3PO_4 provided a low pH to promote/catalyze degradation of cellulose/hemicelluloses/lignin, which subsequently polymerized by dehydration to yield N,P-GCDs (QY: 15.11%). Recently, Durrani et al. [79] used natural green Beans as only precursor to synthesize N,S,P-GCDs (N: 4.1%, S: 0.3%, and P: 0.2%).

3.2.2 ST Synthesis of GCDs

The ST method is a modified version of HT process where an organic solvent (ethanol (C_2H_5OH), dimethylforamide (DMF), acetone, etc.) has been used in place of only

Table 13 Biomass-derived GCDs: synthesis, size, QY, and sensing of Ag^+ ion

Biomass precursor	Synthesis condition	Size distribution/average size (nm)[a]	QY (%)	Linear range (μM)	LOD (μM)	Refs.
Denatured protein	MW (700 W, 5 min)	2.4–5/5.4	14	0.1–10	0.03	[193][b]
Broccoli juice	HT (190 °C, 6 h)	2–6/–	–	0–600	0.5	[194]
Purple *perilla* (*Perilla frutescens L. Britt.*)	HT (260 °C, 5 h)	1–5.5/2.8	9.01	0.0005–0.01, 0.01–3	0.0014	[195][b]
Cryptococcus podzolicus 5–2	HT (200 °C, 6 h)	2.2–8.5/4.8	3.99	0–15	0.11357	[196][b]
Alkali lignin	HT (H_3PO_4, 180 °C, 6 h)	2.3–3.5/2.81	–	100–1600	–	[197]
		2.2–3.2/2.56	"	400–2200	"	
		2.1–2.8/2.32	"	100–2000	"	
		2.15–2.4/2.28	"	100–1600	"	
κ-Carrageenan	HT (GSH, 180 °C, 18 h)	1–3.5/2.1	–	0–100	0.124	[198][b]
Cane molasses	HT (160 °C, 24 h); HT (L/D-cysteine, 120 °C, 2 h)	1.5–6.5/3.6	18.86/17.38	40–220/20–140	3.38/4.09	[199][b]

Note GSH: glutathione
[a] Measured from TEM
[b] Applicability for real sample analyses

water for preparing GCDs (Tables 2, 4, 5, 7, 10, 15). Dried/powdered Lychee dust is mixed in H_2O-C_2H_5OH to heat at 180 °C for 3 h in closed autoclave. After centrifuging and dialyzing the solution, the resultant N-GCDs showed an average size/QY of 3.13 nm/23.5% along with the 1.24% N content [61]. Xu et al. [106] prepared N-GCDs from a mixture of Bran and tartaric acid (TA) in DMF via ST method (150 °C, 8 h) (Fig. 5a). The resultant product is routinely purified to obtain green emitting N-GCDs with QY as high as 46%. Moreover, the quasi-spherical size of N-GCDs and size distribution histogram (2.5–8 nm) can be seen from transmission electron microscopy (TEM)/high-resolution TEM (HRTEM) imaging (Fig. 5b and c). Recently, Zhang et al. [151] demonstrated a one-pot ST method in the preparation of N, Cl-GCDs. They used waste traditional Chinese medicine (*Ligusticum wallichii* and *Corydalis*) as a natural precursor and o-phenylenediamine (OPD) as an N source in C_2H_5OH with 0.25 mL con. HCl to heat at 180 °C for 9 h.

Table 14 Biomass-derived GCDs: synthesis, size, QY, and sensing of Au^{3+} ion

Biomass precursor	Synthesis condition	Size distribution/ average size (nm)[a]	QY (%)	Linear range (μM)	LOD (μM)	Refs.
Peach gum polysaccharide	HT (EDA, 180 °C, 16 h)	2–5/–	28.46	0–50	0.064	[200][b]
Jackfruit seeds	MW (H_3PO_4, 600 W, 1.5 min)	3–7/5	17.91	0–100	0.239	[201]
Gum tragacanth	HT (EDA, 180 °C, 10 h)	0.5–5/~2.2	66.74	0–100	2.69	[202]
Denatured sour milk	HT (160 °C, 3 h)	0.5–3.3/–	13	10–150	0.95	[203][b]
Neem seed kernel	MW (H_3PO_4, 600 W, 100 s)	1.25–3.25/2.2	12.5	0–25	0.016	[204]

Note [a] Measured from TEM
[b] Applicability for real sample analyses

Table 15 Biomass-derived GCDs: synthesis, size, QY, and sensing of Pd^{2+} ion

Biomass precursor	Synthesis condition	Size distribution/ average size (nm)[a]	QY (%)	Linear range (μM)	LOD (μM)	Refs.
Red beetroot	HT (180 °C, 10 h)	5–7/–	27.6	3–43	0.033	[205][d]
Magnolia denudata leaf	ST (Acetone, 120 °C, 5 h); Pluronic F-127 incorporation	–/34[b]	7.97[c]	4–18, 20–200	0.0853	[206][d, e]

Note [a] Measured from TEM
[b] Measured from DLS
[c] Absolute QY
[d] Applicability for real sample analyses
[e] Applicability for analysis in biological system

3.2.3 MW Synthesis of GCDs

MW method offers a homogeneous heating and high reaction rate. Therefore, the synthesis of GCDs can be complete within a few minutes, indicating an energy-saving process. The applicability of MW synthesis for GCDs can be revealed from Tables 2, 3, 4, 7, 9, 10, 12–14. Lily bulb is irradiated with MW in domestic microwave oven (800 W) for 6 min and resultant solid is extracted with H_2O followed by centrifugation/filtration/dialysis to achieve N,P-GCDs (average size/QY: 3.15 nm/17.6%) [100]. Bhatt et al. [225] employed Prickly pear cactus along with *m*-xylylenediamine to prepare N-GCDs through MW irradiation (450 W, 10 min) followed by sequential purification. The N-GCDs (6.76% N) showed a blue fluorescence under UV light with

Table 16 Biomass-derived GCDs: synthesis, size, QY, and sensing of Al^{3+} ion

Biomass precursor	Synthesis condition	Size distribution/average size (nm)[a]	QY (%)	Linear range (μM)	LOD (μM)	Refs.
Pear (*Pyrus pyrifolia*) juice	HT (180 °C, 6 h)	1–3.5/2	10.8	0.005–50	0.0025	[207][c]
Catharanthus roseus leaves	HT (200 °C, 4 h)	–/5	28.2	0–6 0–6[b]	0.5 0.3[b]	[208]
Osmanthus fragrans	HT (180 °C, 10 h)	2–6/3.1	21.9	0.1–100	0.026	[209][c, d]
Arrowroot powder	MW-HT (200 °C, 13 min)	< 10/~5	–	–	37	[210]

Note [a] Measured from TEM
[b] Linear range/LOD for Fe^{3+}
[c] Applicability for real sample analyses
[d] Applicability for analysis in biological system

Table 17 Biomass-derived GCDs: synthesis, size, QY, and sensing of $V^{5+}/Cs^{2+}/Ru^{3+}/Mn^{2+}$ ions

Biomass precursor	Synthesis condition	Size distribution/average size (nm)[a]	QY (%)	Linear range (μM)	LOD (μM)	Refs.
Lemon juice	HT (240 °C, 12 h)	4–5.8/–	21	0–8[b]	3.2[b]	[211][g]
Maple tree leaves	HT (190 °C, 8 h)	2–10/~3.1	–	0.1–100[c]	0.024[c]	[212][g]
Starch	P (300 °C, 6 h, air)	0.5–2.5/1.3	20.7	0–1000[d]	< 20[d]	[213]
Garlic peel extract	HT (180 °C, 36 h)	2–5/3.5	10	10–50[e] 10–50[f]	0.95[e] 0.75[f]	[214]

Note [a] Measured from TEM
[b] Linear range/LOD of V^{5+} in ppm or μg/mL
[c] Linear range/LOD of Cs^{2+}
[d] Linear range/LOD of Ru^{3+}
[e] Linear range/LOD of Mn^{2+}
[f] Linear range/LOD of Fe^{3+}
[g] Applicability for real sample analyses

14.8% QY. Recently, Sandeep et al. [85] transformed human Hair into fluorescent N-GCDs (C:O:N atomic ratio in %: 65:30:5, size range/QY: 3–5 nm/16%) through MW treatment (800 W, 8 min) followed by the sonication and column chromatography.

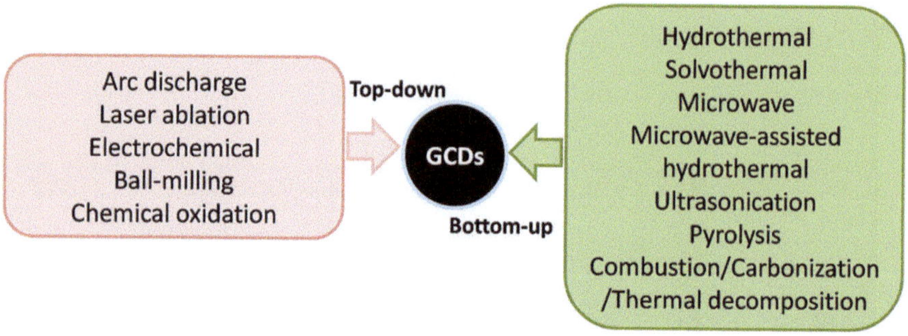

Fig. 3 Various top-down and bottom-up synthetic methods for GCDs/CQDs

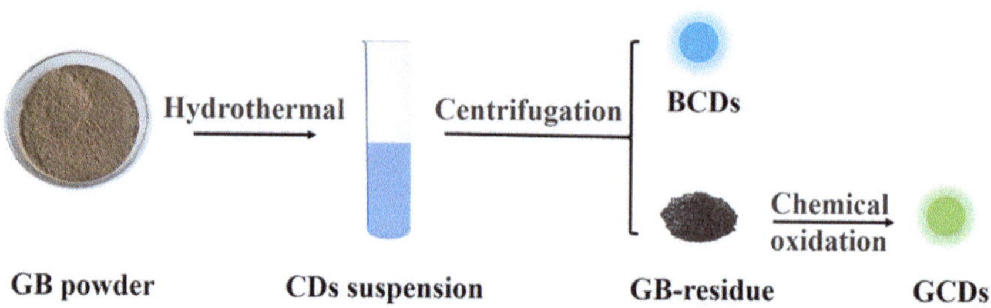

Fig. 4 HT/chemical oxidation syntheses of BCDs/GCDs using GB biomass. Reproduced from ref. [118] with permission from the Royal Society of Chemistry

3.2.4 MW-HT Synthesis of GCDs

In case of MW-HT method, the advantage of HT condition (auto-generated pressure) and MW (homogeneous/fast heating) has been combined to produce a high quality GCDs in short reaction time (Tables 2, 4, 5, 8, 12, 16). For example, Quince fruit (*Cydonia oblonga*) powder is heated in the closed autoclave via MW irradiation (850 W, 30 min). The reaction temperature (220 °C) is attained in 1.5 min, which indicates a quick synthesis process. The N,S-GCDs is obtained after purification, which possessed a small content of N (0.76%) and S (0.41%) elements [189]. Zaman et al. [110] investigated the effect of heating temperature (60–100 °C) in the MW-HT synthesis of GCDs using Empty fruit bunch biochar in isopropyl alcohol (IPA). At 100 °C (reached in only 2 min), the production of GCDs attained maximum yield (15.22 ± 0.84%), which is ascribed to lowering the surface tension of IPA that reduced intermolecular interaction between solvent and GCDs.

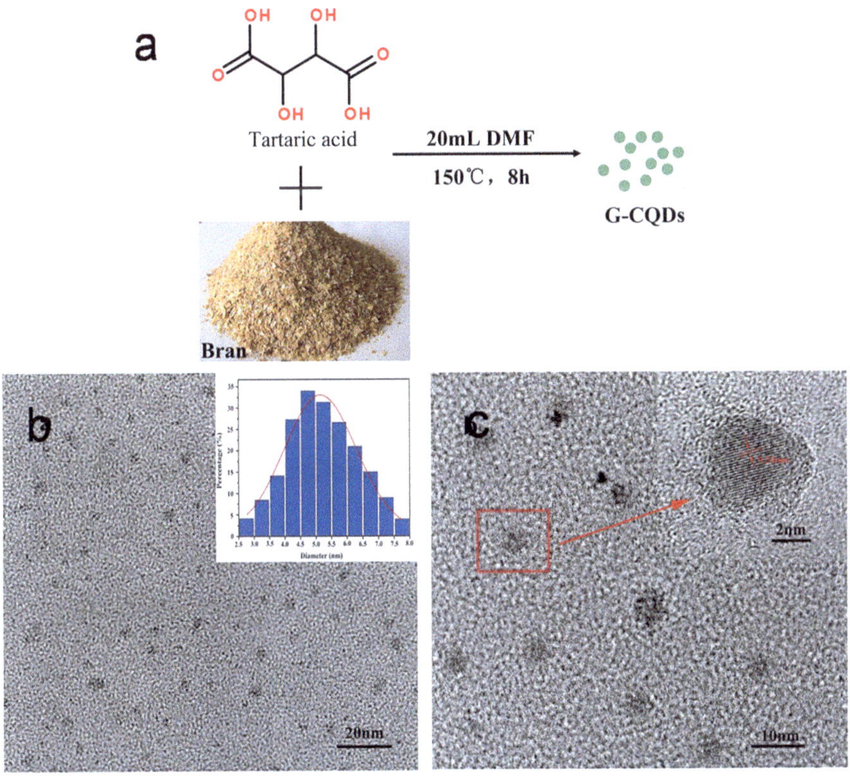

Fig. 5 **a** ST synthesis of N-GCDs using a mixture of bran and TA. **b, c** TEM/HRTEM images of N-GCDs along with the size distribution histogram. Reproduced from ref. [106] with permission from the Royal Society of Chemistry

3.2.5 Ultrasonic (U) Synthesis of GCDs

Some researchers also reported ultrasonic synthesis of GCDs from biomass precursors [74, 139, 142, 164]. The ultrasonic synthesis can be considered as an energy-saving operation because it provides a large amount of localized heat/pressure (5000 K/ > 1000 atm) [142]. Das et al. [142] performed a sonochemical treatment on Jute caddies for 30 min in the presence of con. H_2SO_4. The H_2SO_4 in the reaction medium possesses hydrolysis/oxidant properties and source for S-doping. Consequently, the dehydration, condensation, addition, and aromatization steps led to produce carbonaceous material, which subsequently break into small nanoparticles through the assistance of ultrasonic irradiation. The obtained GCDs are modified with benzalkonium chloride (BZC) to improve their surface activity. Xu et al. [74] synthesized three kinds of N-GCDs via ultrasonic treatment (320 W, 80 Hz, room temperature, 3 h) of Kiwifruit juice in the presence of C_2H_5OH, ethylenediamine (EDA) or acetone. The C_2H_5OH-assisted N-GCDs (~2.7% N, average size: 6.7 nm) showed green emission (473 nm at 360 nm excitation), while EDA-assisted

N-GCDs (~5.7% N, average size: 6.9 nm) and acetone-assisted N-GCDs (~1.7% N, average size: 4.8 nm) exhibited chartreuse (547 nm), and pink (673 nm) fluorescence corresponding to 460 nm and 665 nm excitations, respectively.

3.2.6 Pyrolysis Synthesis of GCDs

In the pyrolysis process, starting precursor is subjected to thermal treatment at moderate/high temperature either in air or in inert atmosphere. Various attempts to synthesize GCDs via simple pyrolysis can be seen in Tables 2, 3, 4, 5, 7, 8, 9, 10, 17. When the extract of Papaya (*Carica papaya*) pulps is pyrolyzed at 200 °C (15 min) in air, it resulted in pale-yellow thick residue containing blue emissive GCDs with 23.7% QY [133]. Shi et al. [115] pyrolyzed Wheat straw at 250 °C (2 h) in air to obtain 58.47% black residue, which subsequently mixed in H_2O via sonication followed by filtration/dialysis to produce N-GCDs (2.81% N).

The pyrolysis product of natural precursor is also combined with HT, MW, ultrasonication, BM, or chemical oxidation treatment to obtain GCDs with narrow size distributions (Tables 2, 3, 5, 9, 10). For example, Hao et al. [84] pyrolyzed blue-green Algae at 250 °C (2 h, N_2 atmosphere) to obtain biochar. Then after, a BM-operation is performed on biochar at optimum condition (800 rpm, 2 h) before alkaline H_2O_2-assisted chemical oxidation and subsequent purification to yield N-GCDs with a narrow particle size (1–2.3 nm) and extensive oxygen-containing functional groups.

3.2.7 Combustion/Carbonization/TD Synthesis of GCDs

Apart from above mentioned bottom-up methods, some reports also showed the synthesis of GCDs via combustion [56], carbonization [57, 67, 150, 187], and TD [105, 166] method. For example, Cheng et al. [56] soaked Willow catkin in a solution of con. H_2SO_4/urea before drying in oven. The dried mass is subjected to grinding/combustion in air to obtain burned ash. Finally, the N,S-GCDs (average size/QY: 7.3 nm/13.3%) is isolated from ash, which contained a high amount of N/S heteroatoms (15.55/7.14%). Preethi et al. [67] carbonized colourless Coconut water at a very low temperature (70 °C, 5 h, pH 7 adjusted by NaOH) to convert into light brown colloidal mixture. After filtration, the suspension is proceeding for sonication/centrifugation to isolate GCDs with average size/QY of 5 nm/14.34%. Ali et al. [105] performed a simple TD operation (200 °C, 6 h) on Grape (*Vitis vinifera*) juice. The large particles are separated by filtration to attain GCDs with 8 nm average size and 32.1% QY.

4 Properties of GCDs in Relevance to Sensing Application

4.1 Absorption and PL Characteristics

GCDs generally show intense absorption in the UV region (below 400 nm) with one/ more than one distinct peaks and extended tail in visible region. The peak at lower wavelength is assigned to $\pi \rightarrow \pi^*$ transitions of the aromatic C = C/C–C bonds and peak at higher wavelength corresponds to $n \rightarrow \pi^*$ transitions associated with the C = O/C–O bonds [64, 111, 79, 81]. *Tagetes patula* derived GCDs showed three strong absorption peaks at 248/273/319 nm. The first two peaks are assigned to the $\pi \rightarrow \pi^*$ transitions of the sp^2 C = C bonds and third is originating from the $n \rightarrow \pi^*$ transitions of C = O bonds [81].

Fluorescence property is one of the most striking characteristics of GCDs, which generally shows an excitation-wavelength dependent (EWD) emission (emission peak shifts according to excitation-wavelength). The origin of PL in GCDs is still not fully understood, but researchers proposed that the quantum confinement in carbogenic core, degree of graphitization, surface/edge defects, emissive trap sites, and various functional groups on the surface mainly determines their PL behaviour [33, 226]. For example, *Miscanthus* grass derived GCDs and their doped counterparts (N-GCDs, P-GCDs, and N,P-GCDs) showed a typical EWD emissions in the excitation-wavelength range of 320–410 nm (red-shift of emission peaks with increasing excitation-wavelength, Fig. 6). The emission-wavelengths with maximum intensities are located at 431/512/455/485 nm for GCDs/N-GCDs/P-GCDs/N,P-GCDs, respectively (Fig. 6a–d). Interestingly, the N-GCDs showed significantly high emission intensity and peak shift towards higher wavelength with gradual increase in their intensity (Fig. 6b). As a result, the N-GCDs exhibited maximum QY (11.65%) followed by N,P-GCDs (9.63%), GCDs (4.71%), and P-GCDs (2.33%). The EWD behaviours of GCDs/doped-GCDs are attributed to the collective effect of carbogenic core, surface states and quantum confinement [57].

Apart from EWD nature, some GCDs showed an excitation-wavelength independent (EWID) behaviour [67, 116, 83]. Lemon peels derived N-GCDs showed a gradual increase of emission intensity without an obvious shift of peak position in the excitation-wavelength range of 325–370 nm, indicating a EWID PL characteristic. The amorphous nature of carbon core along with the uniform and defect-free surface states is responsible for EWID behaviour [83]. *Spirulina* algae derived N-GCDs exhibited both EWD/EWID PL profiles in the excitation-wavelength ranges of 400–440/450–490 nm, respectively. The EWD PL is due to the trapped excitations in surface defects that create different band gaps, while EWID characteristic reflected uniform particle size and surface state distributions in the N-GCDs [227].

Up-conversion PL of GCDs is also noticed, which means emission peak at lower wavelength than that of excitation-wavelength [60, 72, 167, 87]. Such kind of behaviour is usually explained by anti-Stokes PL and multiple-photon activation mechanism [167, 60]. For example, Mango-peel derived N,S-GCDs showed up-conversion emissions in the excitation-wavelength range of 700–800 nm. The

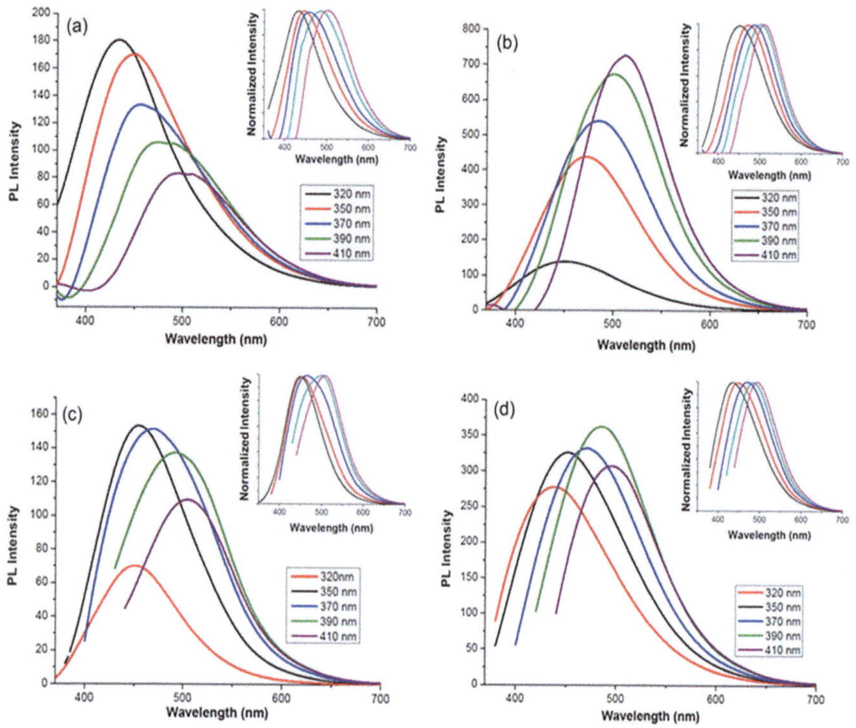

Fig. 6 EWD PL spectra of **a** GCDs, **b** N-GCDs, **c** P-GCDs, and **d** N,P-GCDs derived from *Miscanthus* grass. Reproduced from ref. [57] with permission from the Royal Society of Chemistry

maximum up-conversion emission is located at ~ 422 nm when $\lambda_{ex} = 700$ nm is used [87].

4.2 Photo-Stability/Environmental Stability

A stable optical characteristic of GCDs with the exposure of light represents its photo-stability, while there are several environmental factors such as pH, solvent, temperature, and ionic strength that can affect its optical behaviour. Gardenia fruit derived N,S-GCDs exhibited an excellent PL stability (after 4 months storage), photo-stability (unaffected PL with the exposure of UV light, 3 h), salt resistance (up to 1 M NaCl concentration), and applicability in a wide pH range (1–14) [169]. Bee-pollen derived N-GCDs showed 50% reduction of emission intensity when exposed with UV light (24 h), which is ascribed to the destroy of surface emissive sites via $^{\bullet}$OH radical. The N-GCDs also showed relatively stable PL intensity in 5.5–7.5 pH range in comparison to highly acidic/basic condition. Moreover, different temperature (15–45 °C), salt solution (NaCl), and time (up to 24 h) insignificantly affected the absorption/PL spectra, indicating the suitability of N-GCDs in multiple environments [73]. EWID PL behaviour of N-GCDs (derived from Tea powder and *o*-phenylenediamine) showed two distinct emission peaks in water (351/387 nm) as well as in many organic solvents (C_2H_5OH, ACTN, tetrahydrofuran, DMF, dimethylsulfoxide; 351/499 nm), suggesting multifarious optical nature of N-GCDs. Moreover, the intensity ratio of F_{387}/F_{351} in water exhibited insignificant effect in 7–9 pH, illustrating the applicability of N-GCDs in neutral/slightly-alkaline medium [228].

4.3 Biocompatability and Cytotoxicity

GCDs are found to be a good choice for bio-applications including metal ions detection in living cells due to their non-cytotoxic nature. For example, Spider-silk derived N-GCDs showed only 12% reduction in cell viability for HepG2 cells (12 h incubation) even at a high concentration of 1 mg/mL, suggesting their minimal cytotoxicity [165]. Qu et al. [206] derived N-GCDs from *Magnolia denudata* leaves, which showed ~ 95% cell viability for Li-7 cells in a concentration range of 0.001–0.6 mg/mL, indicating their applicability as a nanoprobe in biological medium.

5 Metal Ions Sensing Mechanism of GCDs

Quenching of fluorescence signal (turn-off) after interaction with metal ions is the most common observation of GCDs, where surface functional groups play a crucial role. Different functional groups existing on the surface of GCDs have tendency to coordinate/complex/chelate with specific metal ions to generate fluorescence-based sensing platform [33, 229, 230]. The changes in fluorescence characteristics are induced by the energy/electron transfer (ET) during interactions between GCDs and analyte, which resulted in modifications in the electron–hole recombination. There are five common mechanistic possibilities for the quenching of fluorescence signal after the introduction of metal ion: (i) photo-induced ET (PET)/ET, (ii) inner filter effect (IFE), (iii) fluorescence/Förster resonance energy transfer (FRET), (iv) static quenching effect (SQE), and (v) dynamic quenching effect (DQE) [33, 229, 230].

5.1 PET/ET

It is a non-radiative ET process between GCDs and metal ions, which is induced by the photo-excitation of sensing nanoprobe [33, 229]. There are several GCDs, which showed PET/ET type fluorescence quenching in the presence of heavy metal ions. For example, Palm-kernel shell derived GCDs showed maximum quenching with Cu^{2+}/Fe^{3+} ions. The Cu^{2+}/Fe^{3+} ions formed coordination bonds with the –COOH functional groups of GCDs to facilitate ET process from GCDs to metal ions [113]. The Au^{3+} ion sensing mechanism of GCDs (derived from Neem-kernels) is explained on the basis of PET from the excited states of GCDs to self-coordinated Au^{3+} [204].

5.2 *IFE*

When excitation/emission spectra of GCDs overlap with the absorption spectrum of metal ion, the fluorescence intensity of GCDs gets suppressed due to its absorption over metal ions [33, 229]. The occurrence of IFE can be endorsed by the nearly invariant average fluorescence lifetime (τ_{av}) of GCDs and absorption peak of metal ions after quenching operation because of the lack of new complex formation [33, 229]. It is observed that IFE is a frequent quenching mechanism for the detection of Cr^{6+} and Co^{2+}. For example, the fluorescence intensity of Seaweed (*Sargassum carpophyllum*) derived N,S-GCDs decreased in the presence of Cr^{6+}. A precise overlap of the excitation/emission peaks of N,S-GCDs (360/430) with the absorption peaks of Cr^{6+} (258 nm, 357 nm, and 430 nm) can be revealed from Fig. 7a. Moreover, τ_{av} did not show an apparent difference before (1.11 ns) and after the addition of Cr^{6+} (1.18 ns) (Fig. 7b). Based on these experimental facts, the quenching process is ascribed to the IFE between N,S-GCDs and Cr^{6+} [144]. Co^{2+} quenched the fluorescence signal of *Sida cordifolia* root derived N-GCDs via IFE [126].

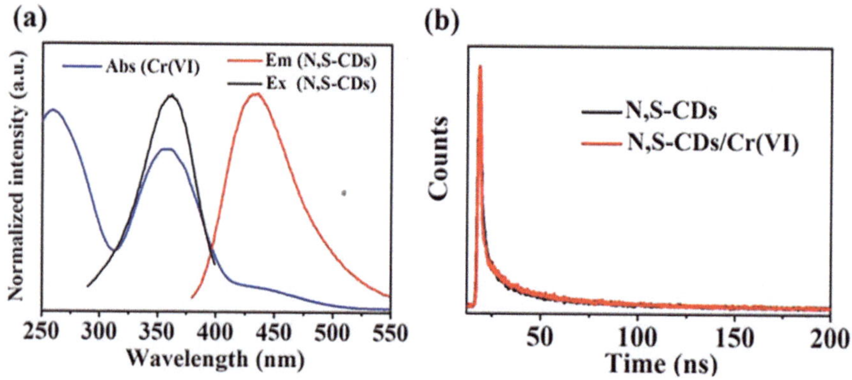

Fig. 7 **a** Excitation/emission spectra of N,S-GCDs along with the absorption spectrum of Cr^{6+}. **b** Lifetime profiles of N,S-GCDs before/after the addition of Cr^{6+}. Reproduced from ref. [144] with permission from the Royal Society of Chemistry

5.3 FRET

The non-radiative energy transfer process can be classified into dipole–dipole interaction-based FRET and electron-exchange-based exchange energy transfer or Dexter energy transfer [33, 229]. The GCDs may exhibit FRET-based fluorescence quenching via non-radiative energy transfer between GCDs (in excited state) and analyte (in ground state) [200, 149]. Therefore, lifetime of GCDs should be sufficient enough along with the adequate distance from the analyte (1–10 nm) for effective FRET [229].

For example, a natural peach gum polysaccharide-EDA synthesized N-GCDs showed the involvement of FRET in the quenching-based sensing of Au^{3+} along with the ET from N-GCDs to Au^{3+} [200]. A surface passivated N,S-GCDs (synthesized from a mixture of banana pseudo-stem cellulose, SO_3H ionic liquid (IL), and PEG400) showed a partial overlap of excitation band (397 nm) with the absorption band of Cr^{6+} (360 nm), while emission band (455 nm) did not show a measurable influence (Cr^{6+} absorption at 455 is very small). Moreover, an τ_{av} of N,S-GCDs (3.21 ns) remarkably decreased after the addition of Cr^{6+} (2.07 ns), indicating the quenching phenomenon via FRET mechanism [149].

5.4 SQE

SQE in the quenching operation occurs due to the formation of non-luminescent complex between GCDs and metal ions in the ground state [229, 230]. SQE can be inferred by the following characteristics: (i) changes in the absorption spectrum of GCDs in the presence of quencher, (ii) insignificant change in τ_{av} after quenching process, (iii) reduction in the stability of generated complex with increasing temperature, which is accessed by the decreasing value of quenching rate constant (k_q) [229, 230].

The Stern–Volmer equation (Eq. 1) is routinely applied to analyze the progress of quenching process [73, 229].

$$\frac{F_0}{F} = 1 + K_{SV}[C] = 1 + k_q\tau_0[C] \tag{1}$$

where F_0, F, and [C] represent the initial emission intensity of GCDs, emission intensity after the addition of metal ion, and concentration of analyte, respectively. The slope (Stern–Volmer quenching constant, K_{SV}) of Stern–Volmer plot (F_0/F versus [C]) can be used to estimate k_q, where τ_0 is the lifetime of GCDs. Various GCDs may involve SQE-type of sensing towards heavy metal ions. For example, Grape seeds/thiourea (TU)-derived N,S-GCDs showed a sensing aptitude of Cu^{2+} via SQE-based quenching mechanism. Based on the significant change in the K_{SV} value (decrease with increasing temperature, Fig. 8a), change in the absorption spectra (before and after quenching, Fig. 8b), and increase in the zeta potential (–35.3/–3.4 mV before/

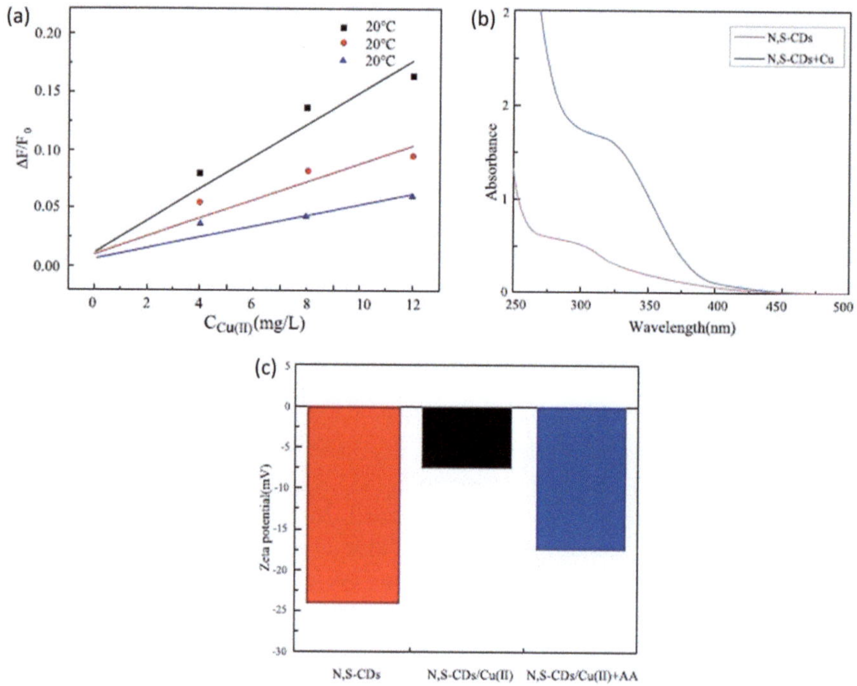

Fig. 8 **a** Stern–Volmer plots of N,S-GCDs at different temperatures. **b** Absorption spectra and **c** Zeta potentials of bare N,S-GCDs and N,S-GCDs along with Cu^{2+}. Reproduced from ref. [112] with permission from the Royal Society of Chemistry

after quenching, Fig. 8c), it is inferred that Cu^{2+} complexes with N,S-GCDs to form non-fluorescent adduct for SQE-based fluorescence quenching [112]. The Fe^{3+} detection of industrial hemp-derived N-GCDs is also explained by SQE-mechanism [231].

5.5 *DQE*

Unlike the involvement of ground state complex in the SQE, the DQE is based on interaction of metal ions and GCDs in the excited state [229, 230]. The photo-excited energy is dissipated from GCDs to metal ions due to dynamic collision between them in the excited state. Following characteristics may be the signatures for DQE: (i) nearly unchanged absorption spectrum of GCDs with/without metal ions, (ii) reduction of τ_{av} after quenching, (iii) increment in the stability of DQE at higher temperatures, which is marked by larger k_q [229, 230]. Wool-keratin derived N,S-GCDs showed a DQE mechanism with Fe^{3+}, which is verified by the insignificant change in the absorption spectrum of N,S-GCDs (Fig. 9a), denying the feasibility

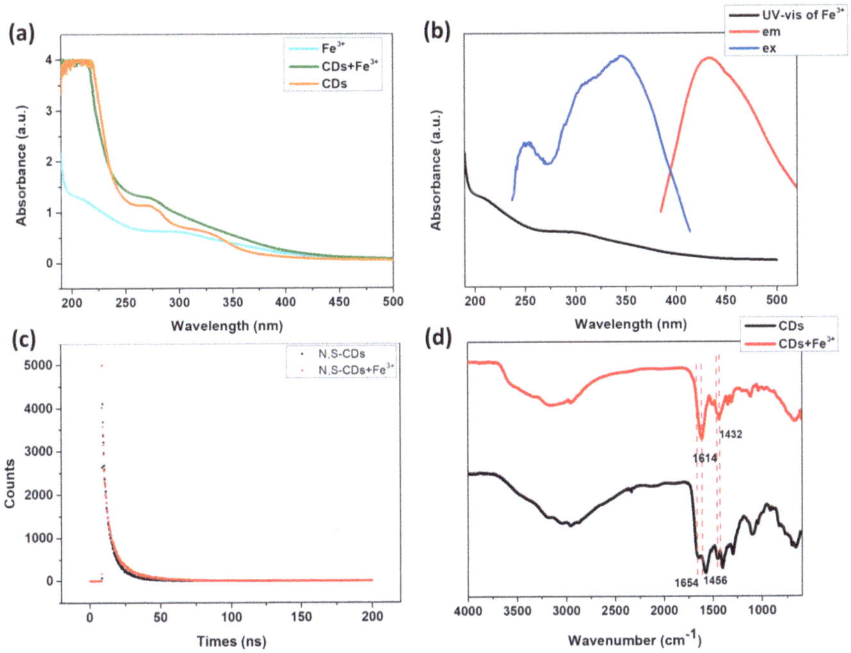

Fig. 9 **a** Absorption spectra of Fe^{3+}, N,S-GCDs, and N,S-GCDs/Fe^{3+}. **b** Excitation/emission profiles of N,S-GCDs and absorption spectrum of Fe^{3+}. **c** Lifetime curves and **d** FTIR spectra of N,S-GCDs in the absence/presence of Fe^{3+}. Reproduced from ref. [146] with permission from the Royal Society of Chemistry

of spectral overlap (Fig. 9b), obvious change in the τ_{av} (Fig. 9c), and red-shift of the C = O peak in the Fourier transform infrared (FTIR) spectrum (Fig. 9d) after quenching operation [146]. It is observed that τ_{av} of the bare N-GCDs (3.57 ns; derived from traditional Chinese medicine) considerably decreased to 2.81 ns after the introduction of Fe^{3+}. Moreover, K_{SV} at 70 °C (3.01×10^4 M^{-1}) is estimated to be larger than at 30 °C (2×10^4 M^{-1}) along with the unobvious change in the absorption profile of N-GCDs/Fe^{3+} (similar like N-GCDs and Fe^{3+}). Therefore, Fe^{3+} quenched the fluorescence signal of N-GCDs via DQE-mechanism [86].

6 GCDs-Based Heavy Metal Ions Sensor

6.1 *Fe^{3+}/Fe^{2+} Ions Sensor*

Iron (Fe) is one of the most essential elements for living organisms to regulate physiological processes such as transport of oxygen/electron, deoxyribonucleic acid (DNA) synthesis, and cellular metabolism [232]. However, the deficiency/overload of Fe may cause functional disorders in organisms [233–235]. Fe commonly exists in Fe^{2+} and Fe^{3+} states. Among them, Fe^{2+} is important for living organism, while Fe^{3+} is toxic and carcinogenic in nature [236]. Therefore, quantitative detection of Fe^{3+}/Fe^{2+} is crucial to avoid potential health risk and understand physiological function. Fluorescence method to detect Fe^{3+}/Fe^{2+} is quite imperative and developed in last decade [237, 238]. Among various fluorescent probes, biomass-based GCDs have shown particular interest for the sensitive and selective detection of Fe^{3+} (Table 2) and Fe^{2+} (Table 3).

6.1.1 Fe^{3+} Ion Sensor

Turn-off based fluorimetric detection of Fe^{3+} using N-GCDs (N content: 2.67%, average size/QY: 3.7 nm/16.4%) derived from biomass (Oriental plant leaves) has begun in 2013 [37]. Subsequently, GCDs or doped counterparts derived from various biomasses are applied in the selective and sensitive determination of Fe^{3+} (Table 2).

Chen et al. [43] synthesized N,S-GCDs (N/S content: 4.32/0.72%) from Garlic with 13% QY and 3.6 nm average size, which showed excellent sensitivity for Fe^{3+} (LOD: 0.22 nM). The sensor is also applied for the Fe^{3+} detection in environmental water samples and RAW264.7 cells. *Lycii fructus* fruit is used as a biosource to synthesize N-GCDs (N content: 8.4 wt%, QY: 17.2%) using HT method (Table 2). The detection ability of N-GCDs for Fe^{3+} is found to be in the linear range of 0–30 μM with a good LOD (0.021 μM). Moreover, the sensor probe showed their potentiality to analyse Fe^{3+} in the real samples (water and urine) and visualization in the living cells [47].

N-GCDs with a high N content (28.8%) and QY of 11.65% is synthesized by the thermal carbonization (180 °C, 4 h) of *Miscanthus* grass extract in the presence of EDA. Benefitted with the amino and carboxyl functional groups in the N-GCDs, the SQE-based Fe^{3+} sensing performance is attributed with a wide linear detection range (0.02–200 μM) and appreciable LOD (0.02 μM) [57].

Wang et al. [66] reported N,S,P-GCDs (average size/QY: 3.2 nm/8.7%), which is synthesize from green Pepper seeds as only starting precursor (Table 2). Among various tested metal ions, the N,S,P-GCDs showed maximum fluorescence quenching (80%) with Fe^{3+}. Due to the appreciable sensitivity of sensor (Table 2), it is also applied for the lake and tap water samples analyses with recovery/standard relative deviation (RSD) ranges of 99.98–103.6/1.99–5.19%, respectively. Additionally, the N,S,P-GCDs could also monitored Fe^{3+} in HeLa cells via decreased intracellular fluorescence intensity.

Senol et al. [70] observed an excellent LOD (0.75 nM) for the detection of Fe^{3+} when N-GCDs is used as a fluorescent nanoprobe. The N-GCDs in this report is obtained by the HT treatment of Cranberry (Table 2), which possessed an average size of 5.0 nm and QY of 14.5%. The sensor is also applied for real water samples (spring and tap water) with recovery/RSD ranges of 99.2–100.5/0.1–0.4%, respectively.

Green bean-derived N,S,P-GCDs (N/S/P content: 4.1/0.3/0.2%, average size: 13 ± 3 nm, QY: 7.2%) via HT method exhibited their potentiality to sense Fe^{3+} with

LOD/limit of quantification (LOQ) as low as 3.6/12.25 nM. The synthesized N,S,P-GCDs showed green emission (440 nm) at 360 nm irradiation. Furthermore, the N,S,P-GCDs could be able to detect Fe^{3+} in the tap water (recovery/RSD: 102.6–115/1.3–5.7%) and mitochondria of A549 cells (suppressed fluorescence intensity), indicating their applicability towards environmental assessment and biosensing in sub-cellular organelles [79].

6.1.2 Fe^{2+} Ion Sensor

Mango peel-derived N,S-GCDs (N/S content: 1.34/1.78%, average size/QY: 3.0 nm/8.5%) is successfully used to detect Fe^{2+} with a good LOD (1.2 μM) [87]. Subsequently, leaves of mango tree (*Mangifera indica*) are applied for the synthesis of N-GCDs (3.15% N), which showed LOD as low as 0.62 ppm in the detection of Fe^{2+} [88].

Mathew et al. [91] employed *Wrightia tinctoria* leaves to synthesize N-GCDs (average size: 3.61 nm). The fluorescence peak of N-GCDs effectively quenched in the presence of Fe^{2+} (LOD: 0.517 μM) due to the coordination-based interaction between Fe^{2+} and functional groups of N-GCDs. Additionally, the N-GCDs modified electrode showed their potentiality for the electrochemical detection of Fe^{2+} in the linear range of 5–25 μM with LOD as low as 2.3 nM.

Recently, Lemon-juice derived N-GCDs are applied for the sensing of Fe^{2+} with excellent performance (linear range/LOD: 2–625/0.063 μM). Authors deduced that a combined effect of IFE and SQE is responsible for fluorescence quenching. Furthermore, the N-GCDs is also used as a nanoprobe to analyze tap water samples with recoveries/RSDs of 98.9–102.3/ < 0.6% [92].

6.2 *Cu^{2+} Ion Sensor*

Cu is another essential heavy metal for the smooth biological functioning of living organisms [239]. Meanwhile, imbalance of Cu^{2+} is known for metabolic disorder and neurodenerative diseases such as Wilson and Alzheimer diseases [240, 241]. The permissible limit of Cu is set to be 02 ppm (~31.5 μM) in drinking water according to the WHO guideline (Table 1).

The N-GCDs obtained from pipe tobacco can be recognized as the first biomass-derived GCDs in the sensing application of Cu^{2+} [93]. Subsequent developments of biomass-based GCDs and their sensing performances for Cu^{2+} are summarized in Table 4.

Rooj et al. [102] applied *Tuberose* petals to prepare GCDs via pyrolysis method (Table 4). The synthesize GCDs (average size: 4.03 nm, QY: 3%) is used for Cu^{2+} sensing (linear range/LOD: 0–70/0.2 μM) due to the binding efficiency of Cu^{2+} with the amino groups of GCDs, and therefore transfer of electron via non-radiative manner.

Chang et al. [107] reported a one-step HT process for the synthesis of GCDs from Smilax China biomass (Table 4). The bright yellow fluorescent GCDs (maximum λ_{em} = 542 nm at λ_{ex} = 470 nm) with QY/average size of 22.37%/2.1 nm is used for the selective/sensitive detection of Cu^{2+}, which showed three linear ranges (0.5–10 μM, 75–225 μM, and 250–350 μM) and satisfactory LOD (0.028 μM). The GCDs also showed intracellular monitoring ability of Cu^{2+} in neuroma PC12 cells.

The GCDs synthesized by pyrolyzing Avocado seeds at 600 °C exhibited linear ranges/LODs of 0.006–0.009, 1.5–4/0.0024 μM and 0.0005–0.0009, 30–200/ 0.00007 μM for Cu^{2+} and Cr^{6+}, respectively [114]. The linear range (0.1–100 μM) for the selective detection of Cu^{2+} using blue-fluorescent GCDs derived from *Plumeria alba* flowers are also remarkable [117].

Xu et al. [119] synthesized N-GCDs with QY as high as 59% by the HT treatment of Wolfberry stems as only biosource (Table 4). Nearly 87% drop in the fluorescence intensity of aqueous N-GCDs in the presence of Cu^{2+} justified a good quenching efficiency. As a result, the N-GCDs showed excellent selectivity and appreciable sensitivity (linear range/LOD: 10–80/2.83 nM) for Cu^{2+}. The quenching process is explained on the basis of coordination between the functional groups of N-GCDs and Cu^{2+} that facilitated ET between them.

6.3 Co^{2+} Ion Sensor

Trace level of cobalt (Co) is an integral part of vitamin B12, which is important for various biological processes such as synthesis/regulation of DNA, functioning of nervous system, and formation of red blood cell [242]. Although Co is necessary for biological function, it can cause neurotoxicity, pneumonia, and risk of cancer in human above tolerance limit [243]. Therefore, development of an efficient platform for the detection of Co^{2+} is an important area of research.

Pig skin derived N-GCDs (QY; 24.1%) via HT method (Table 5) is used in the fluorimetric detection of Co^{2+} with a satisfactory LOD (0.68 μM). Due to the precise overlap of excitation/emission spectra of the N-GCDs with absorption spectra of Co^{2+}, the quenching phenomenon is ascribed to IFE. Moreover, the N-GCDs could recognize Co^{2+} in HeLa cells via gradual quenching of strong green fluorescence with time [120]. Subsequently, GCDs derived from Kelp [121], *Artemisia annua* [122], Flax straw [123], frozen Tofu [124], and Alkali lignin [125] are used for the Co^{2+} sensing with satisfactory performances (Table 5).

Recently, *Sida cordifolia* root/triethylene tetraamine (TETA) derived N-GCDs (13.8 at% N, QY: 17.8%) exhibited LOD of Co^{2+} as low as 0.11 μM in the linear range of 0 to 4.7 μM. Additionally, the N-GCDs exhibited satisfactory recovery ranges/RSDs (99.60–100.22/ < 2.5%) in the quantification of Co^{2+} from lake/river water samples [126].

6.4 Zn^{2+} Ion Sensor

As a micronutrient, zinc (Zn) plays a crucial role for the cellular function, enzymatic activity, and growth/development of living organisms [244]. It is estimated that ~17% of population throughout the world are prone to Zn deficiency, which may cause pregnancy problem, immunity loss, neuronal disorder, and early death [245]. Therefore, monitoring of a trace level of Zn^{2+} is vital to identify its potential risk.

An acid-assisted ultrasonication (HCl/H_2O, 6 h) followed by reflux treatment (90 °C, 6 h) on potato starch produced Si-GCDs (4.60 at% Si) with an average size/QY of < 5 nm/10%. The fluorescence peak of Si-GCDs (~515 nm) selectively quenched with Zn^{2+} via nonspecific interaction and showed a wide linear range of 0–20 μM with LOD as low as 1 nM [127].

Subsequently, *Campomanesia phaea* derived GCDs [128] and Onion extract/EDA derived N-GCDs [129] are used for the detection of Zn^{2+} with satisfactory results (Table 6).

6.5 Cr^{6+}/Cr^{3+} Ions Sensor

Cr commonly exists in Cr^{3+}/Cr^{6+} states in the environment. Among them, Cr^{6+} is considered as a hazardous contaminant for environmental pollution and human health [246]. Industrial processes such as leather tanning, chrome-plating, metal finishing, and wood manufacturing are major sources for Cr^{6+} contamination [247]. According to the latest report of WHO, the maximum permissible limit of Cr^{6+} is 50 μg/L (Table 1). Biomass derived GCDs have shown profound possibility in the detection of Cr^{6+} via fluorimetric method (Table 7).

HT-synthesized N-GCDs (biosource: Jeera, N content: 2.35 mass%, QY: 5.33%) are modified with cystamine via EDC/NHS coupling reaction to develop a high selectivity for Cr^{6+} with minimum LOD of 1.57 μM. The probe also showed their feasibility to detect Cr^{6+} and total Cr in real samples [137].

Interestingly, the N-GCDs synthesized from *Panax notoginseng* along with CA and EDA showed a remarkable LOD (0.185 nM) in the linear range of 0–6 μM for the detection of Cr^{6+} via IFE-based fluorescence quenching. The N-GCDs are featured with 5.69% N element along with a good QY (18.41%) [140].

The N,B-GCDs (N/B content: 3.8/0.8%) synthesized from 3-aminophenylboronic acid (APBA)-assisted acid hydrolysis of alkali lignin followed by HT treatment showed QYs of 0.095/7.5/4.5% corresponding to the 300/330/490 nm excitation. Furthermore, LODs corresponding to these excitations are estimated to be 0.054/0.049/0.077 μM, indicating possibility for the sensitive detection of Cr^{6+} via tri-channel platform [145].

Recently, Kolekar et al. [153] synthesized Zn-GCDs (QY: 7.47%) from *Cinnamon* biomass along with the zinc acetate (ZA) through HT method. Significant drop of

QYs (3.80/2.41%) after the addition of Cr^{6+}/Mn^{7+} indicated the quenching of fluorescence originating from Zn-GCDs. The Zn-GCDs showed maximum quenching within 2 min with appreciable linear ranges/LODs for the selective detection of Cr^{6+}/Mn^{7+} (Table 7) as well as feasibility for real samples analyses.

6.6 Hg^{2+} Ion Sensor

It is known that Hg has adverse effects on the growth and photosynthesis process of phytoplankton/macrophytes [248, 249]. Moreover, oxidative stress induced by Hg^{2+} play a crucial role for Hg toxicity [250]. Therefore, detection of Hg^{2+} with simple platform is reasonable to combat its toxic effect. The journey of Hg^{2+} sensing with biomass-derived GCDs begins in the mid 2012 when pomelo peel-synthesized N-GCDs (QY: 6.9%) are used as a fluorescent nanoprobe. The LOD for the detection of Hg^{2+} is estimated to be as low as 0.23 nM [154]. Subsequent developments of Hg^{2+} sensing using GCDs can be viewed in Table 8.

Mushroom-synthesized N-GCDs showed 4.13 nM LOD for the detection of Hg^{2+}, which is significantly improved by its modification with dihydrolipoic acid (DHLA) (LOD: 17.4 pM). Strong affinity of Hg^{2+} with thiol functional groups present in DHLA is responsible for such a high sensitivity of the sensor [162]. Gardenia fruit derived N,S-GCDs (7.13% N, 0.52% S) showed 0.32 μM LOD for Hg^{2+} in a linear fitted concentration range of 0.2–20 μM [169].

Samota et al. [172] reported an exceptionally high QY of 83% for the N,S-GCDs derived from the HT treatment of an aquatic weed (*Typha angustata* Bory) and TU. The fluorescence intensity of N,S-GCDs quenched with Hg^{2+} in a wide linear range and LOD of 0.01–60 μM and 3.1 nM, respectively, along with the applicability in natural water analyses.

Recently, a photoinduced charge transfer-based fluorescence quenching occurred between N-GCDs (amine-modified, QY: 2.8%) and Hg^{2+} that resulted in LOD of 8.29 μM and linear range of 0–100 μM. Moreover, the fabricated fluorescent hydrogel (by encapsulating N-GCDs into cellulose-based hydrogel) exhibited a feasibility to detect Hg^{2+} with linear range/LOD of 0–220/47.3 μM [175].

6.7 Pb^{2+} Ion Sensor

Pb is another potentially toxic heavy metal ion which needs to detect before health effect [251]. According to the WHO, exposure of Pb up to 10 μg/L is allowed without much toxicity (Table 1).

Tan et al. [176] employed Sago industrial waste to synthesize GCDs for the sensing of Pb^{2+} and Cu^{2+} (Table 9). Subsequently, GCDs derived from various biomasses such as Chocolate [177], Tulsi leaves [178], *Lantana camara* berries [179], *Coccinia indica* [103], and Spinach extract [180] are employed for the selective sensing of Pb^{2+} with appreciable sensitivities (Table 9).

Recently, a ratiometric fluorescence sensor is developed from red emitting N-GCDs (6.24% N) for the recognition of Pb^{2+}. The N-GCDs are synthesized by the MW treatment of Cyano-bacteria under alkaline condition. Interestingly, red emitting N-GCDs showed three emission peaks (452/600/657 nm) at $\lambda_{ex} = 400$ nm. The fluorescence peak intensity ratio (F_{468}/F_{657}) of red emitting N-GCDs is used to detect Pb^{2+}, which showed two linear ranges (0–50 μM and 100–300 μM) with a LOD of 0.85 μM [181].

6.8 Cd^{2+} Ion Sensor

Cd has shown adverse effect on the kidney and bone of human body and can also cause the growth of cancer cells [252]. The WHO has set the permissible limit of Cd within 03 μg/L (Table 1). Therefore, development of an efficient sensor probe to detect Cd^{2+} within toxic limit is an important task. The performances of GCDs-based sensor platforms are summarized in Table 10.

Gu et al. [182] demonstrated Cd^{2+} sensing capability of the N,S-GCDs, which is obtained from scallion biomass (Table 10). The quenching-based fluorescence sensor exhibited two linear ranges (0.1–3 μM, 5–30 μM) with an appreciable LOD (15 nM). Subsequently, a turn-on based sensor platform is developed from coconut-coir derived GCDs (QY; 48%) with LOD as low as 0.18 nM in the detection of Cd^{2+}. Moreover, the sensor probe is applicable for real samples analyses [183].

Recently, *Polyalthia longifolia* leaves are HT treated at 150 °C for 6 h to yield N-GCDs with an average size/QY of 3.33 nm/22%. The interaction of Cd^{2+} with the functional groups of N-GCDs caused aggregation-induced quenching (AIQ) of fluorescence signal to detect Cd^{2+} with an acceptable LOD/LOQ (2.4/7.3 nM) and a wider linear range (7.3 nM–12 μM). Moreover, a recovery range of 98–101% for real-life water (tap water/industrial effluents) samples justified the applicability of sensing method [186].

6.9 Sn^{2+} Ion Sensor

Tin (Sn) is a non-essential element, which possesses a capability of genotoxicity, neurotoxicity, and gastrointestinal irritation in living organisms [253, 254, 255]. GCDs-based nanoprobes are shown their possibilities to detect Sn^{2+} via fluorimetric method (Table 11).

An acid-assisted carbonization of rice husk resulted in fluorescent carbon nanoparticles, which is used for the detection of Sn^{2+} (18.7 μM LOD) [187].

Recently, N,La-GCDs is prepared from a mixture of enzymatic hydrolyzed lignin, melamine, and $La(NO_3)_3.6H_2O$ using HT method (Table 11). The functional groups present on the N,La-GCDs had a great affinity with Sn^{2+}/Fe^{3+} for the ET from N,La-GCDs to vacant d-orbitals of Sn^{2+}/Fe^{3+}. As a result, the nanoprobe exhibited an excellent LOD of 1.1 μM for Sn^{2+} and a satisfactory LOD of 0.99 μM for Fe^{3+}. Moreover, in vivo detection of Sn^{2+} in zebrafish is also demonstrated by N,La-GCDs [188].

6.10 *As^{3+} Ion Sensor*

The presence of As in water body is a serious health threat due to its high toxicity [256]. The WHO has announced the permissible limit of As, which is below 10 μg/L (Table 1). Therefore, the detection of As^{3+} within the permissible limit is an important area of research. GCDs-based optical nanoprobes are utilized for this purpose (Table 12).

Ramezani et al. [189] demonstrated a turn-on based fluorimetric detection of As^{3+} using N,S-GCDs (N/S content: 0.76/0.41%; derived from Quince fruit, Table 12). It is observed that the quenched fluorescence of N,S-GCDs in the presence of MnO_4^- restored when As^{3+} is introduced with MnO_4^- before adding to N,S-GCDs. As a result, the sensing of As^{3+} responded to a good linear range/LOD of 0.1–2/0.02 μg/mL. Subsequently, Prickly pear cactus [190], Banana leaves [191], and *Cynodon dactylon* [192], biomasses are used to derive GCDs for As^{3+} detection (Table 12).

Recently, *Cynodon dactylon* derived GCDs are used for the metal ions sensing application with an appreciable performance (LOD for As^{3+}/Fe^{3+}: 19 nM/0.10 μM). The fluorescence turn-on (enhancement) phenomenon of GCDs with As^{3+} is ascribed to the weak electrostatic interaction between them, while strong interaction of Fe^{3+} with functional groups caused quenching operation [192].

6.11 *Ag^+ Ion Sensor*

Due to wide applicability of silver (Ag) in the field of electronic, photography, and pharmaceuticals, it frequently releases into environment as an industrial waste. It is known that Ag^+ in free form is highly toxic in comparison to bound form such as AgCl, Ag_2S, etc. [257]. GCDs have shown a possibility to sense Ag^+ with various detection ranges/LODs (Table 13).

Denatured protein synthesized N-GCDs (31.04 wt% N) showed an appreciable LOD (0.03 μM) for Ag^+. The N-GCDs also showed credibility to detect Ag^+ in real water samples with satisfactory recoveries/RSDs [193]. A low LOD of 1.4 nM for Ag^+ using *purple perilla* derived N-GCDs is remarkable, which is based on SQE-mechanism [195].

Recently, Fan et al. [199] obtained two kinds of chiral GCDs (L-GCDs/D-GCDs) by the HT processing of primary GCDs liquid (derived from the HT treatment of cane molasses) in the presence of L-cysteine/D-cysteine. The average particle size of L-GCDs/D-GCDs (3.6 nm) is found to be larger than bare GCDs (1.7 nm) along with good QYs (18.86/17.38%). It is observed that both chiral GCDs are selective towards Ag^{+} under fluorimetric detection, however, L-GCDs showed slightly better sensitivity (linear range/LOD: 40–220/3.38 μM) than D-GCDs (linear range/LOD: 20–140/4.09 μM). Moreover, chiral GCDs-based fluorescence sensor showed the capability of tap water samples analyses.

6.12 Au^{3+} Ion Sensor

Ionic form of gold (Au^{+}/Au^{3+}) is more toxic in comparison to metallic form. A report shows that Au^{+}/Au^{3+} possess cytotoxic effect on human cell lines (skin keratinocyte and blood lymphocyte cells) [258]. Therefore, analytical assessment of Au^{3+} is of fundamental importance.

Liao et al. [200] synthesized N-GCDs (14.1 wt%) by the HT treatment of Peach gum polysaccharide in the presence of EDA, which is used for the fluorimetric detection of Au^{3+} with LOD as low as 64 nM. Subsequently, GCDs-derived from other biomasses are utilized to sense Au^{3+} (Table 14).

Raji et al. [204] applied a MW-assisted pyrolysis to synthesize GCDs (QY: 12.5%) from Neem seed kernels. Among various metal ions, the maximum PL intensity (~452 nm) selectively quenched with Au^{3+} and showed LOD as low as 16 nM in the linear range of 0–25 μM. The quenching is inferred due to aggregation caused by the coordination of GCDs–Au^{3+} and donor excited PET from excited state GCDs to coordinated Au^{3+}.

6.13 Pd^{2+} Ion Sensor

A small dose of palladium (Pd) may cause various health risks including allergic reaction, skin/eye irritation, etc. [259]. Therefore, detection of Pd^{2+} is meaningful to find out its presence even up to trace level (Table 15).

Hashemi et al. [205] utilized Red beetroot in the HT synthesis of N,S-GCDs (4.59% N, 2.59% S). The size of green emitting N,S-GCDs is confined in the range of 5–7 nm with a high QY (27.6%). The N,S-GCDs showed an excellent sensitivity in the detection of Pd^{2+} (linear range/LOD: 3–43 μM/33 nM).

Subsequently, *Magnolia denudata* leaves derived near-infrared N-GCDs are mixed with Pluronic F-127 to fabricate a nanoprobe (absolute QY: 7.97%) for the detection of Pd^{2+}. The quenching of 678 nm emission peak of N-GCDs with Pd^{2+} via SQE showed two linear ranges (4–18 μM and 20–200 μM) and LOD of 85.3 nM. Moreover, the sensor probe is also applicable in the environmental water samples (mineral/tap/lake) with recoveries/RSDs of 95.54–107.14/ < 4.6% and monitoring in living cells [206].

6.14 Al^{3+} Ion Sensor

Aluminium (Al) is the most abundant metal (~8%) in the earth's crust and has diverse application from energy generation to packaging, architecture, and fabrication of marine components [260]. However, it can induce toxicity in the form of oxidative stress, metabolic imbalance, immunological/genetic/enzymatic disorder, apoptosis, dysplasia, and necrosis [261]. The utility of GCDs for the sensing of Al^{3+} can be revealed from Table 16.

Pear-juice derived N-GCDs (QY: 10.8%) showed blue fluorescence under UV light (365 nm) and 471 nm emission peak at 390 nm excitation. The enhancement of emission peak intensity via chelating effect of Al^{3+} is used to develop a turn-on sensor for Al^{3+}, which exhibited a wide linear range (0.005–50 μM) and excellent LOD (2.5 nM) [207].

Subsequently, Al^{3+} sensing ability of N-GCDs is reported, which is synthesized by the HT treatment of *Osmanthus fragrans* at 180 °C for 10 h. The QY of N-GCDs is estimate to be 21.9% with respect to quinine sulphate. The Al^{3+} sensing is performed by the restoration of fluorescence signal, which is quenched in the presence of quercetin (turn-off–on). The sensor system showed an excellent linear relationship in the wide Al^{3+} concentration range of 0.1–100 μM with an appreciable LOD/LOQ (26 nM/0.1 μM). Additionally, Al^{3+} detection in human serum and human bladder cancer cells (T24) indicated the possibility of sensor in real and biological environment [209].

6.15 $V^{5+}/Cs^{2+}/Ru^{3+}/Mn^{2+}$ Ions Sensor

Hoan et al. [211] reported lemon juice-derived GCDs for the sensing of V^{5+}. The LOD of V^{5+} is estimated to be 3.2 ppm in a linear range of 0–8 ppm. Furthermore, V^{5+} detection ability of GCDs is also verified in the blood serum samples.

Maple-tree leaves synthesized blue fluorescent N-GCDs with a size range of 2–10 nm is applied as a probe to selectively access the presence of Cs^{2+} in the aqueous solution. The result showed a linear range/LOD of 0.1–100 μM/24 nM for Cs^{2+}. Moreover, Cs^{2+} spiked seawater/tap samples is also analyzed with N-GCDs for practical applicability [212].

Starch-based GCDs (GCDs-300) is prepared by a simple pyrolysis of starch in air at 300 °C, which showed Ru^{3+} sensing capability in a wide linear range of 0–1000 μM with < 20 μM LOD [213].

Recently, Krishnaiah et al. [214] reported an N,S-GCDs based fluorescence sensor for the selective detection of Mn^{2+}/Fe^{3+} ions. Due to good interaction between Mn^{2+}/Fe^{3+} ions and surface/edge functional groups of N,S-GCDs, the fluorescence intensity of N,S-GCDs effectively decreased linearly in the concentration of 10–50 μM. The LOD of sensor is calculated to be 0.95/0.75 μM for Mn^{2+}/Fe^{3+} ions.

7 Conclusion

Biomass and biomass-based waste have shown immense potential to produce GCDs/ doped-GCDs with narrow size distribution and appreciable QY. It is observed that HT treatment is a highly applied synthetic method for that, however, MW/MW-HT methods are advantageous to synthesize GCDs in short times. Besides tunable PL, the GCDs/doped-GCDs are also featured with EWID and up-conversion behaviour. The photo-stability, environmental stability, and biocompatability of GCDs offered them to be used as an excellent nanoprobe for the detection of heavy metal ions in aqueous medium and biological environment. Due to the presence of abundant functional groups, GCDs can bind with various metal ions to respond the quenching of fluorescence signal via different mechanistic pathways. Apart from most common turn-off based detection, some metal ions are recognized by fluorescence enhancement. Therefore, PL-based sensing of hazardous metal ions using GCDs are promising to analyze them in real water bodies as well as tracking in biological cells.

Acknowledgements PD thanks to University of Allahabad, Prayagraj, India for infrastructural facility and Science and Engineering Board (SERB), New Delhi, India for financial support through the fast track grant (*SB/FT/CS-190/2011*). He also thanks to Mr. Shishir Singh, IIT Kanpur to help in getting research papers for this chapter.

References

1. Gaur, N., Sarkar, A., Dutta, D., Gogoi, B. J., Dubey, R., & Dwivedi, S. K. (2022). Evaluation of water quality index and geochemical characteristics of surface water from Tawang India. *Scientific Reports, 12*, 11698.
2. Channegowda, M. (2020). Recent advances in environmentally benign hierarchical inorganic nano-adsorbents for the removal of poisonous metal ions in water: A review with mechanistic insight into toxicity and adsorption. *Nanoscale Advances, 2*, 5529–5554.
3. Crichton, R. R. (2016). *Metal toxicity–an introduction Metal chelation in medicine, Chapter 1*. https://doi.org/10.1039/9781782623892-00001
4. (2022) *Guidelines for drinking-water quality*. World Health Organization, Geneva, Fourth edition incorporating the first and second addenda. https://iris.who.int/bitstream/handle/10665/352532/9789240045064-eng.pdf
5. Wang, S., Yu, J., Zhao, P., Li, J., & Han, S. (2021). Preparation and mechanism investigation of CdS quantum dots applied for copper ion rapid detection. *Journal of Alloys and Compounds, 854*, 157195.
6. Wang, Q., Wei, K. N., Huang, S. Z., Tang, Q., Tao, Z., & Huang, Y. (2021). Turn-off supramolecular fluorescence array sensor for heavy metal ion identification. *ACS Omega, 6*, 31229–31235.
7. Guo, W., Zhang, C., Ma, T., Liu, X., Chen, Z., Li, S., & Deng, Y. (2021). Advances in aptamer screening and aptasensors' detection of heavy metal ions. *Journal of Nanobiotechnology, 19*, 166.
8. Rossi, A., Cuccioloni, M., Magnaghi, L. R., Biesuz, R., Zannotti, M., Petetta, L., Angeletti, M., & Giovannetti, R. (2022). Optimizing the heavy metal ion sensing properties of functionalized silver nanoparticles: the role of surface coating density. *Chemosensors, 10*, 483.

9. Nguyen, K. G., Baragau, I. A., Gromicova, R., Nicolaev, A., Thomson, S. A. J., Rennie, A., Power, N. P., Sajjad, M. T., & Kellici, S. (2022). Investigating the effect of N-doping on carbon quantum dots structure, optical properties and metal ion screening. *Scientific Reports, 12*, 13806.
10. Ju, B., Wang, Y., Zhang, Y. M., Zhang, T., Liu, Z., Li, M., & Zhang, S. X. A. (2018). Photostable and low-toxic yellow-green carbon dots for highly selective detection of explosive 2,4,6-trinitrophenol based on the dual electron transfer mechanism. *ACS Applied Materials & Interfaces, 10*, 13040–13047.
11. Li, H., Yan, X., Kong, D., Jin, R., Sun, C., Du, D., Lin, Y., & Lu, G. (2020). Recent advances in carbon dots for bioimaging applications. *Nanoscale Horizons, 5*, 218–234.
12. Li, W., Wei, Z., Wang, B., Liu, Y., Song, H., Tang, Z., Yang, B., & Lu, S. (2020). Carbon quantum dots enhanced the activity for the hydrogen evolution reaction in ruthenium-based electrocatalysts. *Materials Chemistry Frontiers, 4*, 277–284.
13. Cailotto, S., Negrato, M., Daniele, S., Luque, R., Selva, M., Amadio, E., & Perosa, A. (2020). Carbon dots as photocatalysts for organic synthesis: Metal-free methylene-oxygen-bond photocleavage. *Green Chemistry, 22*, 1145–1149.
14. Jana, P., & Dev, A. (2022). Carbon quantum dots: A promising nanocarrier for bioimaging and drug delivery in cancer. *Materials Today Communications, 32*, 104068.
15. Wu, Q., Wang, L., Yan, Y., Li, S., Yu, S., Wang, J., & Huang, L. (2022). 'Chitosan-derived carbon dots with room-temperature phosphorescence and energy storage enhancement properties. *ACS Sustainable Chem Eng, 10*, 3027–3036.
16. Xu, X., Ray, R., Gu, Y., Ploehn, H. J., Gearheart, L., Raker, K., & Scrivens, W. A. (2004). Electrophoretic analysis and purification of fluorescent single-walled carbon nanotube fragments. *Journal of the American Chemical Society, 126*, 12736–12737.
17. Feng, M., Wang, Y., He, B., Chen, X., & Sun, J. (2022). Chitin-based carbon dots with tunable photoluminescence for Fe^{3+} detection. *ACS Appl Nano Mater, 5*, 7502–7511.
18. Tiwari, A., Walia, S., Sharma, S., Chauhan, S., Kumar, M., Gadly, T., & Randhawa, J. K. (2023). High quantum yield carbon dots and nitrogen-doped carbon dots as fluorescent probes for spectroscopic dopamine detection in human serum. *Journal of Materials Chemistry B, 11*, 1029–1043.
19. Zhao, P., & Zhu, L. (2018). Dispersibility of carbon dots in aqueous and/or organic solvents. *Chemical Communications, 54*, 5401–5406.
20. Zhang, T., Wang, X., Wu, Z., Yang, T., Zhao, H., Wang, J., Huang, H., Liu, Y., & Kang, Z. (2021). Highly stable and bright blue light-emitting diodes based on carbon dots with a chemically inert surface. *Nanoscale Advances, 3*, 6949–6955.
21. Minervini, G., Panniello, A., Madonia, A., Carbonaro, C. M., Mocci, F., Sibillano, T., Giannini, C., Comparelli, R., Ingrosso, C., Depalo, N., Fanizza, E., Curri, M. L., & Striccoli, M. (2022). Photostable carbon dots with intense green emission in an open reactor synthesis. *Carbon, 198*, 230–243.
22. Zhang, Q., Shi, R., Li, Q., Maimaiti, T., Lan, S., Ouyang, P., Ouyang, B., Bai, Y., Yu, B., & Yang, S. T. (2021). Low toxicity of fluorescent carbon quantum dots to white rot fungus *Phanerochaete chrysosporium. Journal of Environmental Chemical Engineering, 9*, 104633.
23. Martindale, B. C. M., Hutton, G. A. M., Caputo, C. A., Prantl, S., Godin, R., Durrant, J. R., & Reisner, E. (2017). Enhancing light absorption and charge transfer efficiency in carbon dots through graphitization and core nitrogen doping. *Angewandte Chemie International Edition, 56*, 6459–6463.
24. Siddique, A. B., Pramanick, A. K., Chatterjee, S., & Ray, M. (2018). Amorphous carbon dots and their remarkable ability to detect 2,4,6-trinitrophenol. *Scientific Reports, 8*, 9770.
25. Yan, F., Jiang, Y., Sun, X., Bai, Z., Zhang, Y., & Zhou, X. (2018). Surface modification and chemical functionalization of carbon dots: A review. *Microchimica Acta, 185*, 424.
26. Anastas, P. T., & Warner, J. C. (1998). *Green chemistry: Theory and practice* (p. 30). Oxford University Press, New York.
27. Chahal, S., Yousefi, N., & Tufenkji, N. (2020). 'Green synthesis of high quantum yield carbon dots from phenylalanine and citric acid: Role of stoichiometry and nitrogen doping. *ACS Sustainable Chemistry & Engineering, 8*, 5566–5575.

28. Wang, L., Choi, W. M., Chung, J. S., & Hur, S. H. (2020). Multicolor emitting N-doped carbon dots derived from ascorbic acid and phenylenediamine precursors. *Nanoscale Research Letters, 15*, 222.
29. Nammahachak, N., Aup-Ngoen, K. K., Asanithi, P., Horpratum, M., Chuangchote, S., Ratanaphan, S., & Surareungchai, W. (2022). Hydrothermal synthesis of carbon quantum dots with size tunability via heterogeneous nucleation. *RSC Advances, 12*, 31729–31733.
30. Wu, Q., Zhang, S., Li, S., Yan, Y., Yu, S., Zhao, R., & Huang, L. (2023). Chitosan-based carbon dots with multi-color-emissive tunable fluorescence and visible light catalytic enhancement properties. *Nano Research, 16*, 1835–1845.
31. Piasek, A., Pulit-Prociak, J., Zielina, M., & Banach, M. (2024). Fluorescence of D-glucose-derived carbon dots: Effect of process parameters. *Journal of Fluorescence, 34*, 1693–1705. https://doi.org/10.1007/s10895-023-03392-z
32. Landa, S. D. T., Bogireddy, N. K. R., Kaur, I., Batra, V., & Agarwal, V. (2022). Heavy metal ion detection using green precursor derived carbon dots. *iScience, 25*, 103816.
33. Singh, P., Arpita, Kumar, S., Kumar, P., Kataria, N., Bhankar, V., Kumar, K., Kumar, R., Hsieh, C. T., & Khoo, K. S. (2023). Assessment of biomass-derived carbon dots as highly sensitive and selective templates for the sensing of hazardous ions. *Nanoscale, 15*, 16241–16267.
34. Wu, Z. L., Zhang, P., Gao, M. X., Liu, C. F., Wang, W., Leng, F., & Huang, C. Z. (2013). One-pot hydrothermal synthesis of highly luminescent nitrogen-doped amphoteric carbon dots for bioimaging from *Bombyx mori* silk–natural proteins. *Journal of Materials Chemistry B, 1*, 2868–2873.
35. Fu, L., Liu, T., Yang, F., Wu, M., Yin, C., Chen, L., & Niu, N. (2022). A multi-channel array for metal ions discrimination with animal bones derived biomass carbon dots as sensing units. *Journal of Photochemistry and Photobiology, A: Chemistry, 424*, 113638.
36. Liu, S., Tian, J., Wang, L., Zhang, Y., Qin, X., Luo, Y., Asiri, A. M., Al-Youbi, A. O., & Sun, X. (2012). Hydrothermal treatment of grass: A low-cost, green route to nitrogen-doped, carbon-rich, photoluminescent polymer nanodots as an effective fluorescent sensing platform for label-free detection of Cu(II) ions. *Advanced Materials, 24*, 2037–2041.
37. Zhu, L., Yin, Y., Wang, C. F., & Chen, S. (2013). Plant leaf-derived fluorescent carbon dots for sensing, patterning and coding. *Journal of Materials Chemistry C, 1*, 4925–4932.
38. Xu, J., Zhou, Y., Liu, S., Dong, M., & Huang, C. (2014). Low-cost synthesis of carbon nanodots from natural products used as a fluorescent probe for the detection of ferrum(III) ions in lake water. *Analytical Methods, 6*, 2086–2090.
39. Yang, X., Zhuo, Y., Zhu, S., Luo, Y., Feng, Y., & Dou, Y. (2014). Novel and green synthesis of high-fluorescent carbon dots originated from honey for sensing and imaging. *Biosensors & Bioelectronics, 60*, 292–298.
40. Vikneswaran, R., Ramesh, S., & Yahya, R. (2014). Green synthesized carbon nanodots as a fluorescent probe for selective and sensitive detection of iron (III) ions. *Materials Letters, 136*, 179–182.
41. Liu, R., Zhang, J., Gao, M., Li, Z., Chen, J., Wu, D., & Liu, P. (2015). A facile microwave-hydrothermal approach towards highly photoluminescent carbon dots from goose feathers. *RSC Advances, 5*, 4428–4433.
42. Sachdeva, A., & Gopinath, P. (2015). Green synthesis of multifunctional carbon dots from coriander leaves and their potential application as antioxidants, sensors and bioimaging agents. *The Analyst, 140*, 4260–4269.
43. Chen, Y., Wu, Y., Weng, B., Wang, B., & Li, C. (2016). Facile synthesis of nitrogen and sulfur co-doped carbon dots and application for Fe(III) ions detection and cell imaging. *Sensors and Actuators, B: Chemical, 223*, 689–696.
44. Shi, L., Zhao, B., Li, X., Zhang, G., Zhang, Y., Dong, C., & Shuang, S. (2016). Eco-friendly synthesis of nitrogen-doped carbon nanodots from wool for multicolor cell imaging, patterning, and biosensing. *Sensors and Actuators, B: Chemical, 235*, 316–324.
45. Song, P., Zhang, L., Long, H., Meng, M., Liu, T., Yin, Y., & Xi, R. (2017). A multianalyte fluorescent carbon dots sensing system constructed based on specific recognition of Fe(III) ions. *RSC Advances, 7*, 28637–28646.

46. Atchudan, R., Edison, T. N. J. I., Aseer, K. R., Perumal, S., Karthik, N., & Lee, Y. R. (2018). Highly fluorescent nitrogen doped carbon dots derived from *Phyllanthus acidus* utilized as a fluorescent probe for label-free selective detection of Fe^{3+} ions, live cell imaging and fluorescent ink. *Biosensors & Bioelectronics, 99*, 303–311.
47. Sun, X., He, J., Yang, S., Zheng, M., Wang, Y., Ma, S., & Zheng, H. (2017). Green synthesis of carbon dots originated from *Lycii fructus* for effective fluorescent sensing of ferric ion and multicolor cell imaging. *Journal of Photochemistry and Photobiology, B: Biology, 175*, 219–225.
48. He, Y., Liang, L., Liu, Q., Guo, J., Liang, D., & Liu, H. (2017). Green preparation of nitrogen doped carbon quantum dot films as fluorescent probes. *RSC Advances, 7*, 56087–56092.
49. Diao, H., Li, T., Zhang, R., Kang, Y., Liu, W., Cui, Y., Wei, S., Wang, N., Li, L., Wang, H., Niu, W., & Sun, T. (2018). 'Facile and green synthesis of fluorescent carbon dots with tunable emission for sensors and cells imaging. *Spectrochimica Acta A, 200*, 226–234.
50. Atchudan, R., Edison, T. N. J. I., Aseer, K. R., Perumal, S., & Lee, Y. R. (2018). Hydrothermal conversion of *Magnolia liliiflora* into nitrogen-doped carbon dots as an effective turn-off fluorescence sensing, multi-colour cell imaging and fluorescent ink. *Colloids and Surfaces B, 169*, 321–328.
51. Murugan, N., & Sundramoorthy, A. K. (2018). 'Green synthesis of fluorescent carbon dots from *Borassus flabellifer* flowers for label-free highly selective and sensitive detection of Fe^{3+} ions. *New Journal of Chemistry, 42*, 13297–13307.
52. Lu, M., Duan, Y., Song, Y., Tan, J., & Zhou, L. (2018). Green preparation of versatile nitrogen-doped carbon quantum dots from watermelon juice for cell imaging, detection of Fe^{3+} ions and cysteine, and optical thermometry. *Journal of Molecular Liquids, 269*, 766–774.
53. Jayaweera, S., Ke, Y., Hu, X., & Ng, W. J. (2019). Facile preparation of fluorescent carbon dots for label-free detection of Fe^{3+}. *Journal of Photochemistry and Photobiology, A: Chemistry, 370*, 156–163.
54. Zhang, Y., Gao, Z., Yang, X., Chang, J., Liu, Z., & Jiang, K. (2019). Fish-scale-derived carbon dots as efficient fluorescent nanoprobes for detection of ferric ions. *RSC Advances, 9*, 940–949.
55. Kailasa, S. K., Ha, S., Baek, S. H., Phan, L. M. T., Kim, S., Kwak, K., & Park, T. J. (2019). Tuning of carbon dots emission color for sensing of Fe^{3+} ion and bioimaging applications. *Materials Science and Engineering C, 98*, 834–842.
56. Cheng, C., Xing, M., & Wu, Q. (2019). A universal facile synthesis of nitrogen and sulfur co-doped carbon dots from cellulose-based biowaste for fluorescent detection of Fe^{3+} ions and intracellular bioimaging. *Materials Science and Engineering C, 99*, 611–619.
57. Picard, M., Thakur, S., Misra, M., & Mohanty, A. K. (2019). *Miscanthus* grass-derived carbon dots to selectively detect Fe^{3+} ions. *RSC Advances, 9*, 8628–8637.
58. Khan, Z. M. S. H., Rahman, R. S., Shumaila, I., & S. and Zulfequar, M. (2019). Hydrothermal treatment of red lentils for the synthesis of fluorescent carbon quantum dots and its application for sensing Fe^{3+}. *Optical Materials, 91*, 386–395.
59. Wu, X., Ma, C., Liu, J., Liu, Y., Luo, S., Xu, M., Wu, P., Li, W., & Liu, S. (2019). In situ green synthesis of nitrogen-doped carbon-dot-based room-temperature phosphorescent materials for visual iron ion detection. *ACS Sustainable Chem Eng, 7*, 18801–18809.
60. Zhang, W., Jia, L., Guo, X., Yang, R., Zhang, Y., & Zhao, Z. (2019). Green synthesis of up- and down-conversion photoluminescent carbon dots from coffee beans for Fe^{3+} detection and cell imaging. *The Analyst, 144*, 7421–7431.
61. Sahoo, N. K., Jana, G. C., Aktara, M. N., Das, S., Nayim, S., Patra, A., Bhattacharjee, P., Bhadra, K., & Hossain, M. (2020). Carbon dots derived from lychee waste: Application for Fe^{3+} ions sensing in real water and multicolor cell imaging of skin melanoma cells. *Materials Science and Engineering C, 108*, 110429.
62. Atchudan, R., Edison, T. N. J. I., Perumal, S., Muthuchamy, N., & Lee, Y. R. (2020). Eco-friendly synthesis of tunable fluorescent carbon nanodots from *Malus floribunda* for sensors and multicolor bioimaging. *Journal of Photochemistry and Photobiology, A: Chemistry, 390*, 112336.

63. Gupta, D. A., Desai, M. L., Malek, N. I., & Kailasa, S. K. (2020). Fluorescence detection of Fe^{3+} ion using ultra-small fluorescent carbon dots derived from pineapple (*Ananas comosus*): Development of miniaturized analytical method. *Journal of Molecular Structure, 1216*, 128343.
64. Sawalha, S., Silvestri, A., Criado, A., Bettini, S., Prato, M., & Valli, L. (2020). Tailoring the sensing abilities of carbon nanodots obtained from olive solid wastes. *Carbon, 167*, 696–708.
65. Christopoulou, N. M., Kalogianni, D. P., & Christopoulos, T. K. (2021). *Posidonia oceanic* (mediterranean tapeweed) leaf litter as a source of fluorescent carbon dot preparations. *Microchemical Journal, 161*, 105787.
66. Wang, W., Chen, J., Wang, D., Shen, Y., Yang, L., Zhang, T., & Ge, J. (2021). Facile synthesis of biomass waste-derived fluorescent N, S, P co-doped carbon dots for detection of Fe^{3+} ions in solutions and living cells. *Analytical Methods, 13*, 789–795.
67. Preethi, M., Viswanathan, C., & Ponpandian, N. (2021). A green path to extract carbon quantum dots by coconut water: Another fluorescent probe towards Fe^{3+} ions. *Particuology, 58*, 251–258.
68. Architha, N., Ragupathi, M., Shobana, C., Selvankumar, T., Kumar, P., Lee, Y. S., & Selvan, R. K. (2021). Microwave-assisted green synthesis of fluorescent carbon quantum dots from *Mexican mint* extract for Fe^{3+} detection and bio-imaging applications. *Environmental Research, 199*, 111263.
69. Zhu, L., Shen, D., Liu, Q., Wu, C., & Gu, S. (2021). Sustainable synthesis of bright green fluorescent carbon quantum dots from lignin for highly sensitive detection of Fe^{3+} ions. *Applied Surface Science, 565*, 150526.
70. Senol, A. M., & Onganer, Y. (2022). A novel "turn-off" fluorescent sensor based on cranberry derived carbon dots to detect iron (III) and hypochlorite ions. *Journal of Photochemistry and Photobiology, A: Chemistry, 424*, 113655.
71. Atchudan, R., Edison, T. N. J. I., Perumal, S., Vinodh, R., Sundramoorthy, A. K., Babu, R. S., & Lee, Y. R. (2022). *Morus nigra*-derived hydrophilic carbon dots for the highly selective and sensitive detection of ferric ion in aqueous media and human colon cancer cell imaging. *Colloids and Surfaces A, 635*, 128073.
72. Gu, L., Zhang, J., Yang, G., Tang, Y., Zhang, X., Huang, X., Zhai, W., Fodjo, E. K., & Kong, C. (2022). Green preparation of carbon quantum dots with wolfberry as on-off-on nanosensors for the detection of Fe^{3+} and L-ascorbic acid. *Food Chemistry, 376*, 131898.
73. Shan, F., Fu, L., Chen, X., Xie, X., Liao, C., Zhu, Y., Xia, H., Zhang, J., Yan, L., Wang, Z., & Yu, X. (2022). 'Waste-to-wealth: Functional biomass carbon dots based on bee pollen waste and application. *Chinese Chemical Letters, 33*, 2942–2948.
74. Xu, J., Cui, K., Gong, T., Zhang, J., Zhai, Z., Hou, L., Zaman, F., & Yuan, C. (2022). Ultrasonic-assisted synthesis of N-doped, multicolor carbon dots toward fluorescent inks, fluorescence sensors, and logic gate operations. *Nanomaterials, 12*, 312.
75. Gracia, K. D. J., Thavamani, S. S., Amaladhas, T. P., Devanesan, S., Ahmed, M., & Kannan, M. M. (2022). 'Valorisation of bio-derived fluorescent carbon dots for metal sensing, DNA binding and bioimaging. *Chemosphere, 298*, 134128.
76. Nagaraj, M., Ramalingam, S., Murugan, C., Aldawood, S., Jin, J. O., Choi, I., & Kim, M. (2022). Detection of Fe^{3+} ions in aqueous environment using fluorescent carbon quantum dots synthesized from endosperm of *Borassus flabellifer*. *Environmental Research, 212*, 113273.
77. Huang, J., Deng, Z., Ding, C., Jin, Y., Wang, B., & Chen, J. (2022). Peroxyoxalate/carbon dots chemiluminescent reaction for fluorescent and visual determination of Fe^{3+}. *Microchemical Journal, 181*, 107782.
78. Zhang, Q., He, S., Zheng, K., Zhang, L., Lin, L., Chen, F., Du, X., & Li, B. (2022). Green synthesis of mustard seeds carbon dots and study on fluorescence quenching mechanism of Fe^{3+} ions. *Inorganic Chemistry Communications, 146*, 110034.
79. Durrani, S., Zhang, J., Mukramin, Wang, H., Wang, Z., Khan, L. U., Zhang, F., Durrani, F., Wu, F. G., & Lin, F. (2023). Biomass-based carbon dots for Fe^{3+} and adenosine triphosphate detection in mitochondria. *ACS Appl Nano Mater, 6*, 76–85.

80. Elizabeth, A. T., James, E., Jesan, L. I., Arockiaraj, S. D., & Vasu, A. E. (2023). Green synthesis of value-added nitrogen doped carbon quantum dots from *Crescentia cujete* fruit waste for selective sensing of Fe^{3+} ions in aqueous medium. *Inorganic Chemistry Communications, 149*, 110427.
81. Patra, S., Singh, M., Subudhi, S., Mandal, M., Nayak, A. K., Sahu, B. B., & Mahanandia, P. (2023). One-step green synthesis of in–situ functionalized carbon quantum dots from *Tagetes patula* flowers: Applications as a fluorescent probe for detecting Fe^{3+} ions and as an antifungal agent. *Journal of Photochemistry and Photobiology, A: Chemistry, 442*, 114779.
82. Pansari, P., Durga, G., & Sharma, R. (2023). Carbon nanoprobe derived from *Nyctanthes arbor-tristis* flower: Unveiling the surface defect-derived fluorescence. *Spectrochimica Acta A, 303*, 123119.
83. Thakkar, H., Bhandary, P., & Thakore, S. (2023). Biogenic carbon dot-embedded chitosan hydrogels as a two-stage fluorescence off–on probe for sequential and ratiometric detection of Fe(III) and glutathione. *ACS Appl Nano Mater, 6*, 16253–16266.
84. Hao, H. C., Chen, S., Tan, Z. X., & Jiang, H. (2023). Preparation of high-yield carbon quantum dots and paper-based sensors from biomass wastes by mechano-chemical method. *Journal of Environmental Chemical Engineering, 11*, 111406.
85. Sandeep, D. H., Krushna, B. R. R., Sharma, S. C., Chandrasekaran, K., Inbanathan, J., George, F. A., Francis, D., Nadar, N. R., Lingaraju, K., & Nagabhushana, H. (2023). Transforming human hair fibers into carbon dots: Utilization in flexible films, fingerprint detection, counterfeit prevention and Fe^{3+} detection. *Materials Today Sustainability, 24*, 100605.
86. Qiu, Y., Xia, L., Shi, R., Yuan, L., Zhang, Y., Chen, A., Zhou, K., Wu, H., Zhang, K., Xia, Z., & Fu, Q. (2024). Cost-efficient and ultrasensitive hydrogel-based visual point-of-care sensor integrated with surface patterning and strongly emissive carbon dots directly from Prunus mume Carbonisatus. *Sensors and Actuators B: Chemical, 401*, 134958.
87. Jiao, X. Y., Li, L. S., Qin, S., Zhang, Y., Huang, K., & Xu, L. (2019). The synthesis of fluorescent carbon dots from mango peel and their multiple applications. *Colloids and Surfaces A, 577*, 306–314.
88. Singh, J., Kaur, S., Lee, J., Mehta, A., Kumar, S., Kim, K. H., Basu, S., & Rawat, M. (2020). Highly fluorescent carbon dots derived from *Mangifera indica* leaves for selective detection of metal ions. *Science of the Total Environment, 720*, 137604.
89. Tejwan, N., Sharma, A., Thakur, S., & Das, J. (2021). Green synthesis of a novel carbon dots from red Korean ginseng and its application for Fe^{2+} sensing and preparation of nanocatalyst. *Inorganic Chemistry Communications, 134*, 108985.
90. Liu, W., Jiang, C., Zhang, L., Li, X., Hou, Q., & Ni, Y. (2022). Valorization of cellulose pulp derived carbon quantum dots by controllable fractionation. *Industrial Crops and Products, 188*, 115560.
91. Mathew, S., Chacko, A. R., Korah, B. K., Punnose, M. S., & Mathew, B. (2022). Green synthesized carbon quantum dot as dual sensor for Fe(II) ions and rational design of catalyst for visible light mediated abatement of pollutants. *Applied Surface Science, 606*, 154975.
92. Zhang, Y., Li, Z., Sheng, L., & Meng, A. (2023). Lemon juice-derived nitrogen-doped carbon quantum dots for highly sensitive and selective determination of ferrous ions and cell imaging. *Colloids and Surfaces A, 657*, 130580.
93. Sha, Y., Lou, J., Bai, S., Wu, D., Liu, B., & Ling, Y. (2013). Hydrothermal synthesis of nitrogen-containing carbon nanodots as the high-efficient sensor for copper(II) ions. *Materials Research Bulletin, 48*, 1728–1731.
94. Wei, J., Zhang, X., Sheng, Y., Shen, J., Huang, P., Guo, S., Pan, J., & Feng, B. (2014). Dual functional carbon dots derived from cornflour via a simple one-pot hydrothermal route. *Materials Letters, 123*, 107–111.
95. Gedda, G., Lee, C. Y., Lin, Y. C., & Wu, H. F. (2016). Green synthesis of carbon dots from prawn shells for highly selective and sensitive detection of copper ions. *Sensors and Actuators, B: Chemical, 224*, 396–403.
96. Niu, X., Liu, G., Li, L., Fu, Z., Xu, H., & Cui, F. (2015). Green and economical synthesis of nitrogen-doped carbon dots from vegetables for sensing and imaging applications. *RSC Advances, 5*, 95223–95229.

97. Shi, L., Zhao, B., Li, X., Zhang, G., Zhang, Y., Dong, C., & Shuang, S. (2017). Green-fluorescent nitrogen-doped carbon nanodots for biological imaging and paper-based sensing. *Analytical Methods, 9*, 2197–2204.
98. Ma, X., Dong, Y., Sun, H., & Chen, N. (2017). Highly fluorescent carbon dots from peanut shells as potential probes for copper ion: The optimization and analysis of the synthetic process. *Materials Today Chemistry, 5*, 1–10.
99. Liu, L., Gong, H., Li, D., & Zhao, L. (2018). Synthesis of carbon dots from pear juice for fluorescence detection of Cu^{2+} ion in water. *Journal of Nanoscience and Nanotechnology, 18*, 5327–5332.
100. Gu, D., Zhang, P., Zhang, L., Liu, H., Pu, Z., & Shang, S. (2018). Nitrogen and phosphorus co-doped carbon dots derived from lily bulbs for copper ion sensing and cell imaging. *Optical Materials, 83*, 272–278.
101. Bhamore, J. R., Jha, S., Park, T. J., & Kailasa, S. K. (2018). Fluorescence sensing of Cu^{2+} ion and imaging of fungal cell by ultra-small fluorescent carbon dots derived from *Acacia concinna* seeds. *Sensors and Actuators, B: Chemical, 227*, 47–54.
102. Rooj, B., Dutta, A., Islam, S., & Mandal, U. (2018). Green synthesized carbon quantum dots from *Polianthes tuberose L.* petals for copper (II) and iron (II) detection. *Journal of Fluorescence, 28*, 1261–1267.
103. Radhakrishnan, K., Panneerselvam, P., & Marieeswaran, M. (2019). A green synthetic route for the surface-passivation of carbon dots as an effective multifunctional fluorescent sensor for the recognition and detection of toxic metal ions from aqueous solution. *Analytical Methods, 11*, 490–506.
104. Murugan, N., Prakash, M., Jayakumar, M., Sundaramurthy, A., & Sundramoorthy, A. K. (2019). Green synthesis of fluorescent carbon quantum dots from *Eleusine coracana* and their application as a fluorescence 'turn-off' sensor probe for selective detection of Cu^{2+}. *Applied Surface Science, 476*, 468–480.
105. Ali, H. R. H., Hassan, A. I., Hassan, Y. F., & El-Wekil, M. M. (2020). Development of dual function polyamine-functionalized carbon dots derived from one step green synthesis for quantitation of Cu^{2+} and S^{2-} ions in complicated matrices with high selectivity. *Analytical and Bioanalytical Chemistry, 412*, 1353–1363.
106. Xu, J., Wang, C., Li, H., & Zhao, W. (2020). Synthesis of green-emitting carbon quantum dots with double carbon sources and their application as a fluorescent probe for selective detection of Cu^{2+} ions. *RSC Advances, 10*, 2536–2544.
107. Chang, D., Shi, L., Zhang, Y., Zhang, G., Zhang, C., Dong, C., & Shuang, S. (2020). Smilax China-derived yellow-fluorescent carbon dots for temperature sensing, Cu^{2+} detection and cell imaging. *The Analyst, 145*, 2176–2183.
108. Liu, Z., Li, B., Shi, X., Li, L., Feng, Y., Jia, D., & Zhou, Y. (2021). Target-oriented synthesis of high synthetic yield carbon dots with tailored surface functional groups for bioimaging of zebrafish, flocculation of heavy metal ions and ethanol detection. *Applied Surface Science, 538*, 148118.
109. Tai, J. Y., Leong, K. H., Saravanan, P., Tan, S. T., Chong, W. C., & Sim, L. C. (2021). Facile green synthesis of fingernails derived carbon quantum dots for Cu^{2+} sensing and photodegradation of 2,4-dichlorophenol. *Journal of Environmental Chemical Engineering, 9*, 104622.
110. Zaman, A. S. K., Tan, T. L., Chowmasundaram, Y. A/P, Jamaludin, N., Sadrolhosseini, A. R., Rashid, U., & Rashid, S. A. (2021). Properties and molecular structure of carbon quantum dots derived from empty fruit bunch biochar using a facile microwave-assisted method for the detection of Cu^{2+} ions. *Optical Materials, 112*, 110801.
111. Chattopadhyay, S., Mehrotra, N., Jain, S., & Singh, H. (2021). Development of novel blue emissive carbon dots for sensitive detection of dual metal ions and their potential applications in bioimaging and chelation therapy. *Microchemical Journal, 170*, 106706.
112. Li, J., Xu, O., & Zhu, X. (2021). A facile green and one-pot synthesis of grape seed derived carbon quantum dots as a fluorescence probe for Cu(II) and ascorbic acid. *RSC Advances, 11*, 34107–34116.

113. Balakrishnan, T., Ang, W. L., Mahmoudi, E., Mohammad, A. W., & Sambudi, N. S. (2022). Formation mechanism and application potential of carbon dots synthesized from palm kernel shell via microwave assisted method. *Carbon Resources Conversion, 5*, 150–166.
114. Avila, J. M., Ayala, M. R., Kumar, Y., Perez-Tijerina, E., Robles, M. A. R., & Agarwal, V. (2022). Avocado seeds derived carbon dots for highly sensitive Cu (II)/Cr (VI) detection and copper (II) removal via flocculation. *Chemical Engineering Journal, 446*, 137171.
115. Shi, J., Zhou, Y., Ning, J., Hu, G., Zhang, Q., Hou, Y., & Zhou, Y. (2022). Prepared carbon dots from wheat straw for detection of Cu^{2+} in cells and zebrafish and room temperature phosphorescent anti-counterfeiting. *Spectrochima Acta A, 281*, 121597.
116. Mohamed, R. M. K., Mohamed, S. H., Asran, A. M., Alsohaimi, I. H., Hassan, H. M. A., Ibrahim, H., & El-Wekil, M. M. (2023). Bifunctional ratiometric sensor based on highly fluorescent nitrogen and sulfur biomass-derived carbon nanodots fabricated from manufactured dairy product as a precursor. *Spectrochima Acta A, 293*, 122444.
117. He, Y., Chen, X., Wang, P., Li, X., Wang, B., Wang, X., Wu, Z., & Wang, W. (2023). 'One-step green synthesis of carbon dots derived from *Plumeria alba* flowers for sensing and bioimaging. *New Journal of Chemistry, 47*, 8877–8884.
118. Wang, B., Lan, J., Ou, J., Bo, C., & Gong, B. (2023). Ganoderma lucidum bran-derived blue-emissive and green-emissive carbon dots for detection of copper ions. *RSC Advances, 13*, 14506–14516.
119. Xu, Y., Lan, J., Wang, B., Bo, C., Ou, J., & Gong, B. (2023). Simple fabrication of carbon quantum dots and activated carbon from waste wolfberry stems for detection and adsorption of copper ion. *RSC Advances, 13*, 21199–21210.
120. Wen, X., Shi, L., Wen, G., Li, Y., Dong, C., Yang, J., & Shuang, S. (2016). Green and facile synthesis of nitrogen-doped carbon nanodots for multicolour cellular imaging and Co^{2+} sensing in living cells. *Sensors and Actuators, B: Chemical, 235*, 179–187.
121. Zhao, C., Li, X., Cheng, C., & Yang, Y. (2019). Green and microwave assisted synthesis of carbon dots and application for visual detection of cobalt(II) ions and pH sensing. *Microchemical Journal, 147*, 183–190.
122. Du, F., Cheng, Z., Kremer, M., Liu, Y., Wang, X., Shuang, S., & Dong, C. (2020). A label-free multifunctional nanosensor based on N-doped carbon nanodots for vitamin B_{12} and Co^{2+} detection, and bioimaging in living cells and zebrafish. *Journal of Materials Chemistry B, 8*, 5089–5095.
123. Hu, G., Ge, L., Li, Y., Mukhtar, M., Shen, B., Yang, D., & Li, J. (2020). Carbon dots derived from flax straw for highly sensitive and selective detections of cobalt, chromium, and ascorbic acid. *Journal of Colloid and Interface Science, 579*, 96–108.
124. Wen, X., Wen, G., Li, W., Zhao, Z., Duan, X., Yan, W., Trant, J. F., & Li, Y. (2021). Carbon dots for specific "off-on" sensing of Co^{2+} and EDTA for *in vivo* bioimaging. *Materials Science and Engineering C, 123*, 112022.
125. Pang, Z., Fu, Y., Yu, H., Liu, S., Yu, S., Liu, Y., Wu, Q., Liu, Y., Nie, G., Xu, H., Nie, S., & Yao, S. (2022). Efficient ethanol solvothermal synthesis of high-performance nitrogen-doped carbon quantum dots from lignin for metal ion nanosensing and cell imaging. *Industrial Crops and Products, 183*, 114957.
126. Mathew, S., & Mathew, B. (2023). Heteroatom doped carbon dots from green sources as metal ion sensor and as fluorescent ink. *Diamond & Related Materials, 139*, 110293.
127. Qiang, R., Yang, S., Hou, K., & Wang, J. (2019). 'Synthesis of carbon quantum dots with green luminescence from potato starch', New. *Journal of Chemistry, 43*, 10826–10833.
128. da Silva Júnior, A. H., Macuvele, D. L. P., Riella, H. G., Soares, C., & Padoin, N. (2021). Novel carbon dots for zinc sensing from *Campomanesia phaea. Materials Letters, 283*, 128813.
129. Dastidar, D. G., Mukherjee, P., Ghosh, D., & Banerjee, D. (2021). Carbon quantum dots prepared from onion extract as fluorescence turn-on probes for selective estimation of Zn^{2+} in blood plasma. *Colloids and Surfaces A, 611*, 125781.
130. Tyagi, A., Tripathi, K. M., Singh, N., Choudhary, S., & Gupta, R. K. (2016). Green synthesis of carbon quantum dots from lemon peel waste: Applications in sensing and photocatalysis. *RSC Advances, 6*, 72423–72432.

131. Shen, J., Shang, S., Chen, X., Wang, D., & Cai, Y. (2017). Highly fluorescent N, S-co-doped carbon dots and their potential applications as antioxidants and sensitive probes for Cr (VI) detection. *Sensors and Actuators, B: Chemical, 248*, 92–100.
132. Bhatt, S., Bhatt, M., Kumar, A., Vyas, G., Gajaria, T., & Paul, P. (2018). Green route for synthesis of multifunctional fluorescent carbon dots from Tulsi leaves and its application as Cr(VI) sensors, bio-imaging and patterning agents. *Colloids and Surfaces. B, Biointerfaces, 167*, 126–133.
133. Pooja, D., Singh, L., Thakur, A., & Kumar, P. (2019). Green synthesis of glowing carbon dots from *Carica papaya* waste pulp and their application as a label-free chemo probe for chromium detection in water. *Sensors and Actuators, B: Chemical, 283*, 363–372.
134. Roshni, V., Misra, S., Santra, M. K., & Ottoor, D. (2019). One pot green synthesis of C-dots from groundnuts and its application as Cr (VI) sensor and *in vitro* bioimaging agent. *Journal of Photochemistry and Photobiology, A: Chemistry, 373*, 28–36.
135. Athika, M., Prasath, A., Duraisamy, E., Devi, V. S., Sharma, A. S., & Elumalai, P. (2019). Carbon-quantum dots derived from denatured milk for efficient chromium-ion sensing and supercapacitor applications. *Materials Letters, 241*, 156–159.
136. Qing, W., Chen, K., Yang, Y., Wang, Y., & Liu, X. (2020). Cu^{2+}-doped carbon dots as fluorescence probe for specific recognition of Cr (VI) and its antimicrobial activity. *Microchemical Journal, 152*, 104262.
137. Roshni, V., Gujar, V., Pathan, H., Islam, S., Tawre, M., Pardesi, K., Santra, M. K., & Ottoor, D. (2019). Bioimaging applications of carbon dots (C. dots) and its cystamine functionalization for the sensitive detection of Cr(VI) in aqueous samples. *Journal of Fluorescence, 29*, 1381–1392.
138. Sakaew, C., Sricharoen, P., Limchoowong, N., Nuengmatcha, P., Kukusamude, C., Kongsri, S., & Chanthai, S. (2020). Green and facile synthesis of water-soluble carbon dots from ethanolic shallot extract for chromium ion sensing in milk, fruit juices, and wastewater samples. *RSC Advances, 10*, 20638–20645.
139. Qiu, Y., Gao, D., Yin, H., Zhang, K., Zeng, J., Wang, L., Xia, L., Zhou, K., Xia, Z., & Fu, Q. (2020). Facile, green and energy efficient preparation of fluorescent carbon dots from processed traditional Chinese medicine and their applications for on-site semi-quantitative visual detection of Cr(VI). *Sensors and Actuators, B: Chemical, 324*, 128722.
140. Zheng, X., Qin, K., He, L., Ding, Y., Luo, Q., Zhang, C., Cui, X., Tan, Y., Li, L., & Wei, Y. (2021). Novel fluorescent nitrogen-doped carbon dots derived from *Panax notoginseng* for bioimaging and high selectivity detection of Cr^{6+}. *The Analyst, 146*, 911–919.
141. Wang, D., Zhang, L., Li, P., Li, J., & Dong, C. (2020). 'Convenient synthesis of carbon nanodots for detecting Cr(VI) and ascorbic acid by fluorimetry', New. *Journal of Chemistry, 44*, 20806–20811.
142. Das, P., Maruthapandi, M., Saravanan, A., Natan, M., Jacobi, G., Banin, E., & Gedanken, A. (2020). Carbon dots for heavy-metal sensing, pH-sensitive cargo delivery, and antibacterial applications. *ACS Appl Nano Mater, 3*, 11777–11790.
143. Das, M., Thakkar, H., Patel, D., & Thakore, S. (2021). Repurposing the domestic organic waste into green emissive carbon dots and carbonized adsorbent: A sustainable zero waste process for metal sensing and dye sequestration. *Journal of Environmental Chemical Engineering, 9*, 106312.
144. Tian, H., Ju, G., Li, M., Fu, W., Dai, Y., Liang, Z., Qiu, Y., Qin, Z., & Yin, X. (2021). Fluorescent "on–off–on" sensor based on N, S-codoped carbon dots from seaweed *(Sargassum carpophyllum)* for specific detection of Cr(VI) and ascorbic acid. *RSC Advances, 11*, 35946–35953.
145. Zhu, L., Shen, D., & Luo, K. H. (2022). 'Triple-emission nitrogen and boron co-doped carbon quantum dots from lignin: Highly fluorescent sensing platform for detection of hexavalent chromium ions. *Journal of Colloid and Interface Science, 617*, 557–567.
146. Song, Y., Qi, N., Li, K., Cheng, D., Wang, D., & Li, Y. (2022). Green fluorescent nanomaterials for rapid detection of chromium and iron ions: Wool keratin based carbon quantum dots. *RSC Advances, 12*, 8108–8118.

147. Goswami, J., Rohman, S. S., Guha, A. K., Basyach, P., Sonowal, K., Borah, S. P., Saikia, L., & Hazarika, P. (2022). Phosphoric acid assisted synthesis of fluorescent carbon dots from waste biomass for detection of Cr(VI) in aqueous media. *Materials Chemistry and Physics, 286*, 126133.
148. John, B. K., John, N., & Mathew, B. (2023). 'Green synthesis of fluorescent carbon dots from *Annona reticulata* leaves as a sensor for chromium (III) ions. *Materials Today: Proceedings, 72*, 169–174.
149. Meng, F., Xu, H., Wang, S., Wei, J., Zhou, W., Wang, Q., Li, P., Kong, F., & Zhang, Y. (2022). One-step high-yield preparation of nitrogen- and sulfur-codoped carbon dots with applications in chromium(VI) and ascorbic acid detection. *RSC Advances, 12*, 19686–19694.
150. Rangel, M., Saluja, S., Barba, V., Perez-Huerta, J. S., & Agarwal, V. (2023). Dual-emissive waste oil based S-doped carbon dots for acetone detection and Cr(VI) detection/reduction/removal. *Journal of Environmental Chemical Engineering, 11*, 109438.
151. Zhang, S., Mao, Y., Sun, J., Song, T., Song, Z., Zhao, X., & Wang, W. (2023). One-pot solvothermal preparation of triple-emission N, Cl doped carbon quantum dots from waste traditional Chinese medicines as a fluorescent sensor for sensing water and Cr (VI). *Colloids and Surfaces A, 669*, 131471.
152. Feng, B., Chen, Z., Li, N., Bi, Y., Kong, F., Wang, Z., & Tan, S. (2023). Novel polysaccharide-derived carbon dots doped with N, S, P: Synthesis, characterization, in vitro antioxidant, fluorescence sensor for chromium (VI) detection and cellular imaging. *Diamond & Related Materials, 139*, 110295.
153. Kolekar, A. G., Nille, O. S., Koparde, S. V., Patil, A. S., Waghmare, R. D., Sohn, D., Anbhule, P. V., Kolekar, G. B., Gokavi, G. S., & More, V. R. (2024). Green, facial zinc doped hydrothermal synthesis of cinnamon derived fluorescent carbon dots (Zn-Cn-CDs) for highly selective and sensitive Cr^{6+} and Mn^{7+} metal ion sensing application. *Spectrochimica Acta A, 304*, 123413.
154. Lu, W., Qin, X., Liu, S., Chang, G., Zhang, Y., Luo, Y., Asiri, A. M., Al-Youbi, A. O., & Sun, X. (2012). Economical, green synthesis of fluorescent carbon nanoparticles and their use as probes for sensitive and selective detection of mercury(II) ions. *Analytical Chemistry, 84*, 5351–5357.
155. Lu, W., Qin, X., Asiri, A. M., Al-Youbi, A. O., & Sun, X. (2013). Green synthesis of carbon nanodots as an effective fluorescent probe for sensitive and selective detection of mercury(II) ions. *Journal of Nanoparticle Research, 15*, 1344.
156. Huang, H., Lv, J. J., Zhou, D. L., Bao, N., Xu, Y., Wang, A. J., & Feng, J. J. (2013). One-pot green synthesis of nitrogen-doped carbon nanoparticles as fluorescent probes for mercury ions. *RSC Advances, 3*, 21691–21696.
157. Gaddam, R. R., Vasudevan, D., Narayan, R., & Raju, K. V. S. N. (2014). Controllable synthesis of biosourced blue-green fluorescent carbon dots from camphor for the detection of heavy metal ions in water. *RSC Advances, 4*, 57137–57143.
158. Yu, J., Song, N., Zhang, Y. K., Zhong, S. X., Wang, A. J., & Chen, J. (2015). Green preparation of carbon dots by Jinhua bergamot for sensitive and selective fluorescent detection of Hg^{2+} and Fe^{3+}. *Sensors and Actuators, B: Chemical, 214*, 29–35.
159. Essner, J. B., Laber, C. H., Ravula, S., Parada, L. P., & Baker, G. A. (2016). Pee-dots: Biocompatible fluorescent carbon dots derived from the upcycling of urine. *Green Chemistry, 18*, 243–250.
160. Ye, Q., Yan, F., Luo, Y., Wang, Y., Zhou, X., & Chen, L. (2017). Formation of N, S-codoped fluorescent carbon dots from biomass and their application for the selective detection of mercury and iron ion. *Spectrochimica Acta A, 173*, 854–862.
161. Li, Z., Zhang, Y., Niu, Q., Mou, M., Wu, Y., Liu, X., Yan, Z., & Liao, S. (2017). A fluorescence probe based on the nitrogen-doped carbon dots prepared from orange juice for detecting Hg^{2+} in water. *Journal of Luminescence, 187*, 274–280.
162. Venkateswarlu, S., Viswanath, B., Reddy, A. S., & Yoon, M. (2018). Fungus-derived photoluminescent carbon nanodots for ultrasensitive detection of Hg^{2+} ions and photoinduced bactericidal activity. *Sensors and Actuators, B: Chemical, 258*, 172–183.

163. Xu-Cheng, F., Xuan-Hua, L., Jin, J. Z., Zhang, J., & Wei, G. (2018). Facile synthesis of bagasse waste derived carbon dots for trace mercury detection. *Materials Research Express, 5*, 065044.
164. Ansi, V. A., & Renuka, N. K. (2019). Exfoliated graphitic carbon dots: Application in heavy metal ion sensing. *Journal of Luminescence, 205*, 467–474.
165. Sahoo, N. K., Das, S., Jana, G. C., Aktara, M. N., Patra, A., Maji, A., Beg, M., Jha, P. K., & Hossain, M. (2019). Eco-friendly synthesis of a highly fluorescent carbon dots from spider silk and its application towards Hg (II) ions detection in real sample and living cells. *Microchemical Journal, 144*, 479–488.
166. Gharat, P. M., Pal, H., & Choudhury, S. D. (2019). 'Photophysics and luminescence quenching of carbon dots derived from lemon juice and glycerol. *Spectrochimica Acta A, 209*, 14–21.
167. Liu, G., Jia, H., Li, N., Li, X., Yu, Z., Wang, J., & Song, Y. (2019). High-fluorescent carbon dots (CDs) originated from China grass carp scales (CGCS) for effective detection of Hg(II) ions. *Microchemical Journal, 145*, 718–728.
168. Zhang, Q., Zhang, X., Bao, L., Wu, Y., Jiang, L., Zheng, Y., Wang, Y., & Chen, Y. (2019). The application of green-synthesis-derived carbon quantum dots to bioimaging and the analysis of mercury(II). *Journal of Analytical Methods in Chemistry, 2019*, 8183134.
169. Sun, D., Liu, T., Wang, C., Yang, L., Yang, S., & Zhuo, K. (2020). Hydrothermal synthesis of fluorescent carbon dots from gardenia fruit for sensitive on-off-on detection of Hg^{2+} and cysteine. *Spectrochimica Acta A, 240*, 118598.
170. Hu, C., Wang, K. H., Chen, Y. Y., Maniwa, M., Lin, K. Y. A., Kawai, T., & Chen, W. (2022). Detection of Fe^{3+} and Hg^{2+} ions through photoluminescence quenching of carbon dots derived from urea and bitter tea oil residue. *Spectrochimica Acta A, 272*, 120963.
171. Kasinathan, K., Samayanan, S., Marimuthu, K., & Yim, J. H. (2022). Green synthesis of multicolour fluorescence carbon quantum dots from sugarcane waste: Investigation of mercury (II) ion sensing, and bio-imaging applications. *Applied Surface Science, 601*, 154266.
172. Samota, S., Tewatia, P., Rani, R., Chakraverty, S., & Kaushik, A. (2022). Carbon dot nanosensors for ultra-low level, rapid assay of mercury ions synthesized from an aquatic weed, *Typha angustata* Bory (Patera). *Diamond & Related Materials, 130*, 109433.
173. Dhandapani, E., Maadeswaran, P., Raj, R. M., Raj, V., Kandiah, K., & Duraisamy, N. (2023). A potential forecast of carbon quantum dots (CQDs) as an ultrasensitive and selective fluorescence probe for Hg (II) ions sensing. *Materials Science and Engineering B, 287*, 116098.
174. Tang, D., Wang, Q., Zhao, S., Yi, J., Lan, M., Zeng, J., & Huang, L. (2023). Environmental synthesis of yellow fluorescent carbon dots for on-off-on detection of mercury and cysteine. *Inorganic Chemistry Communications, 156*, 111160.
175. Li, X., Shi, Q., Li, M., Song, N., Xiao, Y., Xiao, H., James, T. D., & Feng, L. (2024) 'Functionalization of cellulose carbon dots with different elements (N, B and S) for mercury ion detection and anti-counterfeit applications. *Chinese Chemical Letters, 35*, 109021.
176. Tan, X. W., Romainor, A. N. B., Chin, S. F., & Ng, S. M. (2014). Carbon dots production via pyrolysis of sago waste as potential probe for metal ions sensing. *Journal of Analytical and Applied Pyrolysis, 105*, 157–165.
177. Liu, Y., Zhou, Q., Li, J., Lei, M., & Yan, X. (2016). Selective and sensitive chemosensor for lead ions using fluorescent carbon dots prepared from chocolate by one-step hydrothermal method. *Sensors and Actuators, B: Chemical, 237*, 597–604.
178. Kumar, A., Chowdhuri, A. R., Laha, D., Mahto, T. K., Karmakar, P., & Sahu, S. K. (2017). Green synthesis of carbon dots from *Ocimum sanctum* for effective fluorescent sensing of Pb^{2+} ions and live cell imaging. *Sensors and Actuators, B: Chemical, 242*, 679–686.
179. Bandi, R., Dadigala, R., Gangapuram, B. R., & Guttena, V. (2018). Green synthesis of highly fluorescent nitrogen-doped carbon dots from *Lantana camara* berries for effective detection of lead(II) and bioimaging. *Journal of Photochemistry and Photobiology, B: Biology, 178*, 330–338.
180. Gao, Y., Jiao, Y., Zhang, H., Lu, W., Liu, Y., Han, H., Gong, X., Li, L., Shuang, S., & Dong, C. (2019). One-step synthesis of a dual-emitting carbon dot-based ratiometric fluorescent probe

for the visual assay of Pb^{2+} and PPi and development of a paper sensor. *Journal of Materials Chemistry B, 7*, 5502–5509.
181. Yong, C., Lei, Y., Liu, Y., Li, Y., Wang, N., Tong, B., & Tao, L. (2023). Mechanistic regulation of gram-scale synthesis of triple emission cyanobacteria-based carbon dots and visual ratiometric sensing applications. *Applied Surface Science, 623*, 157049.
182. Gu, D., Hong, L., Zhang, L., Liu, H., & Shang, S. (2018). Nitrogen and sulfur co-doped highly luminescent carbon dots for sensitive detection of Cd (II) ions and living cell imaging applications. *Journal of Photochemistry and Photobiology, B: Biology, 186*, 144–151.
183. Chauhan, P., Dogra, S., Chaudhary, S., & Kumar, R. (2020). Usage of coconut coir for sustainable production of high-valued carbon dots with discriminatory sensing aptitude toward metal ions. *Materials Today Chemistry, 16*, 100247.
184. Abidin, N. H. Z., Wongso, V., Hui, K. C., Cho, K., Sambudi, N. S., Ang, W. L., & Saad, B. (2020). The effect of functionalization on rice-husks derived carbon quantum dots properties and cadmium removal. *Journal of Water Process Engineering, 38*, 101634.
185. Huang, Z. Y., Wu, W. Z., Li, Z. X., Wu, Y., Wu, C. B., Gao, J., Guo, J., Chen, Y., Hu, Y., & Huang, C. (2022). Solvothermal production of tea residue derived carbon dots by the pretreatment of choline chloride/urea and its application for cadmium detection. *Industrial Crops and Products, 184*, 115085.
186. Sariga, Kolaprath, M. K. A., Benny, L., & Varghese, A. (2023). A facile, green synthesis of carbon quantum dots from *Polyalthia longifolia* and its application for the selective detection of cadmium. *Dyes and Pigments, 210*, 111048.
187. Ngu, P. Z. Z., Chia, S. P. P., Fong, J. F. Y., & Ng, S. M. (2016). Synthesis of carbon nanoparticles from waste rice husk used for the optical sensing of metal ions. *New Carbon Materials, 31*, 135–143.
188. Zhang, X., Wu, J., Wu, M., Wang, L., Yu, D., Yan, N., Wu, H., Zhu, J., & Chen, J. (2023). Noncytotoxic lanthanum and nitrogen co-doped lignin-based carbon dots for selective detection of ions in biological imaging. *Journal of Environmental Chemical Engineering, 11*, 109881.
189. Ramezani, Z., Qorbanpour, M., & Rahbar, N. (2018). Green synthesis of carbon quantum dots using quince fruit (*Cydonia oblonga*) powder as carbon precursor: Application in cell imaging and As^{3+} determination. *Colloids and Surfaces A, 549*, 58–66.
190. Radhakrishnan, K., & Panneerselvam, P. (2018). Green synthesis of surface-passivated carbon dots from the prickly pear cactus as a fluorescent probe for the dual detection of arsenic(III) and hypochlorite ions from drinking water. *RSC Advances, 8*, 30455–30467.
191. Alam, M. B., Hassan, N., Sahoo, K., Kumar, M., Sharma, M., Lahiri, J., & Parmar, A. S. (2022). Deciphering interaction between chlorophyll functionalized carbon quantum dots with arsenic and mercury toxic metals in water as highly sensitive dual-probe sensor. *Journal of Photochemistry and Photobiology, A: Chemistry, 431*, 114059.
192. Gurung, S., Neha, Arun, N., Joshi, M., Jaiswal, T., Pathak, A. P., Das, P., Singh, A. K., Tripathi, A., & Tiwari, A. (2023) 'Dual metal ion (Fe^{3+} and As^{3+}) sensing and cell bioimaging using fluorescent carbon quantum dots synthesised from *Cynodon dactylon. Chemosphere, 339*, 139638.
193. Liu, X., Li, T., Hou, Y., Wu, Q., Yi, J., & Zhang, G. (2016). Microwave synthesis of carbon dots with multiresponse using denatured proteins as carbon source. *RSC Advances, 6*, 11711–11718.
194. Arumugam, N., & Kim, J. (2018). Synthesis of carbon quantum dots from broccoli and their ability to detect silver ions. *Materials Letters, 219*, 37–40.
195. Zhao, X., Liao, S., Wang, L., Liu, Q., & Chen, X. (2019). Facile green and one-pot synthesis of *Purple perilla* derived carbon quantum dot as a fluorescent sensor for silver ion. *Talanta, 201*, 1–8.
196. Ji, X., Yuan, X., Nian, H., Song, P., Xiang, Y., Wei, Y., Wang, S., Qin, K., Zhang, Q., & Tu, Y. (2020). Yeast *Cryptococcus Podzolicus* derived fluorescent carbon dots for multicolour cellular imaging and high selectivity detection of pollutant. *Dyes and Pigments, 182*, 108621.
197. Zhou, P., Xu, J., Hou, X., Dai, L., Xiao, X., Zhang, C., & Huo, K. (2022). Lignin fractionation-inspired carbon dots to enable trimodule fluorescent sensing of pH, silver ion and cysteine. *Industrial Crops and Products, 185*, 115127.

198. Qi, Q., Sun, L., Xu, J., Guo, X., Zhang, H., & Zhao, X. (2022). Carrageenan-derived sulfur, nitrogen co-doped carbon dots for sequential detection of Ag^+ and lime sulfur with "on-off-on" pattern. *Journal of Alloys and Compounds, 922*, 166129.
199. Fan, X., Jiang, L., Liu, Y., Sun, W., Qin, Y., Liao, L., & Qin, A. (2023). Chiral CQD-based PL and CD sensors for high sensitive and selective detection of heavy metal ions. *Optical Materials, 137*, 113620.
200. Liao, J., Cheng, Z., & Zhou, L. (2016). 'Nitrogen-doping enhanced fluorescent carbon dots: Green synthesis and their applications for bioimaging and label-free detection of Au^{3+} ions', ACS Sustainable Chem. *Eng., 4*, 3053–3061.
201. Raji, K., Ramanan, V., & Ramamurthy, P. (2019). 'Facile and green synthesis of highly fluorescent nitrogen-doped carbon dots from jackfruit seeds and its applications towards the fluorimetric detection of Au^{3+} ions in aqueous medium and in in vitro multicolor cell imaging', New. *Journal of Chemistry, 43*, 11710–11719.
202. Rahmani, Z., & Ghaemy, M. (2019). One-step hydrothermal-assisted synthesis of highly fluorescent N-doped carbon dots from gum tragacanth: Luminescent stability and sensitive probe for Au^{3+} ions. *Optical Materials, 97*, 109356.
203. Sharma, A. S., Xuing, J., Viswadevarayalu, A., Rong, Y., Sabarinathan, D., Ali, S., Agyekum, A. A., Li, H., & Chen, Q. (2020). Facile preparation of fluorescent carbon quantum dots from denatured sour milk and its multifunctional applications in the fluorometric determination of gold ions, *in vitro* bioimaging and fluorescent polymer film. *Journal of Photochemistry and Photobiology, A: Chemistry, 401*, 112788.
204. Raji, K., Thiyagarajan, S. K., Suresh, R., Vadivel, R., Palanivel, D., & Ramamurthy, P. (2022). Neem seed derived green C-dots: A highly sensitive luminescent probe for aqueous Au^{3+} ions and nurtures green gold recovery. *Colloids and Surfaces A, 641*, 128523.
205. Hashemi, N., & Mousazadeh, M. H. (2021). Green synthesis of photoluminescent carbon dots derived from red beetroot as a selective probe for Pd^{2+} detection. *Journal of Photochemistry and Photobiology, A: Chemistry, 421*, 113534.
206. Qu, Y., Li, D., Liu, J., Du, F., Tan, X., Zhou, Y., Liu, S., & Xu, W. (2022). *Magnolia denudata* leaf-derived near-infrared carbon dots as fluorescent nanoprobes for palladium(II) detection and cell imaging. *Microchemical Journal, 178*, 107375.
207. Bhamore, J. R., Jha, S., Singhal, R. K., Park, T. J., & Kailasa, S. K. (2018). Facile green synthesis of carbon dots from *Pyrus pyrifolia* fruit for assaying of Al^{3+} ion via chelation enhanced fluorescence mechanism. *Journal of Molecular Liquids, 264*, 9–16.
208. Arumugham, T., Alagumuthu, M., Amimodu, R. G., Munusamy, S., & Iyer, S. K. (2020). A sustainable synthesis of green carbon quantum dot (CQD) from *Catharanthus roseus* (white flowering plant) leaves and investigation of its dual fluorescence responsive behavior in multi-ion detection and biological applications. *Sustainable Materials and Technologies, 23*, e00138.
209. Yu, C., Qin, D., Jiang, X., Zheng, X., & Deng, B. (2021). N-doped carbon quantum dots from *Osmanthus fragrans* as a novel off-on fluorescent nanosensor for highly sensitive detection of quercetin and aluminium ion, and cell imaging. *Journal of Pharmaceutical and Biomedical Analysis, 192*, 113673.
210. Raveendran, P. T. V., Aswathi, B. S., & Renuka, N. K. (2022). Arrowroot derived carbon dots: Green synthesis and application as an efficient optical probe for fluoride ions. *Materials Today: Proceedings, 51*, 2417–2421.
211. Hoan, B. T., Thanh, T. T., Tam, P. D., Trung, N. N., Cho, S., & Pham, V. H. (2019). A green luminescence of lemon derived carbon quantum dots and their applications for sensing of V^{5+} ions. *Materials Science and Engineering B, 251*, 114455.
212. Chellasamy, G., Arumugasamy, S. K., Govindaraju, S., & Yun, K. (2022). Green synthesized carbon quantum dots from maple tree leaves for biosensing of cesium and electrocatalytic oxidation of glycerol. *Chemosphere, 287*, 131915.
213. Chen, M., Liu, C., Zhai, J., An, Y., Li, Y., Zheng, Y., Tian, H., Shi, R., He, X., & Lin, X. (2022). Preparation of solvent-free starch-based carbon dots for the selective detection of Ru^{3+} ions. *RSC Advances, 12*, 18779–18783.

214. Krishnaiah, P., Atchudan, R., Perumal, S., Gangadaran, P., Manoj, D., Ahn, B. C., Kumar, R. S., Almansour, A. I., Lee, Y. R., & Jeon, B. H. (2024). Multifunctional carbon dots originated from waste garlic peel for rapid sensing of heavy metals and fluorescent imaging of 2D and 3D spheroids cultured fibroblast cells. *Spectrochimica Acta A, 304*, 123422.
215. Joseph, J., & Anappara, A. A. (2017). White-light-emitting carbon dots prepared by the electrochemical exfoliation of graphite. *ChemPhysChem, 18*, 292–298.
216. Ge, L., Hu, G., Shi, B., Guo, Q., Li, L., Zhao, L., & Li, J. (2019). Photoluminescence of carbon dots prepared by ball milling and their application in Hela cell imaging. *Applied Physics A, 125*, 641.
217. Thulasi, S., Kathiravan, A., & Jhonsi, M. A. (2020). Fluorescent carbon dots derived from vehicle exhaust soot and sensing of tartrazine in soft drinks. *ACS Omega, 5*, 7025–7031.
218. Ganesan, K., Hayagreevan, C., Jeevagan, A. J., Adinaveen, T., Sophie, P. L., Amalraj, M., & Bhuvaneshwari, D. S. (2024). Candle soot derived carbon dots as potential corrosion inhibitor for stainless steel in HCl medium. *Journal of Applied Electrochemistry, 54*, 89–102.
219. Sun, Y. P., Zhou, B., Lin, Y., Wang, W., Fernando, K. A. S., Pathak, P., Meziani, M. J., Harruff, B. A., Wang, X., Wang, H., Luo, P. G., Yang, H., Kose, M. E., Chen, B., Veca, L. M., & Xie, S. Y. (2006). Quantum-sized carbon dots for bright and colorful photoluminescence. *Journal of the American Chemical Society, 128*, 7756–7757.
220. Reyes, D., Camacho, M., Camacho, M., Mayorga, M., Weathers, D., Salamo, G., Wang, Z., & Neogi, A. (2016). Laser ablated carbon nanodots for light emission. *Nanoscale Research Letters, 11*, 424.
221. Liu, M., Xu, Y., Niu, F., Gooding, J. J., & Liu, J. (2016). Carbon quantum dots directly generated from electrochemical oxidation of graphite electrodes in alkaline alcohols and the applications for specific ferric ion detection and cell imaging. *The Analyst, 141*, 2657–2664.
222. Zhou, J., Booker, C., Li, R., Zhou, X., Sham, T. K., Sun, X., & Ding, Z. (2007). An electrochemical avenue to blue luminescent nanocrystals from multiwalled carbon nanotubes (MWCNTs). *Journal of the American Chemical Society, 129*, 744–745.
223. Qu, D., & Sun, Z. (2020). The formation mechanism and fluorophores of carbon dots synthesized via a bottom-up route. *Materials Chemistry Frontiers, 4*, 400–420.
224. Muro-Hidalgo, J. M., Bazany-Rodríguez, I. J., Hernández, J. G., Pabello, V. M. L., & Thangarasu, P. (2023). Histamine recognition by carbon dots from plastic waste and development of cellular imaging: Experimental and theoretical studies. *Journal of Fluorescence, 33*, 2041–2059.
225. Bhatt, S., Vyas, G., & Paul, P. (2022). Microwave-assisted synthesis of nitrogen-doped carbon dots using prickly pear as the carbon source and its application as a highly selective sensor for Cr(VI) and as a patterning agent. *Analytical Methods, 14*, 269–277.
226. Dubey, P. (2023). An overview on animal/human biomass-derived carbon dots for optical sensing and bioimaging applications. *RSC Advances, 13*, 35088–35126.
227. Emami, E., & Mousazadeh, M. H. (2023). Nitrogen-doped carbon dots for sequential 'on-off-on' fluorescence probe for the sensitive detection of Fe^{3+} and L-alanine/L-histidine. *Journal of Photochemistry and Photobiology, A: Chemistry, 438*, 114536.
228. Ding, C., Xing, H., Guo, X., Yuan, H., Li, C., Zhang, X., & Jia, X. (2023). Tea-derived carbon dots with two ratiometric fluorescence channels for the independent detection of Hg^{2+} and H_2O. *Analytical Methods, 15*, 1998–2005.
229. Sekar, A., Yadav, R., & Basavaraj, N. (2021). 'Fluorescence quenching mechanism and the application of green carbon nanodots in the detection of heavy metal ions: A review, New. *Journal of Chemistry, 45*, 2326–2360.
230. Lou, Y., Hao, X., Liao, L., Zhang, K., Chen, S., Li, Z., Ou, J., Qin, A., & Li, Z. (2021). Recent advances of biomass carbon dots on syntheses, characterization, luminescence mechanism, and sensing applications. *Nano Select, 2*, 1117–1145.
231. Qi, H., Liu, C., Jing, J., Jing, T., Zhang, X., Li, J., Luo, C., Qiu, L., & Li, Q. (2022). Two kinds of biomass-derived carbon dots with one-step synthesis for Fe^{3+} and tetracyclines detection. *Dyes and Pigments, 206*, 110555.

232. Abbaspour, N., Hurrell, R., & Kelishadi, R. (2014). Review on iron and its importance for human health. *Journal of Research in Medical Sciences: The Official Journal of Isfahan University of Medical Sciences, 19*, 164–174.
233. Haehling, S., Jankowska, E. A., Van Veldhuisen, D. J., Ponikowski, P., & Anker, S. D. (2015). Iron deficiency and cardiovascular disease. *Nature Reviews. Cardiology, 12*, 659–669.
234. Soliman, A. T., Sanctis, V. D., Yassin, M., & Soliman, N. (2017). Iron deficiency anemia and glucose metabolism. *Acta Bio-Medica, 88*, 112–118.
235. Brown, R. A. M., Richardson, K. L., Kabir, T. D., Trinder, D., Ganss, R., & Leedman, P. J. (2020). Altered iron metabolism and impact in cancer biology, metastasis, and immunology. *Frontiers in Oncology, 10*, 476.
236. Zheng, P., Zhao, H., Wang, J., Liu, R., Ding, N., Mao, X., & Lai, C. (2020). Detection and separation of Fe(II) and Fe(III) in aqueous solution by laser-induced breakdown spectroscopy coupled with chelating resin enrichment and pH value adjustment. *Journal of Analytical Atomic Spectrometry, 35*, 3032–3038.
237. Kardar, Z. S., Shemirani, F., & Zadmard, R. (2020). Determination of iron(II) and iron(III) via static quenching of the fluorescence of tryptophan-protected copper nanoclusters. *Microchimica Acta, 187*, 81.
238. Sasan, S., Chopra, T., Gupta, A., Tsering, D., Kapoor, K. K., & Parkesh, R. (2022). Fluorescence "turn-off" and colorimetric sensor for Fe^{2+}, Fe^{3+}, and Cu^{2+} ions based on a 2,5,7-triarylimidazopyridine scaffold. *ACS Omega, 7*, 11114–11125.
239. Bost, M., Houdart, S., Oberli, M., Kalonji, E., Huneau, J. F., & Margaritis, I. (2016). Dietary copper and human health: Current evidence and unresolved issues. *Journal of Trace Elements in Medicine and Biology, 35*, 107–115.
240. Bandmann, O., Weiss, K. H., & Kaler, S. G. (2015). Wilson's disease and other neurological copper disorders. *Lancet Neurology, 14*, 103–113.
241. Zhang, Y., Yang, Y. S., Wang, C. M., Chen, W. C., Chen, X. L., Wu, F., & He, H. F. (2023). Copper metabolism–related genes in entorhinal cortex for Alzheimer's disease. *Scientific Reports, 13*, 17458.
242. Giedyk, M., Goliszewska, K., & Gryko, D. (2015). Vitamin B12 catalysed reactions. *Chemical Society Reviews, 44*, 3391–3404.
243. Leyssens, L., Vinck, B., Straeten, C. V. D., Wuyts, F., & Maes, L. (2017). Cobalt toxicity in humans-a review of the potential sources and systemic health effects. *Toxicology, 387*, 43–56.
244. Kloubert, V., & Rink, L. (2015). Zinc as a micronutrient and its preventive role of oxidative damage in cells. *Food & Function, 6*, 3195–3204.
245. Khan, S. T., Malik, A., Alwarthan, A., & Shaik, M. R. (2022). The enormity of the zinc deficiency problem and available solutions; an overview. *Arabian Journal of Chemistry, 15*, 103668.
246. Shin, D. Y., Lee, S. M., Jang, Y., Lee, J., Lee, C. M., Cho, E. M., & Seo, Y. R. (2023). Adverse human health effects of chromium by exposure route: A comprehensive review based on toxicogenomic approach. *International Journal of Molecular Sciences, 24*, 3410.
247. Hausladen, D. M., Alexander-Ozinskas, A., McClain, C., & Fendorf, S. (2018). Hexavalent chromium sources and distribution in California groundwater. *Environmental Science and Technology, 52*, 8242–8251.
248. Faucheur, S. L., Campbell, P. G. C., Fortin, C., & Slaveykova, V. I. (2014). Interactions between mercury and phytoplankton: Speciation, bioavailability, and internal handling. *Environmental Toxicology and Chemistry, 33*, 1211–1224.
249. Cosio, C., Flück, R., Regier, N., & Slaveykova, V. I. (2014). Effects of macrophytes on the fate of mercury in aquatic systems. *Environmental Toxicology and Chemistry, 33*, 1225–1237.
250. Yang, X., Han, X., Zhang, Y., Liu, J., Tang, J., Zhang, D., Zhao, Y., & Ye, Y. (2020). Imaging Hg^{2+}-induced oxidative stress by NIR molecular probe with "dual-key-and-lock" strategy. *Analytical Chemistry, 92*, 12002–12009.
251. Kumar, A., Kumar, A., Cabral-Pinto, M. M. S., Chaturvedi, A. K., Shabnam, A. A., Subrahmanyam, G., Mondal, R., Gupta, D. K., Malyan, S. K., Kumar, S. S., Khan, S. A., & Yadav, K. K. (2020). Lead toxicity: Health hazards, influence on food chain, and sustainable remediation approaches. *International Journal of Environmental Research and Public Health, 17*, 2179.

252. Godt, J., Scheidig, F., Grosse-Siestrup, C., Esche, V., Brandenburg, P., Reich, A., & Groneberg, D. A. (2006). The toxicity of cadmium and resulting hazards for human health. *Journal of Occupational Medicine and Toxicology, 1*, 22.
253. Bernardo-Filho, M., da Conceicão Cunha, M., de Oliveira Valsa, J., de Araujo, A. C., da Silva, F. C. P., & de Souza da Fonseca, A. (1994). Evaluation of potential genotoxicity of stannous chloride: inactivation, filamentation and lysogenic induction of *Escherichia coli. Food Chem. Toxicol., 32*, 477–479.
254. Elsabbagh, H. S., Moussa, S. Z., & El-Tawil, O. S. (2002). Neurotoxicologic sequelae of tributyltin intoxication in rats. *Pharmacological Research, 45*, 201–206.
255. Boogaard, P., Boisset, M., Blunden, S., Davies, S., Ong, T., & Taverne, J. P. (2003). 'Comparative assessment of gastrointestinal irritant potency in man of tin(II) chloride and tin migrated from packaging. *Food and Chemical Toxicology, 41*, 1663–1670.
256. Balali-Mood, M., Naseri, K., Tahergorabi, Z., Khazdair, M. R., & Sadeghi, M. (2021). Toxic mechanisms of five heavy metals: Mercury, lead, chromium, cadmium, and arsenic. *Frontiers in Pharmacology, 12*, 643972.
257. Eisler, R. (1996). *Silver hazards to fish, wildlife, and invertebrates: A synoptic review* (p. 32). Environmental Protection Agency.
258. Dasari, T. P. S., Zhang, Y., & Yu, H. (2015). Antibacterial activity and cytotoxicity of gold (I) and (III) ions and gold nanoparticles. *Biochemical Pharmacology, 6*, 199.
259. Kielhorn, J., Melber, C., Keller, D., & Mangelsdorf, I. (2002). Palladium-a review of exposure and effects to human health. *International Journal of Hygiene and Environmental Health, 205*, 417–432.
260. (2020) '*Aluminium properties, production and applications: an introduction*', Aalco-ferrous and non-ferrous metals stockist, AZoM. https://www.azom.com/article.aspx? Article ID = 2861.
261. Igbokwe, I. O., Igwenagu, E., & Igbokwe, N. A. (2019). Aluminium toxicosis: A review of toxic actions and effects. *Interdisciplinary Toxicology, 12*, 45–70.

Carbon Quantum Dots for Smart Electronic Devices

V. Arul, D. Senthil Vadivu, K. Radhakrishnan, N. Sampathkumar, S. Jayakumar, and R. Sivagurusundar

Abstract Nanoscale carbon-based materials called carbon quantum dots (CQDs) have become a competitive challenger in the field of smart electronics. A thorough review of the synthesis, characteristics, and uses of CQDs in contemporary electronics is given in this book chapter. Because of their distinctive electrical, optical, and chemical characteristics, CQDs which are often smaller than 10 nm are well-suited for a variety of uses. The chapter begins by exploring the various synthesis methods used to produce CQDs, highlighting their versatility and scalability. It then delves into the intriguing properties of CQDs, including their size-dependent behavior, photoluminescence, high surface area, and biocompatibility. These properties form the foundation upon which their diverse applications in smart electronic devices are built. The core of the chapter elucidates the extensive applications of CQDs, spanning from quantum dot solar cells and light-emitting diodes to sensors, energy storage devices, flexible electronics, and biomedical devices. The discussion outlines how CQDs contribute to improving device efficiency, sustainability, and functionality in each application domain. This chapter addresses issues of scalability, long-term stability,

V. Arul (✉)
Department of Chemistry, Sri Eshwar College of Engineering (Autonomous), Coimbatore, Tamil Nadu 641202, India
e-mail: kvarulchem6@gmail.com

D. S. Vadivu
Department of Chemistry, Sree Saraswathi Thiyagaraja College, Pollachi, Tamil Nadu 642205, India

K. Radhakrishnan
Department of Chemistry, Centre for Material Chemistry, Karpagam Academy of Higher Education, Coimbatore, Tamil Nadu 641021, India

N. Sampathkumar
Department of Chemistry, SSM College of Arts and Science, Dindigul, Tamil Nadu 624002, India

S. Jayakumar
Department of Physics, Government Arts College, Udumalpet, Tamil Nadu 642126, India

R. Sivagurusundar
Sri Paramakalyani Centre of Excellence in Environmental Sciences, Manonmaniam Sundaranar University, Alwarkurichi, Tamil Nadu 627412, India

V. Kumar et al. (eds.), *Green Carbon Quantum Dots*, Engineering Materials,
https://doi.org/10.1007/978-981-97-6203-3_13

and toxicity assessment, offering insights into the practical hurdles that must be overcome for widespread adoption. Additionally, it touches upon the integration of CQDs into existing electronic architectures, a pivotal step for their seamless inclusion in the electronics industry. As we traverse the landscape of Carbon Quantum Dots for Smart Electronic Devices, it becomes evident that these tiny carbon nanoparticles hold immense potential to revolutionize the electronics industry. Their synthesis methods and properties pave the way for innovative solutions in smart electronics, promising a future of devices that are not only smarter but also more efficient, sustainable, and adaptable to the evolving needs of society.

Keywords Carbon quantum dots · Properties · Electronic applications · Smart electronic devices · Challenges

1 Introduction

Carbon quantum dots (CQDs), alternatively known as CQDs, have emerged as highly promising materials with a wide range of applications, particularly in the development of intelligent electronic devices. These nanoscale particles, typically smaller than ten nanometers, consist of carbon atoms arranged in a sp^2 hybridized configuration, similar to graphene structures. Their unique electrical, optical, and chemical properties make them particularly appealing for cutting-edge electronic applications. This chapter delves into the investigation of the potential of carbon quantum dots in the realm of intelligent electronic devices, covering aspects such as their synthesis, properties, and various applications [1].

As we transition to the subsequent chapter, we will explore the intriguing domain of carbon quantum dots in the context of smart electronic devices. We will delve into the synthesis methods responsible for producing these exceptional nanomaterials, explore their distinct properties that differentiate them from traditional materials, and uncover the broad spectrum of applications that have thrust them into the forefront of modern electronics [2]. All of these aspects will be thoroughly examined within the upcoming sections.

Carbon quantum dots epitomize a fusion of scientific creativity and technological advancement, offering a myriad of possibilities across diverse fields, including energy harvesting, storage, and biomedicine. Navigating through the landscape of CQDs, we will delve into their synthesis, structure, and properties, while also delving into the enthralling realm of quantum physics that underlies their remarkable behavior [3].

In the subsequent chapters, we will explore the practical applications of CQDs in various electronic devices, ranging from advanced solar cells and light-emitting diodes to flexible electronics and biomedical devices. Through this exploration, we will gain insights into how CQDs are poised to revolutionize the landscape of intelligent electronic technologies, ushering in a future where devices are not only smarter but also more sustainable, energy-efficient, and user-friendly [4, 5].

The ensuing chapters will provide in-depth exploration into the synthesis, properties, applications, challenges, and future prospects of carbon quantum dots, offering valuable insights into the transformative potential of these minuscule yet powerful nanoparticles. The synthesis of Carbon Quantum Dots for Smart Electronic Devices, and prepare to be enthralled by the exciting possibilities that lie ahead in the realm of electronics.

1.1 Structural Properties of Carbon-Based Quantum Dots

1.1.1 Carbon Dots (CDs)

Amorphous carbon dots, or CDs, can be anywhere from one nanometer in size to twenty nanometers in diameter. An electrical band gap that is size-independent occurs in CDs because, unlike larger structures, CDs do not undergo the quantum confinement phenomenon. Their principal component is a carbon core with a preponderance of sp^3 hybridization, however a small amount of sp^2 hybridization is also present. According to the findings of an X-ray diffractometer analysis, a unique graphite signal predominates due to sp^3 hybridization. The importance of hybridization degree in band gap engineering has been shown by multiple computational experiments. Research like this has also shown how important surface functional groups are. Because of their malleability, these groups can be simply modified utilizing a range of synthesis techniques to manipulate the electrical and optical properties of CDs [6].

1.1.2 Carbon Quantum Dots (CQDs)

Nanoscale crystalline carbon spherical formations exhibiting the quantum confinement phenomena are called carbon quantum dots (CQDs). A lattice constant that falls somewhere in the middle of graphene and graphite is characteristic of chemical quantum dots (CQDs). In terms of size and height, their diameters usually fall somewhere between one and twenty nanometers. Graphene quantum dots (GQDs) have superior crystallinity, but sp^2 carbon is less crystalline, hence they fall short in this regard. Functionalization or alteration of CQD surfaces typically results in the presence of amine functional groups and oxygen-containing carbonyl and carboxyl groups. Interactions between surface functionalization and the quantum confinement effect impact the band gap energy of CQDs. These classifications give light on the structural variety of carbon-based quantum dots by analyzing their creation processes, carbon atom arrangements, crystalline structures, and dimensionalities [7].

1.2 Electronic Properties of Carbon-Based Quantum Dots

Carbon quantum dots (QDs) can have their electrical properties tuned by tweaking a bunch of settings. It is possible to do this change. This class includes features such as surface functional clusters, heteroatom dopants, electron energy levels, and electrical interactions between adjacent nanoparticles [8].

1.2.1 GQDs: Superior Electron Transfer

Research has demonstrated that graphene quantum dots (GQDs) are more capable of transporting electrons than carbon dots (CDs) and carbon quantum dots (C-QDs). The creation of clearly defined and nearly uniform edge sites during size reduction, as well as their highly crystalline structure, may be responsible for this. The ordered core structure of GQDs enables an efficient and effective charge transfer across carbon cores. On the other hand, you must be careful with functionalization because some changes can create electron trap sites, which impede electron transit [9].

1.2.2 C-QDs: Versatile Electron Functions

There are various applications for the p-electron network that forms in Carbon Quantum Dots (C-QDs) due to sp^2 hybridization. They play a variety of tasks in the electron transport chain, including accepting and donating electrons, conducting electrons, and connecting different types of electrons. Employing C-QDs is often crucial for reducing material cracking and maintaining electron conductivity following cycling. Research indicates that C-QDs may find use in optoelectronic devices and supercapacitors due to their two to four times better electron mobilities compared to their hole mobilities. Worse yet, the mobility of charge carriers in C-QDs is influenced by the length of the ligand; as the ligand gets longer, the electron mobility drops exponentially [10].

1.2.3 Band Gap Engineering in GQDs and C-QDs

The energy of the HOMO state in GQDs is 5.3 eV, whereas the energy of the LUMO state is 3.8 eV below the vacuum state. Information of this nature has been provided by the sources. Both the molecular orbital with the highest level of occupancy and the molecular orbital with the lowest level of occupancy are represented by these relative values. To achieve tunability in the band gap energies of GQDs, one can employ a variety of strategies, including size/shape engineering, surface functionalization, doping, and defect manipulation, among others. The density of atomic rings within the GQDs is a good indicator of their energy content. As the aromatic ring concentration of CQDs increases, the energy gap between the p-states decreases.

This phenomenon opens up the possibility of manipulating the band gap energy and of using the quantum capacitance behavior. The band gap energy is hybridization dependent, according to computational studies of CD architectures. The potential for modifying electrical and optical properties is now within reach, thanks to this finding. We provide the groundwork for the potential application of carbon-based quantum dots (QDs) in electronic devices with these new insights into their electrical properties, obtained using selective band gap engineering approaches. By establishing a framework for their possible use, this is achieved [11].

1.2.4 Electrochemical Luminescence (ECL) of Carbon Quantum Dots

In the electrochemical luminescence (ECL) process, carbon quantum dots (CQDs) with different levels of oxidation, referred to as reduced (r-CQDs) and oxidized (o-CQDs), were produced using different ways. This allowed for the discovery of the critically important role that regulated diffusion of o-CQDs onto the electrode surface plays. A clear association between ECL emission and o-CQD oxidation was demonstrated by the fact that the ECL wave displayed particular characteristics and aligned with the oxidation peak that was found in cyclic voltammograms (CVs). The reproducibility and stability of ECL were highlighted by continuous cyclic scanning. On the other hand, the cathodic ECL seen in the o-CQDs/$K_2S_2O_8$ system was attributed to the distinctive "loose shell" structure of o-CQDs, which made it easier for o-CQDs radicals to be generated. Within the system of o-CQDs and $K_2S_2O_8$, the reduction of $S_2O_8{}^{2-}$ resulted in the release of a powerful oxidizing agent known as the SO^{4-} radical. This radical received electrons from the anionic o CQD$^-$, thereby generating emitters that are responsible for the emission of electron–hole pairs (ECL). In addition to offering useful insights into the interplay between composition, morphology, and surface features of CQDs in ECL phenomena, the decreased ECL activity that was seen in r-CQDs provided more support for the association that exists between ECL and the oxidation state of the CQD surface [12].

2 Applications of Carbon Quantum Dots in Smart Electronic Devices

Carbon quantum dots have found applications in various smart electronic devices, revolutionizing the field of electronics and enabling innovative technologies:

2.1 *Recent Progress of Carbon Quantum Dots for Supercapacitors*

Carbon quantum dots, or C-QDs, show great promise when added to the electrodes and electrolytes of supercapacitors, among other components. This section provides a synopsis of the current state of the art regarding the application of bare C-QDs as supercapacitor electrodes. They are covered extensively. Bare C-QDs often exhibit the capacitive behavior seen in Electric Double-Layer Capacitor (EDLC) mechanisms. There has been new research that reveals these materials could have a role in pseudocapacitors. Graphene quantum dots (GQDs) have found extensive use due to their enhanced conductivity and properties associated with edge states. Consequently, a specific capacitance that meets expectations is the result of these characteristics. A lot of hope and promise has been shown by using bare chemical quantum dots (C-QDs) as electrodes in supercapacitors.

To improve the efficiency of supercapacitors, one may use materials with a high conductivity profile in conjunction with carbon quantum dots (C-QDs). Supercapacitors with enhanced performance were made by combining graphene microfibers with enriched carbon dots. There are many other ways this was achieved, but this is just one. Compared to reduced graphene oxide (RGO) fibre, the material's specific capacitance was three times greater because of this. As a whole, the performance is improved because this combination increases both mechanical strength and specific surface area (SSA), which in turn improves the total performance.

Utilisation of Green Quantum Dots in Micro-Supercapacitors It has already been proven that micro-supercapacitors can use bare GQDs as electrodes, meaning no other electrode materials are required. This graphene quantum dots (GQDs), made using the solvothermal method with graphene oxide as the primary material, and displayed a spherical shape. These particles, which were electrophoretic deposited, had a size range of 1.0–5.4 nm and were used to make electrodes. The specific area, rate capability, power response, and durability of these micro-supercapacitors were all above average. The huge specific surface area, numerous active sites, and readily accessible edges of GQDs allowed this to happen. On top of that, they proved to be incredibly durable.

When produced from graphene oxide using hydrothermal or solvothermal techniques, graphene quantum dots (GQDs) exhibit perfect electric double-layer supercapacitor performance. Multiple research have confirmed this. Heat treatment, which removed oxygen-containing functional groups, increased the performance of GQDs by purifying them. As a result, the GQDs' performance was enhanced. The number of electrons passing through the carbon core increased as a result of this. By virtue of their uniform nanoscale dimension, GQDs effectively inhibit the stacking of graphene nanosheets. The edge effect also increases the number of active sites for ion diffusion, which in turn results in unprecedentedly high supercapacitance performance. Carbon dots (CDs), a subset of carbon quantum dots (C-QDs), display properties that are still being discovered today when it comes to the applications of supercapacitors. Surface functionalization of electrodes improves their wettability

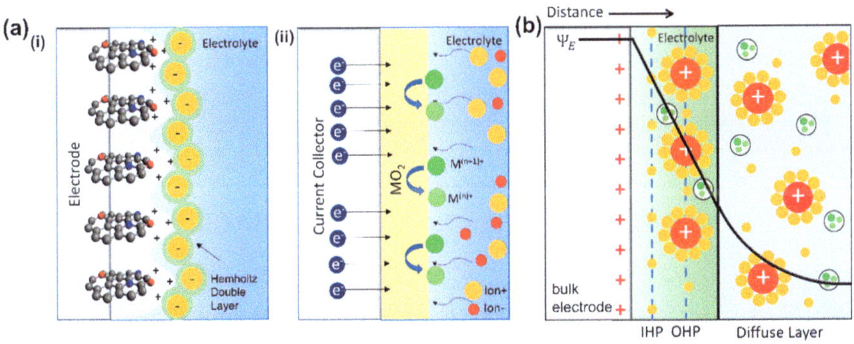

Fig. 1 Shows the Stern model, which describes the electrolyte ions on a charged electrode. A diagram of the ion distribution and mechanism is also included, along with (a) examples of the model's use in electric-double-layer capacitors (EDLCs) and pseudo capacitance supercapacitors [13]

while simultaneously increasing their conductivity, morphological distinctiveness, and active surface area. Because it has been shown that electric capacitors with CDs and $NiCo_2O_4$ nanowires perform much better than those with NiCo2O4 alone in terms of specific capacitance enhancement, it follows that CDs may improve electric capacitor performance. These developments have highlighted the enormous promise of bare carbon-based quantum dots, specifically C-QDs and GQDs, to improve supercapacitor performance by means of various integration and production strategies [13] (Fig. 1).

2.1.1 Wearable Supercapacitors

In contemporary and upcoming applications, there is a growing need for energy storage devices that can adapt to various structures and shapes, including those of human and animal bodies, plants, and soft robots. The demand for flexible and conformal supercapacitors is on the rise, facilitating the development of energy-storing fabrics and bio-integrated IoT sensors. Carbon-based supercapacitors stand out as the preferred choice for such applications due to their biocompatibility.

Advantages of C-QDs for Wearable Supercapacitors: Carbon Quantum Dots (C-QDs) offer several characteristics that make them highly appealing for wearable supercapacitor applications. Their structure allows for flexibility without compromising electrochemical properties. For instance, composite structures like GQDs/$Ni(OH)_2$ on carbon cloths demonstrate remarkable flexibility, bending up to 180 degrees with minimal change in cyclic voltammetry (CV) curves. However, thorough CV measurements under various bending conditions are necessary to validate the flexibility of supercapacitors [14, 15].

Composites with CQD matrices are very pliable; in fact, some of these materials may be bent more than 90 degrees without compromising their electrochemical performance. Among the many advantages of CQD composites over competing materials is this very quality. The wearability of C-QD supercapacitors can be ascertained by testing the electrochemical characteristics at varying bending angles [16].

Preventing Leakage Using Gel Electrolytes: Gel electrolytes are commonly used to protect flexible supercapacitors against leakage during bending testing. An N-GQD/GH/CF asymmetric flexible fibre supercapacitor assembly was built using a polyethylene terephthalate (PET) substrate. For instance, this construction made use of PVA-H_2SO_4.

When it comes to C-QD supercapacitors, transparent and flexible electrodes are both viable options. Common examples of flexible electrodes include conductive polymers, carbon cloth, and nickel foam. Alternatively, micro-supercapacitors constructed from interdigitated graphene show great promise in terms of flexibility; yet, their binding capacity may decrease after charge–discharge cycling. Also, C-QDs have a lot of potential uses because of how transparent they are; for example, transparent wearable devices might use supercapacitors that are almost invisible. The development of wearable supercapacitors utilizing C-QDs opens up avenues for various applications in flexible and conformal energy storage systems. Further research is needed to optimize their performance, durability, and transparency for seamless integration into wearable technologies [17].

2.2 *Use of CQDs for Solar Energy*

Carbon Quantum Dots (CQDs) are gaining attention for their potential application in Energy Downshift (EDS) materials, particularly in solar cells like dye-sensitized solar cells (DSSCs). EDS materials play a vital role in efficiently harnessing the solar spectrum by absorbing harmful ultraviolet (UV) radiation and emitting lower-energy photons, thus protecting solar cells and enhancing their overall efficiency. The use of CQDs in EDS applications is advantageous for a number of reasons, including their cost-effectiveness, their uncomplicated synthesis processes, their strong UV absorption, their emission in the visible spectrum, and their high level of environmental friendliness. The incorporation of CQDs into DSSCs as EDS layers results in the transformation of incident ultraviolet light into visible light. This visible light can then be absorbed by the sensitizer layer, which ultimately leads to an improvement in the external quantum efficiency (EQE) in the key wavelength range of 300–400 nm. Researchers have successfully synthesized highly luminescent CQDs with impressive quantum yields (QYs) through methods such as hydrothermal synthesis. These CQDs, when integrated into DSSCs as EDS layers, have demonstrated notable enhancements in device performance. For example, CQDs synthesized from citric acid and ethylenediamine exhibited increased power conversion efficiency (PCE) by enhancing photon absorption and carrier generation. Similarly, N-doped CQDs synthesized via a one-pot hydrothermal technique showed improved PCE and stability in DSSCs compared to conventional sensitizers. In conclusion, the utilization of CQDs as EDS materials holds significant promise for advancing the efficiency and stability of solar cells, especially DSSCs, by mitigating the adverse effects of UV radiation and enhancing photon absorption. Continued research and development in this field are essential for the advancement of sustainable and efficient solar energy technologies [18, 19].

2.2.1 CQDs as Sensitizers or Co-Sensitizers in Solar Cells

Carbon Quantum Dots (CQDs) are emerging as promising sensitizers or co-sensitizers in solar cells, offering affordability, sustainability, and enhanced performance compared to traditional sensitizers like Ru-based dyes. Synthesized CQDs have been effectively utilized in solar cell applications, leading to notable enhancements in device performance. CQDs with surface functional groups demonstrated broad absorption spectra across the visible range, improving power conversion efficiency (PCE) in TiO_2-based solar cells (Fig. 2).

Nitrogen-doped CQDs exhibited high absorption and efficient electron transfer to the TiO_2 conduction band, contributing to improved device performance. In situ growth of CQDs on TiO_2 surfaces resulted in hybridized CQD/TiO_2 photo-anodes with enhanced electron–hole pair formation and carrier transfer, leading to increased device efficiency. Moreover, nitrogen-doped CQDs served as efficient sensitizers in quantum dot solar cells, facilitating photo carrier injection into TiO_2 and enhancing photocurrent and PCE. Co-sensitization of DSSCs with nitrogen-doped CQDs further improved performance, with exceptional up-conversion luminescence expanding the device's absorption spectrum. Overall, CQDs represent a promising avenue for advancing solar cell technology, offering cost effectiveness, sustainability, and enhanced performance. Continued research in this field is essential for driving further advancements and realizing the full potential of CQDs in solar cell applications [20, 21].

2.2.2 CQDs Serve as an Electron Transport Layer (ETL)

Carbon Quantum Dots (CQDs) are gaining attention as potential Electron Transport Layers (ETLs) in solar cells, playing a crucial role in extracting electrons from the active layer and minimizing recombination losses. In contrast to traditional metal oxide ETLs like TiO_2 and ZnO, CQDs offer unique advantages, including excellent photo-stability, low toxicity, and efficient electron extraction capability. CQDs can either replace or complement existing ETLs, showcasing their versatility and performance enhancement potential. Synthesized through methods like chemical vapor deposition (CVD), CQDs, with an average diameter of 3.5 nm, have been successfully integrated into solution-processed organic solar cells. Yan and colleagues demonstrated their effectiveness in achieving optimized power conversion efficiencies (PCEs) comparable to devices using conventional ETL materials. The incorporation of CQDs improved interfacial connections, reducing resistance and enhancing electron injection. Notably, devices with CQDs-ETLs exhibited enhanced

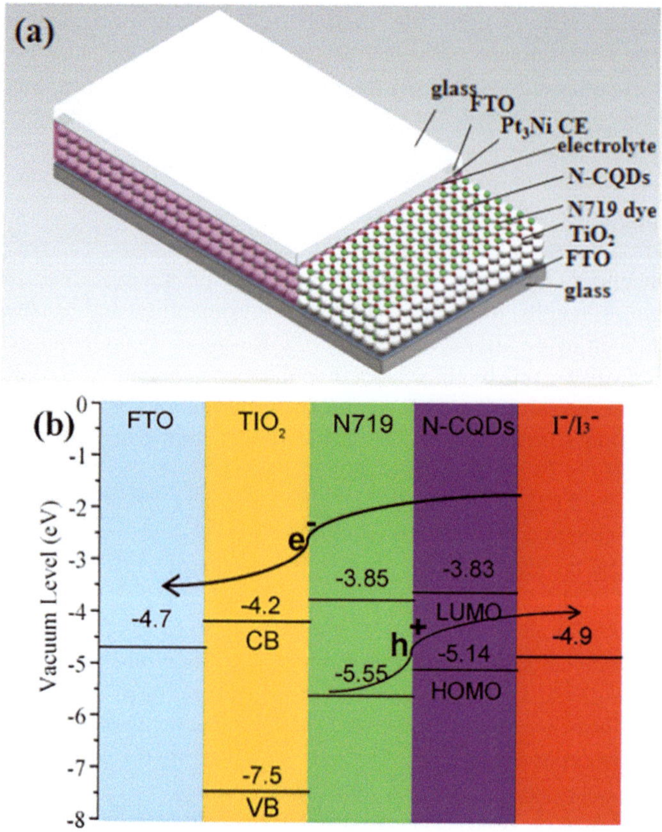

Fig. 2 Presents **a** a schematic diagram depicting the co-sensitized solar cells and **b** the distribution of energy levels along with the processes of charge transfer at the interface [21]

long-term thermal stability due to reduced CQD diffusion into the active layer. This suggests that CQDs hold promise for advancing solar cell technology, providing a pathway for more efficient and stable devices. Further research and development in this direction are crucial for realizing the full potential of CQDs in solar cells [22].

2.2.3 CQDs Serve as the Hole Transit Layer (HTL)

Carbon Quantum Dots (CQDs) are emerging as promising candidates for Hole Transport Layers (HTLs) in solar cells, addressing challenges associated with stability and conductivity while offering compatibility with energy levels crucial for efficient charge transport. Traditionally, HTLs like PEDOT:PSS face limitations in stability and electrical conductivity, prompting the exploration of alternatives. Inorganic semiconductor-based HTLs offer improved stability but at a higher cost, making CQDs an appealing option due to their favorable properties. CQDs exhibit stability comparable to inorganic HTLs but with lower manufacturing costs and higher electrical conductivity, leading to enhanced device performance. Synthesized CQDs demonstrate appropriate energy levels for efficient hole transfer and electron blocking, facilitating improved charge transport within the device. In applications such as perovskite solar cells, CQDs have shown feasibility as HTMs, albeit with modest efficiencies, highlighting the need for further optimization, particularly regarding perovskite coverage. A synergistic approach of nitrogen-doped (N-CQDs) and oxidized (O-CQDs) CQDs, resulting in significant performance improvements attributed to reduced resistance and improved surface characteristics of PEDOT:PSS due to interactions with N-CQDs. In conclusion, CQDs offer stability, conductivity, and energy level compatibility desirable for HTL materials in solar cells, with their incorporation showing potential for enhancing device performance and efficiency. Further research and optimization of CQD-based HTLs are essential for their integration into next-generation solar cell technologies [23].

2.2.4 CQDs as a Donor and Acceptor in Solar Cells

In recent years, carbon quantum dots (CQDs) have gained attention as a possible solar device acceptor material. Fullerene derivatives have been largely supplanted by them in many cases. In their interaction with nearby materials, CQDs can behave as acceptors, taking electrons or donating them, depending on the situation.

Carbon quantum dots (CQDs) display electron acceptor behavior in solar devices, according to studies utilising multiple fabrication strategies. Hydrothermal synthesis of P3HT polymer matrices containing carbon quantum dots (CQDs) with an average diameter of 6.2 nm resulted in a significant decrease in photoluminescence. By establishing the acceptor role of the CQDs in the process of receiving electrons from the polymer, it was found that the quenching was caused by the electron transfer from the photo-excited P3HT to the CQDs. Chemical vapor deposition was used to produce C-CQDs with a 3.5 nm diameter. These C-CQDs, when combined with inverted organic solar cells, acted as acceptor substances, whereas P3HT acted as donor material. A direct correlation was found between the concentration of carbon quantum dots (C-CQDs) in the active layer and performance, with a PCE of 5 watts being the optimal value. These results indicate that carbon quantum dots (CQDs) may contribute to organic solar cell efficiency by acting as efficient electron acceptors. Future work on CQD-based devices and their impact on solar cell technology will hinge on the completion of supplementary research and optimisation processes [24].

2.2.5 Doping with CQDs

Doping is a crucial component in increasing the efficiency of solar cells since it modifies the optoelectronic characteristics of materials. Research demonstrated that solution-processed organic solar cells using carbon quantum dots (CQDs) synthesised via chemical vapor deposition had an improved power conversion efficiency (PCE). As a result, CQDs are being considered as a possible green dopant for commercial solar cells. A notable 11% enhancement was observed with a doping concentration of 0.075 weight percent. The concentrations of the dopants varied between 0.025 and 0.1 weight percent. However, due to improved light absorption and reduced series resistance, a further increase to 0.1% resulted in a reduction of 3.55% PCE. Also, dye-sensitized solar cells (DSSCs) using synthetic N-CQDs as dopants were considered in the study. The integration of N-CQDs led to a significant increase in the PCE, which reached 8.75% under the impact of a single sun. To do this, we reduced recombination between the photo-anode and electrolyte and increased absorbance. Also, N-CQDs improved DSSC performance when utilised as sensitizers and co-sensitizers. Using N-CQDs as the co-active layer resulted in the greatest improvement. This study's findings show that carbon quantum dots (CQDs) could be a dopant that makes solar cells work better. Nevertheless, further investigation into their potential use with different types of solar cells is necessary [25].

2.3 White LEDs (WLEDs) by CQD

With whole width half maxima larger than 80 nm, semiconductor quantum dots' (CQDs') emission characteristics are wide. This goes against the emission characteristics of typical semiconductor quantum dots and commercial rare-earth phosphors. Both the scattered particle size and the strong electron–phonon interaction are to blame for this diffuse emission. As a result of their ability to efficiently cover the emission spectrum range of 400 to 760 nm, carbon quantum dots (CQDs) are a possible material for white light-emitting diodes (WLEDs). In order to make CQD-based WLEDs, two primary technologies are used: phosphor-converted WLEDs, or pc-WLEDs, and electroluminescent WLEDs, or e-WLEDs. In photoconductor-based WLEDs, CQDs act as light-converting phosphors; in electron-based WLEDs, they serve as emitting layers. Research has demonstrated that E-WLEDs outperform pc-WLEDs in terms of efficiency, suggesting a potential new direction for WLED development.

To achieve warm white light with a color temperature below 4000 Kelvin, a suitable red emission component is essential. Red emission CQDs (R-CQDs) play a crucial role in this regard, although their efficient synthesis remains a challenge due to the susceptibility to defect formation. Surface defect state emission and π–π stacking effects are leveraged to realize single component white emission carbon quantum dots (SCWE-CQDs), which emulate sunlight and offer excellent color quality white emission. Various strategies, including heteroatom doping, have been explored to enhance SCWE in CQDs.

The quantum yield and long-wavelength intensity of SCWE-CQDs have been enhanced due to recent advancements in their synthesis, such as the use of white emission carbon dots (CDs) modified with hexadecyltrimethylammonium bromide

(CTAB). Moreover, N-CNDs and S-CNDs, carbon nanodots doped with nitrogen and sulphur, have been synthesised with the aim of attaining SCWE. Improving aggregation size in CQDs and overcoming challenges like self-quenching are ongoing endeavors. Increasing efficiency and widening emission peaks are the targets of these initiatives. Adding heteroatoms and adjusting aggregation behaviors are two promising ways to improve the efficiency of electroluminescent WLEDs. Recent advances have been made in the production of W-CNQDs, which are dual emissive carbon nitride quantum dots that exhibit blue-yellow fluorescence and phosphorescence. With a white emission efficiency of 25%, these quantum dots have outperformed all other white-emitting materials to this point. Carbonyl groups in W-CNQDs play a crucial function in enhancing intersystem crossing and inducing bright yellow phosphorescence. All of these new advancements show how hard people are trying to make efficient, high-quality WLEDs using CQDs. Future solid-state lighting sources may find answers in these WLEDs [26, 27].

2.4 *Electrochemiluminescence (ECL) of CQDs for Heavy Metals Sensor*

Electrochemiluminescence (ECL) merges electrochemistry with luminescence, generating light through high-energy electron transfer reactions at electrodes. There are two basic pathways that are responsible for the ECL phenomena. These processes are the annihilation pathway and the co-reactant pathway. As a result of their high sensitivity, quantum yield (QY), and low background signal, carbon-based quantum dots (C-QDs) have emerged as potentially useful materials for applications in the field of electronics and communications.

Applications of Carbon-Based Quantum Dots in ECL for metal ion detection, phosphorus-doped carbon quantum dots (P-CQDs) have been instrumental in enhancing ECL intensity. By introducing phosphorus doping, P-CQDs create emissive traps within their structure, facilitating electron transfer reactions and resulting in improved sensitivity for detecting metal ions such as Cu^{2+}. This enhancement in ECL intensity enables more precise and reliable detection of metal ions, crucial for various analytical and environmental monitoring applications.

Additionally, a novel ECL sensor has been designed for the purpose of detecting Hg^{2+} ions in a specific manner. This sensor makes use of a sophisticated nanoprobe that is composed of graphene quantum dots (GQDs)-DNA-AuNP, in conjunction with a nanocomposite that is composed of poly(5-formylindole)/reduced graphene oxide (P5FIn/erGO). The operation of the sensor is accomplished by taking advantage of the creation of T-Hg^{2+}-T complexes with thymine-rich single-stranded DNA (ssDNA) that is immobilised on the surface of the electrode. Because of this complex chemical interaction, the sensor is able to achieve high sensitivity and selectivity in the detection of Hg^{2+} ions. As a result, it is an extremely useful instrument for applications in analytical chemistry and environmental monitoring, both of which need precise detection of heavy metal ions [28].

2.5 Bio-Sensing Applications

Electrochemiluminescence (ECL) immunosensors have been developed for the purpose of detecting carcinoembryonic antigen (CEA). This is accomplished through the utilisation of perylenetetracarboxylic acid (PTCA) and carbon quantum dots (CQDs) as dual luminophores in the electrochemiluminescence (ECL) immunosensors. In this sandwich-type configuration, graphene functions as a nano carrier, while $S_2O_8{}^{2-}$ is the co-reactant that is responsible for the reaction. These luminophores have a synergistic effect that increases the intensity of the ECL, which in turn enables the detection of CEA in human serum samples with a high degree of sensitivity.

In order to differentiate cancer cells, carbon dots that have been coupled with nitrogen-doped hydrazide have been produced. These carbon dots have a high quantum efficiency and minimal excitation potential emission. These manufactured carbon dots make it easier to differentiate cancer cells from normal cells by using differing degrees of hydrogen peroxide release as a criterion. Doping carbon quantum dots with elements such as phosphorus and sulphur is one of the strategies that has been utilised in order to improve the efficiency of electron-conducting lasers (ECL) using carbon-based quantum dots. Doping with phosphorus results in the formation of emissive traps, whereas doping with sulphur results in surface state modifications; both of these effects contribute to improved ECL performance. The detection sensitivity of carbon quantum dots in ECL assays can be further improved by utilising the Surface Plasmon Coupling ECL (SPC-ECL) mode. This mode is characterised by the fact that the ECL intensity is increased by the surface plasmon coupling effect of gold nanoparticles (AuNPs). These improvements highlight the potential for carbon-based quantum dots to revolutionise ECL-based sensing platforms. These quantum dots offer increased sensitivity, selectivity, and adaptability in the detection of a wide range of analytes, including metal ions and bio-molecules. The continuation of research and development in this field holds the potential to produce cutting-edge diagnostic and analytical instruments that are based on ECL methodology [29].

2.6 Latent Fingerprint Enhancement Using Carbon Dots

Fingerprint analysis is a fundamental component of forensic investigation, relying on the distinctive ridge patterns found on human fingers. Traditional methods for developing latent fingerprints involve techniques such as silver nitrate, iodine vapor, and cyanoacrylate fuming. However, recent advancements in nanochemistry present exciting prospects for enhancing latent fingerprint visualization and analysis.

Plasmonic nanoparticles, Carbon dots (C-dots), and quantum dots are new materials that are being developed for improving latent fingerprints by nanochemistry techniques. These nanoparticles improve visibility and chemical recognition by strongly sticking to fingerprint residues and providing high-resolution imaging. As an example, it has been demonstrated that anti-cotinine antibody functionalized gold nanoparticles may produce high-quality fingerprint impressions while also detecting the nicotine metabolite cotinine. Insights regarding the individual's lifestyle habits can be gleaned from this approach.

The adaptability of nanomaterial-based techniques has been demonstrated by the successful use of antibody/magnetic particle conjugates for the multiplexed detection of drugs of abuse and their metabolites inside a single latent fingerprint. To improve the accuracy of fingerprint analysis and make background-free photographs easier to get, powders based on carbon dots allow for the color-tunable visualisation of latent fingerprints under various light sources. To tackle the self-quenching effect of carbon dots, various strategies have been suggested, including using a diluent matrix, developing core–shell nanostructures, doping heteroatoms, and taking use of resonance energy transfer (RET) and π-π interactions.

The case study shows that adding carbon dots to a silica matrix makes latent fingerprints visible in high detail and allows for color tuning. These thermally treated carbon dots show astounding detail and color tunability at various excitation wavelengths; they also have a quantum yield of 15%. This method streamlines forensic investigation by doing away with the need for different powders to see fingerprints on different colored backgrounds. The effectiveness of powders based on carbon dots is demonstrated by the fact that automated fingerprint identification system (AFIS) analysis produces similar results with less processing time.

These advancements in carbon dot-based fingerprint enhancement hold promising prospects for forensic investigation, offering enhanced visualization capabilities, molecular recognition, and multiplexed detection within latent fingerprints. Further research and development in this domain have the potential to revolutionize forensic techniques and enhance outcomes in criminal investigations [30] (Fig. 3).

Fig. 3 Schematic representation of CQDs for fingerprint analysis [31]

2.7 *Molecular Sensing Using Carbon Dots in Forensic Detection*

Lab-on-a-chip systems represent an innovative approach in analytical chemistry, integrating microfluidic assays and nanoprobes to achieve rapid, multiplexed detection with minimal sample quantities. These portable devices, designed for simplicity and user-friendliness, hold significant potential across various fields, including medical diagnostics, forensic toxicology, and explosives trace detection. Nanosensors equipped with recognition units capture target molecules, yielding clear signals (optical, electric, mechanical, or acoustic) for precise detection. In medical diagnostics, Nanosensors play a vital role in pre-emptively identifying underlying health conditions. In forensic applications, they detect narcotics, drugs of abuse, explosives, bioterrorism-associated pathogens, and determine the time since death. Wearable biosensors enable continuous monitoring of adverse lifestyle habits, such as alcohol abuse, relevant to criminal behavior. Carbon dots are increasingly utilized in affinity sensors due to their high selectivity and sensitivity in binding specific compounds. Photo-induced charge transfer (PCT), photo-induced electron transfer (PET), inner filter effects (IFE) and Resonance energy transfer (RET),are common sensing strategies. Top-Priority Requests, Police agencies place a premium on the detection of illegal narcotics and explosives. Furthermore, forensic investigation relies heavily on the detection of biofluids and DNA profiles at crime scenes. Critical information on attempts at murder or suicide can be gleaned from evaluating exposure to harmful chemicals and metals [32].

3 Challenges and Future Perspectives

Carbon quantum dots offer promising opportunities for advanced electronic devices; however, several challenges must be addressed to fully harness their potential:

- **Scalability**: Scalable production methods and cost-effective synthesis techniques are imperative to meet the demands of large-scale electronic device manufacturing. Current synthesis routes often involve complex procedures or expensive precursors, hindering mass production. Addressing scalability issues will require the development of innovative synthesis strategies that prioritize efficiency, yield, and cost-effectiveness.
- **Stability**: Ensuring the long-term stability of carbon quantum dots (CQDs) in various electronic environments is crucial for their practical application. CQDs may experience degradation or morphological changes over time, impacting device performance and reliability. Strategies to enhance stability include surface passivation, encapsulation, or engineering robust carbon structures to withstand harsh operating conditions.

- **Toxicity**: While carbon quantum dots generally exhibit favorable biocompatibility, comprehensive toxicity assessments are essential, particularly for biomedical applications. Understanding the potential risks associated with CQDs is critical for their safe use in medical devices, drug delivery systems, and theranostic applications. Addressing toxicity concerns requires rigorous evaluation of CQD formulations, including assessments of cytotoxicity, genotoxicity, and immunotoxicity.
- **Integration**: Efficient integration of carbon quantum dots into existing electronic device architectures is paramount for seamless adoption in the electronics industry. Challenges may arise in achieving compatibility with established fabrication processes, ensuring proper alignment with device interfaces, and optimizing electrical connectivity. Developing versatile integration methods and standardizing protocols for CQD incorporation will facilitate their integration into various electronic platforms.
- **Future Perspectives**: Overcoming these challenges will unlock the full potential of carbon quantum dots (CQDs) in smart electronic devices. Future research directions may focus on:

Exploring novel synthesis routes to enhance scalability, yield, and cost-effectiveness of CQD production. Investigating alternative precursors, reaction conditions, and reactor designs could lead to more efficient synthesis methods, enabling large-scale manufacturing of CQDs for electronic applications.

Developing robust encapsulation methods and engineering stable carbon structures to improve the long-term stability of CQDs. Strategies such as surface passivation, polymer coating, or hybridization with other materials could mitigate degradation mechanisms and enhance the durability of CQDs in electronic environments.

Conducting comprehensive toxicity studies to ensure the safe use of CQDs in biomedical applications and other sensitive areas. In-depth evaluations of cytotoxicity, genotoxicity, immunotoxicity, and long-term biocompatibility are essential for understanding the potential risks associated with CQDs and establishing safe usage guidelines.

Advancing integration techniques to seamlessly incorporate CQDs into diverse electronic device architectures. Developing compatible interfaces, optimizing deposition methods, and refining device fabrication processes will facilitate the integration of CQDs into various electronic platforms, including sensors, displays, and energy storage devices. By addressing these challenges and pursuing innovative research directions, carbon quantum dots hold the promise of revolutionizing the field of electronic devices with their unique properties and versatile applications. Continued interdisciplinary efforts across materials science, chemistry, engineering, and biology will drive the development of CQD-based technologies towards practical implementation and commercialization.

4 Conclusion

In conclusion, carbon quantum dots hold great promise for revolutionizing the field of smart electronic devices. Their unique properties, versatile synthesis methods, and broad range of applications make them a compelling material for future electronic technologies. As researchers continue to overcome challenges and explore new possibilities, CQDs are poised to play a pivotal role in shaping the future of electronics.

In the ever-evolving landscape of electronics, the emergence of carbon quantum dots (CQDs) has opened up new frontiers of possibility. This chapter has illuminated the remarkable journey of CQDs in the realm of smart electronic devices, from their synthesis and unique properties to their diverse applications and future prospects.

The applications of CQDs span a wide spectrum, encapsulating quantum dot solar cells that harness their light-absorbing prowess, light-emitting diodes that utilize their photoluminescence for improved displays, sensors and detectors that capitalize on their surface sensitivity, and energy storage devices that leverage their charge storage capabilities. CQDs have even ventured into the domain of flexible electronics and biomedical devices, promising innovations in wearables, diagnostics, and therapeutics.

While the potential of CQDs is undeniable, there are challenges to address. Scalability remains a pressing concern, necessitating scalable production methods for widespread adoption. Ensuring the long-term stability of CQDs in various electronic environments is essential, as is conducting thorough toxicity assessments, particularly for biomedical applications. Integration of CQDs into existing electronic architectures is a key milestone in their journey toward becoming a staple in the electronics industry.

Carbon Quantum Dots for Smart Electronic Devices represents a promising chapter in the ongoing story of electronic innovation. As we stand on the precipice of a new era, CQDs hold the promise of smarter, more efficient, and more sustainable electronic devices. Their versatility, coupled with the ever-advancing state of research and development, makes them a cornerstone in the foundation of future electronics. As we venture forward, we must remain vigilant in overcoming challenges, exploring new possibilities, and realizing the transformative potential that carbon quantum dots bring to the world of smart electronic devices.

References

1. Wang, X., Feng, Y., Dong, P., & Huang, J. (2019). A mini review on carbon quantum dots: Preparation, properties, and electrocatalytic application. *Frontiers in Chemistry, 7*, 671.
2. Zuo, J., Jiang, T., Zhao, X., Xiong, X., Xiao, S., & Zhu, Z. (2015). Preparation and application of fluorescent carbon dots. *Journal of Nanomaterials, 2015*, 1–13.
3. Ahmad, F., & Khan, A. M. (2017). Carbon quantum dots: Nanolights. *International Journal of Petrochemical Science & Engineering, 2*(7), 247–250.

4. Wang, Y., & Hu, A. (2014). Carbon quantum dots: Synthesis, properties and applications. *Journal of Materials Chemistry C, 2*(34), 6921.
5. Arul, V., Chandrasekaran, P., Sivaraman, G., & Sethuraman, M. G. (2023). Biogenic preparation of undoped and heteroatoms doped carbon dots: Effect of heteroatoms doping in fluorescence, catalytic ability and multicolour in-vitro bio-imaging applications-a comparative study. *Materials Research Bulletin, 162*, 112204.
6. Arul, V., Radhakrishnan, K., Sampathkumar, N., Vinoth Kumar, J., Abirami, N., & Inbaraj, B. S. (2023). Detoxification of toxic organic dye by heteroatom-doped fluorescent carbon dots prepared by green hydrothermal method using Garcinia mangostana extract. *Agronomy, 13*(1), 205.
7. Zhu, S., Song, Y., Zhao, X., Shao, J., Zhang, J., & Yang, B. (2015). The photoluminescence mechanism in carbon dots (graphene quantum dots, carbon nanodots, and polymer dots): Current state and future perspective. *Nano Research, 8*, 355–381.
8. Wang, B., Yu, J., Sui, L., Zhu, S., Tang, Z., Yang, B., & Lu, S. (2020). Rational design of multi-color-emissive carbon dots in a single reaction system by hydrothermal. *Advanced Sciences*, 2001453.
9. Tepliakov, N. V., Kundelev, E. V., Khavlyuk, P. D., Xiong, Y., Leonov, M. Y., Zhu, W., Baranov, A. V., Fedorov, A. V., Rogach, A. L., & Rukhlenko, I. D. (2019). Sp^2–sp^3-Hybridized atomic domains determine optical features of carbon dots. *ACS Nano, 13*, 10737–10744.
10. Zhou, J., Booker, C., Li, R., Zhou, X., Sham, T. K., Sun, X., & Ding, Z. (2007). An electrochemical avenue to blue luminescent nano crystals from multi-walled carbon nanotubes (MWCNTs). *Journal of the American Chemical Society, 129*, 744–745.
11. Chen, G., Wu, S., Hui, L., Zhao, Y., Ye, J., Tan, Z., Zeng, W., Tao, Z., Yang, L., & Zhu, Y. (2016). Assembling carbon quantum dots to alayered carbon for high-density supercapacitor electrodes. *Scientific Reports, 6*, 19028.
12. Liu, Y., Li, W., Wu, P., Ma, C., Wu, X., Xu, M., Luo, S., Xu, Z., & Liu, S. (2019). Hydrothermal synthesis of nitrogen and boron co-dopedcarbon quantum dots for application in acetone and dopamine sensors and multicolor cellular imaging. *Sensors Actuators B Chemical, 281*, 34–43.
13. Permatasari, F. A., Irham, M. A., Bisri, S. Z., & Iskandar, F. (2021). Carbon-based quantum dots for supercapacitors: Recent advances and future challenges. *Nanomaterials, 11*(1), 91.
14. Lee, K., Lee, H., Shin, Y., Yoon, Y., Kim, D., & Lee, H. (2016). Highly transparent and flexible supercapacitors using graphene-graphene quantum dots chelate. *Nano Energy, 26*, 746–754.
15. Hong, Y., Xu, J., Chung, J. S., & Choi, W. M. (2020). Graphene quantum dots/$Ni(OH)_2$ nanocomposites on carbon cloth as a binder-free electrode for supercapacitors. *Journal of Materials Science and Technology, 58*, 73–79.
16. Soram, B. S., Thangjam, I. S., Dai, J. Y., Kshetri, T., Kim, N. H., & Lee, J. H. (2020). Flexible transparent supercapacitor with core-shell Cu@Ni@NiCoS nanofibers network electrode. *Chemical Engineering Journal, 395*, 125019.
17. Li, Z., Wei, J., Ren, J., Wu, X., Wang, L., Pan, D., & Wu, M. (2019). Hierarchical construction of high-performance all-carbon flexible fiber supercapacitors with graphene hydrogel and nitrogen-doped graphene quantum dots. *Carbon, 154*, 410–419.
18. Barr, M. C., Rowehl, J. A., Lunt, R. R., Xu, J., Wang, A., Boyce, C. M., Im, S. G., Bulović, V., & Gleason, K. K. (2011). Direct monolithic integration of organic photovoltaic circuits on unmodified paper. *Advanced Materials, 23*(31), 3500–3505.
19. Izatt, R. M., Izatt, S. R., Bruening, R. L., Izatt, N. E., & Moyer, B. A. (2014). Challenges to achievement of metal sustainability in our high-tech society. *Chemical Society Reviews., 43*(8), 2451–2475.
20. Mirtchev, P., Henderson, E. J., Soheilnia, N., Yip, C. M., & Ozin, G. A. (2012). Solution phase synthesis of carbon quantum dots as sensitizers for nanocrystalline TiO_2 solar cells. *Journal of Materials Chemistry, 22*(4), 1265–1269.
21. Yang, Q., Yang, W., Zhang, Y., Ge, W., Yang, X., & Yang, P. (2020). Precise surface state control of carbon quantum dots to enhance charge extraction for solar cells. *Nanomaterials, 10*(3), 460.

22. Li, H., Shi, W., Huang, W., Yao, E.-P., Han, J., Chen, Z., Liu, S., Shen, Y., Wang, M., & Yang, Y. (2017). Carbon quantum dots/tio x electron transport layer boosts efficiency of planar heterojunction perovskite solar cells to 19%. *Nano Letters, 17*(4), 2328–2335.
23. Wang, Y., Yuan, Z., Shi, G., Li, Y., Li, Q., Hui, F., Sun, B., Jiang, Z., & Liao, L. (2016). Dopant-free spiro-triphenylamine/fluorene as hole-transporting material for perovskite solar cells with enhanced efficiency and stability. *Advanced Functional Materials, 26*(9), 1375–1381.
24. Nguyen, D. N., Roh, S. H., Kim, D.-H., Lee, J. Y., Wang, D. H., & Kim, J. K. (2021). Molecular manipulation of PEDOT: PSS for efficient hole transport by incorporation of N-doped carbon quantum dots. *Dyes and Pigments, 194*, 109610.
25. Cui, B., Yan, L., Gu, H., Yang, Y., Liu, X., Ma, C.-Q., Chen, Y., & Jia, H. (2018). Fluorescent carbon quantum dots synthesized by chemical vapor deposition: An alternative candidate for electron acceptor in polymer solar cells. *Optical Materials, 75*, 166–173.
26. Yuan, F., Yuan, T., Sui, L., Wang, Z., Xi, Z., Li, Y., Li, X., Fan, L., Tan, Z., Chen, A., Jin, M., & Yang, S. (2018). Engineering triangular carbon quantum dots with unprecedented narrow bandwidth emission for multicolored LEDs. *Nature Communications, 9*(1), 2249.
27. Zhu, J., Bai, X., Zhai, Y., Chen, X., Zhu, Y., Pan, G., Zhang, H., Dong, B., & Song, H. (2017). Carbon dots with efficient solid-state photoluminescence towards white light-emitting diodes. *Journal of Materials Chemistry C, 5*(44), 11416–11420.
28. Li, L., Zhao, W., Luo, L., Liu, X., Bi, X., Li, J., Jiang, P., & You, T. (2022). Electrochemiluminescence of carbon-based quantum dots: Synthesis, mechanism and application in heavy metal ions detection. *Electroanalysis, 34*(4), 608–622.
29. Pourmadadi, M., Rahmani, E., Rajabzadeh-Khosroshahi, M., Samadi, A., Behzadmehr, R., Rahdar, A., & Ferreira, L. F. R. (2023). Properties and application of carbon quantum dots (CQDS) in biosensors for disease detection: A comprehensive review. *Journal of Drug Delivery Science and Technology, 80*, 104156.
30. Ding, L., Peng, D., Wang, R., & Li, Q. (2021). A user-secure and highly selective enhancement of latent fingerprints by magnetic composite powder based on carbon dot fluorescence. *Journal of Alloys and Compounds, 856*, 158160.
31. Nugroho, D., Oh, W. C., Chanthai, S., & Benchawattananon, R. (2022). Improving minutiae image of latent fingerprint detection on non-porous surface materials under uv light using sulfur doped carbon quantum dots from *magnolia grandiflora* flower. *Nanomaterials, 12*(19), 3277.
32. Verhagen, A., & Kelarakis, A. (2020). Carbon dots for forensic applications: A critical review. *Nanomaterials, 10*(8), 1535.

MIX
Papier aus verantwortungsvollen Quellen
Paper from responsible sources
FSC® C105338

If you have any concerns about our products,
you can contact us on
ProductSafety@springernature.com

In case Publisher is established outside the EU,
the EU authorized representative is:
Springer Nature Customer Service Center GmbH
Europaplatz 3, 69115 Heidelberg, Germany

Printed by Libri Plureos GmbH
in Hamburg, Germany